PROGRAMMABLE CONTROLLERS

PROGRAMMABLE CONTROLLERS

Theory and Implementation

First Edition

L. A. Bryan
E. A. Bryan

An Industrial Text Co. Publication
Atlanta • Chicago

Editor Lisa DuPree
Preface Kim Ahmed, VP Industrial Text Company
Illustrations Carol Clerk and Kathie Reinisch
Layout Coordination Lisa DuPree
Technical Editing Kim Ahmed
Typesetting Coordination Maxine Maddox
Production Coordination Julie Graham
Cover Coordination Donna Roberson

Published by Industrial Text Co.
 First Edition
Printed and bound in the United States of America
94 93 10 9

Library of Congress Cataloging in Publication Data.

Bryan, L.A. / Bryan, E.A.

PROGRAMMABLE CONTROLLERS: Theory
and Implementation
TJ 223.P76 B795 1988 88-81557
ISBN 0-944107-30-3

Due to the nature of this publication and because of the different applications of programmable controllers, the readers or users and those responsible for applying the information herein contained must satisfy themselves as to the acceptability of each application and the use of equipment therein mentioned. In no event shall the publisher and others involved in this publication be liable for direct, indirect or consequential damages resulting from the use of any technique or equipment herein mentioned.

The illustrations, charts, and examples in this book are intended solely to illustrate the methods used in each application example. The publisher and others involved in this publication cannot assume responsibility or liability for actual use based on the illustrative uses and applications.

No patent liability is assumed with respect to use of information, circuits, illustrations, equipment, or software described in this text.

Preface

In the late 1960's, Programmable Logic Controllers, or PLCs, were first introduced as a means of automating manufacturing processes. Since then, PLCs have evolved into sophisticated and sometimes complex pieces of equipment. Nevertheless, their use and flexibility has reached a point where they are no longer discretionary pieces of equipment, but necessities. Today, a vast array of different types of programmable controllers exists for use on virtually any type of manufacturing process or equipment.

Industrial Text Company continues to recognize the need for publishing textbooks that reflect the valuable information to be learned from programmable controllers. As a result, *Programmable Controllers: Theory and Implementation* was born. This text is a comprehensive guideline to understanding, programming and using PLCs. The text is generic in nature to facilitate an understanding of just about any type of PLC in the marketplace. A very complex subject has been exhaustively discussed in simple, easy to understand language complemented with a plethora of descriptive figures and tables.

It is our sincere hope that this book will serve as a useful tool in helping you to understand how your PLC works. This in turn, will help you apply your PLC based system in more effective ways to realize gains associated with precise and accurate process control.

Kim Ahmed

"A machine is only as effective as its user"

– Wilbur Wright

About The Authors

Luis Bryan received the Bachelor of Science in Electrical Engineering and the Master of Engineering in Electrical Engineering, both from The University of Tennessee. His major area of studies were in the area of digital systems, electronics and computer engineering. Luis was involved, during his graduate studies, on several projects with national governmental agencies including a joint project between the United States, West Germany and Japan in the "Advanced Instrumentation for Reflood Studies" sponsored by the United States Department of Energy at the Oak Ridge National Laboratory.

Luis' s experience in the application of programmable controllers is extensive and goes back many years. He was involved in international marketing activities as well as PLC applications for a major programmable controller manufacturer. He gave many lectures and seminars in Canada, Mexico and South America on the uses and applications of programmable controllers. Luis worked for a consulting firm in the area of programmable controllers, where he was involved in market studies and consulting in the area of PLCs. He continues to give numerous seminars and lectures on PLCs to industry and government entities including the National Aeronautics and Space Administration (NASA). He has co-authored several other books in the area of programmable controllers.

Luis is an active member of several professional organizations including the Institute of Electrical and Electronic Engineers, IEEE's Instrument Society and Computer Society; he is a senior member of the Instrument Society of America and a member of Phi Kappa Phi honor society and Eta Kappa Nu honor electrical engineering society.

Eric Bryan graduated from The University of Tennessee with a Bachelor of Science in Electrical Engineering, specializing in digital design and computer architecture. He received a Master of Science in Engineering from The Georgia Institute of Technology, where he participated in a special program in Computer Integrating Manufacturing (CIM). His major areas of emphasis at Georgia Tech were industrial automation methods, Flexible Manufacturing Systems (FMS) and Artificial Intelligence (AI). He is an advocate of Artificial Intelligence implementation and its application in industrial automation.

Eric worked for a leading company in the area of automatic laser inspection systems, where he was responsible for the application engineering of large inspection systems. He also worked for a programmable controller engineering consulting firm where he helped the engineering of PLC based systems. Eric has coauthored other publications in the area of PLCs and he is a member of several professional and technical societies.

Acknowledgements

Any major work is brought to fruition by the efforts of many people. We would like to thank the editor Ms. Lisa DuPree for her superb editing and layout coordination which made this text easy to understand and a pleasure to read. We believe that the many hours of editing paid off at the end. We are also very thankful to Mr. Kim Ahmed's help for the many comments and suggestions made during the technical editing. Thanks are also extended to Ms. Carol Clerk and Ms. Kathie Reinisch for their excellent art work and illustrations.

As is the case in any major technical textbook, we are greatly thankful to the corporations listed who provided us with the necessary literature on their products and the photographs used in the text. We extend our thanks to Prof. G. Vachtsevanos and Prof. J. Gilmore, of Georgia Tech, for their help and suggestions in the area of Artificial Intelligence. Special thanks to Mr. Denny R. Zarafis of Texas Instruments for his kind cooperation and suggestions on the AI information he provided.

We also extend our sincere thanks Ms. Diane Olah for the initial cover suggestions, Ms. Donna Roberson for her cover coordination, Ms. Julie Graham for the production coordination and to Maxine Maddox for her long hours during the typesetting output stages.

Allen Bradley Company–Industrial Computer Group
B & R Industrial Automation
Bailey Controls Company
Divelbiss Corp.
Furnas Electric Co.
General Electric Company
Gould Inc.–Industrial Automation System Group
Omron Electronics Inc.
Schaevitz Engineering Co.
Square D Company
Texas Instrument
Thermometrics Inc.
Toshiba/Houston International Corp.
Westinghouse Electric Corp.
WRB Associates (Ladders Inc.)
Xicor

Contents

Chapter 10: PLC SYSTEM DOCUMENTATION

Chapter 11: IMPLEMENTING AND PROGRAMMING THE PLC SYSTEM

Chapter 12: DATA MEASUREMENTS AND TRANSDUCERS

Chapter 13: ARTIFICIAL INTELLIGENCE AND PLC SYSTEMS

Chapter 14: LOCAL AREA NETWORKS

Chapter

–1–

INTRODUCTION TO PROGRAMMABLE CONTROLLERS

"The most important thing in science is not so much to obtain new facts as to discover new ways of thinking about them."

—Sir William Bragg

In this chapter we will discuss the programmable controller from ways it can be defined to introductory applications on how it is used. This introduction will give you an inside look on the philosophy of design behind the creation of the programmable controller as well as its 20-year evolution. Comparisons with other types of control are presented and discussed to provide insight and perspective. Market segmentation of PLCs are pinpointed to clarify the categories where each PLC works best.

1-1 DEFINITION

A programmable controller, formally called a *programmable logic controller* or *PLC*, can be defined as a solid state device member of the computer family. It is capable of storing instructions to implement control functions such as sequencing, timing, counting, arithmetic, data manipulation and communication to control industrial machines and processes. Figure 1-1 illustrates a conceptual diagram of a PLC application.

Many definitions have been used to describe a programmable controller. However, a PLC can be thought of in simple terms as an industrial computer which has a specially designed architecture in both its central unit (the PLC itself) and its interfacing circuitry to field devices (input/output connections to the real world).

As it will be seen throughout this book, a PLC is a mature industrial controller that has its roots of uses and design based on the assumption of simple understanding and practical application.

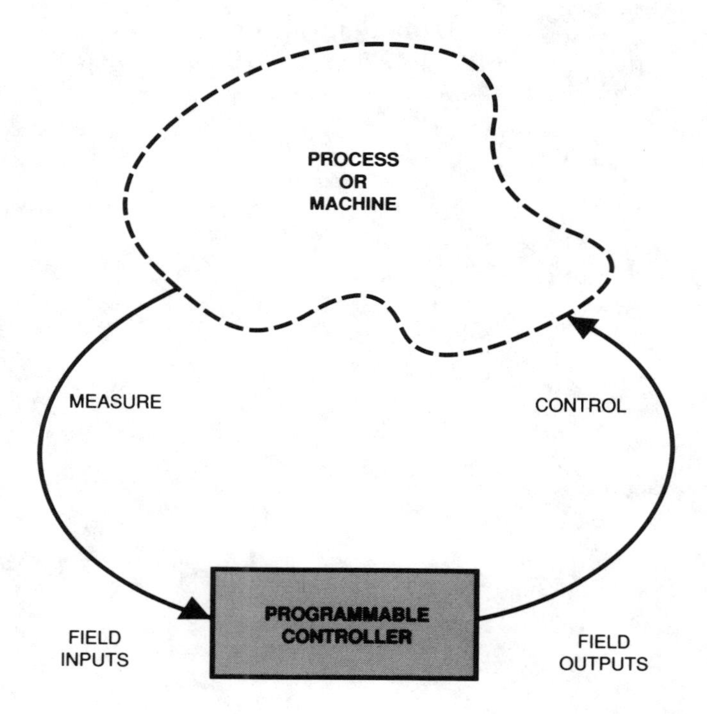

Figure 1-1. PLC Conceptual Application Diagram.

1-2 AN HISTORICAL BACKGROUND

The design criteria for the first programmable controller were specified in 1968 by the Hydramatic division of the General Motors Corporation. The primary goal was to eliminate the high cost associated with inflexible, relay-controlled systems. The specifications required a solid-state system with computer flexibility able to survive in an industrial environment, to be easily programmed and maintained by plant engineers and technicians and, last but not least, to be reusable. Such a control system would reduce machine downtime and provide expandability for the future.

The Conceptual Design of the PLC

The first programmable controllers were more or less just relay replacers. Their primary function was to perform the sequential operations that were previously implemented with relays. These operations included ON/OFF control of machines and processes that required repetitive operation, such as transfer lines and grinding and boring machines. On the other hand, programmable controllers were an improvement over relays. They were easily installed, used considerably less space and energy, had diagnostic indicators that aided trouble-shooting and, unlike relays, were reusable if a project were scrapped. The initial design not only met Hydramatic requirements, but led to further improvements which would spread the use of programmable controllers to other industries.

The programmable controller can be considered a newcomer when it is compared to its elder predecessors in traditional control equipment technology, such as old hardwired relay systems, analog instrumentation and other types of early solid-state logic. The PLC has been improved through the years when it comes to speed of operation, types of interfaces, and data processing capabilities; however, its design requirements still hold to the original intentions—it is simple to use and maintain.

The origins of the programmable controller are largely given to the automotive industry because of its need for a control system that would replace hardwired relay systems. The primary objectives of such a controller were to eliminate the high cost associated with downtime related to machine control problems (hardwired systems), to possess full logic capabilities, and to provide expandability for the future. Some of the initial specifications included the following:

- The new control system had to be price competitive with the use of relay systems.
- The system had to be capable of sustaining an industrial environment.
- The input and output interfaces had to be easily replaceable.
- The controller had to be designed in modular form (architecture) so that sub-assemblies could be removed easily for replacement or repair.

- The control system needed the capability to pass data collection to a central system.
- The system had to to be re-usable.
- The method used to program the controller had to be simple, so that it could be easily used and understood by plant personnel.

The First Programmable Controller

The product implementation to satisfy Hydramatic's specifications was underway and by 1969 the programmable controller had its first offspring. The early controllers met the original specifications and opened the doors to the development of a new control technology.

The first PLCs offered relay functionality, thus replacing the original hardwired relay logic. Modularity, expandability, programmability, and use in an industrial environment were achieved. These controllers were easily installed, used less space and were re-usable. The controller programming, although a little tedious, had a recognizable plant standard: *the ladder diagram format.*

In a short period, programmable controller use started to spread to other industries. By 1971, PLCs were being used to provide relay replacement as the first steps of control automation in other industries such as food and beverage, metals, manufacturing, and pulp and paper.

Early Innovations

During the early 1970s, the advent of microprocessor technology created a dramatic change for what would become the programmable controller. These new microprocessors, or micros, added greater flexibility and intelligence to the PLC. In addition to performing relay replacing functions, the new PLCs were now capable of performing arithmetic and data manipulation functions, operator communication and interaction, and computer communications.

The Cathode Ray Tube (CRT) used in large computers for programming was now becoming the programming tool for interaction between the PLC programmer and the controllers. The use of a CRT offered an alternative to the tedious process of inserting the program by using a manual loading process that varied in complexity depending on the controller.

With the CRT, the original ladder diagram format could now be seen on the screen in the same form as on relay drawings, thus further aiding the trouble-shooting process. The new intelligence added to the PLC allowed the evolution of the ladder programming language. The familiar symbols of the ladder diagram were used to further implement new instructions. These instructions would provide programming access to the micro-created functions. New symbols were added to represent operations such as comparisons, data transfers and arithmetic functions.

The addition of arithmetic functions and improved instruction sets expanded the application of programmable controllers by allowing them to be used with instrumentation devices that provided numerical input data. Logic and sequencing tasks could now be enhanced with the ability to perform calculations based on measured data. That same data could be used as the basis for corrective action. This newly found intelligence was the beginning of many advancements to come.

Later Innovations

Hardware and software enhancements between 1975 and 1979 added even greater flexibility to the programmable controller. Improvements included larger memory capacity, remote input/output capability, analog and positioning control, operator communications, machine fault detection, and software enhancement. These advancements made the programmable controller suitable for an even wider range of applications and contributed greatly to the reduction of wiring and installation cost.

Expanded memory systems allowed storage of larger application programs and amounts of data. Added memory allowed control programs to include not only logic and sequencing tasks, but also data acquisition and manipulation. The ability to store more data meant that resident control information or recipes could be stored and retrieved automatically. For example, if a certain event occurred, all timer preset values could be changed, or high limit presets altered. This flexibility eliminated the need for operators to stop the process to change parameters.

Wiring costs were significantly reduced with the new ability to locate input/output subsystems in remote locations away from the Central Processing Unit (CPU) and near the equipment being controlled. Instead of bringing hundreds of wires back to the CPU, the signals from the subsystems could be multiplexed over two twisted pairs of wire. Remote input/output systems also allowed large systems to be divided into smaller subsystems, which greatly improved maintainability and allowed a gradual start-up of major subsystems.

With the development of analog control, the programmable controller bridged the gap between ON/OFF control systems and instrumentation controls. Up until then, the programmable controller was only capable of ON/OFF control, which limited its control of applications such as chemical batching, water and waste treatment, and mineral processing. Applications such as these required a combination of ON/OFF and variable (analog) control functions.

Another hardware development during this period established provisions to perform positioning control, using stepper output and encoder input feedback. The input interface counts a train of incoming pulses which provide a numeric value to the controller for verification of a move. Using data sent to the module by the controller, the output interface produces a pulse train to be interpreted by the stepper motor translator. Early application of these interfaces included grinders, transfer lines, and paint spray lines.

Enhanced communication allowed PLCs to communicate with other devices that, in turn, helped to improve operator interface. The CRT hardware and software were further developed to aid program entry and monitoring. Production summaries, management reports, and maintenance data could be provided on hard copy through a printer. In late 1979, high-speed communication local area networks evolved. These networks allowed the control tasks of an entire plant to be distributed among several controllers, all communicating as one. With machines and processes communicating to one another, human communications could be eliminated from the control scheme. The local area network opened new, more advanced applications for the programmable controller such as machine transfer lines and material handling and tracking.

Software enhancements brought about computer-like statement instructions. These new instructions allowed easy utilization of the many hardware enhancements. Improvements included statements for manipulating and handling large amounts of data, and for communicating with analog and peripheral devices. The relay-type instructions would have made these tasks very cumbersome or even impossible. Other software enhancements included system routines to improve on-line monitoring of the process. Menu driven routines simplified operator interfacet by merely selecting a numbered function.

With the developments during this period, the programmable controller took the first step towards replacing the minicomputer in many industrial applications.

Today's Controllers

The 1980s have brought about many technological advances in the programmable controller industry that still continue to this date and into the 1990s. These changes not only affect the programmable controller design, but also the philosophical approach to control system design. The following list describes some of the enhancements that have occurred:

Hardware Enhancements:

- Faster scan times were achieved using new advance microprocessor technology.
- Small, low cost PLCs (see Fig. 1-2a) could replace 4 to 10 relays and also reduce space requirements.
- High density I/O systems (see Fig. 1-2b) provided space-efficient interfaces at low cost.
- Intelligent microprocessor based I/O interfaces expanded distributed processing. Typical interfaces are PID (proportional, integral, and derivative control), ASCII communication, positioning, host computer, and language modules (e.g. BASIC, Pascal).
- Special interfaces allowed certain devices to be connected directly to the controller. Typical interfaces included thermocouplers, strain gauges, and fast response inputs.
- Mechanical design improvements included rugged I/O enclosures and I/O systems that made the terminal an integral unit.
- Peripheral equipment improved operator interface techniques and system documentation methods.

Figure 1-2. a) Small PLC with built-in I/O and detachable, hand-held programming unit (Courtesy Omron Electronics, Inc., Schaumburg, IL).

Figure 1-2. b) PLC system with high density I/O (128 points per rack enclosure). Footprint of 19.2W x 9.9H x 6.4D inches (Courtesy of Toshiba Corp., Houston, TX).

Figure 1-2. c) Programmable controller family concept portraying several PLCs: the Minicontrol (96 I/O), Midicontrol (192 I/O) and Multicontrol (1536 I/O) (Courtesy of B & R Industrial Automation, Stone Mountain, GA).

A significant hardware enhancement was the development of programmable controller families like those shown in Figure 1-2c. These families consisted of a product line that ranged from very small single board "micro-controllers", with as few as 10 I/O points, to very large and sophisticated PLCs, with as many as 8000 I/O points and 128,000 words of memory. These family members, utilizing common I/O systems and programming peripherals, typically could be interfaced to a local communication network. The family concept was an important cost savings development for users.

Software Enhancements:

- High-level languages, such as BASIC and Pascal were implemented in some controllers' modules to provide greater programming flexibility when communicating with peripheral devices.
- Advanced functional block instructions were implemented for ladder diagram instruction sets to provide enhanced software capability using simple programming commands.
- Diagnostics and fault detection were expanded from "system" diagnostics, which diagnose controller malfunctions, to include machine diagnostics, which diagnose failures or malfunctions of the controlled machine or process.
- Floating point math made it possible to perform complex calculations for control applications that required gauging, balancing, and statistical computation.
- Data handling and manipulation instructions were improved and simplified to accommodate complex control and data acquisition applications that involved storage, tracking, and retrieval of large amounts of data.

The programmable controller is now a mature control system offering much more than was ever anticipated. It is now capable of communicating with other control systems, providing production reports, scheduling production, and diagnosing its own failures and those of the machine or process. These enhancements have made the programmable controller an important contributor in meeting today's demands of higher quality and productivity. In spite of the fact that the programmable controller has become much more capable, it retains the simplicity and ease of operation that was originally intended.

Programmable Controllers and the Future

The future of programmable controllers relies not only on the continuation of new product developments, but also in the integration of programmable controllers with other control and factory management equipment. There is no doubt that PLCs play a substantial role in the factory of the future. They are presently being incorporated into Computer Integrated Manufacturing (CIM) systems, combining its power and resources with numerical controls, robots, CAD/CAM systems, management information systems and hierarchical computer based systems.

New advances in PLC technology will include features such as better operator interfaces, more human oriented man-machine interfaces, development of more interfaces that will allow communication with other equipment, and hardware and software that will support artificial intelligence systems.

Software advances will provide better interconnections among different types of equipment utilizing communication standards such as the Manufacturing Automation Protocol (MAP). The development of new PLC instructions will be based on the need to add intelligence to the system by introducing knowledge based or process learning type of instructions.

The control philosophy of the future will be determined by the user's concept of the Flexible Manufacturing System (FMS). It is almost certain that the future will cast programmable controllers as an important player in producing the factory of the future. Control strategies will be distributed with "intelligence" instead of being centralized. Super PLCs will be used in applications requiring complex calculations, network communication and supervision of other smaller PLCs and machine controllers.

1-3 PRINCIPLES OF OPERATION

A programmable controller is composed primarily of two basic sections: the Central Processing Unit (CPU) and Input/Output (I/O) interface system. These sections are illustrated in Figure 1-3.

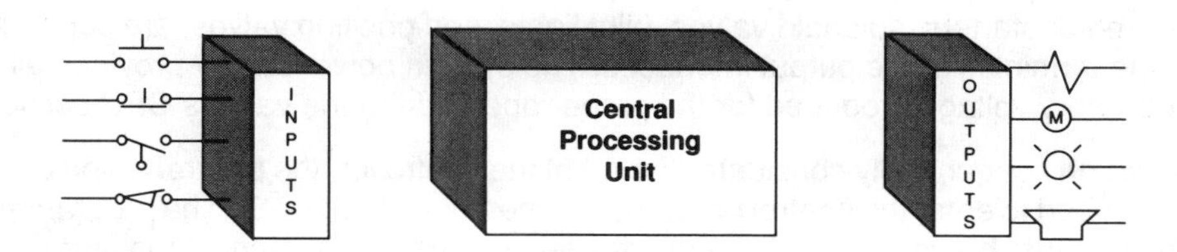

Figure 1-3. Programmable controller block diagram.

The CPU section of a PLC is formed by three components: the processor, the memory system, and the system power supply. Figure 1-4 shows the three sections that comprise the CPU.

The operation of a programmable controller is relatively simple. First, note that the input/output system is physically connected to the field devices encountered in a machine, or that are used in the control of a process. These field devices may be limit switches, pressure transducers, pushbuttons, motor starters, solenoids, etc. The I/O interfaces provide the connection between the CPU and the information providers (inputs) and the controllable devices (outputs).

During the operation, the CPU reads or accepts the input data or status of the field devices via the input interfaces, executes the control program stored in the memory system, and writes or updates the output devices via the output interfaces. This process of sequentially reading the inputs, executing the program in memory, and updating the outputs is known as *scanning*.

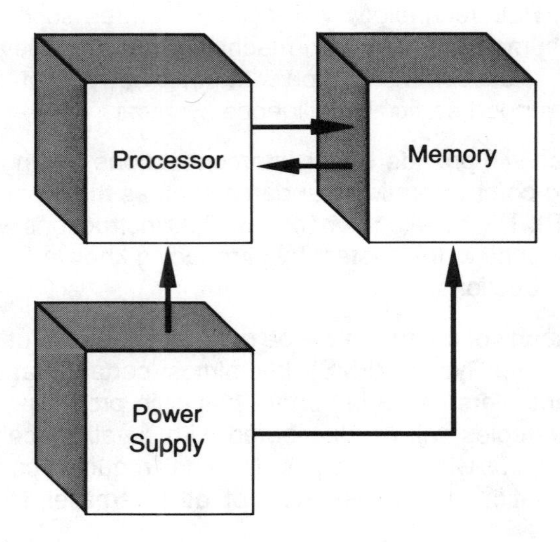

Figure 1-4. Block diagram of major CPU components.

The Input/Output system forms the interface by which field devices are connected to the controller. The main purpose of the interface is to condition the various signals received from or sent to external field devices. Incoming signals from sensors such as pushbuttons, limit switches, analog sensors, selector switches, and thumbwheel switches are wired to terminals on the input interfaces. Devices that will be controlled, like motor starters, solenoid valves, pilot lights, and position valves, are connected to the terminals of the output interfaces. The system power supply provides all the necessary voltages required for the proper operation of the various CPU sections.

Although not generally considered a part of the controller, the programming device is required to enter the control program into memory (Figure 1-5). The programming device must be connected to the controller only when entering or monitoring the program. A CRT is commonly used for program entry and display, but alternate equipment such as personal computers are also available.

In Chapter 4 and 5, we present a more detailed discussion of the Central Processing Unit and how it interacts with memory and the Input/Output interface. Chapters 6, 7, and 8 are dedicated to the discussion of the I/O system.

1-4 PLCs VERSUS OTHER TYPES OF CONTROLS

PLCs vs Relay Control

For years, the question many engineers, plant managers, and Original Equipment Manufacturers (OEM) had asked was, "Should I be using a programmable controller?" At one time, much of a systems engineer's time was spent trying to determine the cost effectiveness of a PLC over relay control. Even today, many panel builders

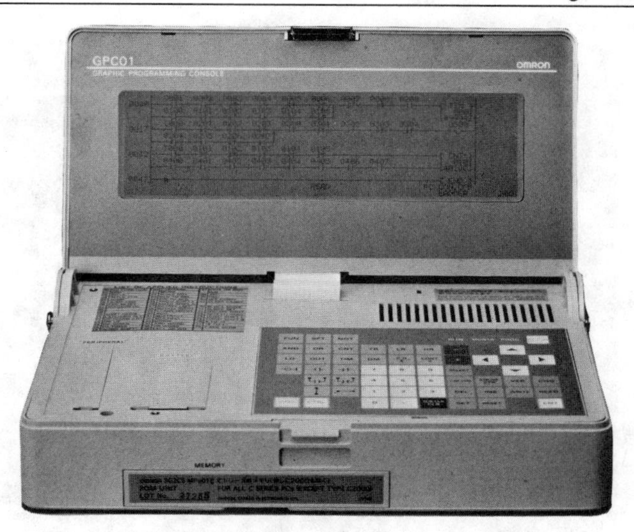

Figure 1-5. PLC's dedicated programming device (Courtesy of Omron Electronics, Schaumburg, IL).

and control systems designers still think that they are faced with this decision. One thing is certain: today's demand for high quality and productivity can hardly be fulfilled economically without electronic control equipment. With rapid developments and increasing competition, the cost for programmable control has been driven down to the point where the old PLC versus relay study is no longer necessary or valid. Programmable controller application can now be evaluated on its own merits.

When deciding whether to use a PLC based system or hardwired relay system, several questions must be asked by the designer. Some of these questions may be:

- Is there a need for flexibility in control logic changes?
- Is there a need for high reliability?
- Are space requirements important?
- Are increased capability and output required?
- Are there data collection requirements?
- Will there be frequent control logic changes?
- Will there be a need for rapid modification?
- Must similar control logic be used on different machines?
- Is there a need for future growth?
- What are the overall costs?

The merits of programmable controller systems make them especially suitable for applications in which the requirements listed above are particularly important for the economic viability of machine or process operation. A case which speaks for itself, shown in Figure 1-6, shows why PLCs are easily favored over relays.

Figure 1-6. Illustration of an installed PLC based system.

If system requirements call for flexibility or future growth, a programmable controller brings returns that will outweigh any initial cost advantage that a relay-control system might have. Even the case in which no flexibility or future expansion is required, a large control system can benefit tremendously from the trouble-shooting and maintenance aids a PLC provides. The extremely short cycle (scan) time of a PLC would allow the productivity of machines, which were previously under electro-mechanical control, to increase considerably.

Although relay-control may be thought to initially cost less, this advantage would be lost if production downtime due to failures is high. These and other considerations for selecting a programmable controller are discussed later in Chapter 16.

PLCs vs Computer Controls

The architecture of the programmable controller CPU is basically the same as that of a general purpose computer. However, some important characteristics distinguish PLCs from general purpose computers.

First, unlike computers, the PLC was specifically designed to survive the harsh conditions of the industrial environment. A well-designed PLC can be placed in areas with substantial amounts of electrical noise, electro-magnetic interference, mechanical vibration, or extreme temperatures and non-condensing humidity.

A second distinction of PLCs is that the hardware and software is designed for easy use by plant electricians and technicians. The hardware interfaces for connecting the field devices are actually a part of the PLC and are easily connected. The modular and self-diagnosing interface circuits are able to pinpoint malfunctions and are easily removed and replaced. The software programming uses conventional relay ladder symbols, or other easily learned languages which are familiar to plant personnel.

The final distinction is one that is changing as PLCs become more intelligent. Whereas computers are complex computing machines capable of executing several programs or tasks simultaneously and in any order, the PLC executes a single program in an orderly and sequential fashion from first to last instruction. In recent years, however, flexibility has been added to the PLC by including instructions that allow subroutine calling, interrupt routines, and a means for bypassing or jumping certain instructions.

PLCs vs Personal Computers

With the proliferation of the personal computer, or PC, many engineers are finding that the personal computer is not a direct competitor of the PLC in control applications, but an ally in the implementation of the control solution. The personal computer possesses similar characteristics as far as CPU architecture is concerned; however, the most distinct difference is in the way the two connect to field devices. While new rugged personal computers can sometimes sustain mid-range industrial environments, interconnection to field devices still presents difficulties. These computers must communicate with I/O interfaces not necessarily designed for them. The programming language may not meet the standards of ladder diagram programming. This presents a problem to people familiar with this plant standard while trouble-shooting and when making changes to the system.

The personal computer is being used more and more as a programming device for PLCs in the market rather than as a controlling device. Personal computers are also employed to get process data from the PLC to perform data gathering and data collection to hard disk storage media. Because of the number crunching capability, the personal computer is well suited to help the PLC and sometimes to bridge the communication gap between a PLC system and other mainframe computers.

1-5 TYPICAL AREAS OF PLC APPLICATIONS

Since its inception, the programmable controller has been successfully applied in virtually every segment of industry including steel mills, paper and pulp plants, food processing plants, chemical and petrochemical plants, and automotive and power plants. PLCs perform a great variety of control tasks, from repetitive ON/OFF control of a simple machine to sophisticated manufacturing and process control. Table 1-1 lists a few of the major areas in which PLCs have been applied and some of the typical applications.

Table 1-1. Typical Programmable Controller Applications

CHEMICAL/PETROCHEMICAL	**GLASS/FILM**
Batch Process	Process
Materials Handling	Forming
Weighing	Finishing
Mixing	Packaging
Finished Product Handling	Palletizing
Water/Waste Treatment	Materials Handling
Pipeline Control	Lehr Control
Off-Shore Drilling	Cullet Weighing
MANUFACTURING/MACHINING	**FOOD/BEVERAGE**
Energy Demand	Bulk Materials Handling
Tracer Lathe	Brewing
Material Conveyors	Distilling
Assembly Machines	Blending
Test Stands	Container Handling
Milling	Packaging
Grinding	Filling
Boring	Weighing
Cranes	Product Handling
Plating	Sorting Conveyors
Welding	Accumulating Conveyors
Painting	Load Forming
Injection/Blow Molding	Palletizing
Metal Casting	Warehouse Storage/Retrieval
	Metal Forming Loading/Unloading
MINING	**METALS**
Bulk Material Conveyors	Blast Furnace Control
Ore Processing	Continuous Casting
Loading/Unloading	Rolling Mills
Water/Waste Management	Soaking Pit
PULP/PAPER/LUMBER	**POWER**
Batch Digesters	Coal Handling
Chip Handling	Burner Control
Coating	Flue Control
Wrapping/Stamping	Load Shedding
	Sorting
	Winding/Processing
	Woodworking
	Cut-to-Length

Although the uses of the programmable controller are far too extensive to list, the application areas shown here are frequently found within the listed industries. Table 1-2 is just a sample list of how PLCs are being applied. The potential benefits of PLC application can no longer be simply neglected or reduced to a cost comparison with relays.

Table 1-2. Examples of PLC Applications

RUBBER AND PLASTIC

Tire Curing Press Monitoring. PLCs perform individual press monitoring for time, pressure, and temperature during each press cycle. Information concerning machine status is stored in tables for later use and alerts the operator of any press malfunctions. Report generation printout for each shift includes a summary of good cures and press downtime due to malfunctions.

Tire Manufacturing. Programmable controllers can be used for tire press/cure systems to control the sequencing of events which must occur to transform the raw tire into a tire fit for the road. This control includes molding the tread pattern and curing the rubber to obtain the road-resistant characteristics. This PLC application substantially reduces the space required and increases reliability of the system and quality of the product.

Rubber Production. Dedicated programmable controllers provide accurate scale control, mixer logic functions, and multiple formula operation of carbon black, oil, and pigment used in the production of rubber. The system maximizes utilization of machine tools during production schedules, tracks in-process inventories, and reduces time and personnel required to supervise the production activity and the manually produced shift-end reports.

Plastic Injection Molding. A PLC system controls variables, such as temperature and pressure, which are used to optimize the injection molding process. The system provides closed-loop injection such that several velocity levels can be programmed to maintain consistent filling, reduce surface defects and stresses, and shorten cycle time. The system can also accumulate production data for future use.

CHEMICAL AND PETROCHEMICAL

Ammonia and Ethylene Processing. Programmable controllers monitor and control large compressors that are used during the manufacture of ammonia, ethylene, and other chemicals. The PLC monitors bearing temperatures, operation of clearance pockets, compressor speed, power consumption, vibration, discharge temperatures, pressure, suction flow, and gas composition.

Dyes. PLCs monitor and control the dye processing used in the textile industry. They provide accurate processing of color blending and matching to predetermined values.

Chemical Batching. The PLC controls the batching ratio of two or more materials in a continuous process. The system determines the rate of discharge of each material, in addition to providing inventory records and other useful data. Several batch recipes can be logged and retrieved automatically or on command from the operator.

Fan Control. PLCs automatically control fans based on levels of toxic gases in a chemical production environment. This system provides effective measures of exhausting gases when a preset level of contamination is reached. The PLC controls the fan start/stop, cycling, and speeds so that safety levels are maintained while energy consumption is minimized.

Gas Transmission and Distribution. Programmable controllers monitor and regulate pressures and flows of gas transmission and distribution systems. Data is gathered and measured in the field and transmitted to the PLC system.

Oil Fields. PLCs provide on-site gathering and processing of data pertinent to characteristics such as the depth and density of drilling rigs. The PLC controls and monitors the total rig operation and alerts the operator of any possible malfunction.

Pipeline Pump Station Control. PLCs control mainline and booster pumps for the distribution of crude oil. Measurement of flow, suction, discharge, and tank low and high limits are some of the functions they fulfill. Possible communications with SCADA (Supervisory Control And Data Acquisition) systems can provide total supervision of the pipeline.

POWER

Plant Power System. The programmable controller regulates the proper distribution of available electricity, gas, or steam. In addition, the PLC monitors power house facilities, schedules distribution of energy, and generates distribution reports. The PLC controls the loads during operation of the plant, as well as the automatic load shedding or restoring during power outages.

Energy Management. Through the reading of inside and outside temperatures, the PLC controls heating and cooling units in a manufacturing plant. The PLC system controls the loads, cycling them during predetermined cycles and keeping track of how long each load should be on or off during the cycle time. The system provides scheduled reports on the amount of energy used by the heating and cooling units.

Coal Fluidization Process. The controller can monitor how much energy is generated from a given amount of coal and regulate the coal crushing and mixing with crushed limestone. The PLC monitors and controls burning rates, temperatures generated, sequencing of valves, and analog control of jet valves.

Compressor Efficiency Control. PLCs control several compressors located at a typical compressor station. The system handles safety interlocks, start-up/shutdown sequences, and compressor cycling and keeps the compressors running at maximum efficiency using the non-linear curves of the compressors.

METALS

Steelmaking. The PLC controls and operates furnaces and produces the metal in accordance with preset specifications. The controller also calculates oxygen requirements, alloy additions, and power requirements.

Loading and Unloading of Alloys. Through accurate weighing and loading sequences, the system controls and monitors the quantity of coal, iron ore, and limestone to be melted. The unloading sequence of the steel to a torpedo car can also be controlled.

Continuous Casting. PLCs direct the molten steel transport ladle to the continuous-casting machine, where the steel is poured into a water-cooled mold for solidification.

Cold Rolling. PLCs are used to control the conversion of semi-finished products into finished goods through cold rolling mills. The system controls the speed of the motors to obtain correct tension and provide adequate gauging of the rolled material.

Aluminum Making. Controllers monitor the refining process in which impurities are removed from bauxite by heat and chemicals. The system can grind and mix the ore with chemicals and then pump them into pressure containers, where they are heated, filtered, and combined with more chemicals.

PULP AND PAPER

Pulp Batch Blending. The PLC controls sequence operation, quantity measurement of ingredients, and storage of recipes for the blending process. The system allows operators to modify batch entries of each quantity if necessary and provides hard copy print-outs for inventory control and for accounting of ingredients used.

Batch Preparation for Paper Making Process. Applications include control of the complete stock preparation system for paper manufacturing. Recipes for each batch tank are selected and adjusted via operator entries. The system can also control the feedback logic for chemical addition based on tank level measurement signals. At the completion of each shift, the PLC system provides management reports for material usage.

Paper Mill Digester. PLC systems provide complete control of pulp digesters for the process of making pulp from wood chips. The system calculates and controls the amount of chips based on density and the digester volume; the percent of cooking liquors is calculated, and the required amounts are added in sequence. The PLC ramps and holds the cooking temperature until the cook is completed. All data concerning the process is then transmitted to the PLC for reporting.

Paper Mill Production. The controller regulates the average basis weight and moisture variable for paper grade. The system manipulates the steam flow valves, adjusts the stock valves to regulate weight, and monitors and controls total flow.

GLASS PROCESSING

Annealing Lehr Control. PLCs control the Lehr used to remove the internal stress from glass products. The system controls the operation by following the annealing temperature curve during the reheating, annealing, straining, and rapid cooling processes through different heating and cooling zones. Improvements are made in the ratio of good glass to scrap, reduction in labor cost, and energy utilization.

Glass Batching. PLCs control the batch weighing system according to stored glass formulas. The system also controls the electromagnetic feeders for infeed to and outfeed from the weigh hoppers, manual shut-off gates, and other equipment.

Cullet Weighing. PLCs direct the cullet system by controlling the vibratory cullet feeder, weight-belt scale, and shuttle conveyor. All sequences of operation and inventory of quantities weighed are kept by the PLC for future use.

Batch Transport. PLCs control the batch transport system including reversible belt conveyors, transfer conveyors to cullet house, holding hoppers, shuttle conveyors, and magnetic separators. The controller takes action after discharge from the mixer and transfers the mixed batch to the furnace shuttle, where it is discharged to the full length of the furnace feed hopper.

MATERIALS HANDLING

Storage and Retrieval Systems. A PLC is used to load parts and carry them in totes in the storage and retrieval system. The controller keeps tracking information such as storage lane number where parts are stored, the parts assigned to specific lanes, and quantity of parts in any particular lane. This PLC arrangement allows rapid changes of parts loaded or unloaded from the system. The controller also provides inventory printouts and informs the operator of any malfunctions.

Automatic Plating Line. The PLC controls a set pattern for the automated hoist which can traverse left, right, up, and down through the various plating solutions. The system knows at all times where the hoist is located.

Conveyor Systems. The PLC controls all the sequential operations, alarms, and safety logic necessary to load and circulate parts on a main line conveyor as well as sorting product to the correct lane. The PLC can also schedule lane sorting to optimize palletizer duty. Records are kept on a shift basis for production of good parts and part rejections if required.

Table 1-2 con't.

Automated Warehousing. The PLC controls and optimizes the movement of stacking cranes and provides high turnaround of materials requests in an automated high cube vertical warehouse. The PLC also controls aisle conveyors and case palletizers to significantly reduce manpower requirements. Inventory control figures are maintained and can be provided on request.

AUTOMOTIVE

Internal Combustion Engine Monitoring. The system acquires data recorded from sensors located at the internal combustion engine. Measurements taken include water temperature, oil temperature, RPMs, torque, exhaust temperature, oil pressure, manifold pressure, and timing.

Carburetor Production Testing. PLCs provide on-line analysis of automotive carburetors in a production assembly line. The systems significantly reduce the test time, while providing greater yield and better quality carburetors. Some of the variables that are tested are pressure, vacuum, and fuel and aire flow.

Monitoring Automotive Production Machines. The system monitors total parts, rejected parts, parts produced, machine cycle time, and machine efficiency. All statistical data are available to the operator and also at the end of each shift.

Power Steering Valve Assembly and Test. The PLC system controls a machine to insure proper balance of the valves and maximize left and right turning ratios.

MANUFACTURING/MACHINING

Production Machines. The PLC controls and monitors automatic production machines at high efficiency rates. The piece-count production and machine status are also monitored. Corrective action can be taken immediately if a failure is detected by the PLC.

Transfer Line Machines. PLCs monitor and control all transfer line machining station operations and the interlocking between each station. The system receives input from the operator to check the operating conditions of line-mounted controls and reports any malfunctions. This arrangement provides greater machine efficiency, higher quality products, and lower scrap levels.

Wire Machine. The controller monitors the time and ON/OFF cycles of a wire drawing machine. The system provides ramping control and synchronization of electric motor drives. All cycles are recorded and reported on demand to obtain the machine's efficiency as calculated by the PLC.

Tool Changing. The PLC controls a synchronous metal cutting machine with several tool groups. The system keeps track of when each tool should be replaced, based on the number of parts it manufactures. It also displays the count and replacements of all the tool groups.

Paint Spraying. PLCs control the painting sequences in auto manufacturing. Style and color information is entered by the operator or a host computer and is tracked through the conveyor until the part reaches the spray booth. The controller decodes the part information and then controls the spray guns to paint the automotive part. The spray gun movement is optimized to conserve paint and to increase part throughput.

1-6 PLC PRODUCT APPLICATION RANGES

Until approximately 1979, the choice in programmable controllers was limited to two types. The first type was small, relatively inexpensive, and not too versatile. These controllers were designed for the purpose of converting small electro-mechanical relay systems to solid-state control systems. Programming was rather tedious, and software functions were limited. The second type of controller was large, expensive, and designed to bridge the gap between small controllers and minicomputers. The functional capability spread between these two classes of controllers made the choice limited, and somewhat difficult. Buying the inexpensive system may have meant trying to live within its limitations, while purchasing the expensive controller could result in wasteful extravagance.

With the wide range of products available today, the chance of closely matching a product to the application is much better, but the task of selecting the best product is much more difficult. This situation is largely related to the fact that conventional classification of products, based on input/output and memory capacity, is no longer valid.

In the early 1980s, PLCs were categorized into three major segments—small, medium, and large—each with distinct characteristic features. The standard features inherent to a particular segment were once a definite function of size (i.e. I/O, memory). As the market continues to grow and new capabilities are implemented, the three major segments have become less distinct. For example, a small PLC may now have some features once found only in large PLCs. Thus, I/O and memory specifications alone do not provide enough information for selecting a product.

Today's programmable controller product ranges can be graphically illustrated as shown in Figure 1-7. This chart is not definitive, but for practical purposes is valid. Segmentation of the market can be viewed now in five groups. Micro-controllers form the first group, being used in applications using generally up to 32 I/O, 20 I/O being the norm. The micro PLCs are followed by the small PLC category which controls up to 128 I/O. The medium (up to 1024 I/O), large (up to 4096 I/O) and very large (up to 8192 I/O) complete the segmentation. Figure 1-8 shows several PLCs that fall into this category classification.

The A, B, and C overlapping areas reflect enhancements of the standard features of PLCs within a particular segment by adding options. These options allow a product to be closely matched to the application without having to purchase the next largest unit.

The major distinction between segments and similarities among overlapping areas will be covered in detail in Chapter 16. These differences are based on I/O count, memory size, programming language, software functions, and other factors. An understanding of the PLC product ranges and their characteristics will allow the user to identify properly the controller that will satisfy a particular application.

1-7 LADDER DIAGRAMS AND THE PLC

The ladder diagram has been, and will be to a certain extent, the traditional way of representing electrical sequences of operations. These diagrams are used to represent the interconnection of field devices in such a way that the activation, or turning ON, of one device would turn ON another one according to a predetermined sequence of events. Figure 1-9 illustrates a simple electrical ladder diagram.

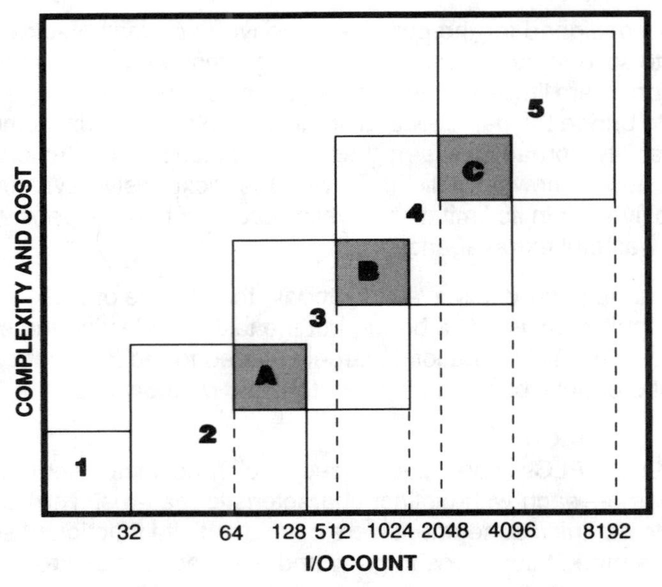

Figure 1-7. Illustration of the PLC product range.

Figure 1-8. Counterclockwise, Toshiba EX-14B 8 In/6 Out (Toshiba, Houston, TX); PC-1100 64 In/64 Out (Westinghouse Electric, Pittsburgh, PA); PC-90 1000 I/O (Bailey Controls, Wickliffe, OH); PLC-3/10 2048 I/O and PLC/3 8192 I/O (Allen Bradley, Highland Heights, OH).

The original ladder diagrams were established to represent hardwired logic circuits used for the control of a machine or equipment. Due to wide industry use, it became a standard way of providing control information from the designers to the users of equipment. As programmable controllers were being introduced, this type of circuit representation was also desirable; not only was it easy to use and interpret, but it was also widely accepted in industry.

The programmable controller can implement all the "old" ladder diagram conditions and much more. The idea is to perform the control operation in a more reliable manner at a lower cost. The PLC implements in its CPU all the hardwired interconnections similar to *wiring,* using the PLC's software instructions. This is accomplished by using familiar ladder diagrams but in a manner that is transparent to the engineer or programmer. As will be seen throughout the book, the knowledge of PLC operation, scanning, and instructions programming will be vital to the proper implementation of a control system.

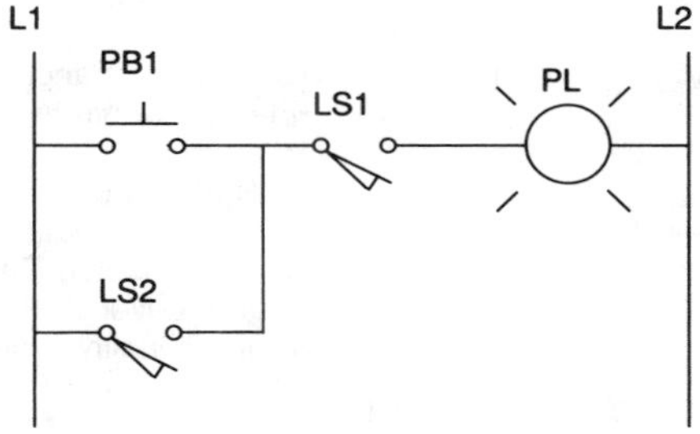

Figure 1-9. Simple electrical ladder diagram.

Figure 1-10 illustrates the connections or transformation to a PLC format of the simple diagram shown in Figure 1-9. Note that the "real" I/O field devices are connected to input or output interfaces, while the ladder program is implemented in a similar manner (as hardwired) inside the programmable controller. As previously mentioned, the CPU will read the status of inputs, energize or turn ON its corresponding circuit element according to the program, and control a real output device via the output interfaces.

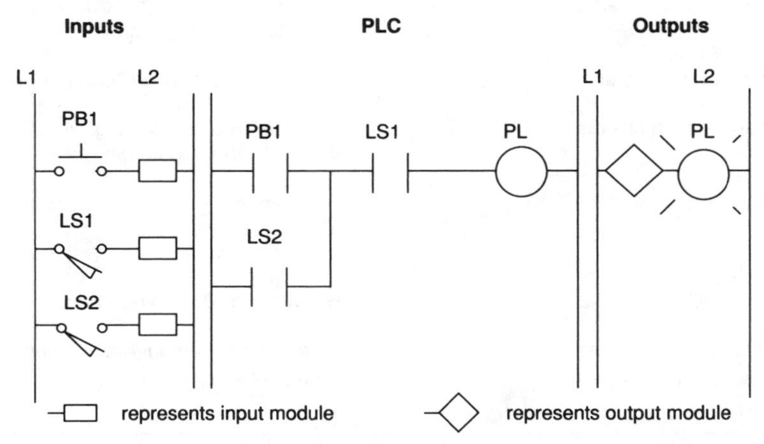

Figure 1-10. PLC implementation of Figure 1-9.

1-8 THE BENEFITS OF USING PLCS

In general, PLC architecture is modular and flexible, allowing hardware and software elements to expand as the application requirements change. In the event that an application outgrows the limitations of the PLC, the unit can easily be replaced with a unit having greater memory and I/O capacity, and the old hardware can be reused for a smaller application. A PLC system provides many benefits to a control solution, from its reliability and repeatability to its programming. The benefits accomplished with a PLC will grow with the individual using these controllers; the more you learn about them, the more you will be able to solve other control problems. Table 1-3 lists some of the many features and benefits obtained with a PLC.

Table 1-3. Typical Programmable Controller Features/Benefits

INHERENT FEATURES	BENEFITS
Solid-State Components	•High Reliability
Programmable Memory	•Simplifies Changes •Flexible Control
Small Size	•Minimal Space Requirements
Microprocessor Based	•Communication Capability •Higher Level Of Performance •Higher Quality Products •Multi-functional Capability
Software Timers/Counters	•Eliminate Hardware •Easily Changed Presets
Software Control Relays	•Reduce Hardware/Wiring Cost •Reduce Space Requirements
Modular Architecture	•Installation Flexibility •Easily Installed •Hardware Purchase Minimized •Expandability
Variety Of I/O Interface	•Controls a Variety Of Devices •Eliminates Customized Control
Remote I/O Stations	•Eliminate Long Wire/Conduit Run
Diagnostic Indicators	•Reduce Trouble-shooting Time •Signal Proper Operation
Modular I/O Interface	•Neat Appearance of Control Panel •Easily Maintained •Easily Wired
Quick I/O Disconnects	•Service w/o Disturbing Wiring
All System Variables Stored In Memory Data	•Useful Management/Maintenance •Can Be Output In Report Form

Without question, the "programmable" feature provides the single greatest benefit for the use and installation of a programmable controller. Eliminating hardwired control in favor of programmable control is the first step towards achieving a flexible control system. Once installed, the control plan can be manually or automatically altered to meet the day-to-day control requirements without changing the field wiring. This easy alteration is possible since there are no physical connections between the field input devices and output devices (see Fig. 1-11), as in a hardwired system. The only connection is through the control program, which can be easily altered.

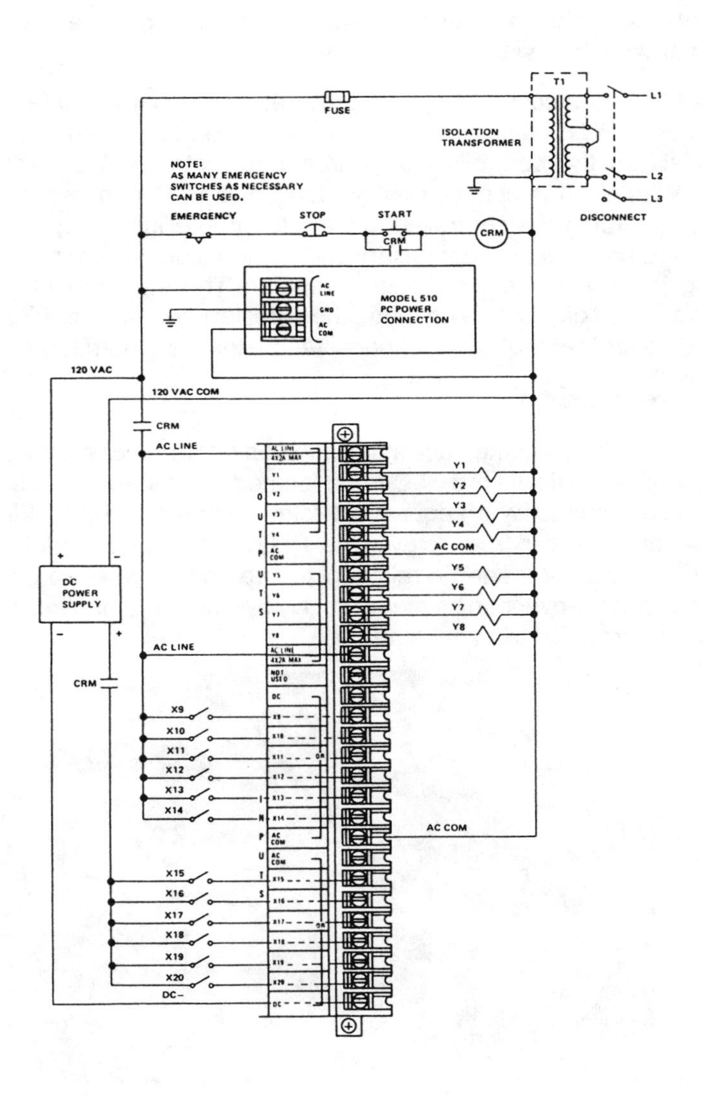

A typical example might involve a solenoid that is controlled by two limit switches connected in series. To change the solenoid operation by placing the two limit switches in parallel or combining a third switch to the existing circuitry could take less than one minute. In most cases, this simple program change can be made without shutting the system down. The same change to a hardwired system would have taken as much as thirty to sixty minutes of downtime. Even a half hour of downtime could mean a costly loss of production. A similar situation exists if there is a need to change a timer preset value or some other constant. A software timer in a PLC can be changed in as little as five seconds. A set of thumbwheel switches and a pushbutton could be configured to input new preset values to any number of software timers very easily. The time-savings benefit of altering software timers as opposed to altering several hardware timers is obvious.

Similar flexibility and cost savings are provided by the hardware features of PLCs. The intelligent CPU is capable of communicating with other intelligent devices. This capability allows the controller to be integrated into local or plantwide control schemes. With such a control configuration, the PLC can send useful English messages regarding the controlled system to an intelligent display. On the other hand, the PLC can receive supervisory information, such as production changes or scheduling information, from a host computer. The standard I/O system could include a variety of digital, analog, and other special interface modules that allow sophisticated control without using expensive, customized interface electronics.

Easy Installation

Several PLC attributes make each installation an easy and cost effective project. Its relatively small size allows the PLC to be located conveniently in usually less than half the space required by an equivalent relay control panel (see Fig. 1-12). On a small scale changeover from relays, the PLC's small and modular construction allows it to be mounted near the relay enclosure and prewired to existing terminal strips. Actual changeover can be made quickly by simply connecting the input/output devices to the prewired terminal strips.

Figure 1-12. PLC installation takes less space than other control systems (Courtesy of Omron Electronics, Schaumburg, IL).

In large installations, remote input/output stations are placed at optimum locations. The remote station is connected to the CPU via a coaxial cable or by a twisted pair of wires. This configuration results in a considerable reduction of material and labor costs that would have been associated with running multiple wires and installing large conduits. The remote subsystem approach also means that various sections of a total system could be completely prewired by an OEM or PLC vendor prior to reaching the installation site. This approach will mean a considerable reduction in the time spent by an electrician during an on-site installation.

Maintenance and Trouble-shooting

From the beginning, programmable controllers have been designed with ease of maintenance in mind. With virtually all components being solid-state, maintenance is reduced to the replacement of a modular, plug-in type component. Fault detection circuits and diagnostic indicators, incorporated in each major component, can tell if the component is working properly, or if it is malfunctioning. With the aid of the programming device, any programmed logic can be viewed to see if inputs or outputs are ON or OFF. Programmed instructions can also be written to annunciate certain failures.

These and several other attributes of the PLC make it a valuable part of any control system. Once installed, its contribution will be quickly noticed and payback will be readily realized. The potential benefits of the PLC, like any intelligent device, will depend on the creativity with which it is applied.

It is obvious from the preceeding discussion that the potential benefits which stem from applying the programmable controller in an industrial application are substantial. The list of benefits shown here has implications greater than first suggested. The bottom line is that through the use of programmable controllers the user will achieve high performance and reliability that will result in higher quality at a reduced cost.

Chapter

—2—

NUMBER SYSTEMS AND CODES

"When you measure what you are speaking about and express it in numbers, you know something about it; but when you cannot measure it, when you cannot express it in numbers, your knowledge is of a meager and unsatisfactory kind."

—William Thomson, Lord Kelvin

This chapter introduces the number systems and digital codes that are most often encountered in programmable controller applications. Binary, octal, decimal, and hexadecimal number systems are most frequently used during input/output address assignments and program preparation. Binary Coded Decimal and Gray codes used in input/output applications are also described and the ASCII character set along with several PLC register formats are presented.

2-1 INTRODUCTION

A familiarity with number systems will prove quite useful when working with either programmable controllers or any digital computer. This is true since a basic requirement of these devices is to represent, store, and operate on numbers, even to perform the simplest of operations. In general, PLCs work on binary numbers in one form or another to represent various codes or quantities. Although number operations are transparent for the most part and of little importance, there will be occasion to utilize a knowledge of number systems.

First let's review some basics. The following statements will apply to any number system:

- Each system has a base or radix.
- Each system can be used for counting.
- Each system can be used to represent quantities or codes.
- Each system has a set of symbols.

The *base* of a number system determines the total number of unique symbols used by that system. The largest-valued symbol always has a value of one less than the base. Since the base defines the number of symbols, it is possible to have a number system of any base. However, number systems are typically chosen for their convenience. The number systems usually encountered while using programmable controllers are base 2, base 8, base 10, and base 16. These systems are also labelled binary, octal, decimal and hexadecimal respectively. To demonstrate the common characteristics of number systems, let's first turn to the familiar decimal system.

2-2 NUMBER SYSTEMS

Decimal Number System

The decimal system, which is most common to us, was undoubtedly adopted as a result of man having ten fingers and ten toes. The base of the decimal number system is 10. The symbols or digits are 0,1,2,3,4,5,6,7,8, and 9. As noted earlier, the total number of symbols is the same as the base, and the largest-valued symbol is one less than the base. Because the decimal system is so commonly used, rarely do we stop to think how we express a number greater than 9. It is, however, important to note that the technique of representing a value greater than the largest symbol is the same for any number system.

To express numbers greater than nine, a place value or weight is assigned to each position that a digit would hold starting from right to left. The first position, starting from the right-most position, is position 0, the second is position 1, and so on, up to the last position (n). The weighted value of each position can be expressed as the base (10 in this case) raised to the power of (n), the position. For the decimal system then, the position weights from right to left are 1,10,100,1000, etc.

Let's take for example, the number 9876:

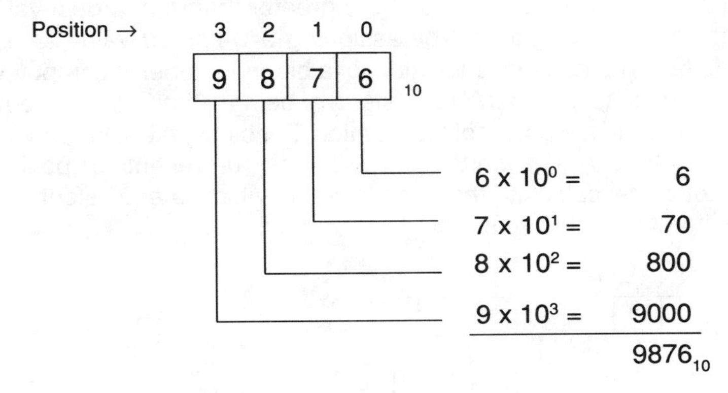

The value of the decimal number is computed by multiplying each digit by the weight of its position and summing the results. As we will see in other number systems, the decimal equivalent of any number can be computed by multiplying the digit times its base raised to the power of the digit's position. This is shown below.

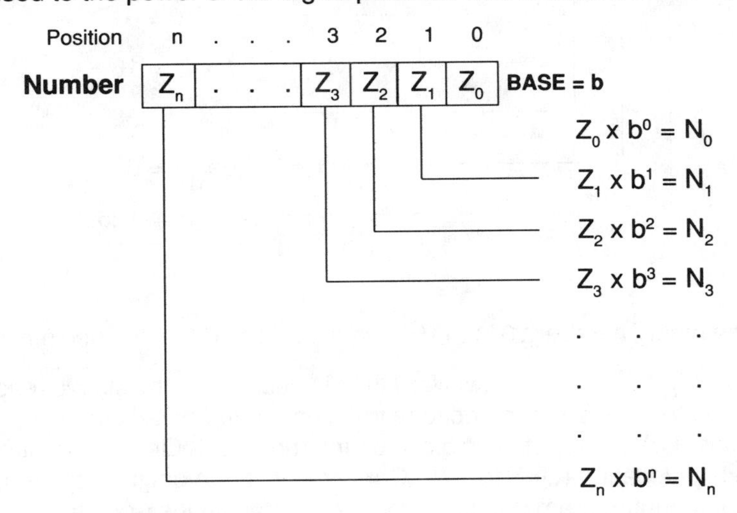

The sum of N_0 through N_n will be the decimal equivalent of the number in base "b".

Binary Number System

The binary numbering system uses the number 2 as the base. The only allowable digits are 0 and 1. There are no 2s, 3s, etc. The binary system is most accommodating for devices such as programmable controllers and digital computers. It was adopted for convenience, since it would be easier to design machines that distinguish between only two entities or numbers rather than ten as in decimal. Most physical elements have two states only: a light bulb is ON or OFF, a valve is open or closed, a switch is ON or OFF, a door is open or closed, and so on. With digital circuits, it is possible to distinguish between two voltage levels (e.g. +5V, 0V), which makes binary very applicable.

As with the decimal system, expressing numbers greater than the largest-valued symbol (in this case 1) is accomplished by assigning a weighted value to each position from right to left. The decimal equivalent of a binary number is computed in the same way as for a decimal number, only instead of being 10 raised to the power of the position, it is 2 raised to the power of the position. For binary then, the weighted values from right to left are 1,2,4,8,16,32,64, etc., representing positions 0,1,2,3,4,5,6, etc. Let's now calculate the decimal value which is equivalent to the value of the binary number 10110110:

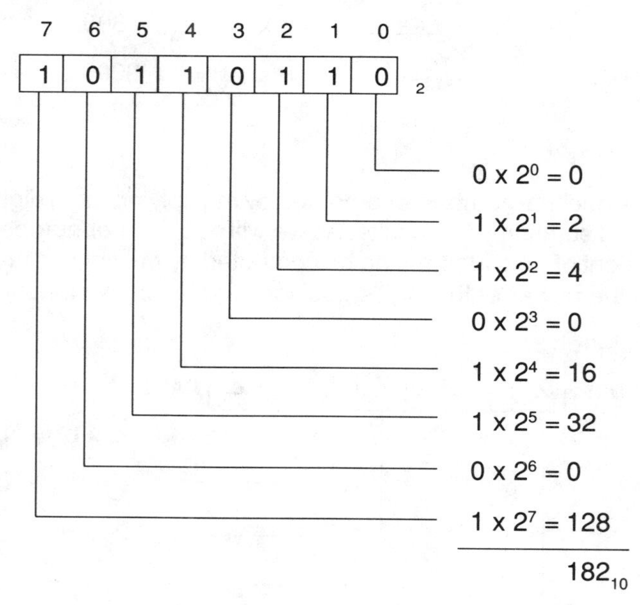

Therefore, the binary number 10110110 is equivalent to 182 in the decimal system.

Counting in binary is a little more awkward than in decimal for the simple reason that we are not used to it. This is true because the binary number system uses only two digits (zero and one), making it possible to count from 0 to 1. Only one change takes place in one digit location (OFF to ON). Conversely, in the decimal system we are accustomed to counting from 0 to 9. There are ten transitions or changes before a new digit position is added.

In binary, just like in decimal, we add another digit position once we run out of transitions. So, when we count in binary 0, 1 the next digit will be 10 (one zero, not ten) just like when we count in decimal 0,1,2...9, the next one will be 10 (ten). Table 2-1 shows a count in binary from 0_{10} to 10_{10}.

Decimal	Binary
0	0
1	1
2	10
3	11
4	100
5	101
6	110
7	111
8	1000
9	1001
10	1010

Table 2-1. Binary and Decimal counting

Each digit of a binary number is known as a "bit", therefore this particular binary number 10110110 (182 decimal) has 8 bits or positions.

A group of 4 bits is known as a nibble, usually a group of 8 bits is a byte, and a group of one or more bytes is a word. Figure 2-1 presents a binary number composed of 16 bits, with the least and most significant bits.

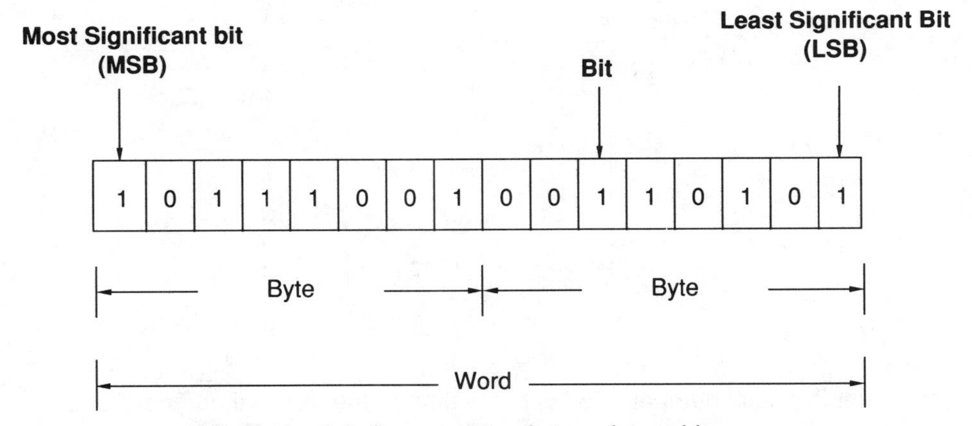

Figure 2.1. One word, two bytes, sixteen bits

Octal Number System

Expressing a number in binary obviously requires substantially more digits than in the decimal system. For example, $91_{10} = 1011011_2$. Too many binary digits can become cumbersome to read or write when using large numbers, especially for human readers or writers. This system uses the number 8 as its base. The eight digits are 0,1,2,3,4,5,6, and 7. The octal numbering system was brought into use because of its distinctive way of representing binary numbers using less digits. An octal count representation of the numbers 0 through 15 (decimal) is shown in Table 2-2.

Decimal	Binary	Octal
0	000	0
1	001	1
2	010	2
3	011	3
4	100	4
5	101	5
6	110	6
7	111	7
8	1000	10
9	1001	11
10	1010	12
11	1011	13
12	1100	14
13	1101	15
14	1110	16
15	1111	17

Table 2-2. Binary and related Octal Code

Like all other number systems, each digit in an octal number has a weighted decimal value according to its position. For example:

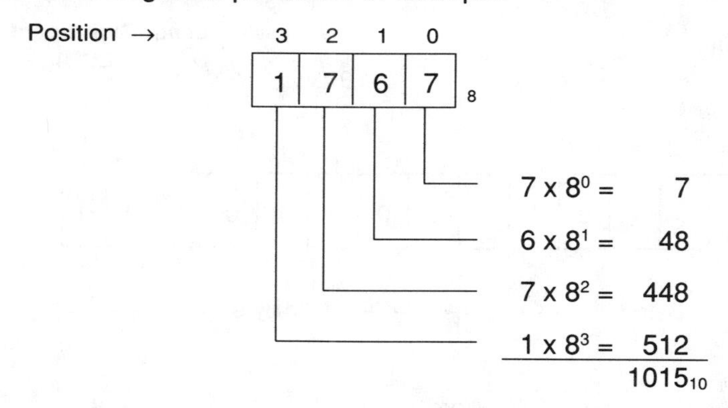

Therefore, the octal number 1767 is equivalent to the decimal number 1015.

As noted, octal is used as a convenient means of writing or handling a binary number. If a binary number is very large (many 1s and 0s), it can be represented by an equivalent octal number.

The octal system has a base of 8, or 2 to the 3rd power, making it possible to represent any binary number in octal by grouping binary bits in groups of three. In this manner, a very large binary number can be easily represented by an octal number with significantly fewer digits. For example:

1	1	1	0	0	0	1	1	1	1	1	0	1	0	1	1	**Binary Number**

1	1 1 0	0 0 1	1 1 1	1 0 1	0 1 1	**3-Bit Groups**

1	6	1	7	5	3	**Octal Digits**

As we will see later, most programmable controllers use the octal number system for referencing input/output and memory addresses. So, the 16-bit binary number can be represented directly by 6 digits in octal.

Hexadecimal Number System

The hexadecimal (hex) numbering system uses 16 as the base. It consists of 16 digits, numbers 0 thru 9, and the letters A through F, which are substituted for the numbers 10 to 15 respectively. The hexadecimal system is used for the same reason the octal system is employed, to represent binary numbers with less digits. The hex numbering system uses one digit to represent four binary digits (or bits) instead of three as in the octal system. Table 2-3 shows a hexadecimal count example of the number 0 through 15 with its decimal and binary equivalents.

Hexadecimal	Binary	Decimal
0	0000	0
1	0001	1
2	0010	2
3	0011	3
4	0100	4
5	0101	5
6	0110	6
7	0111	7
8	1000	8
9	1001	9
A	1010	10
B	1011	11
C	1100	12
D	1101	13
E	1110	14
F	1111	15

Table 2-3. Hexadecimal code related to binary and decimal.

As with the other number systems, hexadecimal numbers can be represented by their decimal equivalents by using the sum of the weights method. Please refer to the example on the following page.

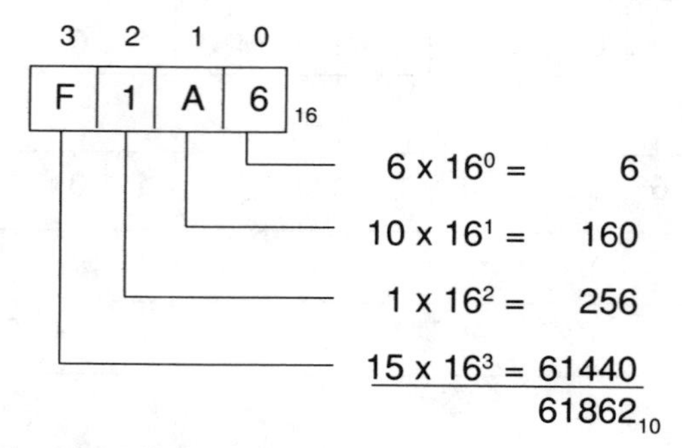

Therefore, the hexadecimal number F1A6 is equivalent to the decimal number 61,862.

The decimal values of the letter-represented digits A through F are used when computing the decimal equivalent; that is 10 for A, 11 for B, and so on. Therefore, the value of A in the example will be 10 times 16^1, while F will be 15 times 16^3. Like octal numbers, hexadecimal numbers can be easily converted to binary numbers without any mathematical conversion. Conversion is accomplished by writing the 4-bit binary equivalent of the hex digit of each position. For example:

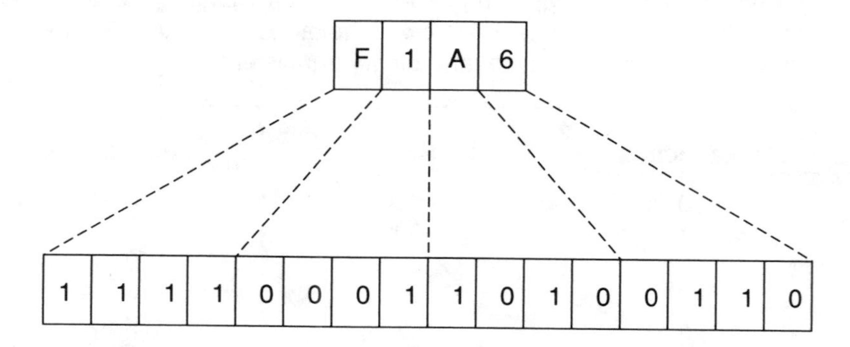

2-3 NUMBER CONVERSIONS

In the previous section we saw how a number of any base can be converted to our familiar decimal system using the sum of the weights method. In this section, we show how a decimal number can be converted to binary, octal or any number system.

To convert a decimal number to its equivalent in any base, a series of divisions by the desired base must be performed. The conversion process starts by dividing the decimal number by the base; if there is a remainder, it is placed in the least significant digit of the new base number. If there is no remainder, a 0 is placed in the next digit position (from right to left). The result is brought down, and the process repeated until the final result of the successive division has been reduced to 0. The methodology used may be a little cumbersome; however, this method proves to be the easiest to understand and employ.

As a generic example (refer to Figure 2-2), let's take the number "Z" and find its base 5 equivalent. The first division ($Z \div 5$) gives an N_1 result and a remainder R_1. The remainder R_1 becomes the first digit of the base 5 number. To obtain the next base 5 digit, the N_1 result is divided again by 5 giving a N_2 result and a R_2 remainder which becomes the second base 5 digit. This process is repeated until the result of the division is 0 ($N_n \div 5$) giving the last remainder R_n, which becomes the most significant digit of the base 5.

Divisions	Remainder
$Z \div 5 = N_1$	R_1
$N_1 \div 5 = N_2$	R_2
$N_2 \div 5 = N_3$	R_3
$N_3 \div 5 = N_4$	R_4
.	
.	
.	
$N_n \div 5 = 0$	R_n
New base 5 number is $(R_n \ldots R_4\, R_3\, R_2\, R_1)_5$	

Figure 2-2. Representation of how to obtain a number Z into its equivalent base 5 number.

Let's now convert the decimal number 35_{10} to its binary (base 2) equivalent using the above rule:

Division	Remainder
$35 \div 2 = 17$	1
$17 \div 2 = 8$	1
$8 \div 2 = 4$	0
$4 \div 2 = 2$	0
$2 \div 2 = 1$	0
$1 \div 2 = 0$	1
Therefore, the base 2 equivalent of 35_{10} would be 100011.	

As a second exercise let's convert the number 1355_{10} to hexadecimal (base 16):

Division	Remainder
1355 ÷ 16 = 84	11
84 ÷ 16 = 5	4
5 ÷ 16 = 0	5

Therefore, the hexadecimal equivalent of 1355_{10} will be $54B_{hex}$ (remember that 11 is represented as B in hexadecimal).

There are other ways to compute a binary number from its decimal equivalent which would tend to be a bit faster. This method employs division by eight instead of by two to convert the number to octal first and then to binary from octal (3 bits at a time). For instance, let's take the number 145_{10}:

Division	Remainder
145 ÷ 8 = 18	1
18 ÷ 8 = 2	2
2 ÷ 8 = 0	2

The octal equivalent of 145_{10} is 221_8. From Table 2-2 we can see that 221_8 is 010010001.

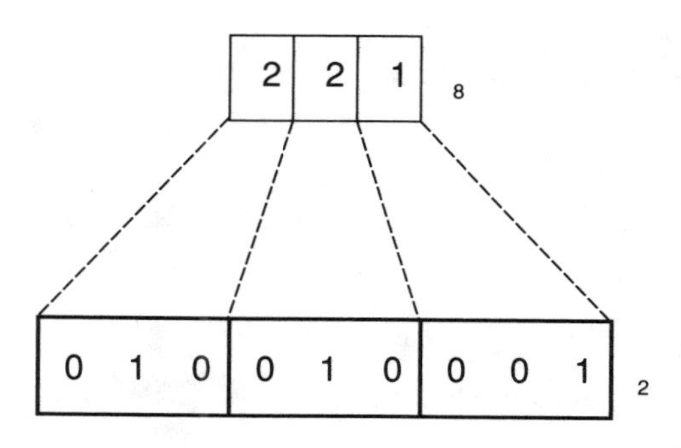

2-4 ONE'S AND TWO'S COMPLEMENT

One's and two's complement of a binary number is an operation used by programmable controllers, as well as computers, to perform internal mathematical calculations. To complement a binary number is to change it to a negative number. This allows the basic arithmetic operations of subtraction, multiplication, and division to be performed through successive addition. For example, to subtract the number 20 from 40 is done by first complementing 20 to obtain -20, and then performing an addition.

The intention here is to introduce the basic concepts of complementing rather than to provide a thorough analysis of arithmetic operations. However, several of the references listed in the back of this book cover this material.

One's Complement

Let us assume that we have a 5-bit binary number that we wish to represent as a negative number. The number is decimal 23.

$$(10111)_2$$

There are two ways of representing this number as a negative number. The first method is simply to place the minus sign in front of the number, as with decimal numbers.

$$-(10111)_2$$

This method is suitable for us, but impossible for programmable controllers or computers, since the only symbols they use are binary 1s and 0s. To represent negative numbers then, digital computing devices use what is known as the complement method. First, the complement method places an extra bit (sign bit) in the most significant position (left-most), and lets this bit determine whether the number is positive or negative. The number is positive if the sign bit is 0 and negative if the sign bit is 1. Using the complement method, +23 would be represented as shown here:

$$0\ 10111_2$$

The negative representation of binary 23 using the 1s complement method, is simply obtained by inverting each bit, or changing 1s to 0s, and 0s to 1s. The one's complement of binary 23 then is shown here.

$$1\ 01000_2$$

If a negative number was given in binary, its complement would be obtained in the same fashion. Let's take the number -15:

$$1\ 0000 \rightarrow -15$$

$$0\ 1111 \rightarrow +15$$

Two's Complement

Two's complement is similar to 1s complement in the sense that 1 extra digit is used to represent the sign. The two's complement computation, however, differs slightly. In 1s complement, all bits are inverted; but in two's complement, each bit from right to left is inverted, but only after the first "1" is detected. Let's use the number +22 as an example:

0 10110 → +22

Its two's complement would be:

1 01010 → -22

Note that in the positive representation of the number 22, starting from the right, the first digit is a 0 so it is not inverted; the second digit is a 1, so all digits after this one are inverted.

If a negative number is given in two's complement, its complement (a positive number) is found in the same fashion.

1 10010 → -14

0 01110 → +14

All bits from right to left are inverted after the first 1 is detected. Other examples of two's complement are shown here:

0 10001 → +17   0 00111 → +7

1 01111 → -17   1 11001 → -7

0 00001 → +1

1 11111 → -1

The two's complement of 0 is not found, since no first 1 is ever encountered in the number. The two's complement of 0 then is still 0.

Two's complement is the most common arithmetic method used in computers as well as programmable controllers.

2-5 BINARY CODES

An important requirement of programmable controllers is to communicate with various external devices that supply information to the controller or receive information from the controller. This input/output function involves the transmission, manipulation, and storage of binary data that, at some point, must be interpreted by humans. Although the machine can easily handle this binary data, we require that the data be converted to a more interpretable form.

One way of satisfying this requirement is to assign a unique combination of 1s and 0s to each number, letter, or symbol that must be represented. This technique is called binary coding. In general, there are 2 categories of codes—those used to represent numbers only, and those that represent letters, symbols, and decimal numbers. Several codes for representing numbers, symbols, and letters have been instituted and are standard throughout the industry. Among the most common are the following:

- ASCII
- BCD
- GRAY

ASCII

Alphanumeric (letters, symbols, decimal numbers) codes are used when information processing equipment, such as printers or CRTs, must handle the alphabet as well as numbers and special symbols. These characters—26 letters (uppercase), 10 numerals (0-9), plus mathematical and punctuation symbols—can be represented using a 6 bit code (i.e. $2^6 = 64$). The most common code for alphanumeric representation is the *American Standard Code for Information Interchange* (ASCII). The acronym is pronounced "askey".

The ASCII code can be 6,7, or 8 bits. Even though the basic alphabet, numbers, and special symbols can be accommodated in a 6 bit code (64 possible characters), standard ASCII character sets use a 7-bit code ($2^7 = 128$ possible characters), which allows lower case and control characters for communication links, in addition to the characters already mentioned. This 7-bit code provides all possible combinations of characters used when communicating with peripherals or interfaces. The 8 bit ASCII code is used when parity check (see Chapter 4) is added to the standard 7-bit code for error checking. Note that all 8 bits can still be placed in one byte. Figure 2-3a shows the binary ASCII code representation of the letter "Z" (132_8). This letter is generally sent/received in serial form between the PLC and other communicating equipment. A typical ASCII transmission is illustrated in Figure 2-3b using the character "Z" as an example. Note that two more bits are added to the character to signify the start and stop of the ASCII character transmission. Appendix E shows a standard ASCII table, while more serial communication explanation is covered in Chapter 8.

BCD

Binary Coded Decimal was introduced as a convenient means for humans to handle numbers that had to be input to digital machines and to interpret numbers output from the machine. The best solution to this problem is a means of converting a code readily handled by man (decimal) to a code readily handled by the equipment (binary). The result is BCD.

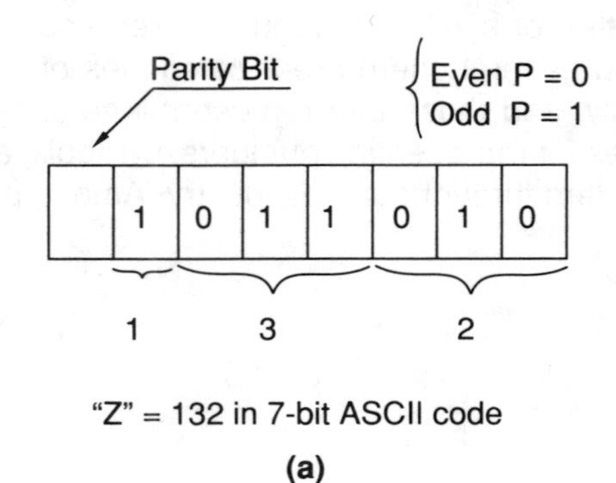

(a)

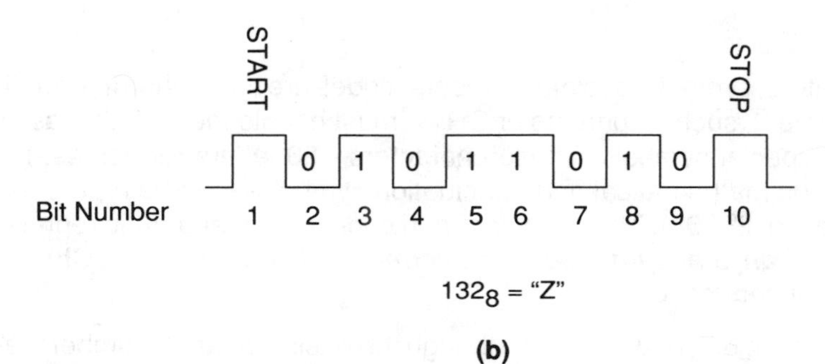

(b)

Figure 2-3. ASCII representation of character "Z" (a); ASCII transmission representation of the letter "Z" (b).

In decimal, we have the numbers 0 thru 9, whereas in BCD, each of these numbers is represented by a 4-bit binary number. Table 2-4 illustrates the relationship between the BCD code and the binary and decimal number systems.

Decimal	Binary	BCD
0	0	0000
1	1	0001
2	10	0010
3	11	0011
4	100	0100
5	101	0101
6	110	0110
7	111	0111
8	1000	1000
9	1001	1001

Table 2-4. BCD code with binary and decimal equivalents.

The BCD representation of a decimal number is obtained simply by replacing each decimal digit by its BCD equivalent. The BCD representation of decimal 7493 is shown here as an example.

0111	**0100**	**1001**	**0011**
7	4	9	3

Typical PLC applications of BCD codes include data entry (time, volume, weights, etc.) via thumbwheel switches (TWS), data display via 7-segment displays, input from absolute encoders, and uses in analog input/output instructions. Nowadays, the necessary circuitry to convert from decimal to BCD and BCD to 7-segment is found already built into the TWS and 7-segment LED devices. This BCD data is then taken by the PLC and converted internally into the binary equivalent of the input data through some instruction. Input or output of BCD data requires 4 lines to an input/output interface for each decimal digit. Figure 2-4 shows a thumbwheel switch and a 7-segment indicator.

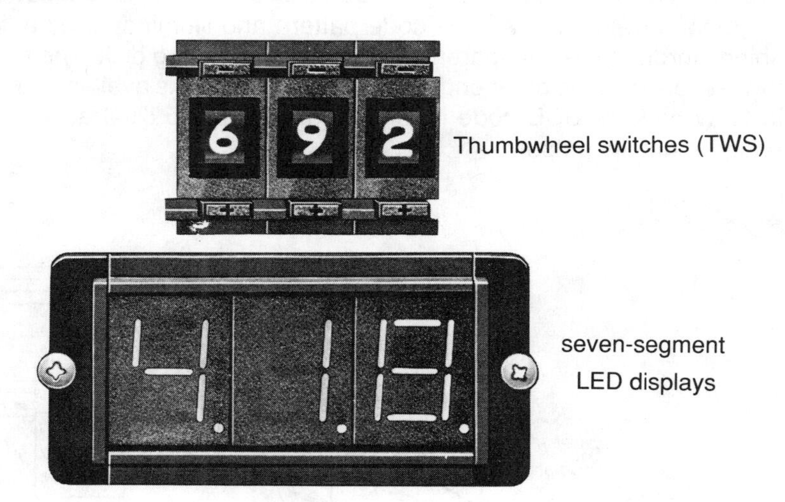

Figure 2-4.

Gray

The Gray code is one of a series of cyclic codes known as reflected codes and suited primarily for position transducers. It is basically a binary code that has been modified in such a way that only 1 bit changes as the counting number increases. In standard binary, as many as 4 digits could change when counting with as few as four binary digits. This drastic change is seen in the transition from binary 7 to 8. Such a change allows a greater chance for error, which would be unsuitable for positioning applications. Most encoders use this code to determine angular position. Table 2-5 shows the Gray code and binary equivalent for comparison.

Gray code	Binary
0000	0000
0001	0001
0011	0010
0010	0011
0110	0100
0111	0101
0101	0110
0100	0111
1100	1000
1101	1001
1111	1010
1110	1011
1010	1100
1011	1101
1001	1110
1000	1111

Table 2-5. Gray Code.

In an optical absolute encoder, the rotor disk consists of an opaque and transparent segment, arranged in a Gray code pattern and illuminated by a light source that shines through the transparent sections of the rotating disk. The transmitted light is then received at the other end in Gray code form and is available for input to the PLC in Gray code or BCD code if converted. Figure 2-5 illustrates a typical absolute encoder and its output.

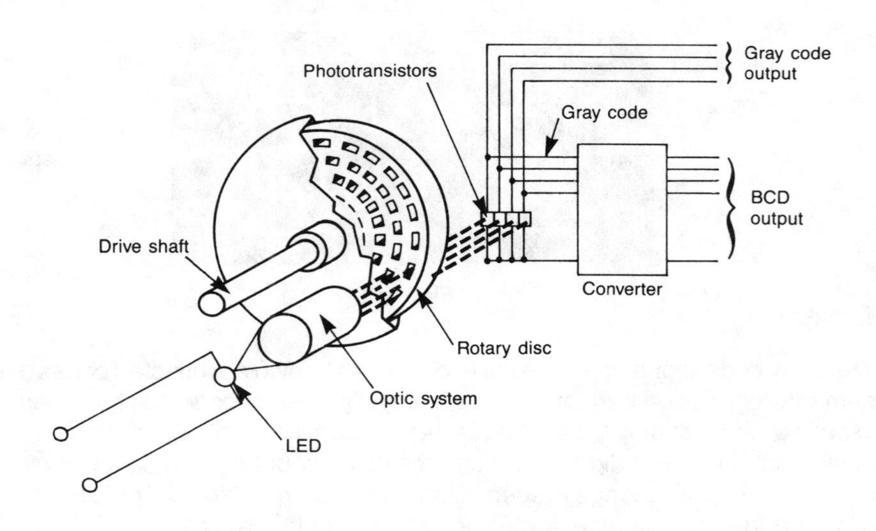

Figure 2-5. Illustration of an absolute encoder with BCD and Gray outputs.

2-6 REGISTER WORD FORMATS

As previously mentioned, the programmable controller performs all of its internal operations in binary using 1s and 0s. In addition, the status of I/O field devices is also read or written to or from the PLC's CPU in binary form. Generally, these operations are performed using a group of 16 bits that represent numbers and codes. Recall that the grouping (unit) of bits on which a particular machine operates is called a *word*. A PLC word is also called a "register" or "location". Figure 2-6 illustrates a 16-bit register, comprised of 2 bytes.

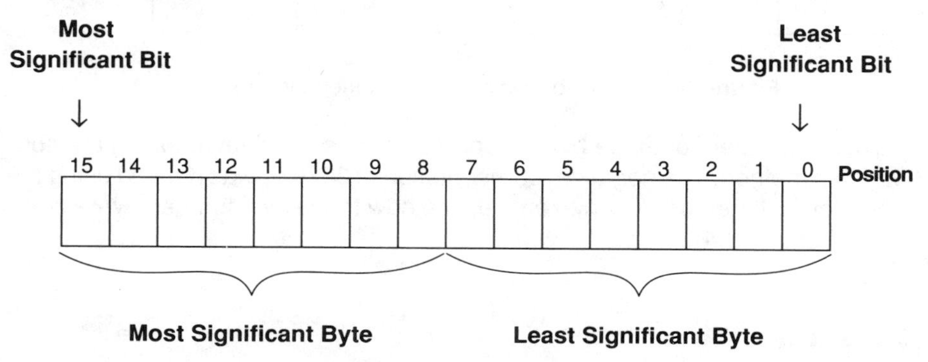

Figure 2-6. A 16-bit register/word

Although the data stored in any register is represented in binary 1s and 0s, the format in which it is stored may differ from one controller to another. Generally, data is represented in straight (not coded) binary, or Binary Coded Decimal (BCD). Let's examine these two formats.

Binary Format

Data stored in this format can be directly converted to its decimal equivalent without any special restrictions. In a 16-bit register then, a maximum value of 65535_{10} can be represented. Figure 2-7 shows the value 65535_{10} in binary format (all bits are 1). As a preview, remember that in using binary we are able to represent the 1 or 0 status, which can be (and will be) interpreted by the PLC as something ON or OFF. All these statuses are stored in registers or words.

15	14	13	12	11	10	9	8	7	6	5	4	3	2	1	0
1	1	1	1	1	1	1	1	1	1	1	1	1	1	1	1

Figure 2-7. A 16-bit register containing $65,535_{10}$ in binary.

If the most significant bit of the register in Figure 2-8 is used as a sign bit, then the maximum decimal value that can be stored is +32767, or -32767.

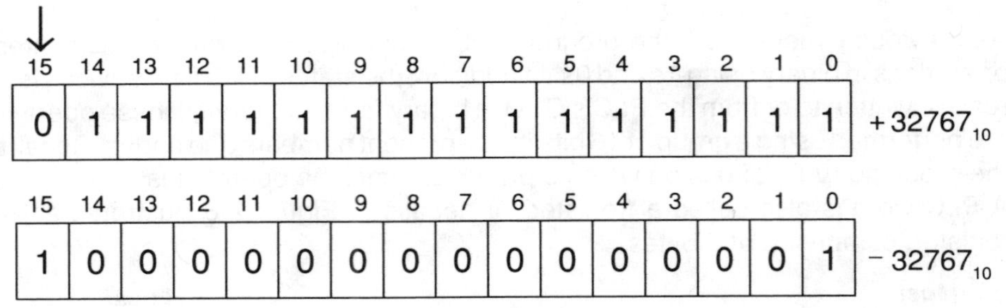

Figure 2-8. Two 16-bit registers with signed bit (MSB).

The decimal equivalent of these binary representations is achieved using the sum of the weights method. The negative representation of 32767_{10} was derived using the two's complement method. As an exercise, see how these two numbers were computed (see Section 2-4).

BCD Format

If data is stored in a BCD format, then 4 bits are used to represent a single decimal digit. The only decimal numbers that these 4 bits can represent are 0 through 9. There are programmable controllers that operate and store data in several of their software instructions such as arithmetic and data manipulation using the BCD format.

Our 16-bit register then can hold up to a 4 digit decimal equivalent, and the decimal values that can be represented are 0000-9999. The binary representation for BCD 9999 is shown in Figure 2-9.

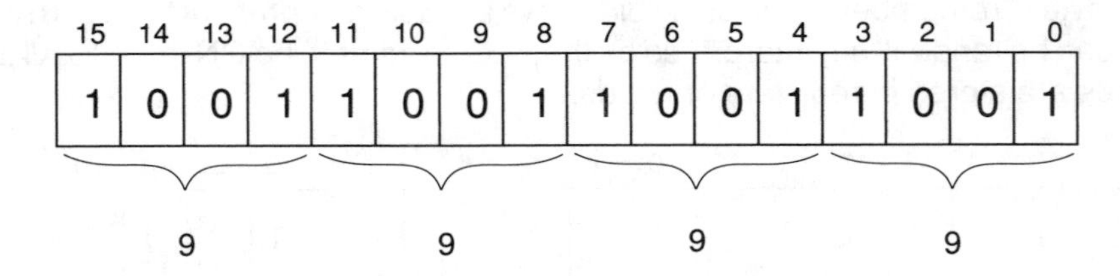

Figure 2-9. Register containing BCD 9999.

Chapter

—3—

LOGIC CONCEPTS

"If it was so, it might; and if it were so, it would be; but as it isn't, it ain't. That's logic."

—Lewis Carroll

A fundamental prerequisite for understanding programmable controllers and their applications is a working knowledge of logic operations. In this chapter and in later chapters, we will show how three basic logic functions—AND, OR, and NOT—can be combined to make very simple to very complex control decisions.

3-1 THE BINARY CONCEPT

The binary concept is not a new idea; in fact, it is a very old one. It simply refers to the idea that many things can be thought of as existing in one of two predetermined states. For instance, a light can be ON or OFF, a switch OPEN or CLOSED, or a motor RUNNING or STOPPED. In digital systems, these two-state conditions can be thought of as a signal that is PRESENT or NOT PRESENT, ACTIVATED or NOT ACTIVATED, HIGH or LOW, ON or OFF, etc. This two-state concept can be the basis for making decisions, and since it is very adaptable to the binary number system, it is a fundamental building block for programmable controllers and digital computers.

Here, and throughout this book, binary "1" represents the presence of a signal, or the occurrence of some event, while binary "0" represents the absence of the signal, or non-occurrence of the event. In digital systems, these two states are actually represented by two distinct voltage levels, as shown in Table 3-1. One voltage is more positive (or at a higher reference) than the other. Often, binary 1 (or logic 1) will be interchangeably referred to as TRUE, ON, or HIGH. Binary 0 (or logic 0) will be referred to as FALSE, OFF, or LOW.

1 (+V)	0 (0V)	EXAMPLE
OPERATED	NOT OPERATED	LIMIT SWITCH
RINGING	NOT RINGING	BELL
ON	OFF	LIGHT BULB
BLOWING	SILENT	HORN
RUNNING	STOPPED	MOTOR
ENGAGED	DISENGAGED	CLUTCH
CLOSED	OPEN	VALVE

Table 3-1. Binary Concept Using Positive Logic

It should be noted in Table 3-1 that the more positive (HIGH) voltage, represented as logic 1, and the less positive (LOW) voltage, represented as logic 0, were arbitrarily chosen. The use of binary logic to represent "1" as the more positive voltage level and the occurrence of some event is referred to as *positive logic.*

Negative logic, as illustrated in Table 3-2, is the use of binary logic so that the more positive voltage level or the occurrence of the event is represented by "0". One "1" represents non-occurrence of the event or the less positive voltage level. Although positive logic is the more conventional of the two, negative logic sometimes may be more convenient.

1 (0V)	0 (+V)	EXAMPLE
NOT OPERATED	OPERATED	LIMIT SWITCH
NOT RINGING	RINGING	BELL
OFF	ON	LIGHT BULB
SILENT	BLOWING	HORN
STOPPED	RUNNING	MOTOR
DISENGAGED	ENGAGED	CLUTCH
OPEN	CLOSED	VALVE

Table 3-2. Binary Concept Using Negative Logic

3-2 LOGIC FUNCTIONS

The binary concept shows how physical quantities (binary variables), that can exist in one of two states, could be represented as "1" or "0". Now, we will see how statements that combine two or more of these binary variables can result in a true or false condition, represented by "1" and "0". The programmable controller will eventually make decisions based on the results of these statements.

Operations performed by digital equipment, such as programmable controllers, are based on three fundamental logic operations—AND, OR, and NOT. These operations are used to combine binary variables to form statements. Each function has a rule that will determine the statement outcome (true or false) and a symbol that represents the operation. For the purpose of this discussion, the result of a statement is called an output (Y), and the conditions of the statement are called inputs (A,B). Both the inputs and outputs represent two-state variables such as those discussed earlier in this section.

The AND Function

The symbol shown in Figure 3-1 is called an AND gate and is used to represent graphically the AND function. The AND output is True(1) only if all inputs are True(1).

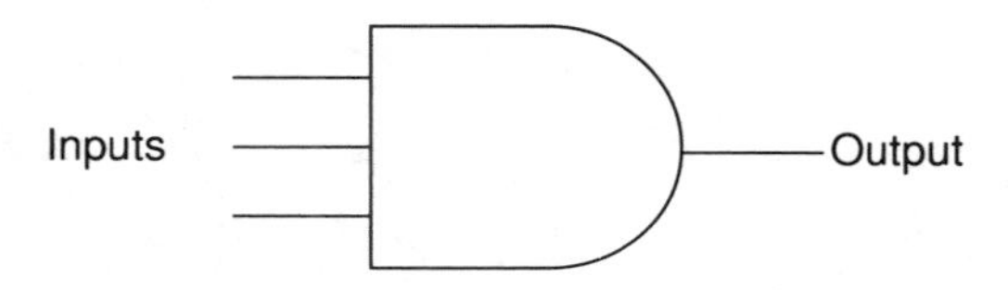

Figure 3-1. Symbol for the AND Function.

The number of inputs to the AND gate function are unlimited, but there is only one output. Figure 3-2 shows the resulting output (Y), based on all possible input combinations for a two-input gate. This mapping of outputs according to its predefined inputs is called a *Truth Table*.

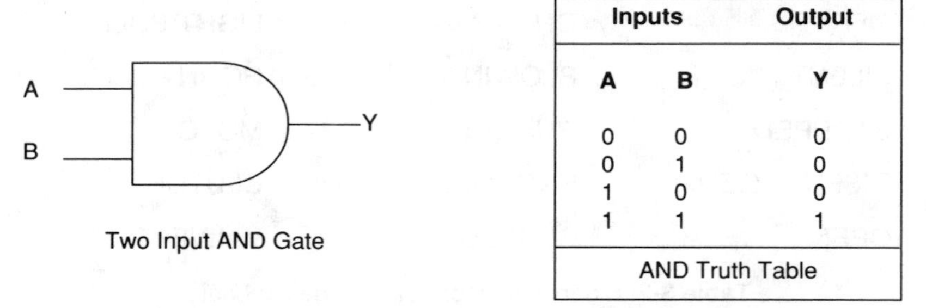

Inputs		Output
A	**B**	**Y**
0	0	0
0	1	0
1	0	0
1	1	1

AND Truth Table

Figure 3-2. Two input AND gate and its truth table.

The following example shows applications of the AND function. A and B represent inputs to the controller.

Example 3-1

Alarm horn will sound if pushbutton PB1 and PB2 are 1 (ON, or depressed) at the same time.

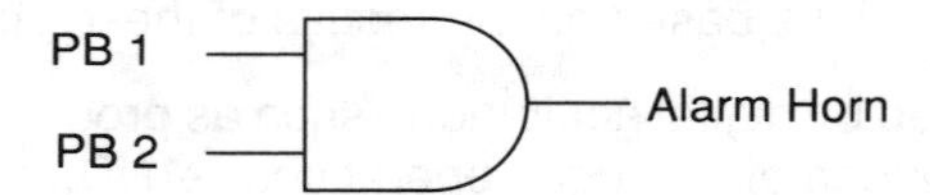

PB1	PB2	Alarm Horn
Not-Pushed (0)	Not-Pushed (0)	Silent (0)
Not-Pushed (0)	Pushed (1)	Silent (0)
Pushed (1)	Not-Pushed (0)	Silent (0)
Pushed (1)	Pushed (1)	Sound (1)

AND Logic Representation

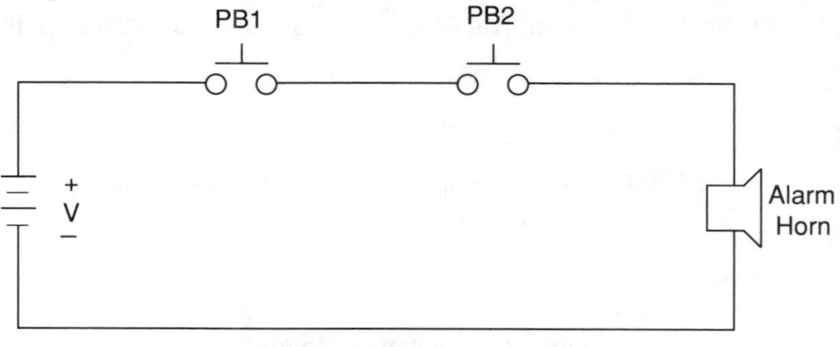

Electronic Representation

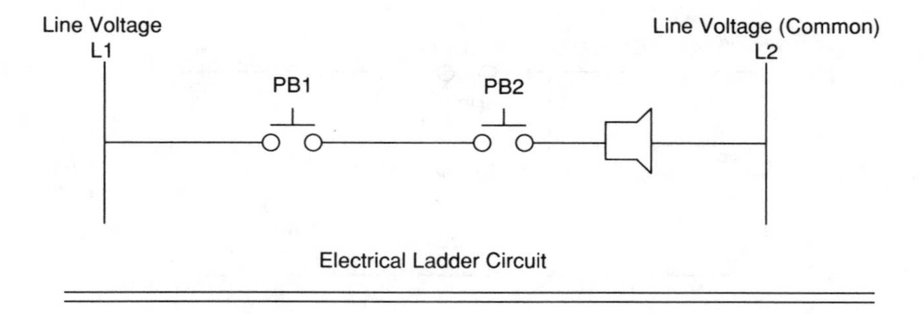

Electrical Ladder Circuit

The OR Function

The symbol shown in Fig. 3-3 , called an OR gate, is used to represent graphically the OR function. The OR output is True(1) if one or more inputs are True(1).

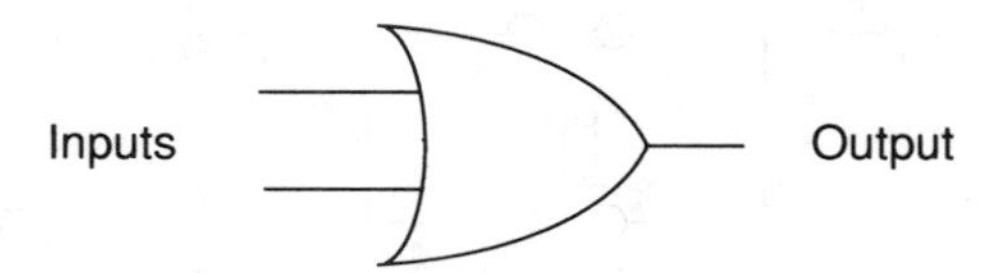

Figure 3-3. Symbol for the OR Function

As for the AND function, the number of inputs to an OR gate function are unlimited, but there is only one output. The truth table (mapping of outputs to predetermined inputs) shows the resulting output(Y), based on all possible input combinations (Fig. 3-4).

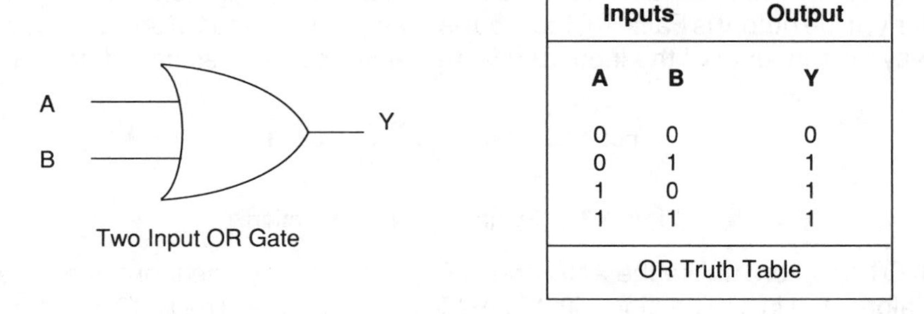

Inputs		Output
A	**B**	**Y**
0	0	0
0	1	1
1	0	1
1	1	1

OR Truth Table

Figure 3-4. Two input OR gate and its truth table.

The next example shows an application of the OR function. A and B are inputs to the controller.

Example 3-2

Alarm horn will sound if either pushbutton PB1 or PB2 is 1 (ON, or depressed).

PB1

PB2

Alarm Horn

Logic Representation

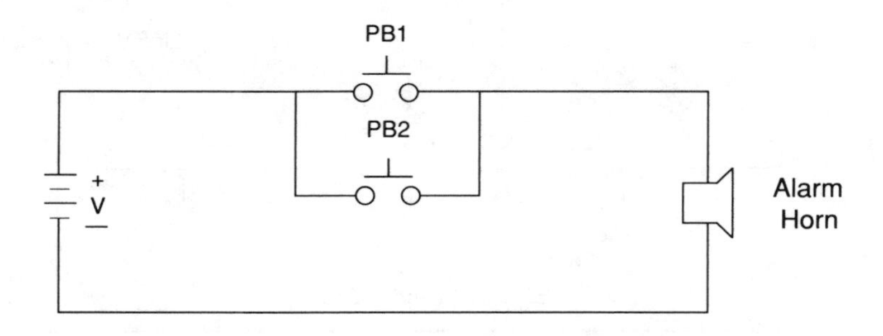

Electronic Representation

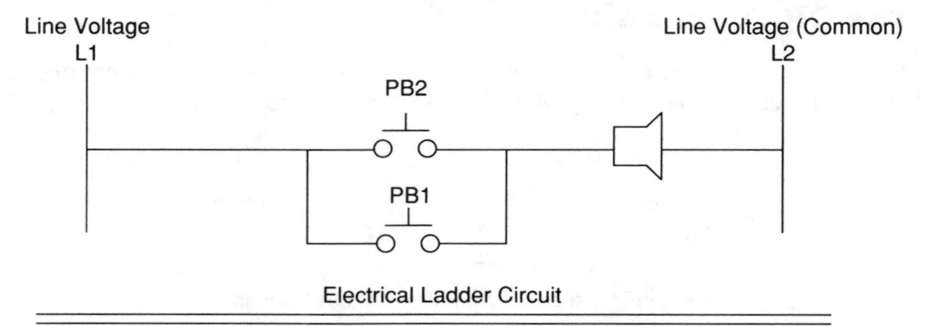

Electrical Ladder Circuit

The NOT Function

The symbol shown in Figure 3-5 is the NOT symbol and is used to represent graphically the NOT function. The NOT output is True (1) if the input is False (0). Similarly, if the output is False (0) the input is True (1). The result of the NOT operation is always the inverse of the input and is, therefore, sometimes called an *inverter.*

Figure 3-5. Symbol for NOT Function.

The NOT function, unlike the AND and OR, can have only one input and is seldom used alone, but in conjunction with the AND or the OR gate. The NOT operation and truth table are shown in Figure 3-6; note that a NOT A can be represented as an A with a bar on top.

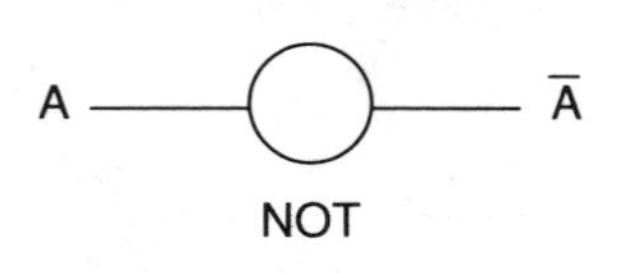

Input	Output
A	$\bar{\mathbf{A}}$
0	1
1	0
NOT Truth Table	

Figure 3-6. NOT gate representation and its truth table.

At first glance, application of the NOT function is not as easily visualized as the AND and OR functions. However, a closer examination of the NOT function shows it to be simple and quite useful. At this point, it would be helpful to recall three things that have been discussed.

1. Assigning 1 or 0 to a condition is arbitrary.
2. A "1" is normally associated with TRUE, HIGH, ON, etc.
3. A "0" is normally associated with FALSE, LOW, OFF, etc.

Examining 2 and 3 shows that logic 1 is normally expected to activate some device (e.g. if Y=1, motor runs), and an output of logic 0 is normally expected to deactivate some device (e.g. if Y=0, motor stops). If these conventions were to be reversed (see Section 3-1), such that a logic 0 is expected to activate some device(e.g. if Y=0, motor runs), and a logic 1 is expected to deactivate some device(e.g. Y=1, motor stops), the NOT function would then have useful application.

1. A NOT is used when a "0"(Low condition) is to activate some device.
2. A NOT is used when a "1"(High condition) is to deactivate some device.

The following two examples show applications of the NOT function. Although the NOT is normally used in conjunction with the AND and OR functions, for clarity the first example shows the NOT used alone.

Example 3-3

The solenoid valve V1 will be open (ON) if the selector switch S1 is ON and if the level switch L1 is NOT ON (liquid has not reached level).

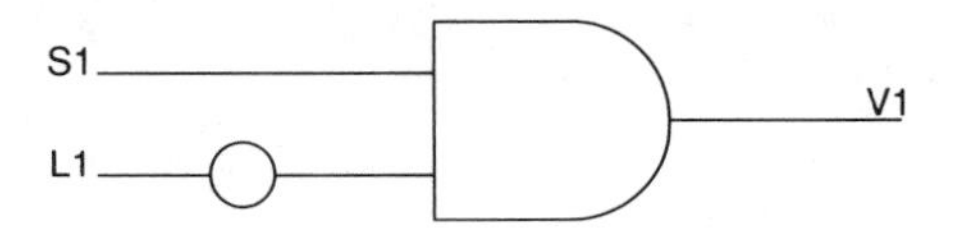

S1	L1	$\overline{(L1)}$	V1
0	0	1	0
0	1	0	0
1	0	1	1
1	1	0	0

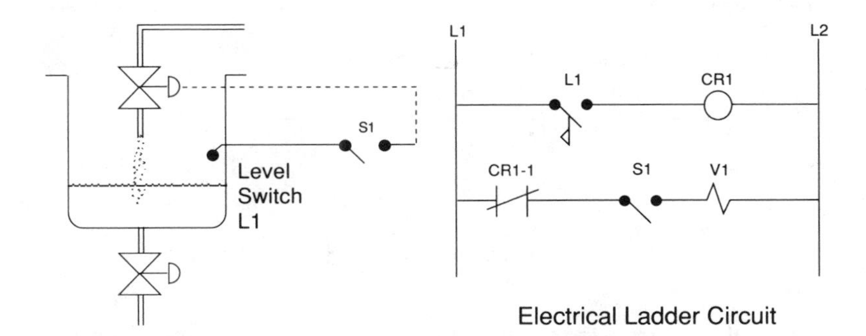

Electrical Ladder Circuit

Note: In this example the level switch L1 is normally open, which closes when the liquid level reaches L1. The ladder circuit requires an auxiliary control relay (CR1) to implement the not-normally open L1 signal. When L1 closes (ON), CR1 is energized, thus opening the normally closed CR1-1 contacts thus deactivating V1. S1 is ON when the system operation is enabled.

Example 3-4

Alarm horn will sound if pushbutton PB1 is 1 (ON, or depressed) and PB2 is NOT 0 (not depressed).

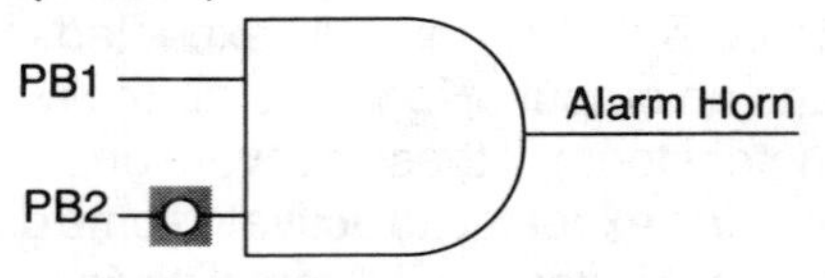

Logic Representation

PB1	PB2	Alarm Horn
Not-Pushed (0)	Not-Pushed (0)	Silent (0)
Not-Pushed (0)	Pushed (1)	Silent (0)
Pushed (1)	Not-Pushed (0)	Sounding (1)
Pushed (1)	Pushed (1)	Silent (0)

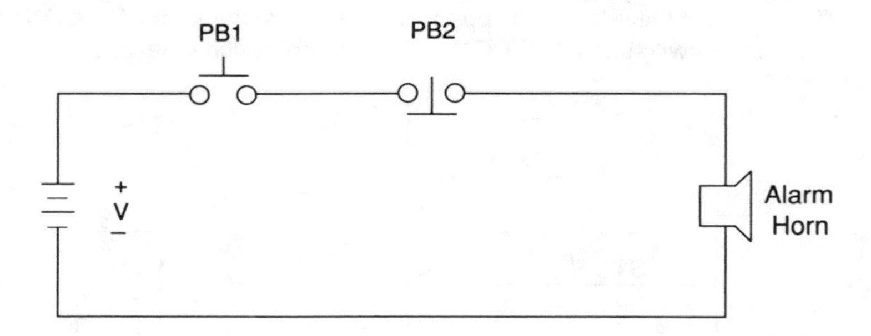

Electronic Representation

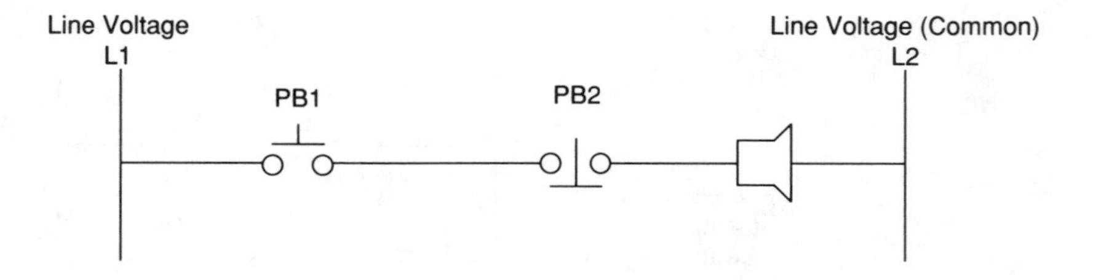

Electrical Ladder Circuit

Note: In this example the physical representation of a field device element that signifies the NOT function is represented as a "normally closed" or "not normally open" switch (PB2). In the logical representation section of this example, the pushbutton switch is represented as NOT OPEN by the symbol —○— .

The two examples showed the NOT symbol placed on inputs to a gate. The NOT symbol placed at the output of an AND gate would negate or invert the normal output result. A negated AND gate is called a NAND gate. The logic symbol and truth table are shown in Figure 3-7.

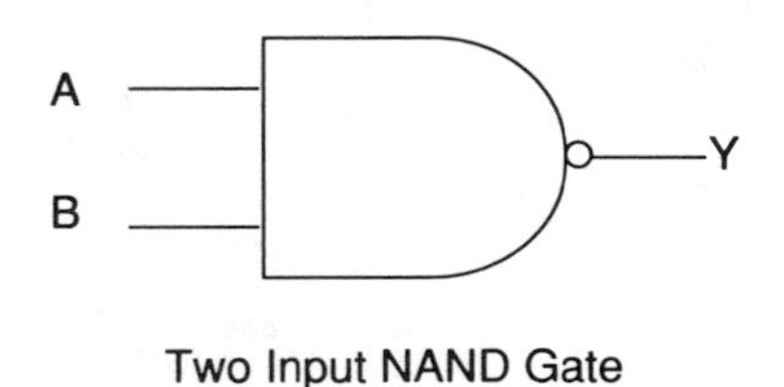

Two Input NAND Gate

Inputs		Output
A	**B**	**Y**
0	0	1
0	1	1
1	0	1
1	1	0

NAND Truth Table

Figure 3-7. Two input NAND gate and its truth table.

The same principle applies if a NOT symbol is placed at the output of the OR gate. The normal output is negated, and the function is referred to as a NOR. The symbol and truth table are shown in Figure 3-8.

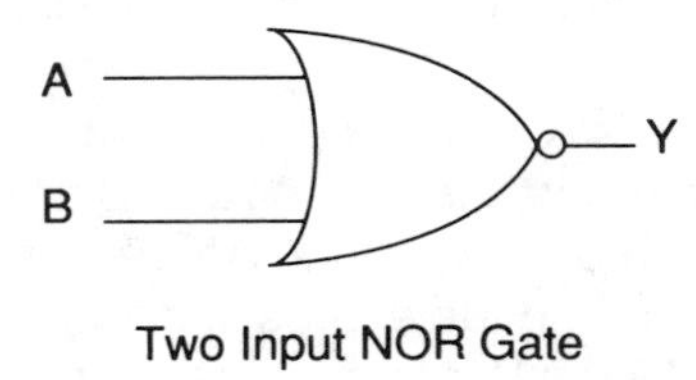

Two Input NOR Gate

Inputs		Output
A	**B**	**Y**
0	0	1
0	1	0
1	0	0
1	1	0

NOR Truth Table

Figure 3-8. Two input NOR gate and its truth table.

3-3 PRINCIPLES OF BOOLEAN ALGEBRA AND LOGIC

An in-depth discussion of Boolean algebra is not required for the purpose of this book and is beyond the book's scope. However, an understanding of the technique for writing shorthand expressions for complex logical statements could be a useful tool when creating a control program of Boolean statements or conventional ladder diagrams.

Boolean algebra was developed in 1849 by an Englishman named George Boole. The intent of this algebra was to aid in the logic of reasoning, an ancient form of philosophy, rather than to implement the purposes of digital logic, a technique not developed at the time. It was meant to provide a simple way of writing complicated combinations of "logical statements," defined earlier as statements that can be either TRUE or FALSE.

When digital logic finally came to be, Boolean algebra already existed as a simple way to analyze and express logic statements. All digital systems are based on this TRUE/FALSE or two-valued logic concept, where 1 represents TRUE and 0 represents FALSE. Because of this relationship between digital logic and Boolean logic, we will occasionally hear logic gates referred to as Boolean gates, or several interconnected gates called a Boolean network, and even a PLC language called Boolean language.

Figure 3-9 summarizes the basic operators of Boolean algebra as they are related to the basic digital logic functions AND, OR, and NOT. A capital letter represents the wire label of an input signal, a multiplication sign (•) relates to the AND operation, while the addition sign (+) relates to the OR operation. A bar over the letter ($\overline{A}$) represents the NOT operation.

Logical Symbol	Logical Statement	Boolean Equation
A, B — Y	Y is "1" if A AND B are "1"	$Y = A \bullet B$ or $Y = AB$
C, D — Y	Y is "1" if A OR B is "1"	$Y = C + D$
A — Y	Y is "1" if A is "0" Y is "0" if A is "1"	$Y = \overline{A}$

Figure 3-9. Summary of Boolean Algebra as related to the AND,OR, and NOT functions.

In Figure 3-9 the AND gate is shown with input signals A and B and an output signal Y. The output can be expressed by the logical statement: Y is "1" if A AND B are "1". The Boolean expression is written $Y = A \bullet B$ and is read *Y equals A ANDed with B.* The Boolean symbol (•) for AND could be removed and the expression written as $Y = AB$. Similarly, if Y is the result of ORing C and D, the Boolean expression is $Y = C+D$ and is read *Y equals C ORed with D.* In the NOT operation, Y is the inverse of A, the Boolean expression is $Y=\overline{A}$ and is read Y equals NOT A. The basic Boolean operations of ANDing, ORing, and inversion are illustrated in Table 3-3. The table also illustrates how these three functions can be combined to obtain any desired logic combination.

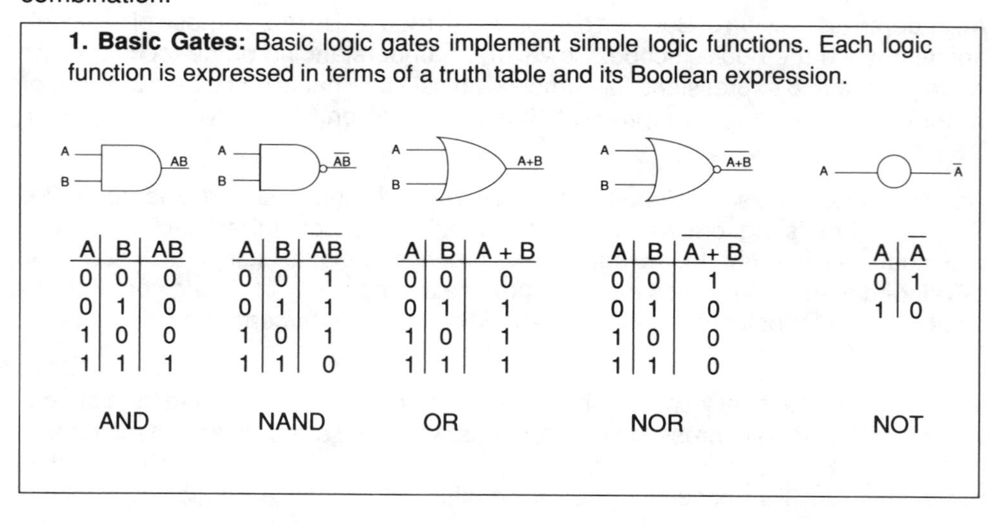

1. Basic Gates: Basic logic gates implement simple logic functions. Each logic function is expressed in terms of a truth table and its Boolean expression.

AND

A	B	AB
0	0	0
0	1	0
1	0	0
1	1	1

NAND

A	B	$\overline{AB}$
0	0	1
0	1	1
1	0	1
1	1	0

OR

A	B	A + B
0	0	0
0	1	1
1	0	1
1	1	1

NOR

A	B	$\overline{A+B}$
0	0	1
0	1	0
1	0	0
1	1	0

NOT

A	$\overline{A}$
0	1
1	0

Table 3-3. Logic Operations Using Boolean Algebra

Table 3-3 continued.

2. Combined Gates. Any combination of control functions can be expressed in Boolean terms using three simple operators: (•), (+), ($\overline{\ }$).

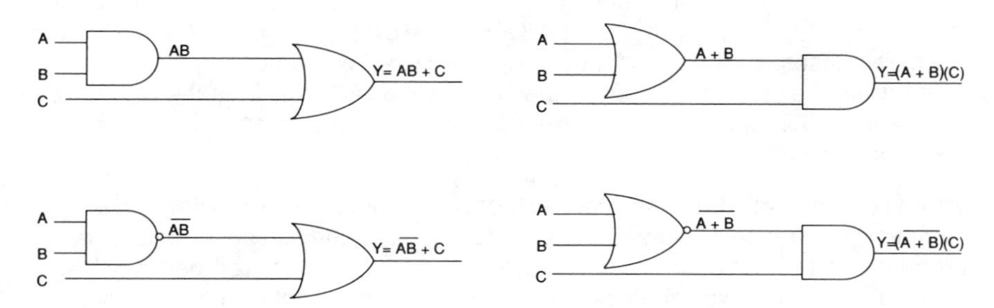

3. Boolean Algebra Rules. Control logic functions can vary from very simple to very complex combinations of input variables. However simple or complex the functions may be, they satisfy these basic rules. The rules are a result of simple combination of the basic truth tables and may be applied to simplify logic circuits.

$A + B = B + A$	*COMMUTATIVE LAWS*
$AB = BA$	
$A + (B + C) = (A + B) + C$	*ASSOCIATIVE LAWS*
$A(BC) = (AB)C$	
$A(B + C) = AB + AC$	*DISTRIBUTIVE LAWS*
$A + BC = (A + B)(A + C)$	
$A(A + B) = A + AB = A$	*LAW OF ABSORPTION*
$\overline{(A + B)} = \overline{A}\ \overline{B}$	*DE MORGAN'S LAWS*
$\overline{(AB)} = \overline{A} + \overline{B}$	
$\overline{\overline{A}} = A\ ,\ \overline{1} = 0\ ,\ \overline{0} = 1$	
$A + \overline{A}\,B = A + B$	
$AB + AC + B\,\overline{C} = AC + B\,\overline{C}$	

4. Order of Operations and Grouping Signs. The order in which Boolean operations (AND,OR,NOT) of an expression are performed is important. This order will affect the resulting logic value of the expression. Consider the three input signals A,B,C. Combining them in the expression Y = A + B•C can result in misoperation of the output device(Y), depending on the order in which the operations are performed. Performing the OR operation prior to the AND operation is written (A + B)•C, and performing the AND operation prior to the OR is written A + (B•C). The result of these two expressions is not the same.

Table 3-3 continued.

The order of priority in Boolean expressions is NOT (inversion) first, AND second, and OR last, unless otherwise indicated by grouping signs such as parentheses, brackets, braces, or the vinculum. According to these rules, the previous expression A + B•C, without any grouping signs, will always be evaluated only as A + (B•C). With the parentheses, it is obvious that B is ANDed with C prior to ORing the result with A. Knowing the order of evaluation then makes it possible to write the expression simply as A + BC, without fear of misoperation. As a matter of convention, the AND operator is usually omitted in Boolean expressions.

When working with Boolean logic expressions, misuse of grouping signs is a common occurrence. However, if signs occur in pairs, they do not generally cause problems if they have been properly placed according to the desired logic. Enclosing within parentheses two variables that are to be ANDed is not necessary, since the AND operator would normally be performed first. If, however, B is to be ORed with C prior to ANDing it with A, then B + C must be placed within parentheses.

When using grouping signs to insure proper order of evaluation of an expression, parentheses ( ) are used first. If additional signs are required, brackets [ ], and finally braces { } are used. An illustration of the use of grouping signs is shown here.

$$Y1 = Y2 + Y5\ [X1(X2 + X3)] + [Y4\ Y3 + X2(X5 + X6)]$$

5. Application of DeMorgan's Laws. DeMorgan's Laws are frequently used to simplify inverted logic expressions or simply to convert an expression into a useable form. See Appendix B for other conversions using DeMorgan's Laws.

According to DeMorgan's Laws:

$$\overline{AB} = \overline{A} + \overline{B} \quad \text{and} \quad \overline{A + B} = \overline{A}\ \overline{B}$$

3-4 PLC CIRCUITS AND LOGIC CONTACT SYMBOLOGY

Hardwired logic refers to logic control functions (timing, sequencing, and control) that are determined by the way devices are interconnected. In contrast to programmable controllers in which logic functions are programmable and easily changed, hardwired logic is fixed and changeable only by altering the way devices are connected or interwired. A prime function of the PLC is to replace existing hardwired control logic or to implement control functions for new systems. Figure 3-10 shows a typical hardwired relay logic circuit and its PLC ladder diagram implementation. The important point of Figure 3-10 is not to understand the process of changing from one circuit to another, but to see the similarities in the representation. The ladder circuit connections in the PLC are implemented via software instructions. All the wiring can be thought of as being inside the CPU (softwired as opposed to hardwired).

Relay logic implemented in PLCs is based on the three basic logic functions (AND, OR, NOT) that were discussed in the previous sections. These functions are used either singly or in combinations to form instructions that will determine if a device is to be switched ON or OFF. How these instructions are implemented to convey commands to the PLC is called the language. The most widely used languages for implementing ON/OFF control and sequencing are ladder diagrams, Boolean mnemonics, and others. These languages are discussed at length in Chapter 9.

The most conventional of these languages are ladder diagrams. Ladder diagrams are also called contact symbology, since its instructions are relay-equivalent contact symbols (i.e. normally-open and normally-closed contacts and coils). Contact symbology is a very simple way of expressing the control logic in terms of symbols that are used on relay control schematics. If the controller language is ladder diagrams, the translation from existing relay logic to programmed logic is a one-step translation to contact symbology. If the language is Boolean mnemonics, conversion to contact symbology is an unrequired step, yet still useful and quite often taken to provide an easily understood piece of documentation. Examples of simple translation from hardwired logic to programmed logic are shown later in Table 3-6 and thoroughly explained later in Chapter 11.

The complete ladder diagram or circuit in Figure 3-10 can be thought of as being formed by individual circuits, each circuit having one output. Each of these circuits is known as a rung or network, therefore, a rung is the contact symbology required to control an output in the PLC. Some controllers allow a rung to have multiple outputs, but one output per rung is the convention. A complete PLC ladder diagram program then consists of several rungs, each controlling an output interface which is connected to an output field device. Each rung is a combination of input conditions (symbols) connected from left to right between two vertical lines, with the symbol that represents the output at the far right. The symbols that represent the inputs are connected in series, parallel, or some combination to obtain the desired logic; these input symbols represent the input devices that are connected to the PLC's input interfaces. When completed, a ladder diagram control program consists of several rungs, with each rung having a specific output that it controls.

The programmed rung concept is a direct carryover from the hardwired relay ladder rung, in which input devices are connected in series and parallel to control various outputs. When activated, these input devices either allow current to flow through the circuit or cause a break in current flow, thereby switching a device ON or OFF. The input symbols on a ladder rung can represent signals generated from connected input devices, connected output devices (see Table 3-4), or from outputs internal to the controller.

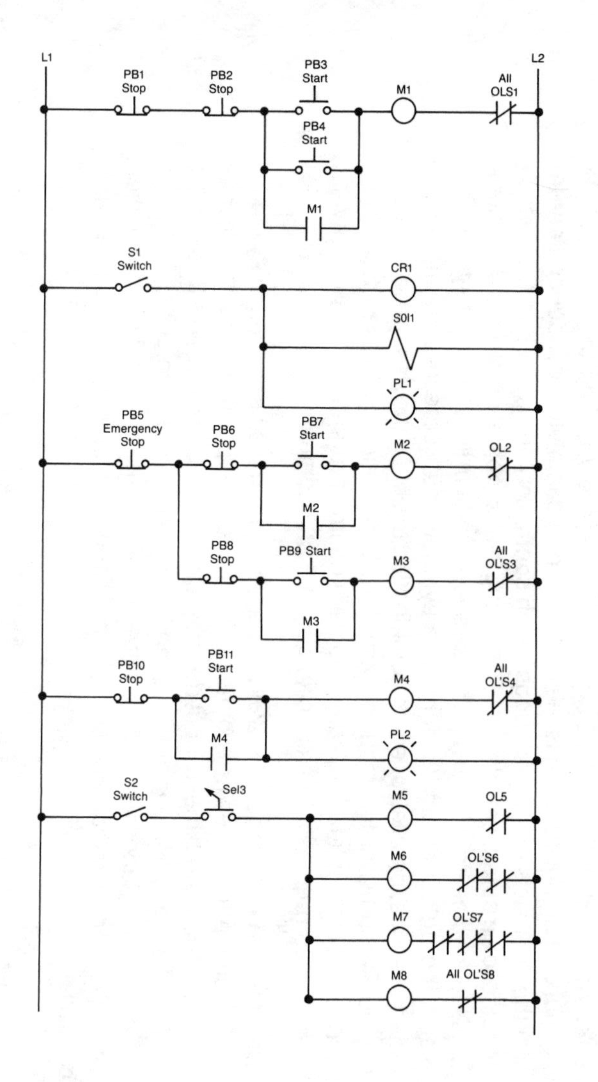

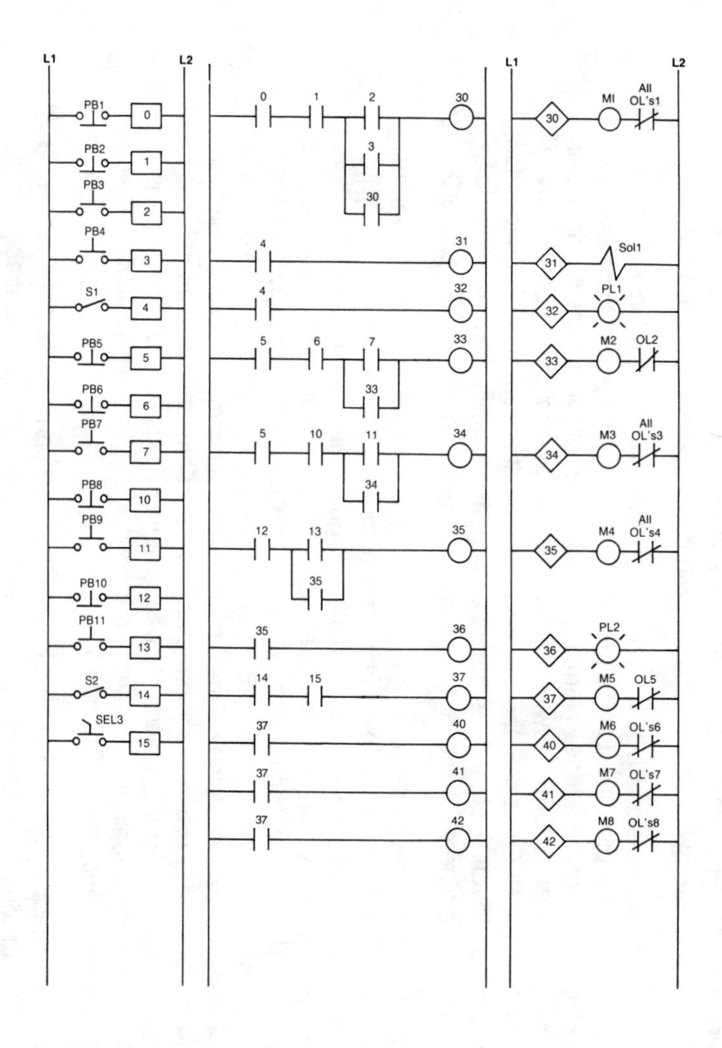

Figure 3-10. Circuit translation from hardwired relay logic (left) to PLC ladder logic (right).

Input Devices	Output Devices
Pushbutton	Pilot Light
Selector Switch	Solenoid Valve
Limit Switch	Horn
Proximity Switch	Control Relay
Timer Contacts	Timer

Table 3-4. ON/OFF input and output devices

Address

Each symbol on the rung will have a *reference number*, which is the *address* in memory where the current status (1 or 0) for the referenced input is stored. When the field signal is connected to an input or an output interface, the address will also be related to the terminal where the signal wire is connected. The address for a given input/output can be used throughout the program as many times as required by the control logic. This programmable controller feature is quite an advantage when compared to relay-type hardware where additional contacts often mean additional hardware. More on I/O interaction and its relationship with the PLC's memory and enclosure placement is described in Section 5-4 and 6-2 respectively.

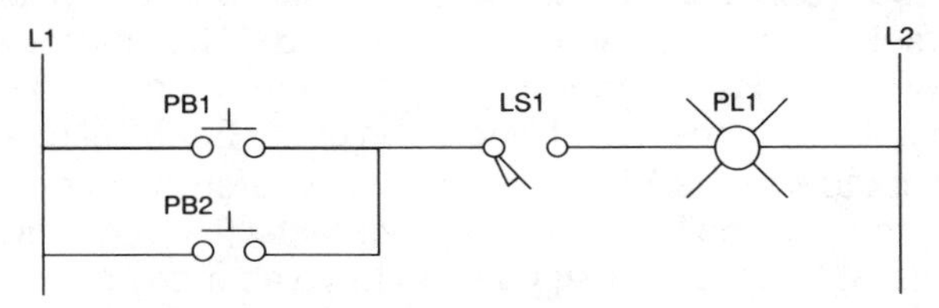

Electromechanical Circuit

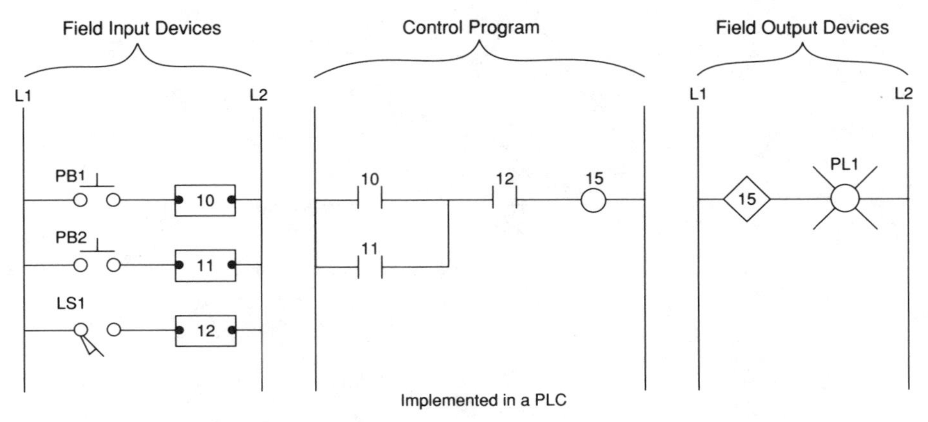

Implemented in a PLC

Note: PB1 would be wired to input 10, PB2 would be wired to input 11, and LS1 would be wired to input 12. PL1 is wired to output 15.

Figure 3-11. Electromechanical relay circuit and equivalent PLC implementation.

Figure 3-11 illustrates a simple elctromechanical relay circuit and its equivalent PLC implementation. Each "real" field device, i.e. pushbuttons PB1 and PB2 and limit switch LS1, are connected to the PLC's input modules which have a reference number—the address. How these symbols are normally referenced is dependent on the controller, but most are referenced using numeric addresses with octal (base 8) or decimal (base 10) numbering. Note that any complete path (all contacts closed) from left to right will energize the output (pilot light PL1).

Power has to flow through 10 and 12 or through 11 and 12 to turn ON output 15 which in turn will energize the light PL1 that is connected to the interface with address 15. The same logic that applies to the electromechanical circuit applies to the PLC circuit. In order to provide power (in the PLC circuit) to the addresses 10, 11 or 12, the devices connected to these input interfaces addressed 10, 11 and 12 must be turned ON (activated) by pressing the pushbutton or closing the limit switch.

Contact Symbols

Programmable controller contacts and relay contacts operate in a very similar fashion. As an example, let's take relay A (Figure 3-12) which has two sets of contacts, one normally open (A-1) and one normally closed (A-2). If relay coil A is not energized, or OFF, contacts A-1 will remain open and contacts A-2 will remain closed. Conversely, if coil A is energized, or turned ON, contacts A-1 will close and contacts A-2 will open. Remember that when a set of contacts closes it provides power flow or continuity in the circuit where it is used. Each set of available coils and its respective contacts in the PLC have a unique reference address by which it is identified. For instance, coil 10 will have contacts (NO and NC) with the same address 10 as its coil (Figure 3-13); you can have as many NO and NC contacts in a PLC as desired.

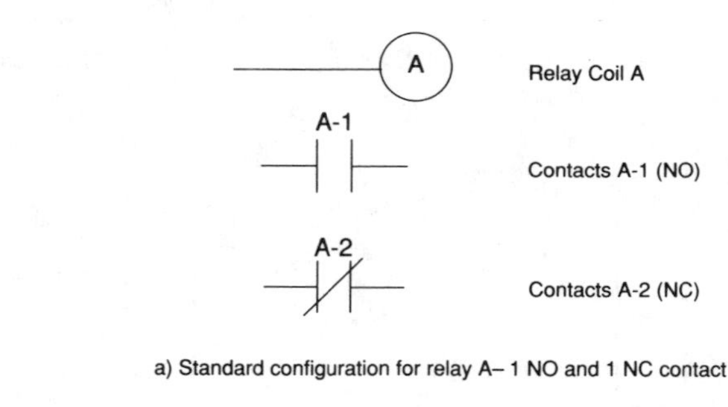

a) Standard configuration for relay A– 1 NO and 1 NC contact

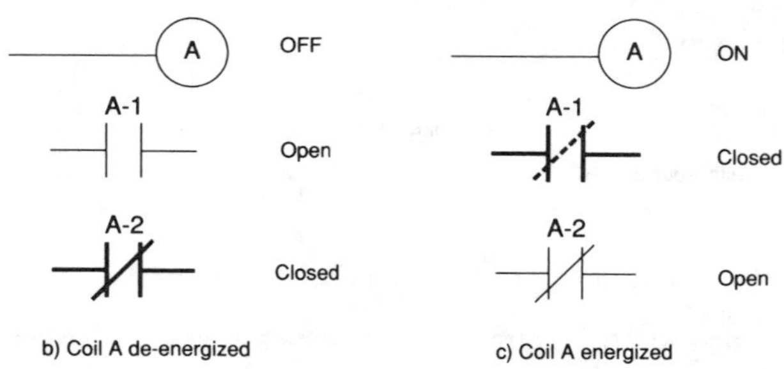

b) Coil A de-energized

c) Coil A energized

Note: The bolded contacts signify power flow representing a closed condition in b) and c).

Figure 3-12. Relay and PLC contact representation.

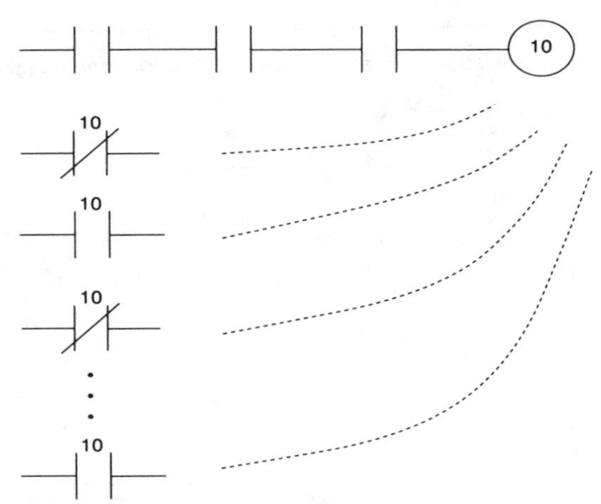

Figure 3-13. Multiple contacts from a PLC output coil.

The following symbols are used to translate relay control logic to contact symbolic logic. These symbols are also the basic instruction set for the Ladder Diagram, excluding timer/counter instructions. These and more advanced instructions are explained in Chapter 9. Table 3-5 lists several translating examples from hardwired relay logic to PLC logic using contact symbology.

Symbol **Definition and Symbol Interpretation**

Normally-opened contact. Represents any input to the control logic. An input can be a connected switch closure or sensor, a contact from a connected output, or a contact from an internal output. When interpreted, the referenced input or output is examined for an "ON" condition. If its status is "1", the contact will close and allow current to flow through the contact. If the status of the referenced input/output is "0", the contact will open and not allow current to flow through the contact.

Normally-closed contact. Represents any input to the control logic. An input can be a connected switch closure or sensor, a contact from a connected output, or a contact from an internal output. When interpreted, the referenced input/output is examined for an "OFF" condition. If its status is "0", the contact will remain closed, thus allowing current to flow through the contact. If the status of the referenced input/output is "1", the contact will open and not allow current to flow through the contact.

Output. Represents any output that is driven by some combination of input logic. An output can be a connected device or an internal output. If any left-to-right path of input conditions is true (all contacts closed), the referenced output is energized (turned ON).

NOT output. Represents any output that is driven by some combination of input logic. An output can be a connected device or an internal output. If any left-to-right path of input conditions are true (all contacts closed), the referenced output is de-energized (turned OFF).

Table 3-5. Example of translation from hardware logic to programmed logic.

Relay Ladder Diagram	Contact or Ladder Diagram	Boolean Equation	Boolean Statements
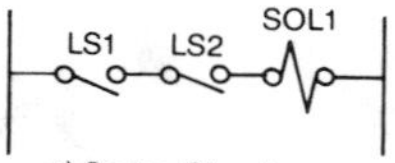 LS1 LS2 SOL1 a) Series Circuit	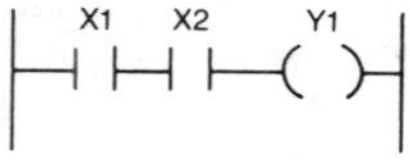 X1 X2 Y1	$Y1 = X1 \bullet X2$	STR X1 AND X2 OUT Y1
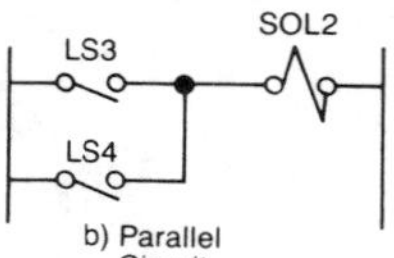 LS3 SOL2 LS4 b) Parallel Circuit	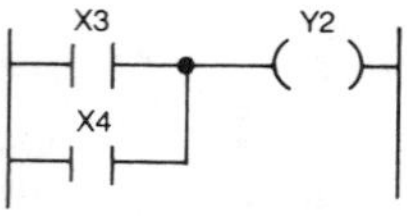 X3 Y2 X4	$Y2 = X3 + X4$	STR X3 OR X4 OUT Y2
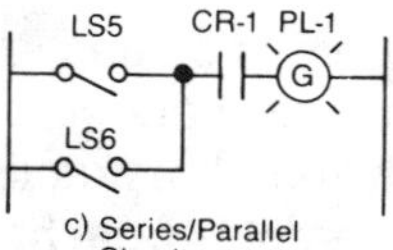 LS5 CR-1 PL-1 G LS6 c) Series/Parallel Circuit	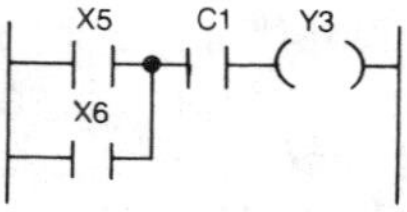 X5 C1 Y3 X6	$Y3 = (X5 + X6) \bullet C1$	STR X5 OR X6 AND C1 OUT Y3
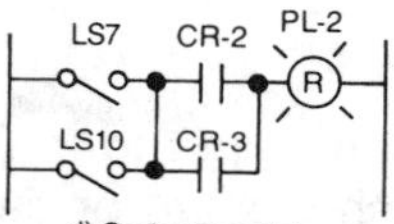 LS7 CR-2 PL-2 R LS10 CR-3 d) Series/Parallel Circuit	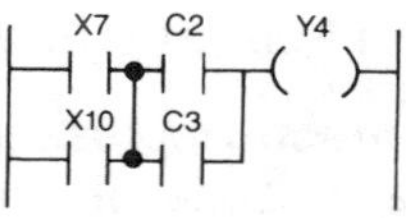 X7 C2 Y4 X10 C3	$Y4 = (X7 + X10) \bullet (C2 + C3)$	STR X7 OR X10 STR C2 OR C3 OUT Y4
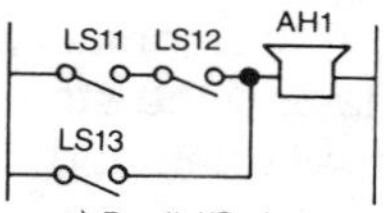 LS11 LS12 AH1 LS13 e) Parallel/Series Circuit	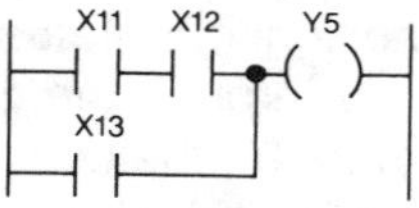 X11 X12 Y5 X13	$Y5 = (X11 \bullet X12) + X13$	STR X11 AND X12 OR X13 OUT Y5
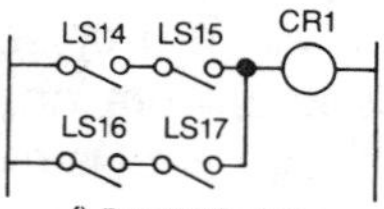 LS14 LS15 CR1 LS16 LS17 f) Parallel/Series Circuit	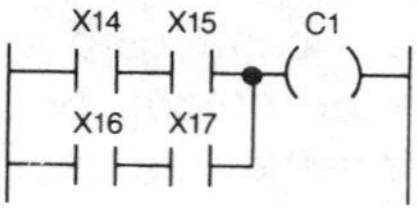 X14 X15 C1 X16 X17	$C1 = (X14 \bullet X15) + (X16 \bullet X17)$	STR X14 AND X15 OR X16 AND X17 OUT C1
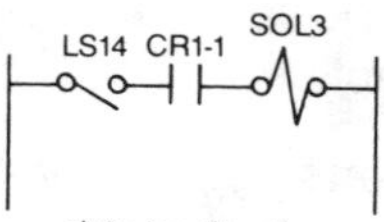 LS14 CR1-1 SOL3 g) Series Circuit	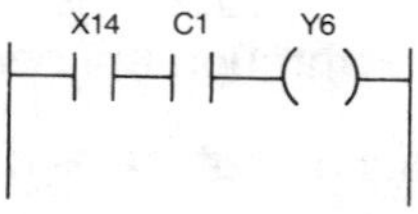 X14 C1 Y6	$Y6 = X14 \bullet C1$	STR X14 AND C1 OUT Y6
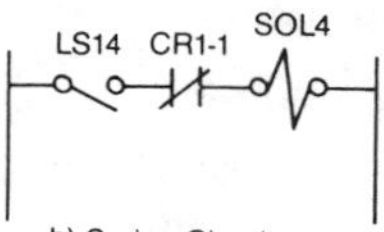 LS14 CR1-1 SOL4 h) Series Circuit	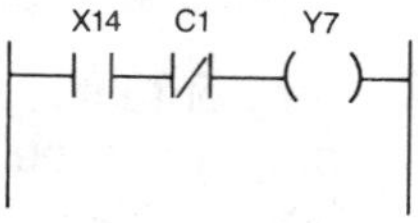 X14 C1 Y7	$Y7 = X14 \bullet \overline{C1}$	STR X14 AND NOT C1 OUT Y7

The following seven points describe guidelines for translating from hardwired logic to programmed logic using contact symbols.

- **The normally-open contact.** When evaluated by the program, this symbol is examined for a "1" to close the contact; therefore, the signal referenced by the symbol must be ON, CLOSED, ACTIVATED, etc.
- **The normally-closed contact.** When evaluated by the program, this symbol is examined for a "0" to keep the contact closed; thus, the signal referenced by the symbol must be OFF, OPEN, DE-ACTIVATED, etc.
- **Outputs.** An output on a given rung will be energized if any left-to-right path has all contacts closed, with the exception of power flow going in reverse before continuing to the right. An output can control a connected device if the reference address is also a termination point, or an internal output used exclusively within the program.
- **Inputs.** Contact symbols on a rung can represent input signals sent from connected inputs, contacts from internal outputs, or contacts from connected outputs.
- **Contact Addresses**. Each program symbol is referenced by an address. If the symbol is referencing a connected input/output device, then the address is determined by the point at which the device is connected.
- **Logic format.** Contacts may be programmed in series or in parallel, depending on the logic required to control the output. The number of series contacts or parallel branches to a rung is dependent on the controller.
- **Repeated use of contacts.** A given input, output, or internal output can be used throughout the program as many times as required.

The circuits in Table 3-5 show how simple hardwired series and parallel circuits can be translated to programmed logic. The series circuit is equivalent to the Boolean AND operation; therefore, all inputs must be ON to activate the output. The parallel circuit is equivalent to the Boolean OR operation; therefore, any one of the inputs must be ON to activate the output; the STR and OUT Boolean statements stand for START (of a new rung) and OUTPUT (of a rung) respectively. Table 3-5 is explained in the following paragraphs.

Series Circuit (a). In this circuit, if both switches LS1 AND LS2 are closed, the solenoid SOL1 will energize. According to (1) of the summary, a normally-opened contact symbol must be programmed.

Parallel Circuit (b). In this circuit, if either of the two switches LS3 OR LS4 close, the solenoid SOL2 will energize. According to (1) of the summary, a normally-opened contact symbol must be programmed.

Series/Parallel circuits (c) and (d). In a series/parallel circuit, the result of ORing two or more inputs is ANDed with one or more series or parallel inputs. In both of these examples, all of the relay circuit elements are normally-opened and must be closed to activate the pilot lights. Normally-open contacts are used in the program.

Parallel/Series circuits (e) and (f). In a parallel/series circuit, the result of ANDing two or more inputs is ORed with one or more series inputs. In both of these examples, all of the relay circuit elements are normally-opened, and must be closed to activate the output device. Normally-open contacts are used in the program.

Internal outputs. Circuit **(f)** controls an electromechanical control relay. Control relays do not normally drive output devices, but drive other relays. They are normally used to provide additional contacts for interlocking logic. The internal output provides the same function in software; however, the number of contacts are unlimited and can be either normally-open or normally-closed.

Use of normally-open contacts. Note in the series circuit (g) the solenoid will energize if LS14 closes and CR1-1 is energized. CR1-1 is a contact from the control relay CR1 in circuit (f) and closes whenever CR1 is energized. In the program, CR1 was replaced by the internal output C1; therefore, the program uses a normally-open contact from the internal output C1. SOL3 will energize when LS14 closes and C1 is energized.

Use of normally-closed contacts. In circuit (h), the solenoid will energize if LS14 closes, and CR1-1 is NOT energized. The program uses a normally-closed contact from the internal output C1. SOL3 will stay energized as long as the limit switch is closed, AND C1 is NOT energized.

Chapter

—4—

PROCESSORS, THE POWER SUPPLY SYSTEM, AND PROGRAMMING DEVICES

"Give us the tools and we will finish the job."

—Winston Churchill

4-1 INTRODUCTION

As mentioned in the first chapter, the Central Processing Unit, CPU for short, is undoubtedly the most important element in a programmable controller. The CPU forms what can be considered the brain of the system. The three components that form the CPU are:

- The Processor
- The Memory System
- The Power Supply System

Figure 4-1 illustrates a simplified block diagram of the CPU. The general CPU architecture may differ from one manufacturer to another, but in general, most of them follow this typical three component organization. Although the power supply is shown inside the CPU block enclosure, it may be a separate unit that is normally mounted next to the enclosure which contains the processor and memory. The programming device, not regarded as part of the CPU per se, completes the total central architecture as the medium of communication between the programmer and the CPU. Figure 4-2 shows a CPU with a built-in power supply.

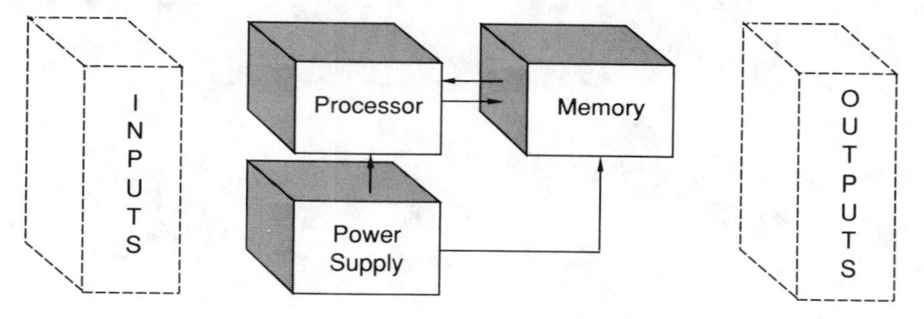

Figure 4-1. CPU Block Diagram.

The term CPU is often used interchangeably with *processor*; however, the CPU term encompasses all the necessary elements that form the intelligence of the system. There are definite relationships between the sections that form the CPU and constant interaction among them. Figure 4-3 illustrates the functional interaction of the basic components that make up the programmable controller. The processor *executes* the control program stored in the memory system in the form of ladder diagrams, while the system power supply provides all the necessary voltage levels to insure proper operation of all the processor and memory components.

This chapter will cover the important aspects of the processor and power supply that comprise the CPU. An inside look at these sections will provide you with a better understanding of the function and operation of a programmable controller. This will be useful when specifying a product and defining the CPU requirements for specific PLC applications. We will also cover some of the most commonly used programming devices that are employed when entering and editing the control program. The next chapter is dedicated to the memory system and the interactions between input and output field devices with the memory and therefore the PLC.

Figure 4-2. CPU with built-in power supply (Courtesy Westinghouse Electric, Pittsburgh, PA).

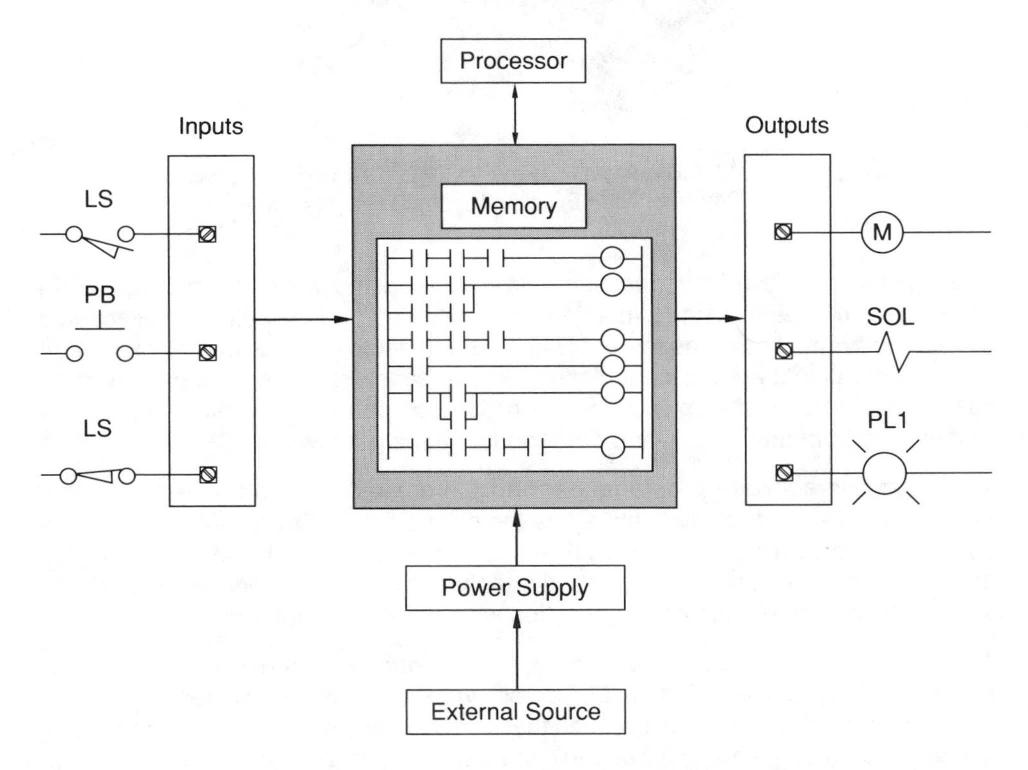

Figure 4-3. Functional interaction of a PLC system.

4-2 PROCESSORS

The intelligence of today's programmable controllers is formed by very small *microprocessors* (micros)—integrated circuits with tremendous computing and control capability. They perform all mathematical operations, data handling, and diagnostic routines that were not possible with relays or their predecessor, the hardwired logic processor. Figure 4-4 illustrates a processor module which contains the microprocessor, its supporting circuity, and the memory system.

Figure 4-4. Allen Bradley's PLC model 5/12, 5/15 and 5/25 processors (Courtesy Allen Bradley, Highland Heights, OH).

The principle function of the processor is to command and govern the activities of the entire system. It performs this function by interpreting and executing a collection of system programs known as the *executive*. The executive is a collection of supervisory programs that are permanently stored and considered a part of the controller itself. By executing the executive, the processor can perform all of its control, processing, communication, and other housekeeping functions.

The executive software programs perform the communication between the PLC system and the user or programmer via the programming device. Other peripheral communication is also supported by the executive; such communication includes the monitoring of field devices, reading diagnostic data from the power supply, I/O system and memory, and communication with an operator interface.

The CPU of a PLC system may contain more than one processor (or micro) to execute the systems duties and/or communications. The basic reason for this arrangement is the speed of operation that can be achieved. The approach of using several microprocessors to divide the control and communication tasks is known as *multi-processing.*

Another multi-processor arrangement places the microprocessor intelligence away from the CPU. This technique involves intelligent I/O interfaces that contain a microprocessor, built-in memory, and a mini executive that performs independent control tasks. A typical intelligent module is the Proportional-Integral-Derivative (PID) control module, which performs closed loop control independent of the CPU.

Microprocessors used in PLCs are categorized according to the word size, or the number of bits that they use simultaneously to perform operations. Standard word lengths are 4, 8, 16 and 32 bits. This word length affects the speed in which most operations are performed. For example, a 16-bit microprocessor can manipulate data faster than an 8-bit micro since it manipulates twice as much data in one operation. The difference in word length is, of course, associated with the capability and degree of sophistication of the controller.

4-3 PROCESSOR SCAN

The basic function of the programmable controller is to read all field input devices and to execute the control program which, according to the logic programmed, will turn the field output devices ON or OFF. In reality, this last process of turning the output devices ON or OFF occurs in two steps. As the processor executes the internal programmed logic, it will turn each of its *internal* output coils that have been programmed ON or OFF. The energizing or de-energizing of these internal outputs will not, at this exact time, turn the output devices ON or OFF. When the processor has finished the evaluation of the control logic program, turning the internal coils ON or OFF, it will perform an *update* to the output interface modules, thereby turning the field devices connected to each interface terminal ON or OFF. This process of reading the inputs, executing the program and updating the outputs is known as the *scan*.

The scan may be represented graphically in Figure 4-5. This process is repeated over and over in the same fashion, making the operation sequential from top to bottom. Sometimes, for the sake of simplicity, some PLC manufacturers call the

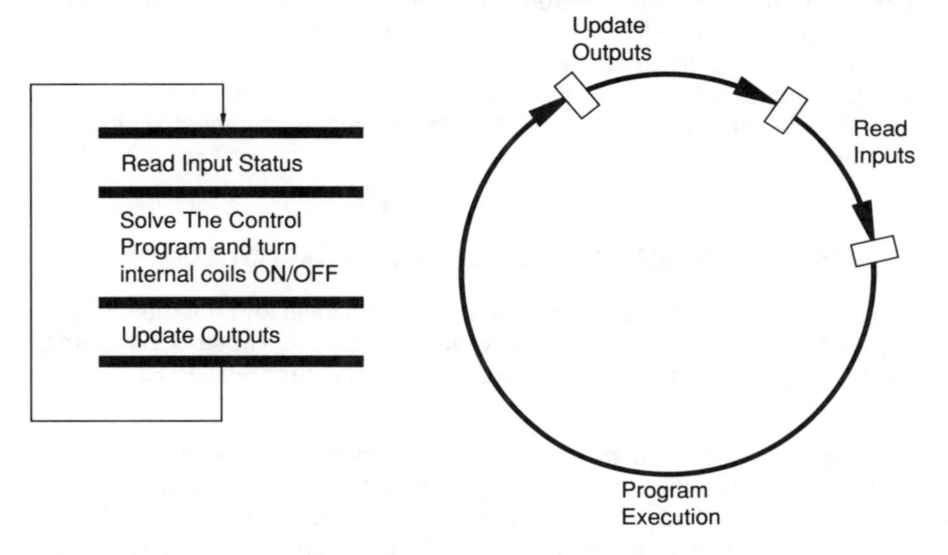

Figure 4-5. PLC total scan representation.

solving of the control program *program scan* and the reading of inputs and updating of outputs the *I/O update*. Nevertheless, the total scan of the system is composed of both. The internal processor signal that indicates that the program scan has ended is called *end of scan (EOS signal)*.

The time it takes to implement a scan is called *scan time*. The scan time is composed of the program scan time and the I/O update time. The program scan time generally depends on the amount of memory taken by the control program and the type of instructions used in the program (time used to execute instructions). The time required to make a single scan can vary from 1 msec to 100 msec. PLC manufacturers generally specify the scan time based only on the amount of application memory used (e.g. 10 msec/1K of programmed memory). The scan time, however, is affected by other factors. The use of remote I/O subsystems increases the scan time as a result of having to transmit/receive the I/O update to remote subsystems. Monitoring of the control program also adds time to the scan, because the micro has to send the status of coils and contacts to the CRT or another monitoring device.

As we just learned, the scan is normally a continuous and sequential process of reading the status of inputs, evaluating the control logic, and updating the outputs. A processor will be able to read the inputs as long as the input signal is not faster than the scan time, i.e. the input signal does not change state (ON to OFF to ON or vice versa) twice during the processor's scan time. For instance, if a controller has a total scan time of 10 msec and needs to monitor an input signal that can change states twice during an 8 msec period (less than the scan), the PLC will never be able to "see" the signal, thus resulting in a possible machine or process malfunction. This scan consideration must always be taken into account when reading discrete input signals and ASCII characters (see ASCII module Chapter 8). The scan specification indicates how fast the PLC can react to inputs and correctly solve the control logic. More on scan evaluation is described in Chapter 9.

The common scan method of monitoring the inputs at the end of each scan may be inadequate for reading certain extremely fast inputs. Some PLCs provide software instructions that will allow the interruption of the continuous program scan in order to receive an input or to update an output immediately. Figure 4-6 illustrates how immediate instructions operate during a normal program scan. These immediate instructions are very useful when the PLC must react instantaneously to a critical input or output.

4-4 SUBSYSTEMS, ERROR CHECKING, DIAGNOSTICS

The PLC's processor constantly communicates with local and remote subsystems (see Chapter 6) or racks as they may also be called. These subsystems are connected via I/O interfaces to field devices located close to the CPU or at a remote location.

The subsystem communication involves data transfer exchange at the end of each program scan, where the processor sends the latest status of outputs to the I/O subsystem and receives the current status of inputs and outputs. The actual communication between the processor and subsystem is performed via an I/O

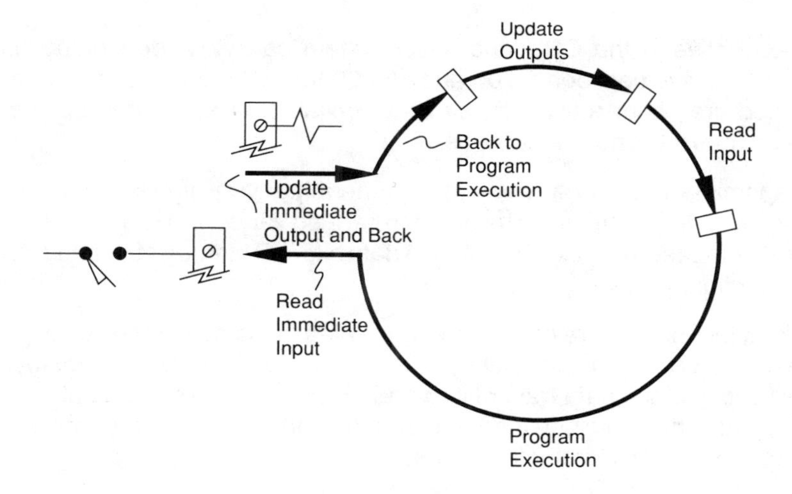

Figure 4-6. PLC scan with immediate I/O update.

subsystem adapter module, located in the CPU, and a remote I/O processor module, located in the subsystem chassis or rack. Figure 4-7 illustrates a typical PLC subsystem configuration.

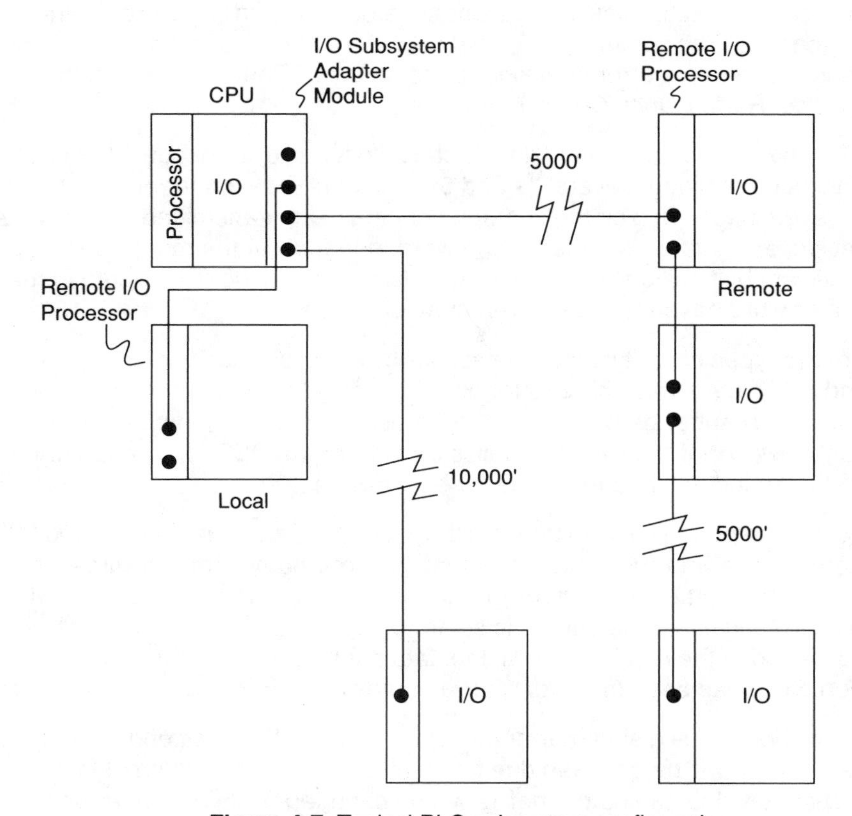

Figure 4-7. Typical PLC subsystem configuration.

The distance between the CPU and a subsystem can vary, depending on the controller, and ranges between 1,000 and 15,000 feet. The communication media generally used are either a twisted pair or twinaxial, coaxial, or fiber optic cable, depending on the PLC and the distance.

The data transmission rate to subsystems is generally at very high speeds, but does vary depending on the controller. The data format also varies, but is normally a serial binary format composed of a fixed number of data bits (I/O status), start and stop bits, and error detection codes.

Error checking techniques are normally incorporated in the continuous communication between the processor and subsystem. These techniques are employed to confirm the validity of the data transmitted and received. The level of sophistication of error checking varies from one manufacturer to another, as does the type of errors reported and the resulting protective action.

Error Checking

The processor uses error-checking techniques to monitor the functional status of the memory, communication links between subsystems and peripherals, and its own operation. Common error-checking techniques include *parity* and *checksum*.

Parity. Parity is perhaps the most common error detection technique, and it is used primarily in communication link applications to detect errors on long, error-prone data transmission lines. The communication between the CPU and subsystems is a prime example in which parity error-checking might be used. Parity check is sometimes called *Vertical Redundancy Check (VRC)*.

When using parity, the transmitted data is checked for an even or odd number of 1s. When data is transmitted, several 1s and 0s are sent. The number of 1s can either be odd or even, depending on the character or data being transmitted. In parity data transmission, an extra bit is added to the word, generally in the most significant or least significant bit position. This extra bit, called the *parity bit (P)*, is used to make each byte or word have an odd or even number of 1s.

There are two types of parity checks: *even parity*, which checks for an even number of 1s, and *odd parity*, which checks for an odd number of 1s. If the transmission of data is sent to a subsystem, the parity type used will be already defined by the controller. However, if the data transmission is from the PLC to a peripheral, the parity method used must be the same for both devices.

Suppose that the processor is to transmit the 7-bit ASCII character "C" (1000011) to a peripheral device, and odd parity is required. The total number of 1s is three, or odd. If the most significant bit is used for the parity bit (P), the transmitted data will be P1000011. To achieve odd parity, P is set to "0" to obtain an odd number of 1s. An error is detected at the receiving end if the data does not contain an odd number of 1s. If even parity had been the error-checking method, P would have been set to 1.

Parity error checking is a single-error detection method. If one bit of data in a word changes, an error will be detected due to the change in the bit pattern. However, if two bits change value, the number of 1s will be changed back, and an error will not be detected even though there is a mis-transmission.

Some processors do not use parity when transmitting information, although it may be required by some peripherals. In this case, parity generation can be accomplished through application software. The parity bit can be set for odd or even parity, with a short routine using functional blocks or a high level language. If a non-parity oriented processor receives data that contains parity, a routine can also be used to mask-out or strip the parity bit.

Checksum. The extra bit of data added to each word, using parity error detection, is often too wasteful to be desirable. In data storage, for example, error detection is desirable, but storing one extra bit for every 8 bits means a 12.5% loss of data storage capacity. For this reason, a data block error-checking method known as *checksum* is used.

Checksum error detection spots errors in blocks of many words instead of in individual words as parity does. This check is accomplished by taking all the words in a data block into consideration and adding to the end of the block one word that reflects a characteristic of the block. This last word, known as the *Block Check Character (BCC)*, is shown in Figure 4-8. This type of checking is appropriate for memory checks and is usually done at power up.

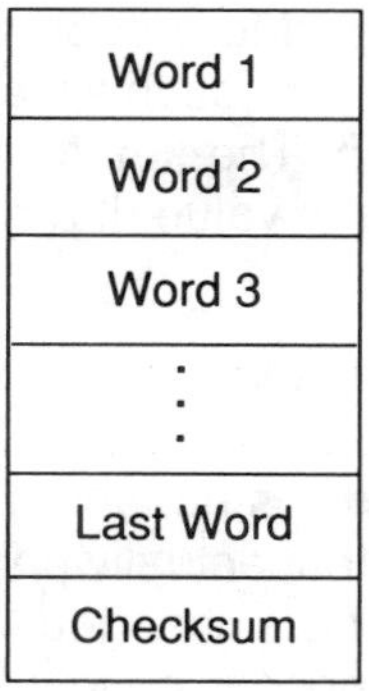

Word 1
Word 2
Word 3
⋮
Last Word
Checksum

Figure 4-8. Block check character at end of data block.

There are several methods of checksum computation. *Cyclic Redundancy Check (CRC)* and *Longitudinal Redundancy Check (LRC)* are two of the most common methods used. Other variations of checksum include the *Cyclic Exclusive-OR Checksum.*

The CRC performs an addition of all the words in the data block, and the resultant sum is stored in the last location (BCC). This summation process can rapidly reach an overflow condition. One variation of CRC will allow the sum to overflow and store only the remainder bits in the BCC word. Typically, the resulting word is complemented and written in the BCC location. During the error check, all words in the block are added together, and the addition of the final BCC word will turn the result to 0. A valid block can be detected by simply checking for 0 sum. Another type of CRC generates the BCC by taking the remainder after dividing the sum by a preset binary number.

The LRC is an error-checking technique based on the accumulation of the result of performing an Exclusive-OR of each of the words in the data block. The operation here is simply the logical Exclusive-OR of the first word with the second word, the result with the third word, and so on. The final Exclusive-OR operation is stored at the end of the block as the BCC.

The Cyclic Exclusive-OR checksum is similar to LRC with some slight variations. The operation starts with a checksum word containing 0s, and an Exclusive-OR operation is done with the first word of the block. This is followed by a left rotate of the bits in the checksum word. The next word in the data block is Exclusive-ORed with the checksum word and then rotated left. This procedure is repeated until the last word of the block is logically operated on. The checksum word is then appended to the block to become the BCC.

Most checksum error detection methods are usually performed by a software routine in the executive program. Typically, the processor performs the checksum computation on memory at power-up and also during the transmission of data. Some controllers perform the checksum on memory during the execution of the control program. This continuous on-line error checking lessens the possibility of the processor operating on invalid data.

Error Detection and Correction. More sophisticated programmable controllers may have an error detection and correction scheme that provides greater reliability than conventional error detection. The key to this type of error correction is the multiple representation of the same value. If a single bit changes, the value would remain the same.

The most common error detection and correction code is the *Hamming Code*. This code relies on parity bits interspread with data bits in a data word. By combining the parity and data bits according to a strict set of parity equations, a small byte is generated that contains a value that actually points to the bit in error. An error can be detected and corrected if any bit is changed in any value. The hardware used to generate and check Hamming Codes is quite complex and essentially implements a set of error-correcting equations.

Error-correcting codes offer the advantage of being able to detect two or more bit errors; however, they can only correct one bit errors. They also present a disadvantage in that they are bit-wasteful. Nevertheless, this scheme will be seen in the future primarily in data communication of hierarchical systems that are completely unmanned, sophisticated, and automatic.

CPU Diagnostics

The processor is responsible for detecting communication as well as other failures that may be encountered during system operation. It must alert the operator or system in case of a malfunction. To achieve this goal, the processor performs error checks during its operation and sends status information to indicators that are normally located on the front of the CPU. Typical diagnostics include memory OK, processor OK, battery OK, and power supply OK. Depending on the controller, a set of fault relay contacts may be available. The fault relay is controlled by the processor and is activated when one or more specific fault conditions (CPU diagnostics) occur. The fault relay contacts can be used in an alarm circuit to signal a failure.

The relay contacts that are generally provided with the controller operate in a "watch-dog" timer fashion. The processor sends a pulse at the end of each scan indicating a correct system operation. If there is a failure, the processor would not send a pulse, the timer would time-out, and the fault relay would activate.

In some controllers, certain CPU diagnostics are available to the user for use during the execution of the control program. These diagnostics use internal outputs that are controlled by the processor, but can be referenced by the user program (e.g. loss of scan, back-up battery low, etc.).

4-5 THE SYSTEM POWER SUPPLY

The system power supply plays a major role in the total system operation. It can very well be considered as the first line manager of system reliability and integrity, for its responsibility is not only to provide internal DC voltages to the system components (i.e. processor, memory, and input/output), but also to monitor and regulate the supplied voltages and warn the CPU if all is not well. The power supply then, has the function of supplying well-regulated power and protection for other system components.

The Input Voltage

Usually, PLC power supplies require input from an AC power source; however, some PLCs will accept input from a DC source. Those that will accept a DC source are quite appealing for applications such as off-shore drilling operations, where DC sources are commonly used. The most common requirement, however, is for 120 VAC or 220 VAC, while a few controllers will accept 24 VDC.

Since it is quite normal for industrial facilities to experience fluctuations in line voltage and frequency, an important specification for the PLC power supply is to tolerate a 10% to 15% variation in line conditions. For example, a power supply with a line voltage tolerance of ±10% when connected to a 120 VAC source will continue to function properly as long as the voltage remains between 108 VAC and 132 VAC. A 220 VAC power supply with ±10% line tolerance will function properly as long as the voltage remains between 198 VAC and 242 VAC. When the line voltage exceeds the upper or lower limits for a specified duration (usually 1 to 3 AC cycles), most power supplies are designed to issue a shutdown command to the processor. Line voltage variations in some plants could eventually become disruptive and may result in frequent loss of production. Normally, in such a case, a constant voltage transformer can be installed to stabilize line conditions.

Constant Voltage Transformers. Good power supplies are designed to tolerate what can be considered normal fluctuations in line conditions, but even the best designed power supply cannot compensate for the especially unstable conditions of the line voltage found in some industrial environments. Conditions that cause the line voltage to drop below proper levels vary depending on applications, and even plant locations. Some of the causes are:

- Startup/Shutdown of nearby heavy equipment, such as large motors, pumps, welders, compressors, and air conditioning.
- Natural line losses that vary with distance from utility substations.
- Intra-plant line losses caused by poorly made connections.
- Brownout situations in which line voltage is intentionally reduced by the utility company.

A constant voltage transformer compensates for voltage changes at its input (primary) to maintain a steady voltage at its output (secondary). When operated at less than the rated load, the transformer can be expected to maintain approximately ±1% regulation with an input voltage variation of as much as 15%. The percent regulation changes as a function of the operated load (PLC power supply and input devices). The constant voltage transformer then must be properly rated to provide ample power to the load. The rating of the constant voltage transformer, in units of volt-amperes (VA), should be selected based on the worst-case power requirements

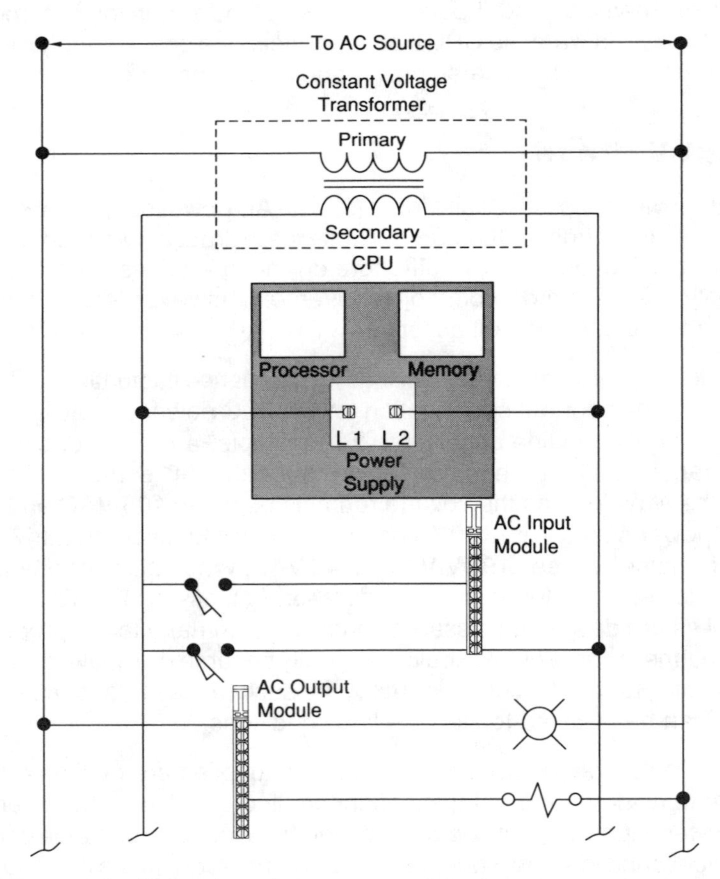

Figure 4-9. Illustration of a constant voltage transformer connection and the PLC system (CPU and modules).

of the load. A recommended rating for the constant voltage transformer should be obtained from the PLC manufacturer. Figure 4-9 illustrates a simplified connection of a constant voltage transformer when applied to programmable controllers.

The Sola* CVS, "standard sinusoidal", or equivalent constant voltage transformer is suitable for PLC applications. This type of transformer incorporates line filters to remove high harmonic content and provide a clean sinusoidal output. Constant voltage transformers that do not filter high harmonics are not recommended for PLC applications. Figure 4-10 illustrates the relationship between the output voltage and input voltage for a typical Sola CVS transformer operated at different loads.

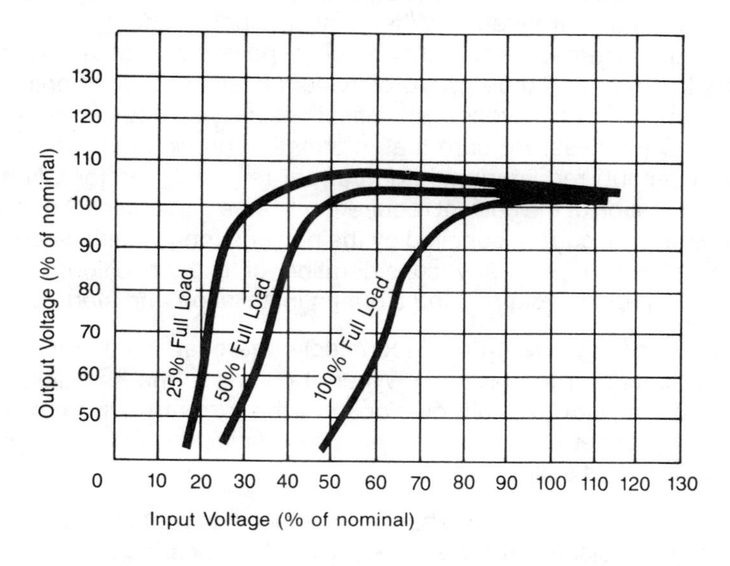

Figure 4-10. Input versus output voltage relationship for a Sola unit.

Isolation Transformers. Often, a programmable controller will be installed in an area where the AC line is stable, however, surrounding equipment may generate considerable amounts of electromagnetic interference (EMI). Such an installation could result in intermittent misoperation of the controller, especially if the controller is not electrically isolated (on a separate AC power source) from the equipment generating the EMI.

Placing the controller on a separate isolation transformer from the potential EMI generators will increase system reliability. The isolation transformer need not be a constant voltage transformer; but it should be connected between the controller and the AC power source.

Loading Considerations

The system power supply provides the DC power for the logic circuits of the CPU and the I/O circuits. Each power supply has a maximum amount of current that it can provide at a given voltage level (e.g. 10 amps at 5 volts). The amount of current that a given power supply is capable of providing is not always sufficient to supply a particular mix of I/O modules. Such a case would cause undercurrent conditions which could result in unpredictable operation of the I/O system.

*Sola Basic Industries, Elk Grove Village, IL

The undercurrent situation in most circumstances is unusual, since most power supplies are designed to accommodate a general mix of the most commonly used I/O modules. The undercurrent condition normally arises in some applications where an excessive number of special purpose I/O modules are used (e.g. power contact outputs, analog inputs/outputs). These special purpose modules usually have higher current requirements than most commonly used digital I/O modules.

Power supply overloading can be an especially annoying condition, since the problem is not always easily detected. The overload condition is often a function of the combination of outputs that are ON at a given time, which means that the overload conditions can appear intermittently. When power supply loading limits have been exceeded, the normal remedy is to add an auxiliary power, or to obtain a supply with a larger current capability. To be aware of system loading requirements ahead of time, users can obtain from the vendor specifications on typical current requirements of the various I/O modules. Be sure that information provided includes per point (single input or output) requirements and current requirements for ON and OFF states. If the summation of the current requirements for a particular I/O configuration is greater than the total current supplied by the power supply, then a second power supply will be required. An early consideration of line conditions and power requirements will help to avoid problems during installation and start-up.

Power Supply Loading Example. Undoubtedly, the best solution to a potential problem is anticipation of the problem. When it comes to the selection of power supplies, the first thing that is usually overlooked is the current loading requirements, which may or may not get you later depending on your luck. For this reason, let's go over an example of load estimation.

Consider an application where a PLC has been selected to control 50 discrete inputs and 25 outputs. Each discrete input and output module is capable of connecting up to sixteen and eight field devices, respectively. In addition to this discrete configuration, the application also requires a special real time clock module, and 5 power contact outputs. The system also uses 3 analog inputs and 3 analog outputs.

The configuration of this PLC application is illustrated in Figure 4-11. The first plug-in module is the power supply, then the processor module, and then the I/O modules. The first step that must be done is the computation of how many modules are required and the total current requirement of these modules to operate properly. Table 4-1 shows a listing of the modules, current requirements for all inputs or outputs ON at the same time, and the power supplies available for this PLC system.

The total power supply current required by the I/O system totals 4655 mA, or 4.655 Amps. This current added to the 1.2 Amps required by the processor indicates that a total of 5.855 Amps will be the minimum current the power supply must provide to insure the proper operation of the system. This total current indicates a worst case condition since it assumes that all I/Os are operating in the ON condition which requires more current than when in the OFF condition.

Module Type	I/O Devices Connected	Connection per Module	No. of Modules Required	Modules' Current @ 5V	Total Current Required
Discrete In	50	16	4	250mA	1000mA
Discrete Out	25	8	4	220mA	880mA
Contact	5	64	1	575mA	575mA
Analog In	3	4	1	600mA	600mA
Analog Out	3	4	1	1200mA	1200mA
Clock	1	1	1	400mA	400mA
				TOTAL	4655mA

Processor's current:	1200mA		
Power Supplies Available:	Type A	3 Amps	digital input
	Type B	5 Amps	
	Type C	6 Amps	
Auxiliary Power Supply:	Type AA	3 Amps	
	Type BB	5 Amps	

Table 4-1. Listing of modules required and current needed by each module.

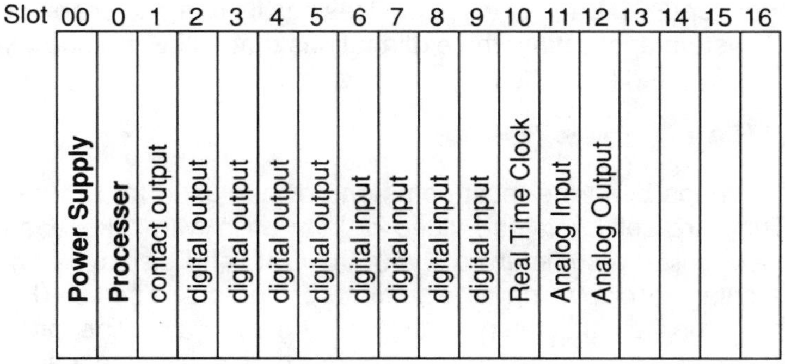

Slot	00	0	1	2	3	4	5	6	7	8	9	10	11	12	13	14	15	16
	Power Supply	**Processer**	contact output	digital output	digital output	digital output	digital output	digital input	digital input	digital input	digital input	Real Time Clock	Analog Input	Analog Output				

Power Supply requires one slot (slot 00).
Processor requires one slot (slot 0).
Twelve I/O slots used; 4 spare.
Auxiliary Power Supplies, if required, must be placed in slot 8.

Figure 4-11. Configuration of PLC.

There are several power supply options which can be used. These include using the 6 Amp supply or a combination of a smaller supply with an auxiliary source. If no expansion is expected, the 6 Amp source will do the job. Conversely, if there is a slight possibility for more I/O requirements, then an auxiliary supply will most likely be needed. The addition of the auxiliary supply can be done at the beginning or when required; however, note that the auxiliary source must be placed in the 8th slot, and for this PLC configuration (Figure 4-11), the I/O addresses will change. Therefore, the reference addresses in the program will have to be reprogrammed to reflect this change. Also remember that the larger the power supply, most likely the higher the price. You must keep all these factors in mind when configuring your system and assigning I/O addresses to field devices.

4-6 PROGRAMMING DEVICES

Since the onset of programmable controllers, developments and advances in the design of programming devices have been in the minds of PLC manufacturers. New and better methods of entering, retrieving and monitoring the internal PLC activity have proved beneficial of programmable controllers in virtually all industries. Due to simple means of entering programs, the user does not have to spend a lot of time learning how to *enter* a program but rather spend time programming and solving the control problem.

In this section we are going to discuss the primary devices that are used in the entering and monitoring of the PLC programs. The most important programming devices include the CRT (cathode ray tube), the mini-programmers, and the personal computer (PC). Each of these devices has its role and scope of utilization within a PLC system application; the explanations that follow will show you where they are generally used and why they are selected.

Cathode Ray Tubes (CRTs)

CRTs have perhaps been the most common devices used for programming the controller. They are self-contained video display units with a keyboard and the necessary electronics to communicate with the CPU and display data. The CRT offers the advantage of displaying large amounts of logic on the screen (Figure 4-12), which greatly simplifies the interpretation of the program. The program logic displayed on the terminal can be ladder diagrams or whatever language is used by the controller.

CRTs are generally classed into two groups: "dumb" and "intelligent." These two types differ greatly in capabilities and in price. Some CRTs are portable and are easily transported. There are also desk-top types that are used primarily in the office or development laboratory.

Dumb CRT. Although the dumb CRT has been used extensively for years as a relatively inexpensive CRT programming device, their use is lessening with new advances in personal computer technology. As the name implies, this CRT is not microprocessor-based, or "not smart"; all the software needed for creating the program and displaying and updating the screen is contained in the controller's executive memory. This terminal must be connected to the PLC to enter or edit the control program. This requirement of constant communication with the processor for programming is known as "on-line" programming. In spite of not being microprocessor based, this dumb video terminal can be used for monitoring and debugging purposes, if the PLC executive provides this option.

The dumb CRT offers the advantage of being usable with different makes of PLCs. However, all dumb CRTs may not be compatible with a user's particular controller. Manufacturers generally provide a list of dumb video terminal models that are recommended for use with their equipment. Program storage is not available with these terminals; however, it can be accomplished using digital cassette recorders specially designed for digital data storage and retrieval. Using these storage devices, the user may load the PLC with new programs or new revisions without having the tedious task of entering it again with the CRT.

Intelligent CRT. The intelligent or smart CRT, shown in Figure 4-13, is a microprocessor-based device that displays logic networks and provides program editing capabilities and other functions independent of the controller's CPU. Contrasted with the dumb CRT, the intelligent terminal has, internal to its own memory, the software that is required to create, alter, and monitor programs. Intelligent CRTs provide a powerful tool when programming, since all the logic, or control programs, can be edited and stored without being connected to the controller. This capability is known as *off-line* programming.

These intelligent devices typically cost more than the dumb terminals; however, most manufacturers' designs allow intelligent CRT's to operate with several PLCs of their programmable controller family. This intercompatibility among family members makes the purchase of an intelligent CRT more justifiable. In general, these CRTs come with a built-in tape or disk device that permits the permanent storage of one or more programs from the different members of the PLC family.

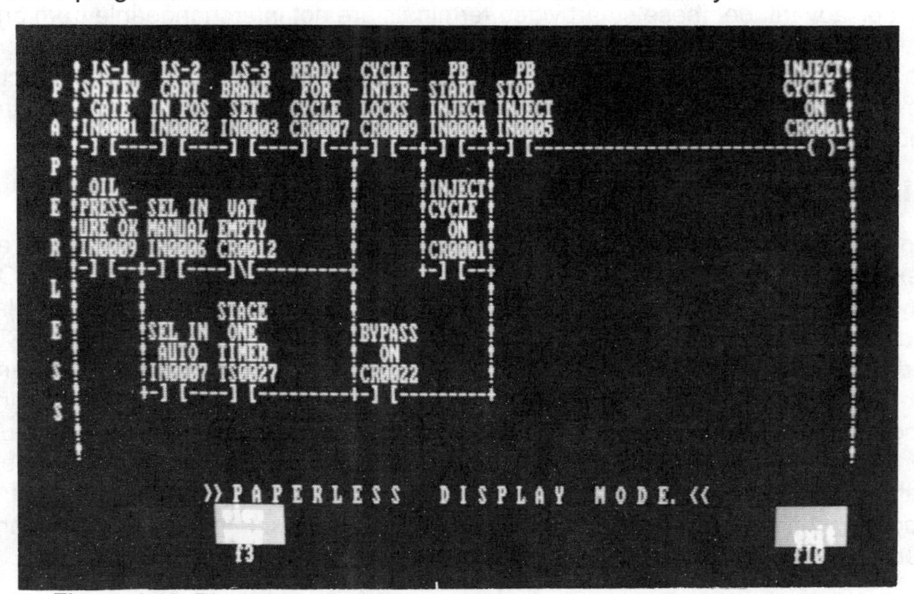

Figure 4-12. Display of a PLC ladder diagram on a CRT programming screen (Courtesy Westinghouse Electric, Pittsburgh, PA).

Figure 4-13. Intelligent CRT (Courtesy of Square D, Milwaukee, WI).

More sophisticated smart CRTs have additional features which make them even more appealing. One feature is a network interface that allows the CRT to connect to the manufacturer's Local Area Network (LAN). This arrangement allows the terminal to access any PLC in the network, change parameters or programs, and monitor any elements (i.e. coils, contacts, inputs, or outputs) without connecting directly to any specific PLC. If the software is available, this arrangement will also allow a means for centralized data collection and display, since it can receive data from the different controllers in the network. Some newer intelligent CRTs provide special graphic keys (on the keyboard) which allow the user to construct various displays that can be recalled on command. Under program control, variable data can be superimposed on the fixed graphic display. Some intelligent CRTs also have software documentation capabilities which eliminate the need for purchasing this additional equipment.

As a disadvantage, these smart video terminals are not interchangeable from one manufacturer's PLC family to another. However, with the increasing number of products in the manufacturers' product lines and user standardization of products, these intelligent devices may be a good choice, especially if the user has standardized with one brand of PLCs.

Mini-Programmers

Mini-programmers, also known as hand-held or manual programmers, are an inexpensive and portable means for programming small PLCs (up to 128 I/O). Physically, these devices resemble hand held calculators, but have a larger display and a somewhat different keyboard. The display is usually LED or dot matrix LCD, and the keyboard consists of numeric keys, programming instruction keys, and special function keys. Instead of the hand-held unit, some controllers have the mini-programmer built-in. In some instances, these built-in programmers are detachable from the PLC. Figure 4-14 shows a small PLC with this built-in and detachable feature. Even though they are used mainly for editing and inputting a control program, mini-programmers can also be a useful tool for starting-up, changing, and monitoring the control logic.

As with smart CRTs, most mini-programmers are designed so that they are compatible with two or more controllers of the product family. The mini-programmer is generally used with the smallest member of the PLC family or in some cases with the next larger member, which is normally programmed with a CRT. With this programming option, small changes or monitoring may be required in the larger controller, and can be accomplished without carrying the CRT to the PLC location. This compatibility feature, offered by some mini-programmers, can be translated into cost savings in the case where an expensive CRT cannot be justified.

Mini-programmers, like CRTs, can be non-smart or smart. The non-smart hand-held programmer can be used for entering and editing the program with limited (limited by memory and display size) on-line monitoring and editing capability. The smart mini-programmer is microprocessor-based and provides the user with many of the features offered by the CRT. These smart devices can often perform system diagnostic routines (memory, communication, display, etc.) and even serve as an operator interface device which could display English messages concerning the controlled machine or process. Figure 4-15 shows a typical mini-programmer and a small PLC in which they are generally used.

Figure 4-14. Small programmable controller with a built-in and detachable hand-held programmer (Courtesy Gould, Andover, MA).

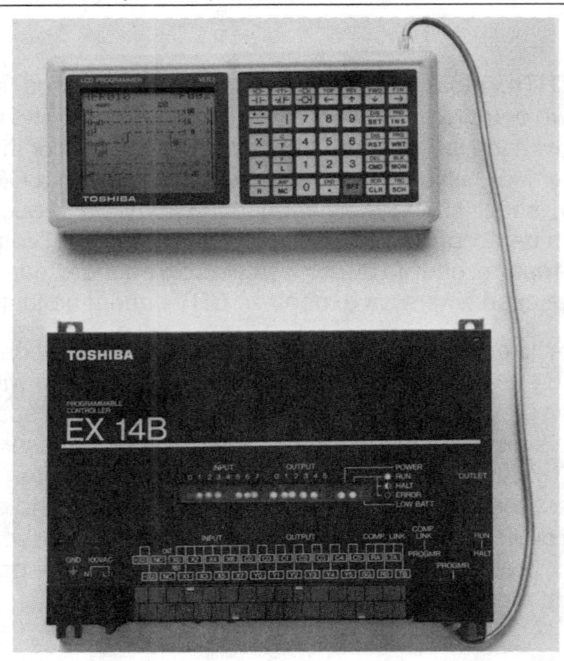

Figure 4-15. Mini-programmer unit used to program a micro PLC (Courtesy of Toshiba, Houston, TX).

Small mini-programmers do not provide storage capability of the program in cassette form; however, a memory module may be connected to a smart hand-held unit and used for storing the program. The type of memory that is generally used is an EEPROM (see next chapter on memory types) whose contents can be altered at a later date. This type of storage may be useful in applications where the control program of one machine needs to be duplicated and easily transferred to other machines (i.e. OEM applications).

Personal Computers

Common usage of the personal computer (PC) in our daily lives has led to a new breed of PLC programming device. Due to the PC's general purpose architecture and de facto standard operating system, several PLC manufacturers and other independent suppliers provide the necessary software to implement the ladder program entry, editing, and real time monitoring of the PLC's control program.

Personal computers will soon be the programming device of choice not so much because of its PLC programming capabilities, but because these PCs may already be present at the location where the user may be performing the programming. The different types of desk-top, lap-top and portable PCs have allowed the programmer to utilize a programming device that can be used in applications other than PLC programming. For instance, a personal computer may also be connected to the PLC's local area network (LAN) (Figure 4-16) to gather and store, on a hard disk, process information which could be vital for future product enhancements. The PC could also communicate through the RS-232C serial port with the PLC and serve as a data handler and supervisor of the PLC control.

In addition to the programming and data collection activity of a PC, documentation of the PLC is sometimes included in the software which provides the ladder programming capability. This documentation capability allows the programmer to define the purpose and function of each I/O address that is being used in the program. General software programs, similar to those obtained for the office environment (i.e. spreadsheet, etc.), will be evolving to allow standard programming packages to be used for report generation. This would be invaluable in using the PC as a man/machine interface, providing a window to the inner workings of the machine or process being controlled by the PLC.

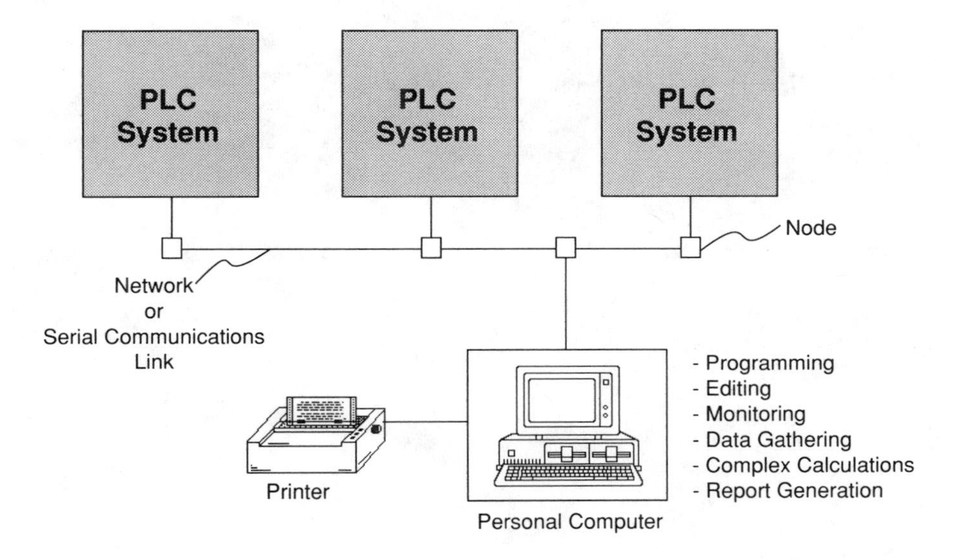

Figure 4-16. Example of a PC connected to a PLC's Local Area Network.

Chapter

—5—

THE MEMORY SYSTEM AND I/O INTERACTION

"Nor will the thought of Dido be bitter to me so long as I have memory, and breath ontrols these limbs."

—Virgil

5-1 MEMORY OVERVIEW

The most important characteristic of a programmable controller is the user's ability to make changes to a control program quickly and easily. This programmability feature is possible because of the architecture of a PLC. The memory system is the area in the CPU where all the sequences of instructions, or programs, are stored and executed by the processor to provide the desired control of field devices. The sections of memory that contain the control programs can be changed, or reprogrammed, to correct for changes in a manufacturing line procedure or during the start-up of a new system.

An understanding of what is stored in the PLC's memory will aid in understanding why certain things are stored as they are and why certain considerations are taken during the I/O assignment in subsystem racks, I/O addresses and addressing methods, and the memory capacity requirement for a particular application. As will be seen later, the interaction between part of the memory system and the I/O modules will help you understand the general operation of a PLC and its I/O scanning method.

Memory Sections

The total memory system in a PLC is composed of two virtual memories, the first one is called the *Executive*, the second one is referred to as *Application* memory (see Figure 5-1). The executive memory system is composed of a collection of permanently stored programs that are considered part of the PLC system itself. These supervisory programs direct all system activities such as execution of the control program, communication with peripheral devices, and other system activities. The executive section is the part of memory where the system's available instruction software is stored, i.e. relay instructions, block transfer, math instructions, etc. This area of memory is not accessible to the user.

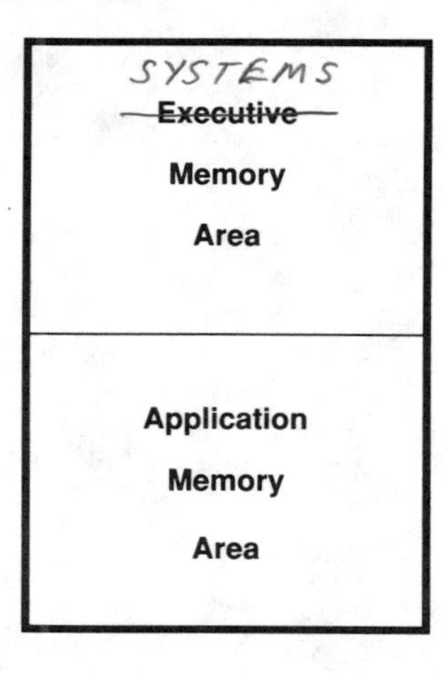

Figure 5-1. Simplified block diagram of the total PLC memory system.

The application memory provides storage area for programmed instructions entered by the user which forms the applications program. The application memory area for a PLC system is composed of several sections, each having a specific function and usage. Section 5-4 (Memory Organization) will cover in detail the executive and application memory areas.

5-2 MEMORY TYPES

The storage and retrieval requirements are not the same for the executive memory and the application memory; therefore, they are not always stored in the same types of memory. For example, the executive requires a memory that permanently stores its contents and cannot be deliberately or accidentally disturbed or altered by loss of electrical power or by the user.This type of memory for the application program may prove to be unsuitable.

Although there are several types, memory can generally be placed into the following two categories: *volatile and nonvolatile.*

Volatile memory will lose its programmed contents if all operating power is lost or removed, whether it is normal power or some form of back-up power. Volatile memory is easily altered and quite suitable for most applications, when supported by battery back-up and possibly a disk copy of the program.

Nonvolatile memory will retain its programmed contents, even during a complete loss of operating power, without requiring a back-up source. Nonvolatile memory generally is unalterable, yet there are special nonvolatile memory types that are alterable. Today's PLCs include those that use nonvolatile memory and volatile memory with battery back-up, as well as those that offer both.

There should be two major concerns regarding the type of memory where the application program is stored. Since this memory is responsible for retaining a control program that will be run each day, volatility should be the prime consideration. Without the application program, production may be delayed or forfeited, and the outcome is usually unpleasant. A second concern should be the ease with which the program stored in memory can be altered. Ease in altering the memory is important, since the memory is ultimately involved in any interaction between the user and the controller. This interaction begins with program entry and continues with program changes made during program generation, system start-up, and ultimate on-line changes such as changing a timer or counter preset value.

The following discussion describes various types of memory and how their characteristics affect the manner in which programmed instructions are retained or altered within the programmable controller.

Read-Only Memory (ROM)

A Read-Only Memory is designed to permanently store a fixed program which is not alterable under ordinary circumstances. It acquires its name from the fact that its contents can be examined or read, but not written into or altered once the data or program has been stored. This is in contrast to memory types that can be read from and written to (see RAM, Read/Write). Because of their nature, ROMs are generally immune to changes due to electrical noise or loss of power. Executive programs are often stored in a ROM.

Generally, programmable controllers rarely use ROM for their application memory. However, in applications that require fixed data, the Read Only Memory offers advantages when speed, cost, and reliability are factors. Generally, the creation of ROM-based PLC programs is accomplished at the factory by the manufacturer. Once the original set of instructions is programmed, it can never be altered by the user. The typical approach to the programming of ROM-based controllers assumes that the program has already been debugged and will never be changed. This debugging would have to be accomplished using a Read/Write-based PLC or possibly a computer. The final program can then be entered into ROM. ROM application memory is typically found only in very small, dedicated programmable controllers.

Random Access Memory (RAM or R/W)

Random Access Memory, often referred to as *Read/Write Memory*, is designed so that information can be written into or read from any unique location. There are two types of RAM: the volatile RAM, which does not retain its contents if power is lost, and non-volatile RAM (see NOVRAM, Core), which retains its contents if power is lost. Volatile RAM normally has a battery backup to sustain its contents during power outages.

Today's controllers, for the most part, use RAM with battery support for application memory. RAM provides an excellent means for easily creating and altering a program, as well as allowing data entry. In comparison to other memory types, RAM is a relatively fast memory. The only noticeable disadvantage of battery supported

Figure 5-2. 4K words by 8-bit RAM memory (4K x 1 bit/chip).

RAM is the fact that it requires a battery that may eventually fail. Battery supported RAM is sufficient for most applications. If a battery backup is not feasible, a controller with a nonvolatile memory option can be used in combination with the RAM (e.g. RAM, EPROM). This type of memory arrangement provides the advantages of both volatile and nonvolatile memory. A RAM chip is shown in Figure 5-2.

Programmable Read Only Memory (PROM)

The PROM is a special type of ROM in that it can be programmed. Less than 1% of today's programmable controllers use PROM for application memory. When it is used, this type of memory will most likely be a permanent storage backup to some type of RAM. Although a PROM is programmable and like any other ROM has the advantage of nonvolatility, it has the disadvantage of requiring special programming equipment, and once programmed it cannot be easily erased or altered. Any program change would require a new set of PROM chips. A PROM memory might be suitable for storing a program that has been thoroughly checked while residing in RAM and will not require further changes or on-line data entry.

Erasable Programmable Read Only Memory (EPROM)

The EPROM is a specially designed PROM that can be reprogrammed after being entirely erased with the use of an ultra-violet (UV) light source. Complete erasure of the contents of each chip requires that the window of the chip (see Figure 5-3) be exposed to the UV-light for approximately 20 minutes. The EPROM can be considered a semi-permanent storage device in that it permanently stores a program until it is ready to be altered. Program changes can be made after the EPROM chip is completely erased.

Figure 5-3. 4K by 8-bit EPROM memory chip.

EPROM provides an excellent storage medium for an application program in which nonvolatility is required, but program changes or on-line data entry is not required. Many Original Equipment Manufacturers (OEMs) use controllers with EPROM type memories to provide permanent storage of the machine program after it has been developed, debugged, and is fully operational. EPROM is probably used because most machines supplied by OEMs will not require any changes or data entry by the user.

An application memory composed of EPROM alone would be unsuitable, if on-line changes and/or data entries are a requirement. However, many controllers offer EPROM application memory as an optional backup to battery supported RAM. EPROM, with its permanent storage capability combined with the easily altered RAM, makes a suitable memory system.

Electrically Alterable Read Only Memory (EAROM)

Electrically Alterable ROMs are similar to EPROMs, but instead of requiring an ultraviolet light source to erase them, an erasing voltage on the proper pin of an EAROM chip would wipe it clean.

Very few controllers use EAROM as application memory, but like EPROM, it provides a nonvolatile means of program storage and can be used as a backup to RAM type memories.

Electrically Erasable Programmable Read Only Memory (EEPROM)

EEPROM is a integrated circuit memory storage device, which was developed in the mid-1970s. Like ROMs or EPROMs, it is a nonvolatile memory, yet it offers the same programming flexibility as the RAM does.

Several of today's small and medium-sized controllers use EEPROM as the only memory within the system. It provides permanent storage of the program and can be easily changed with the use of the standard CRT or manual programming unit. These two advantages will certainly help to eliminate downtime, or delays associated with programming changes, and will also help to lessen, as we will see below, two disadvantages of the EEPROM.

One disadvantage of EEPROM is that writing to a byte of memory is possible only after erasing that byte. The erase/write process takes approximately 10 to15 milli-seconds. This delay period is noticeable when an on-line program change is being made. Another characteristic of the EEPROMs is a limitation on the number of times that a single byte of memory can undergo the erase/write operation (approximately 10,000). These disadvantages are negligible when compared to the remarkable advantages that the EEPROM offers.

Core

Core memory is a nonvolatile type of memory. It received its name from the fact that it stores individual bits by magnetizing small toroidal ferrite cores in the 1 or 0 direction, through a write-current pulse (Figure 5-4). Each core represents a bit and can hold a 1 or 0 even in the event of power loss. Core memory units are nonvolatile because power is not necessary to keep the cores magnetized. The state of each core is also electrically alterable, which makes it a nonvolatile RAM.

Core was used in many of the first programmable controllers and is still used in a few controllers today. It provides an excellent means for permanent storage and is easily altered. Distinct disadvantages of using core memory in programmable controllers are its slow speed, relatively high cost, and somewhat large physical space requirements.

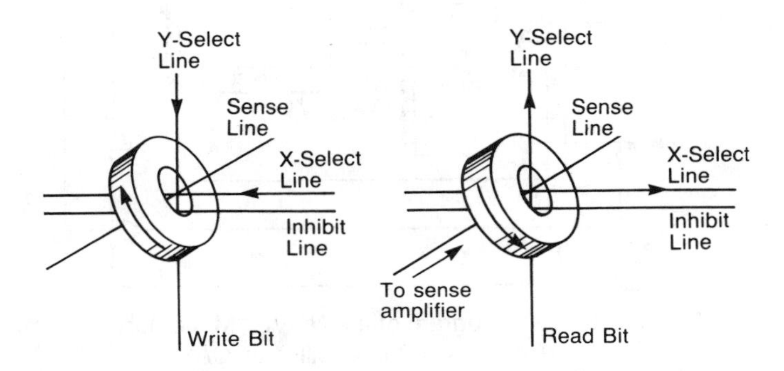

Figure 5-4. Magnetic toroidal illustration used in core memories.

Non-Volatile Random Access Memory (NOVRAM)

NOVRAM, a newer memory fabrication, employs both the conventional RAM and the nonvolatile EEPROM on a single chip. Each bit within the RAM section of the chip has a corresponding adjacent bit within the EEPROM (see Figure 5-5). Nonvolatile data can be stored in the EEPROM, and at the same time, independent data can be written to or accessed from the RAM. Data can be transferred back and forth between the RAM and the EEPROM at any time.

A NOVRAM application memory is the perfect solution to the application requirements of easy reprogrammability and nonvolatility. When entered, the control program is stored directly onto the RAM and is also executed from the RAM. An exact replica of the executed program is dumped onto the nonvolatile EEPROM, without any user intervention, when the STORE operation takes place. When power is removed from the controller, the RAM contents are lost, but are restored by the RECALL (from EEPROM to RAM) command on each power-up. With this memory arrangement, the convenience of easy changes and a permanent copy of the program are provided in a single memory system, as opposed to two separate memory systems (e.g. RAM, EPROM).

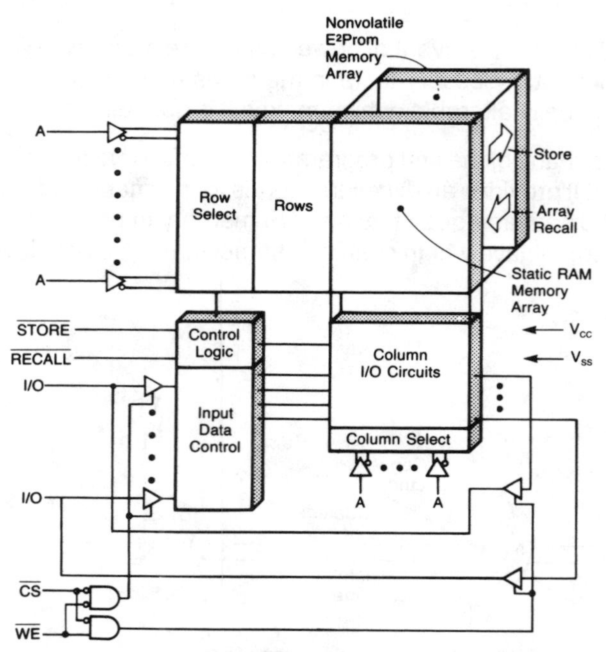

Figure 5-5. Block diagram of the NOVRAM memory system (Courtesy of Xicor, Milpitas, CA).

NOVRAM can be considered the state of the art in application memory for programmable controllers, and will probably replace the current memory technologies used in PLCs. Figure 5-6 shows an actual NOVRAM chip.

Figure 5-6. Magnified photograph of the NOVRAM 2210 memory chip (left) and actual NOVRAM chip (Courtesy of Xicor, Milpitas, CA).

5-3 MEMORY STRUCTURE AND CAPACITY

Basic Structural Units

Programmable controller memories can be visualized as a large two-dimensional array of single unit storage cells, each of which can store a single piece of information in the form of "1" or "0". It is obvious, then, that the binary numbering system is used to represent any information stored in memory. Since BIT is the acronym for BInary digiT, and each cell can store a bit, each cell is called a bit. A bit then is the smallest structural unit of memory and stores information in the form of 1s and 0s. Ones and zeroes are not actually present in each cell; instead, each cell has a voltage charge present (indicating a 1) or not present (indicating a 0). The bit is considered ON if the stored information is 1, and OFF if the stored information is 0. The ON/OFF information stored in a single bit is referred to as *bit status*.

Sometimes it is necessary for the processor to handle more than a single bit. For example, it would be more efficient to handle a group of bits when transferring data to and from memory. To store numbers and codes, as we have discussed, also requires a grouping of bits. A group of bits handled simultaneously is a "byte." A byte is more accurately defined as the smallest group of bits that can be handled by the processor at one time. Although byte size is normally 8 bits, the size can vary depending on the specific controller.

The third and final structural information unit used within the programmable controller is called a "word". A word is also a fixed group of bits that varies according to the controller. In general, a word is the unit that the processor uses when data is to be operated on, or when instructions are to be performed. Word length is usually 1 byte in length or more. For example, a 16 bit word consists of 2 bytes. Typical word lengths used in PLCs are 8, 12, and 16 bits. Figure 5-7 illustrates the structural units of a typical programmable controller memory.

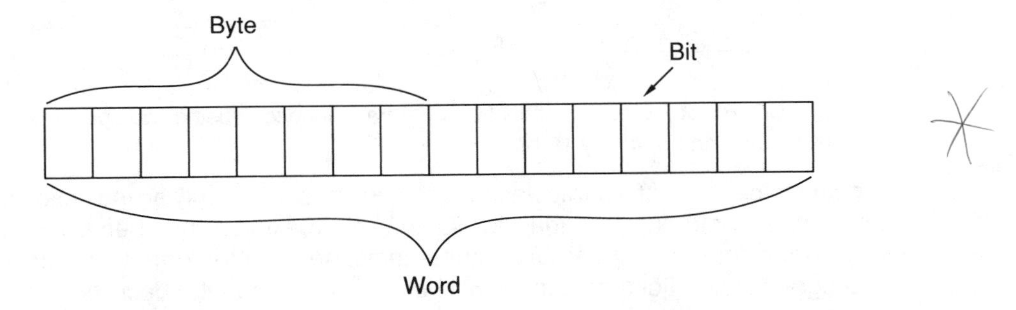

Figure 5-7. Basic Stuctural Units of PLC Memory.

Memory Capacity and Utilization

Memory capacity is a vital concern when considering a programmable controller application. Specifying the right amount of memory can mean a savings in hardware and time in case additional memory capacity is required later. Knowing the memory capacity requirements ahead of time could also mean avoiding the possibility of purchasing a controller that does not have adequate capacity or that is not expandable.

Memory capacity in general is non-expandable in some controllers (small, less than 64 I/O capacity), and expandable in the larger PLCs. Small PLCs have a fixed amount of memory because the available memory is usually more than enough to provide program storage for most small applications. Larger controllers allow memory expandability since the scope of the application and amount of I/O control have less definition.

The amount of application memory is specified in terms of K units where each K unit represents 1024 word locations. A 1K memory contains 1024 storage locations, 2K is 2048 locations, 4K is 4096, and so on. Figure 5-8 illustrates two memory arrays of 4K each; however, one set is of one byte word (8 bits) and the other of two byte words (16 bits).

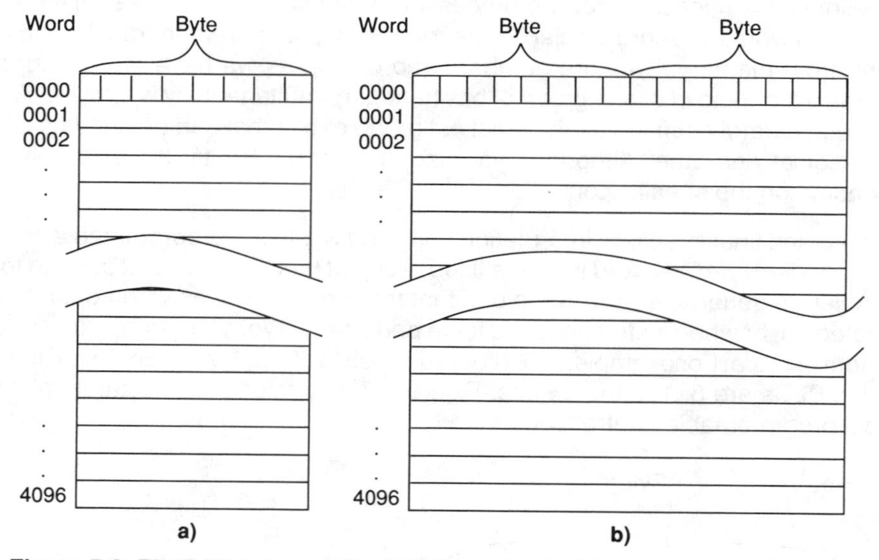

Figure 5-8. Block illustration of 4K (4096) storage locations using two specifications; a) 4K by 8-bit and b) 4K by 16-bits.

The memory capacity of a particular controller in units of K is only an indication of the total number of storage locations available. This maximum number alone is not enough when determining memory requirements. Additional information concerning how program instructions are stored will help in making a better decision. The term "memory utilization" refers to the amount of data that can be stored in one location or, more specifically, to the number of memory locations required to store each type of instruction. This data can be obtained from the manufacturer if not found in the product literature.

Suppose 16 bits were required to store each normally-open and normally-closed contact instruction. With this information, it becomes quite clear that the effective storage area of the memory system in Fig. 5-8a is one half that of Fig.5-8b. This means that to store the same size control program would require 8K memory capacity instead of the specified 4K.

After becoming familiar with how memory is utilized in a particular controller, users can begin a determination of the maximum memory requirements for the application. Although several rules of thumb have been used over the years, no one simple rule has emerged as being the most accurate. However, with a knowledge of the number of outputs, some idea of the number of program contacts needed to drive the logic of each output and information concerning memory utilization, an approximation can be reduced to a simple multiplication.

As an example, let's take an application that requires 70 outputs, where each output may be driven by logic composed of an average of 10 contact elements. In addition to these outputs, we know that 11 timers and 3 counters will be required, each having approximately 8 and 5 contact elements respectively. Furthermore, we estimate that there will be approximately 20 instructions that include addition, subtraction and comparisons, each being driven by about 5 contact elements. For this example, let's use the memory utilization information shown in Table 5-1.

Instruction	Words of Memory Required
Examine ON or OFF (contacts)	1
Output Coil	1
Add/Subtract/Compare	1
Timer/Counter	3

Table 5-1. Memory Utilization.

The first approximation attempt will give us the following estimation of memory:

a) Control logic = 10 contact elements/output rung
   Number of output rungs = 70
b) Control logic = 8 contact elements/timer
   Number of timers = 11
c) Control logic = 5 contact elements/counter
   Number of counters = 3
d) Control logic = 5 contact elements/math and compare
   Number of math and compare = 20

Based on memory utilization the total number of words would be:

a)	Total contact elements	(10 x 70)	700
	Total outputs	(1 x 70)	70
	Total Words		770
b)	Total contact elements	(11 x 8)	88
	Total timers	(11 x 3)	33
	Total Words		121
c)	Total contact elements	(3 x 5)	15
	Total counters	(3 x 3)	9
	Total Words		24
d)	Total contact elements	(5 x 20)	100
	Total math and compare	(1 x 20)	20
	Total Words		120

The total number words of memory required for the storage of these instructions would be 1035 words, or just over 1K of memory.

The calculation performed in the previous example is considered an approximation, because other factors must be considered before the final decision is made. Such considerations may include future expansion. We should keep in mind that memory requirements are also affected by the sophistication of the control program. If the application requires data manipulation and data storage, additional memory will be required. Normally, the enhanced instructions that perform mathematical and data manipulation operations will have greater memory requirements. Depending on the PLC's manufacturer, the amount of application memory may also include the data table and I/O table sections (see next section). If this is the case, then the amount of "real" user application memory would be less than that specified. Exact usage can be determined by consulting the manufacturer's memory utilization specifications.

After determining the minimum memory requirements for the application, it would be wise to add an additional 25% to 50% more memory. This increase will allow for changes, modifications, or future expansion.

5-4 MEMORY ORGANIZATION AND I/O INTERACTION

The memory system, as mentioned before, is composed of two major sections, the system memory and the application memory, which in turn are formed by other areas. This memory organization is known as the *memory map* and can be illustrated as shown in Figure 5-9. Although the two main sections, system memory and application memory, are shown one after the other, they are not necessarily next to each other, physically or by address. The memory map shows not only what is stored in memory, but also where data is stored according to specific locations called memory addresses. The understanding of the memory map may prove to be very useful when creating the control program and defining the data table.

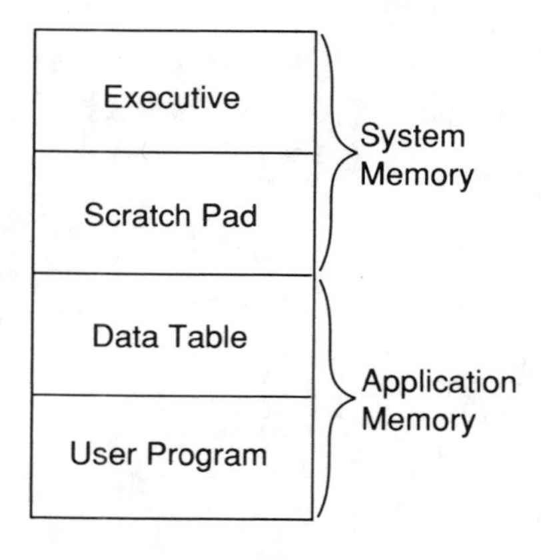

Figure 5-9. A simplified memory map.

It is unlikely that two different controllers will have identical memory maps, but a generalization of memory organization is still valid in light of the fact that all programmable controllers have similar storage requirements. In general, all programmable controllers must have memory allocated for the four items described as follows:

Executive: A permanently stored collection of programs that are considered a part of the system itself. These supervisory programs direct system activities such as execution of the control program, communication with peripheral devices, and other system housekeeping activities.

Scratch Pad: A temporary storage used by the CPU to store a relatively small amount of data for interim calculations or control. Data that is needed quickly is stored in this area to avoid the access time that would be involved if it were stored in the main memory.

Data Table: This area stores any data associated with the control program, such as timer/counter preset values, and any other stored constants or variables that are used by the control program or the CPU. This section also retains the status information of the system inputs once they have been read and the system outputs once they have been set by the control program.

User Program: This area provides storage for any programmed instructions entered by the user. The control program is stored in this area.

The Executive and the Scratch Pad are transparent to the user and thus can be considered a single area of memory that for our purpose is labeled "System Memory". On the other hand, the Data Table and the User Program areas are all accessible and are required by the user for the control application and is therefore called "Application Memory."

It is important to note here that the total memory specified for a controller may include system memory and application memory. For example, a controller with a maximum of 64K may have executive routines that use 32K and a system work area (scratch pad) of 1/4K. This arrangement would leave a total of 31 3/4K for application memory (data table and user memory). Although it is not always the case, the maximum memory normally specified for a given controller will only include the total application memory. Other controllers may specify the available user memory for use by the control program; this may assume a fixed data table area defined by the manufacturer. Now let's take a closer look at the application memory and understand how it really interacts with the user and the program.

The Application Memory

The application memory stores programmed instructions and any data that will be utilized by the processor to perform its control functions. A mapping of the typical elements of this area is shown in Figure 5-10. Each controller has a maximum amount of application memory, which varies depending on the size of the controller. All data is stored in what is called the data table, while programmed instructions are stored in the area allocated for the user program.

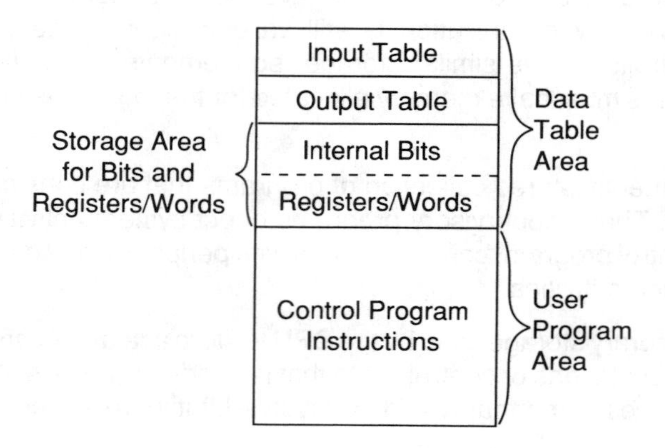

Figure 5-10. Block Diagram of a typical application memory map.

Data Table Memory Area. The data table area of the PLC's application memory is composed of several sections (see Figure 5-10) which are:

- The input table area
- The output table area
- Storage area

These sections contain information in binary form, representing status (ON or OFF), numbers, or codes. Remember that the memory structure is composed of cell areas, or bits, where this binary information is stored. The following is an explanation of each of these areas.

Input Table. The input table is an array of bits that stores the status of digital or discrete inputs, which are connected to input interface circuits. The maximum number of bits in the input table is equal to the maximum number of field inputs that can be connected to the PLC. For instance, a controller with a maximum of 64 inputs would require an input table of 64 bits. Each connected input has a bit in the input table that corresponds exactly to the terminal to which the input is connected. The address of the input device can be interpreted as a bit and word location of its corresponding mapping or position in the input table. Referencing Figure 5-11, the limit switch connected to the input interface shown has an address of 13010_8 as its corresponding bit in the input table. This address comes from the word location 130_8 and the bit number 10_8, both of which are related to the rack position where the module is installed and the module's terminal connected to the field device. If the limit switch is ON (closed) the corresponding bit (13010_8) will be 1; if the limit switch is OFF (open) its corresponding bit will be 0.

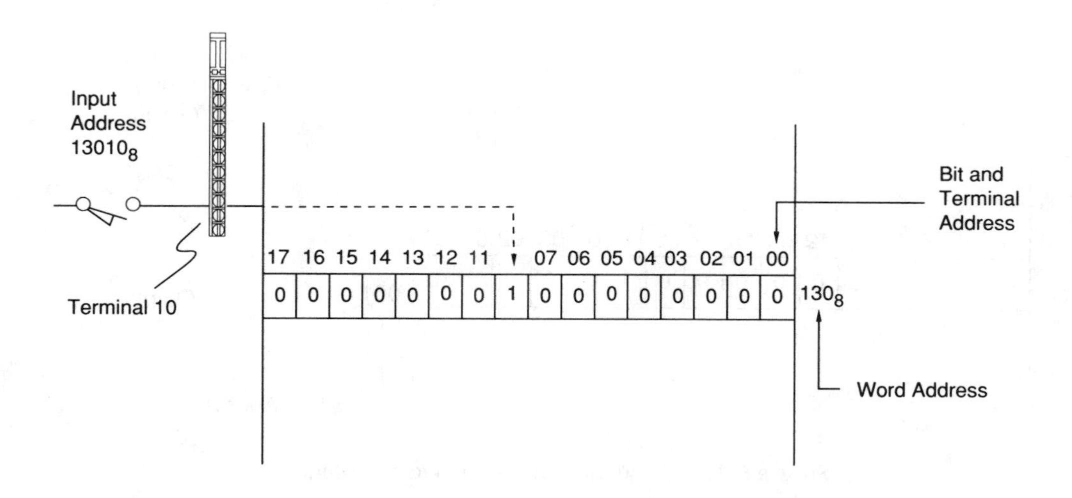

Figure 5-11. Input device mapping to input table.

During PLC operation, the processor will read the status of each input in the input module, and place this value (1 or 0) in the corresponding address in the input table. The input table is constantly changing to reflect the changes of the input module and its connected field devices. These changes in the input table take place during the reading part of the I/O update.

Output Table. The output table is an array of bits that controls the status of digital output devices, which are connected to output interface circuits. The maximum number of bits available in the output table is equal to the maximum number of output field devices that can be interfaced to the PLC. For instance, a controller with a maximum of 128 outputs would require an output table of 128 bits.

Each connected output has a bit in the output table that corresponds exactly to the terminal to which the output is connected. The bits in the output table are controlled (ON/OFF) by the processor as it interprets the control program logic and turns the I/O modules ON or OFF accordingly during the I/O update scan. If a bit in the table is turned ON (logic 1), then the connected output is switched ON. If a bit is cleared or turned OFF (logic 0), the output is switched OFF (refer to Figure 5-12). Remember that the turning ON or OFF of the field devices occurs during the update of outputs after the end of the scan.

Storage Area. This section of the data table may be subdivided in two parts consisting of an *internal bit storage* and a *register* or *word storage area* (see Figure 5-13). The purpose of this data table section, as its name implies, is to store data that can change whether it is one bit or a word (16 bits).

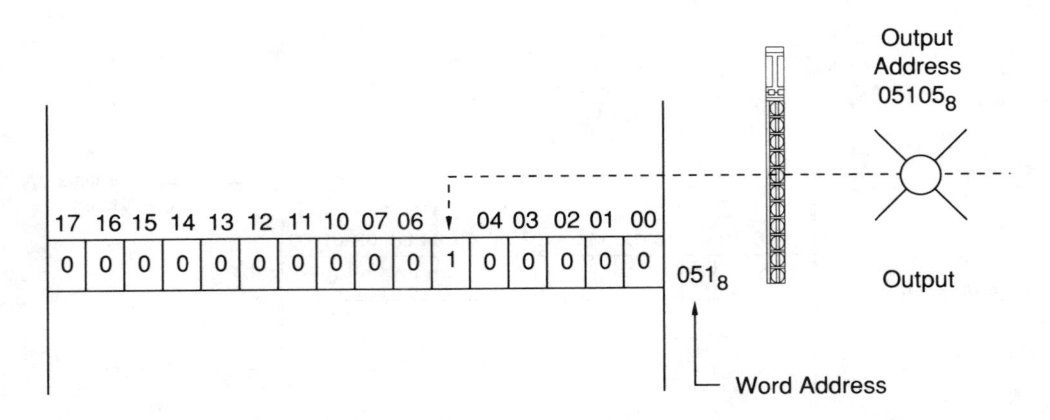

Figure 5-12. Output device mapping to output table.

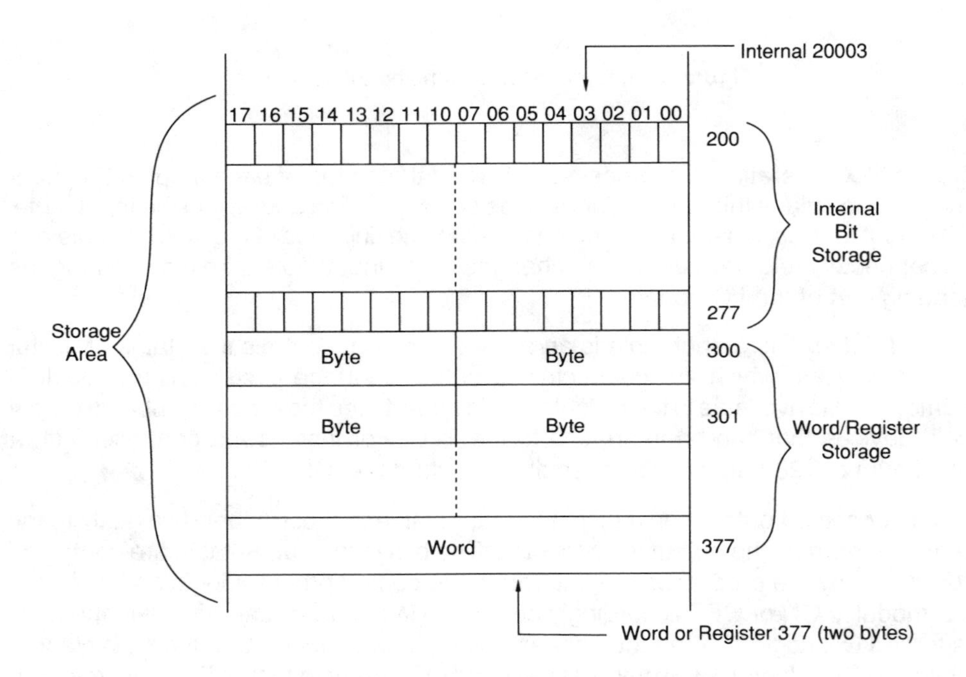

Figure 5-13. Storage area section of the data table.

Storage bits are also referred to as *internal outputs, internal coils, internal (control) relays,* or just simply as *internals.* These internals are generally used to provide an output, for interlocking purposes, of ladder sequences in the control program. The internal outputs do not directly control an output device.

Note that these bits are located in a different address which does not map the output table, and therefore, any output devices. When the processor evaluates the control program, and any of these outputs is energized (logic 1), its referenced contacts (contacts with this address) will change state; the normally open will close and the normally closed will open (see I/O interaction section). These contacts are used in conjunction with other internals or *real* input contacts to form interlocking sequences that are driving an output device or another internal output.

The register/word storage area is used to store groups of bits at a time (bytes, words). This information is in binary and represents quantities or codes. If decimal quantities are stored, the binary pattern of the register represents an equivalent decimal number (see Chapter 3); if a code is stored, the binary pattern may represent a BCD number or an ASCII code character (one character per byte).

Values placed in this register storage area may represent input data from thumbwheel switches, shaft encoders, analog input devices and other types of variables. The data stored may also represent constants that are used in the control program. In addition to input values, these registers can contain output values which are destined to go to output interface modules connected to devices such as analog meters, seven segment LED indicators (BCD), control valves, drive speed controllers and others. Storage registers are also used to hold information whether it is a fixed constant such as a preset timer/counter value, or a changing value such as arithmetic results or accumulated timer/counter values. Depending on the use of the registers they may also be referred to as *input registers, output registers,* and *holding registers*. Table 5-2 shows some typical constants and variables stored in these storage areas.

CONSTANTS	VARIABLES
Timer Preset Value	Timer Accumulated Value
Counter Preset Value	Counter Accumulated Value
Loop Control Setpoints	Resultant Values from Math Operations
Other Compare Setpoints	Analog Input Values
Decimal Tables (recipes)	Analog Output Values
ASCII Characters	BCD Inputs
ASCII Messages	BCD Outputs
Other Numerical Tables	

Table 5-2. Constants and variables generally stored in registers.

User Program Memory Area. The user program memory is an area reserved in the application memory for the storage of the control logic. All the PLC instructions which control the machine or process are stored here. These instructions have been stored by the processor's executive software language which represents each of the PLC instructions.

When the controller is in the RUN mode, and the program is executed, the processor interprets these memory locations and controls the bits in the data table which correspond to a real or internal I/O. The interpretation of the User program is accomplished by the processor's execution of the Executive program.

The maximum amount of User program memory available is normally a function of the controller size (i.e. I/O capacity). In medium and large controllers, the user program area is made flexible by altering the size of the data table so that it meets the minimum data storage requirements. In small controllers, however, the user program area is normally fixed. The amount of memory used in the user's program is directly proportional to the amount of instructions required by the control program. Estimation of this memory usage is generally accomplished by the method described in Section 5-3 which explains memory utilization.

5-5 MEMORY MAP EXAMPLE AND I/O ADDRESSING

The understanding of the memory organization and especially the interaction of the data table's I/O mapping and storage area definitely helps to totally comprehend the functional operation of the programmable controller. Although the memory map is generally taken for granted by some users of PLCs, its thorough understanding usually provides a better perception of not only the PLC's operation, but also of how the control software program will be organized and developed.

Organization of the data table*, or configuration as it is sometimes called, is very important. Not only will the discrete device addresses be defined, but also what registers will be used for numerical and analog control, and for the basic PLC's timing and counting operations. The intention of this section is not to go intodetail about the onfiguration, but to review what has been learned about the memory map, making sure that we know how the memory and I/O interact.

Let's first try to come up with an illustration of an application memory map for a programmable controller. This PLC has 4K words of16 bits of total application memory. The system is capable of connecting 256 I/O devices (128 inputs and 128 outputs) and has 128 internal outputs available. There is capability of having up to 256 storage registers which are selectable in groups of eight word locations, eight being the minimum number of registers possible (32 groups of 8 registers each). The numbering system used by this PLC system is octal (base 8) and the word length is two bytes (16 bits).

To illustrate this memory map may seem trivial to some people; however, at this point no one really knows the starting address of the control program; not that it matters as far as the program is concerned, but it does matter concerning the register address references to be used. These register addresses are the ones referred to in the control program, i.e. timer's preset and accumulated values. With this in mind, let's set the I/O table boundaries. Assuming the inputs are first in the I/O mapping,

* In some controllers, the boundary between the data table and user program is selectable. If the data table is increased, more space is available for storage registers.

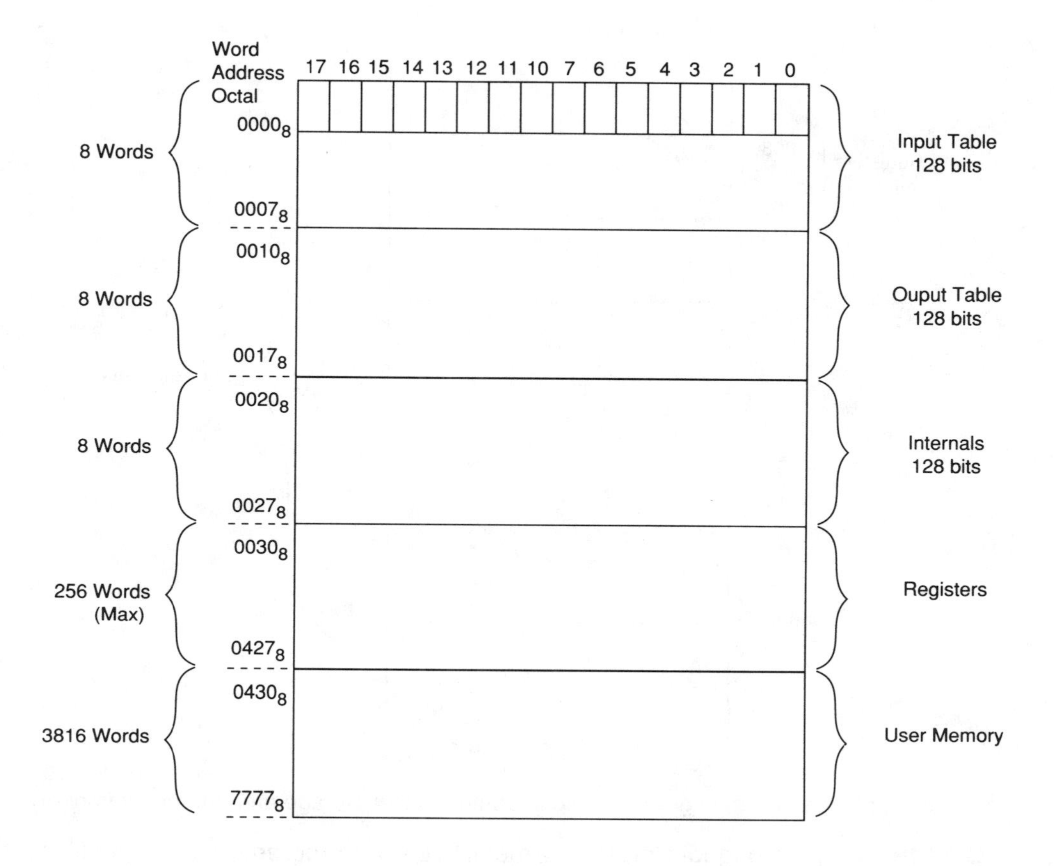

Figure 5-14. I/O table and user memory boundaries.

the input table will start at address 0000_8 and end at 0007_8. The outputs will start at 0010_8 and end at 0017_8 (see Figure 5-14). Since there are 16 bits per memory word, eight input table words comprise the 128 inputs and likewise for the outputs.

The starting address for the internal output storage is at memory location number 0020_8 through 0027_8. Eight words of 16 bits each, for a total of 128 internal output bits. Address 0030_8 indicates the beginning of the register/word storage area. The first eight registers (the minimum possible) end at address 0037_8. Any other 8-register increments will start from 0040_8. The last possible address for a register is 0427_8 which provides for a total of 256 registers (see Figure 5-15).

If all available registers are to be utilized, then the starting memory address for the control program will be 0430_8. This configuration will leave 3816 (decimal) locations to store the control software. This maximum configuration is shown in Figure 5-14.

Although most controllers allow the user to change the range of register boundaries without any concern about the starting memory addresses for the program, the user should know beforehand the number of registers needed. This will prove useful when assigning register addresses in the program.

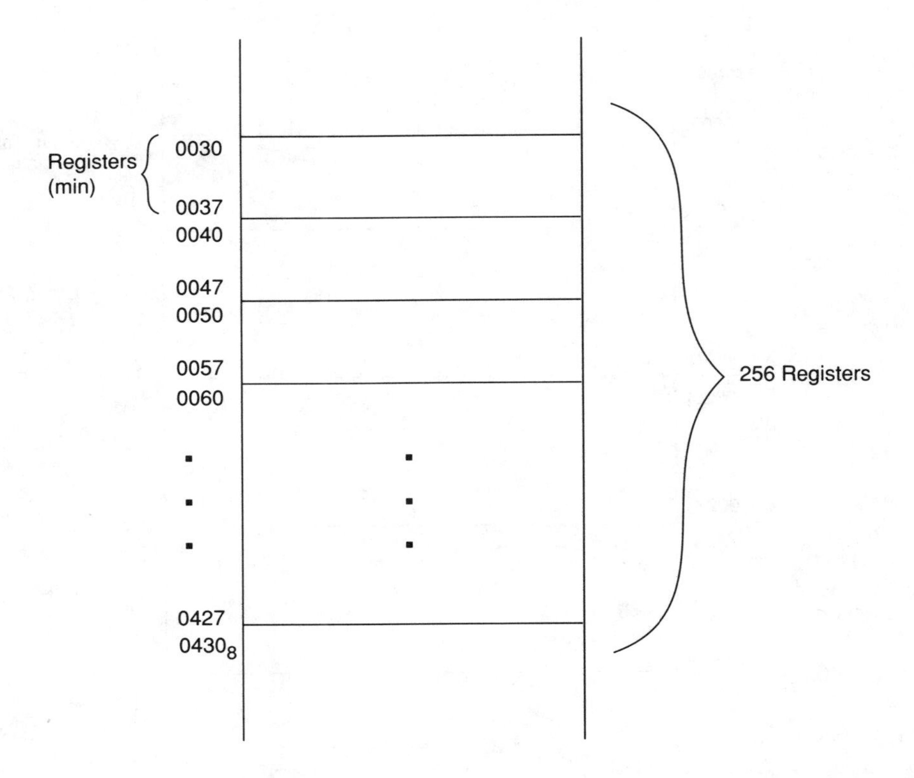

Figure 5-15. Breakdown, in groups of eight, of the register storage area for the example.

I/O Addressing. Throughout this text we mention that the programmable controller's operation simply consists of reading inputs, solving the ladder logic in the user's memory, and updating the outputs. As we get more and more into PLC programming, and into the application of I/O modules, it is worthwhile to pause and review the I/O address relationship with the I/O table and how it is used in the program.

The input/output structure of the programmable controller has been designed with one thing in mind: *simplicity*. The I/O locations in the *racks* or *housings* are mapped to the I/O table where the I/O module placement may define the address of the devices connected to the module. Other PLCs may use internal module switches to define the addresses to be used by the devices connected to the module. In the end, however, all the input and output connections are mapped to the I/O table.

Assume that we have a simple relay circuit that contains a limit switch driving a pilot light (Figure 5-16). This circuit is to be connected to a PLC input module and an output module as shown in Figure 5-17. For the purpose of our discussion, let's say that each module contains sixteen possible input or output channels and that this PLC has a memory map similar to the one shown previously in Figure 5-14. The limit switch is connected to the thirteenth (octal) terminal of the input module while the light is connected to the sixth (octal) terminal of the output module.

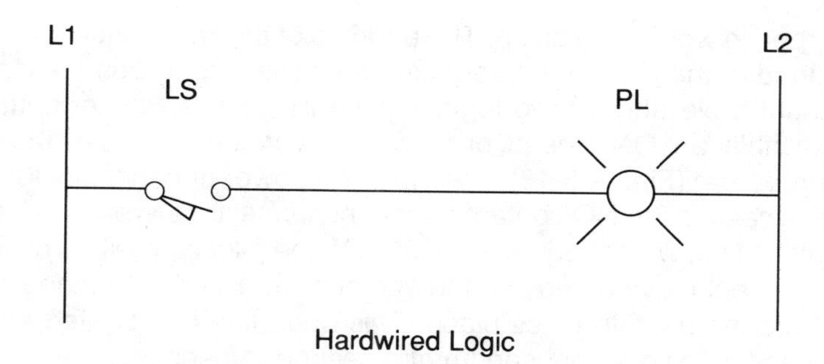

Figure 5-16. Hardwired Logic.

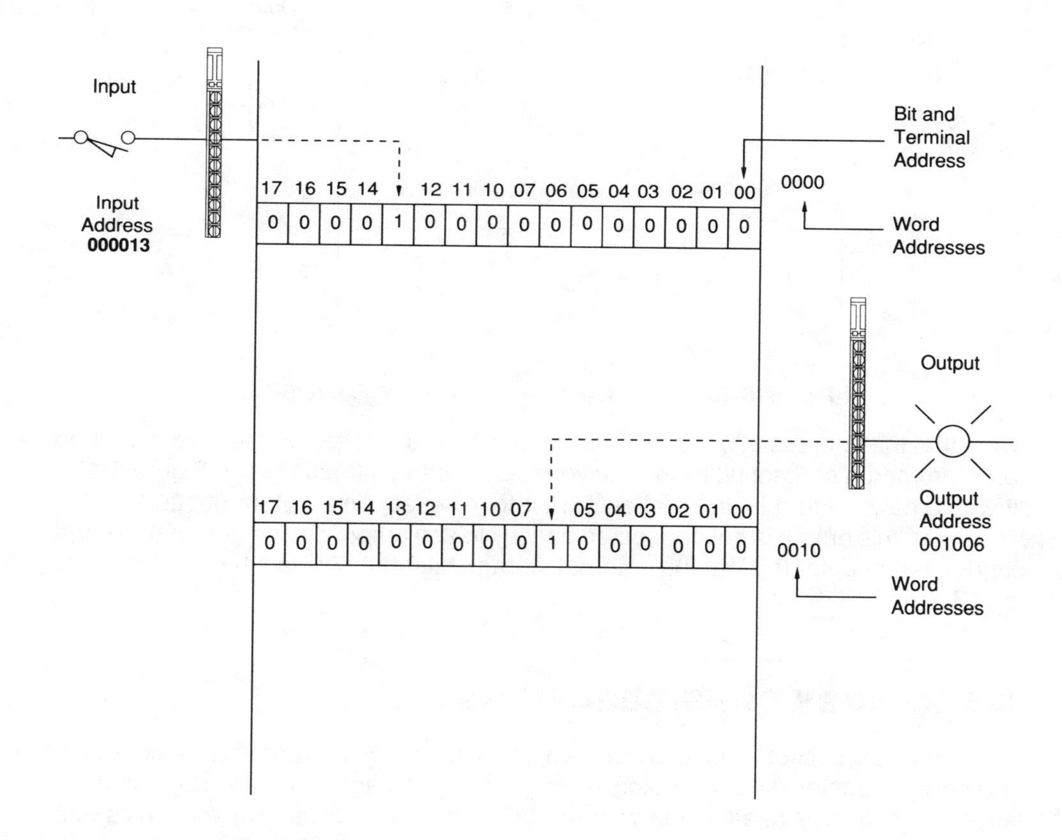

Figure 5-17. Input and Output table and field devices interaction.

Let's say that the I/O modules,due to the placement inside the enclosure, map address (word) 0000 for the input module and address (word) 0010 for the output module. Therefore, the limit switch will be known by the processor as input 000013 and the light as output 001006 i.e. the input is mapped in word 0000 bit 13 and the output is mapped in word 0010 bit 06. The address of the limit switch and the output light are mapped to the I/O table. Every time the processor reads the inputs it will update the input table and turn to logic 1 (ON) those bits whose input devices, mapped to the table, are ON, closed, or 1. When the processor begins the execution of the ladder program (Figure 5-18), it will provide power or continuity to the ladder element, in this case to the NO contact switch, because its reference address is "1" or "ON". At this time, it will set output 001006 ON; the pilot light will turn ON after *all* instructions have been evaluated and the end of scan (EOS) is reached, where the output update to the module takes place. This operation is repeated every scan, which can be as fast as one thousandth of a second (1msec).

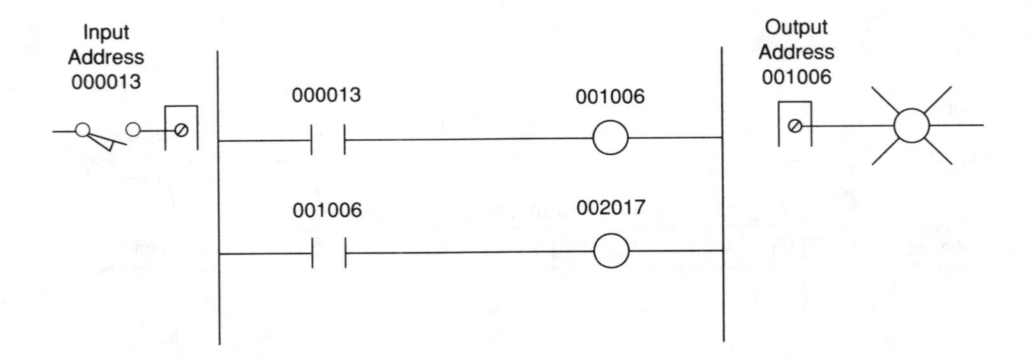

Figure 5-18. PLC ladder implementation of Figure 5-16.

Note that the address 000013 or 001006 can be used as often as required. If we had programmed a contact 001006 to drive internal outputs 002017 (see Figure 5-18), it would have to turn its internal output bit to "1" every time output 001006 is ON. However, this output is not connected directly to any output device (internal storage bit or internal output). Note that internal storage bit 002017 is located in word 0020 bit 17.

5-6 MEMORY CONSIDERATIONS

The previous sections have presented an analysis of programmable controller memory characteristics regarding memory type, storage capacity, organization, structure and their relationship with the I/O addressing. Particular emphasis has been placed on the application memory, which stores the control program and data. Careful consideration must be given to the type of memory, since certain applications require frequent change, while others require permanent storage once the program is debugged. A RAM with battery support may be adequate in most cases, but in others a RAM and optional nonvolatile type memory may be required.

It is important to remember that the memory capacity for a particular controller may not be totally available for application programming. The specified memory capacity may include memory utilized by the Executive routines or the scratch pad, as well as the user program area.

The application memory varies in size, depending on the size of the controller. The total area available for the control program also varies according to the size of the data table. In small controllers, the data table is usually fixed, which means that the user program area will be fixed. In the larger controllers, however, the data table size is usually selectable, according to the data storage requirements of the application. This flexibility allows the program area to be adjusted to meet the application requirements.

When selecting a controller, care must be given to any limitations that may be placed on the use of the available application memory. One controller, for instance, may have a maximum of 256 internal outputs with no restrictions on the number used for timers, counters, or various types of internal outputs. Another controller, however, may have 256 internal outputs available which are restricted to 50 timers, 50 counters, and 156 of any combination of various types of internal outputs. A similar type of restriction could also be placed on data storage registers.

One way to be sure to satisfy memory requirements is first to understand the application requirements for the programming and data storage and the flexibility required for program changes or on-line data entry. This understanding will allow the user to make a decision on the memory type. Creating the program on paper first will help when evaluating capacity requirements. With the use of a memory map, users should learn what is available for the application and then, how the application memory is to be configured for their use. It is also good to know ahead of time if the application memory is expandable.

Chapter

—6—

THE DISCRETE I/O SYSTEM

"Fundamental progress has to do with the reinterpretation of basic ideas."

—Whitehead

6-1 INTRODUCTION

The discrete input/output (I/O) system provides the physical connection between the digital outside world (field equipment) and the central processing unit (Figure 6-1). This is the only "real" connection between the PLC's CPU and discrete field devices. Through various interface circuits and the use of field devices (limit switches, transducers, etc.), the controller can sense and measure physical quantities regarding a machine or process such as proximity, position, motion, level, temperature, pressure, current, and voltage. Based on status of the devices sensed or process values measured, the CPU issues commands that control various devices such as valves, motors, pumps, and alarms. In short, the input/output interfaces are the sensory and motor skills required by the CPU to exercise control over a machine or process.

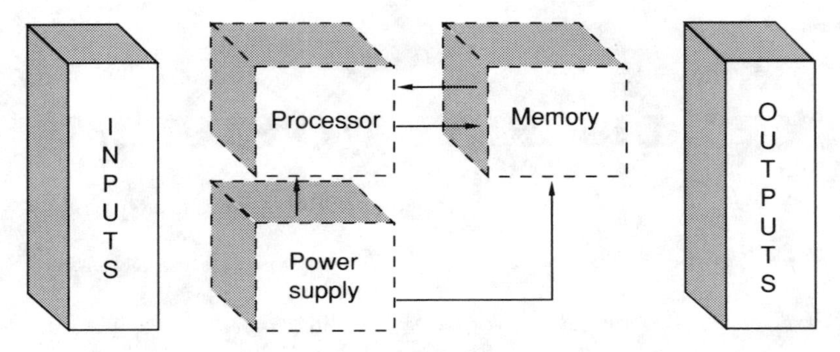

Figure 6-1. I/O System

The early predecessors of today's PLCs were limited to discrete input/output interfaces which allowed only ON/OFF type of devices to be connected. This limitation allowed the PLC only partial control of many processes. These process applications required analog measurement and manipulation of numerical values to control analog and instrumentation devices. Today's controllers, however, have a complete range and variety of discrete and analog interfaces that allow them to be applied to practically any type of control. This chapter will introduce the discrete interfaces, explain their physical, electrical, and functional characteristics, and explore their methods of providing interface to the outside world. The next two chapters will cover the operation and installation of analog I/O interfaces and special function I/O. Figure 6-2 shows a typical discrete I/O system.

6-2 I/O RACK ENCLOSURES AND I/O TABLE MAPPING

Regardless of the type of interface used, I/O modules must be placed or inserted in a *rack enclosure*, usually referred to simply as a "rack". The location of where the module is inserted, for most PLCs, defines the address to reference each connected device. Several PLC manufacturers allow the user to select or set the addresses (to be mapped to the I/O table) for each module by setting internal switches. A rack, in general, recognizes the type of module connected to it, whether input or output, and the class of interface (discrete, analog, numerical, etc.). This module recognition is decoded on the back plane of the rack.

Figure 6-2. Typical PLC discrete I/O system along with the processor and power supply (Courtesy of Furnas Electric, Batavia, IL).

The PLC's rack configuration is an important detail to keep in mind throughout the system configuration. Remember that each of the devices connected is referenced in the control program; therefore, a misunderstanding of the I/O location or addresses will create confusion during and after the programming stages. Figure 6-3 illustrates a typical I/O rack enclosure.

Figure 6-3. Typical rack enclosure (Courtesy of Allen Bradley, Highland Heights, OH).

Generally speaking, there are three categories of enclosures: the *master rack, local rack*, and *remote rack*. A master rack refers to the enclosure containing the CPU or processor module. This rack may or may not have slots available for the insertion of I/O modules. The larger the PLC system, in terms of I/O, the less likely the master rack will have I/O housing capability or space. A local rack is an enclosure which is placed in the same location or area where the master rack is housed. If a master rack contains I/O, it can also be considered a local rack. In general, a local rack (if not master) contains a "local I/O processor" which receives and sends data to and from the CPU. This bi-directional information consists of diagnostics, communication error checking, input status and output updating. This rack also has its I/O mapped to an I/O image or mapping table.

As the name implies, remote racks are enclosures containing I/O modules located far away from the CPU. This remote rack contains an I/O processor (referred to as a remote I/O processor) which communicates I/O information and diagnostic status just like the local rack. The I/O addresses here are also mapped to the I/O table.

The rack concept really emphasizes physical location of the enclosure and the type of processor (local, remote, or main CPU) that will be used in each particular rack. The important point however, is that each and every one of the I/O modules in a rack, whether discrete, analog or special, will have an address by which it is referenced. Therefore, as in the case of discrete I/O, each terminal point connected to the module will have a particular address. This connection point which ties the real field devices to the I/O module, will identify each I/O device by the module's address and the terminal point where it is connected. This will be the *address* used in the control program that identifies the programmed input or output device.

Examples of I/O Rack and I/O Table Mapping

Each PLC manufacturer specifies the rules the user must conform to concerning the I/O module placement. There are modules that accommodate anywhere from 2 to 16 field connections, and the user must follow certain regulations to assure proper I/O addressing. It is not our intention to go over all the different manufacturer's rules, but to interpret and point out in a generic manner how the I/O maps each rack and what could be some of the restrictions imposed on our generic PLC example.

For our example, let's take the PLC's I/O placement specifications shown in Table 6-1. As can be seen in Figure 6-4, several factors determine the address location of each module. The type of module, input or output, determines the first address location from left to right, (0 for O of output, 1 for I of input). The next two address numbers are determined by the rack number and the slot location where the module is placed. The last digit is represented by the terminal connected to the I/O module (0 through 7). Figure 6-5 graphically illustrates a mapping of the I/O table and the modules placed in rack 0 (master rack). Note that the outputs are mapped from word addresses 000 through 077 and the inputs from word addresses 100 through 177.

• There can be up to 7 I/O racks; the first rack (0) is the master rack. Racks 1 through 7 may be local or remote. Each rack has eight slots available for I/O modules.

• PLC discrete I/O modules are available in 4 or 8 points (connections) per module (modularity). Maximum I/O capability of 512 points (any mix).

• I/O Image table is eight bits wide (see I/O table figure).

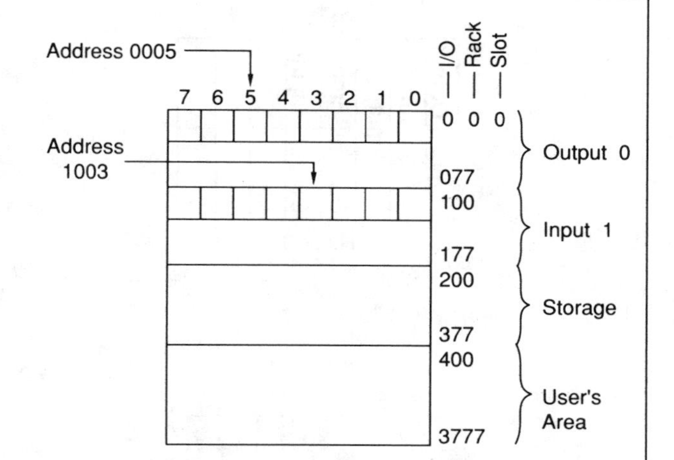

• Type of module, input or output, is detected by the rack's back plane circuitry. If the module is an input, a 1 is placed in front of it three-digit address. If the module is an output, a 0 is placed in front of its three-digit address.

• The number system used is octal.

Table 6-1. Specifications for the I/O rack enclosure used in the example.

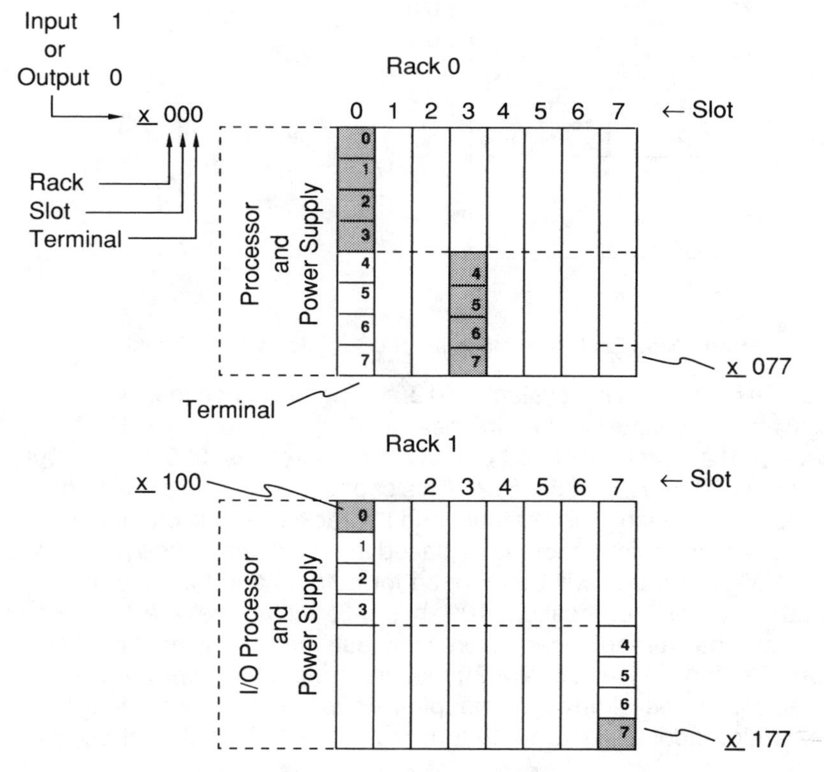

Figure 6-4. Enclosure with available slots.

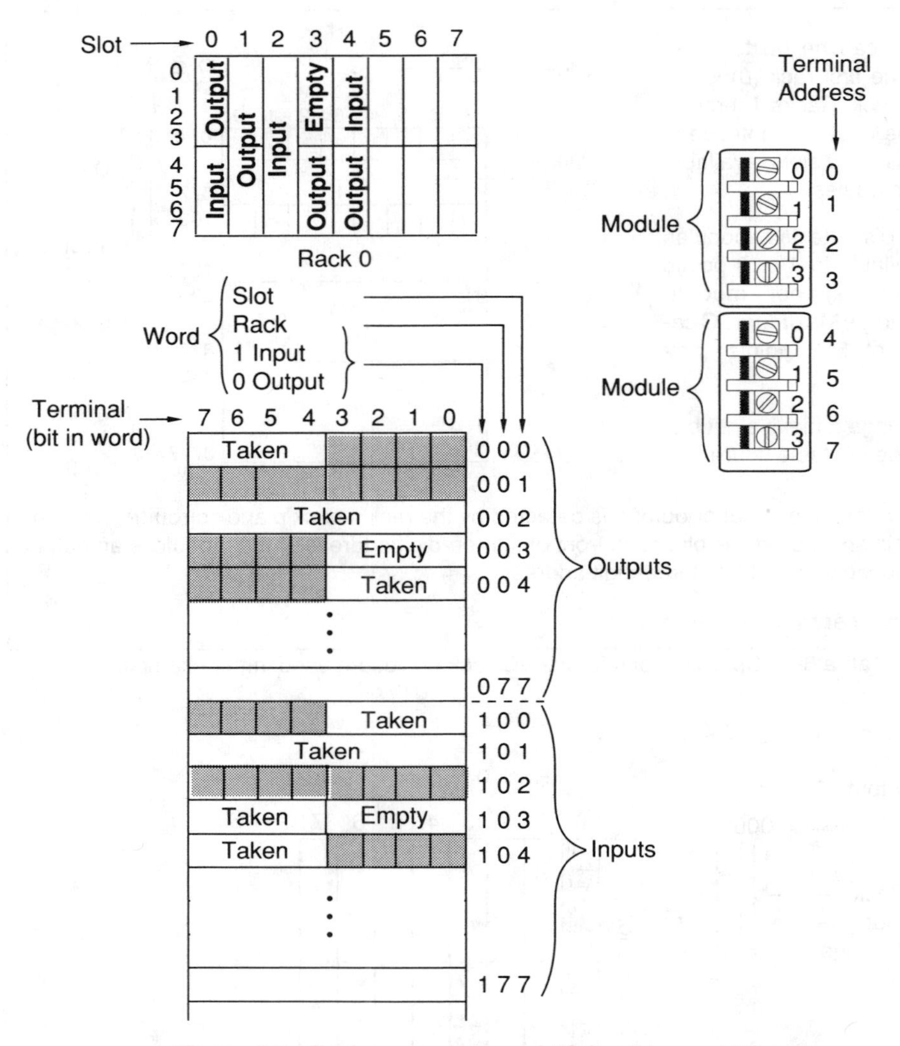

Figure 6-5. Enclosure mapping of I/O modules and I/O table.

The maximum capacity of the system is 512_{10} I/O point connections which is derived by having 64 eight-bit input word addresses, 1000_8, word 100 bit 0, to1777_8, word 177 bit 7,(64 x 8 = 512) and 64 eight-bit output word addresses, 0000_8, word 000 bit 0, to 0777_8, word 077 bit 7,(64 x 8 = 512). This capacity is for *both* inputs and outputs together, not 512 of each. The I/O table and the rack locations cannot overlap. For instance, if a 4-point output module is placed in rack 0, slot 0 (top 0-3), the output table image 000_8, bits 0-3, will be mapped for outputs and the input table image corresponding to this slot location, 100_8 bits 0-3, will not have a reference input mapped since it has already been taken by outputs. If an 8-bit input module is used in location 102_8 (input 1, rack 0, slot 2), the whole eight bits (word address102_8) of the input table would be taken by the mapping, while the corresponding output image to the same slot, address 002, (output 0, rack 0, slot 2) would not be able to be mapped. The bits from the output table that do not have a mapping due to the use of input modules could be used as internal outputs since they cannot be physically connected output field devices (e.g. bits 4-7 of word 000).

6-3 REMOTE I/O SYSTEMS

Large PLC systems (normally upward of 512 I/O) will allow input/output subsystems to be remotely located from the central processing unit. A remote subsystem is usually a rack-type enclosure in which the I/O modules are installed. The rack generally includes a power supply to drive the logic circuitry of the interfaces and a remote I/O adapter or processor module which allows communication with the main processor (CPU). The communication between the I/O adapter modules and the CPU takes place in serial binary form at speeds of up to 1 megabaud (1,000,000 bits transmitted per second). The information packet really contains 1s and 0s representing the status of the I/O and diagnostic information about the remote rack. Capacity of a single subsystem is normally 32, 64, 128, or 256 I/O points. A large system with a maximum capacity of 1024 I/O points might have subsystem sizes of either 64 or 128 points, in which case there could be either eight racks with 128 I/O, sixteen racks with 64 I/O, or some combination of both sizes. In the past, most controllers with remote I/O would allow only discrete interface modules to be placed in the rack. Today analog and special function interfaces are available at remote locations allowing process monitoring and control.

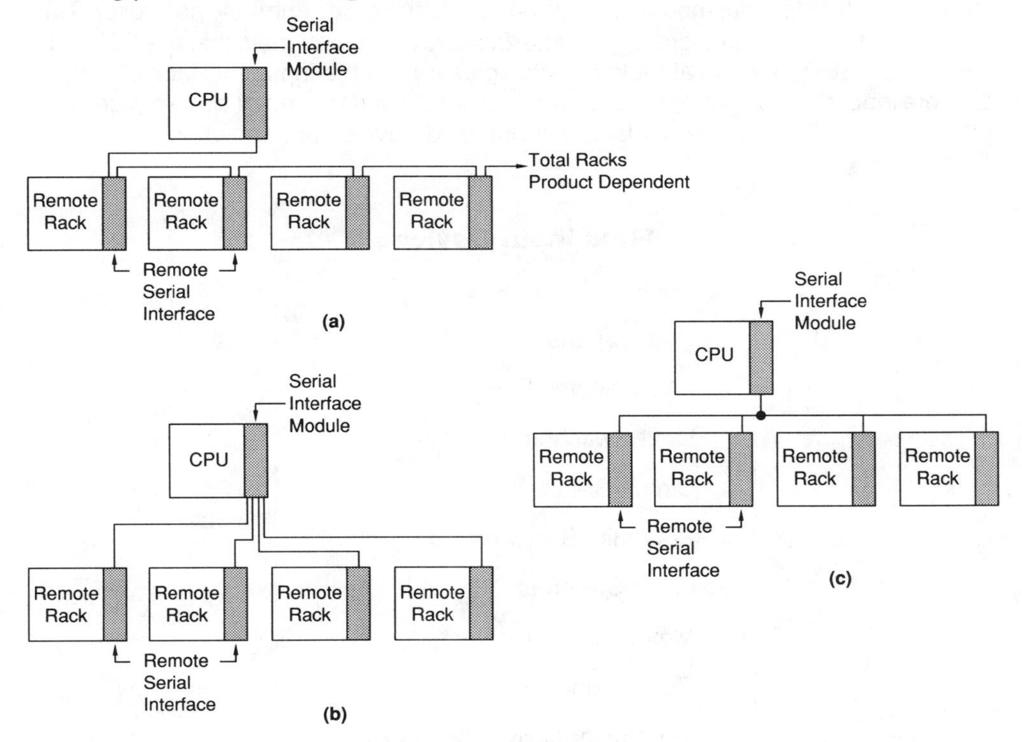

Figure 6-6. Remote I/O Configuration, **a)** Daisy Chain, **b)** Star, and **c)** Multidrop.

Individual racks are normally connected to the CPU using a *daisy chain, star*, or *multi-drop* configuration (Figure 6-6), via one or two twisted-pair conductors or a single coaxial cable. The distance the remote rack can be placed away from the CPU varies among products, but can be as far as two miles. Another approach used is a fiber-optic data link which allows greater distances and has higher noise immunity.

Remote I/O offers tremendous savings on wiring materials and labor costs for large systems in which the field devices are in clusters at various distant locations. With the CPU in a main control room or some other central area, only the communication link is brought back to the processor, replacing hundreds of field wires. When the I/O is distributed, it also offers the advantage of allowing subsystems to be installed and started up independently, as well as allowing maintenance on individual subsystems while others continue to operate. Trouble-shooting and checking connections becomes much easier since hundreds of wires do not need to be checked all the way back to the master rack.

6-4 DISCRETE INPUTS

The most common class of input interface is the digital or discrete type. This interface connects field input devices which will provide an input signal that is separate and distinct in nature to input modules and therefore to the programmable controller. This characteristic limits the discrete input interfaces to sensing signals that are ON/OFF, OPEN/CLOSED, or equivalent to a switch closure. To the input interface circuit, all discrete input devices are essentially a switch that is either open or closed, signifying a 1 (ON) or 0 (OFF). Several discrete input field devices are shown in Table 6-2.

Field Input Devices
Selector Switches
Pushbuttons
Photoelectric Eyes
Limit Switches
Circuit Breakers
Proximity Switches
Level Switches
Motor Starter Contacts
Relay Contacts
Thumbwheel Switches (TWS)

Table 6-2. Discrete Inputs.

Discrete input interfaces receive their necessary voltage and current for proper operation from the back plane of the rack enclosure where they are inserted (see *Loading Considerations* in Chapter 4). The signal that they receive from input field devices can be of different types and/or magnitude (e.g. 120 VAC, 12 VDC). For this reason, discrete input interface circuits are available at various AC and DC voltage ratings. Table 6-3 lists the standard ratings generally encountered for discrete inputs.

Input Interfaces
24 Volts AC/DC
48 Volts AC/DC
120 Volts AC/DC
230 Volts AC/DC
TTL level
Non-Voltage
Isolated Input
5-50 Volts DC (Sink/Source)

Table 6-3. Standard Ratings for Discrete Input Interfaces

During our discussion of input modules it will be helpful if you try to keep in mind the relationship between the interface signals (ON/OFF), where they are inserted (rack and module location), and their I/O table mapping and addressing (used in the program). When in operation, if an input signal is energized (ON), the input interface senses the field device's supplied voltage and converts it to a logic-level signal acceptable to the CPU to indicate the status of that device. A logic 1 indicates ON or CLOSED, and a logic 0 indicates OFF or OPEN. The field status information provided to the standard input module (ON/OFF) is placed into the input table through PLC instructions which include the normally open (—| |—) and normally closed (—|/|—) symbolic instructions. Modules that receive multiple inputs (multibit) such as thumbwheel switches used in register (BCD) interfaces get the input value into the data table (register storage area) via *block transfer* or *get data* instructions. These instructions are covered in detail in Chapter 9.

To properly apply these input interfaces, it is important to have a general understanding of how they operate internally and to have an awareness of certain operating specifications. These specifications are discussed in Section 6-8. Several start-up and maintenance procedures concerning the I/O system are covered in Chapter 15. Now, let us look at the various types of discrete input interfaces and their operation and connections.

AC/DC Inputs

A block diagram of a typical AC/DC input interface circuit is shown in Figure 6-7. Input circuits vary widely among PLC manufacturers, but in general all AC/DC interfaces operate in a manner similar to that described in this diagram. The input circuit is composed of two primary parts: the *power section* and the *logic section*. The power and logic sections of the circuit are normally, but not always, coupled with a circuit, which electrically separates the two thus providing *isolation*.

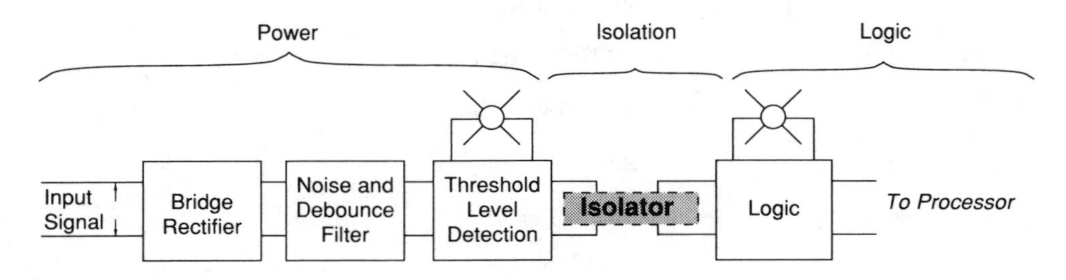

Figure 6-7. Block Diagram for AC/DC Input Circuit.

The power section of an input interface basically performs the function of converting the incoming voltage (230 VAC, 115 VAC, etc.) from an input sensing device, such as those described in Table 6-2, to a DC logic-level signal to be used by the processor during its *read input* section of the scan. The bridge rectifier circuit converts the incoming signal (AC or DC) to a DC level that is passed through a filter circuit, which will protect against signal bouncing and electrical noise on the input power line. This filter causes a signal delay that is typically 9-25 msec. The threshold circuit detects whether the incoming signal has reached the proper voltage level for the specified input rating. If the input signal exceeds and remains above the threshold voltage for a duration of at least the filter delay, the signal will be recognized as a valid input.

Figure 6-8 illustrates a typical input circuit. When a valid signal has been detected, it is passed through the isolation circuit, which completes the electrically isolated transition from AC to logic level. The DC signal from the isolator is used by the logic circuit and made available to the processor via the rack's back plane data bus. Electrical isolation is provided so that there is *no electrical connection* between the field device (power) and the controller (logic). This electrical separation will help prevent large voltage spikes from damaging the logic side of the interface (or the controller). The coupling between the power and logic sections is normally provided by an optical-coupler or a pulse transformer.

Most input circuits will have an LED (power) indicator to signify that the proper input voltage level is present. In addition to the power indicator, an LED may also be available to indicate the presence of logic 1 at the logic section. If the input voltage is present, and if the logic circuit is functioning properly, this LED will be lit. When both indicators are available and the input signal is ON, the two LEDs must illuminate to indicate that the power and logic sections of the module are operating correctly. A device connection diagram is shown in Figure 6-9.

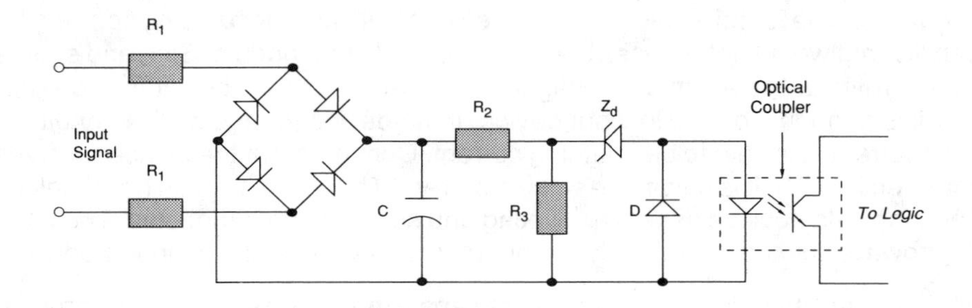

Figure 6-8. Typical input circuit.

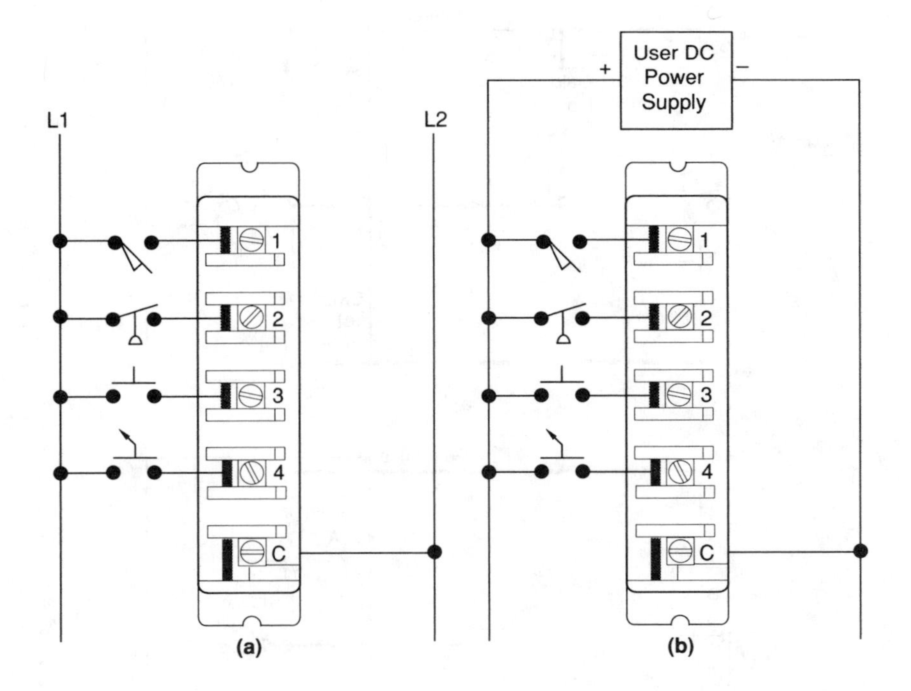

Figure 6-9. Device connection for **a)** an AC input module; **b)** a DC input module.

DC Inputs (Sink/Source)

The DC input module allows interfacing of field input devices which provide a DC output voltage. The difference between the DC input interface and the AC/DC input is that the DC input does not contain a bridge circuit since it does not need to convert an AC signal to a DC level. The range of input voltage for the DC input module varies between 5 VDC and 30 VDC. The module generally recognizes an input signal being ON if the input voltage level is at 40% (or other similar percentage specified by the manufacturer) of the supplied reference voltage. The OFF condition is detected when the input voltage falls under 20% (or other percentage) of the reference DC voltage.

This module can be found to interface field devices in *sinking* or *sourcing* operations, a capability that would not be possible in the AC/DC input module. Sink and source operations refer to the electrical configuration of an electronic circuit in a device, whether it is a module or a field input device. If the device (during its ON condition) *provides* current, it is said to be sourcing current. Conversely, if the device *receives* current when it is ON, it is said to be sinking current. Therefore, we can have sinking and sourcing field devices as well as sinking and sourcing input modules. The most familiar however, is the sourcing field input device and the sinking input module.

Depending on the module, the sink or source capability may be selected using rocker switches inside the module. Figure 6-10 graphically depicts the sink and source operation and current direction.

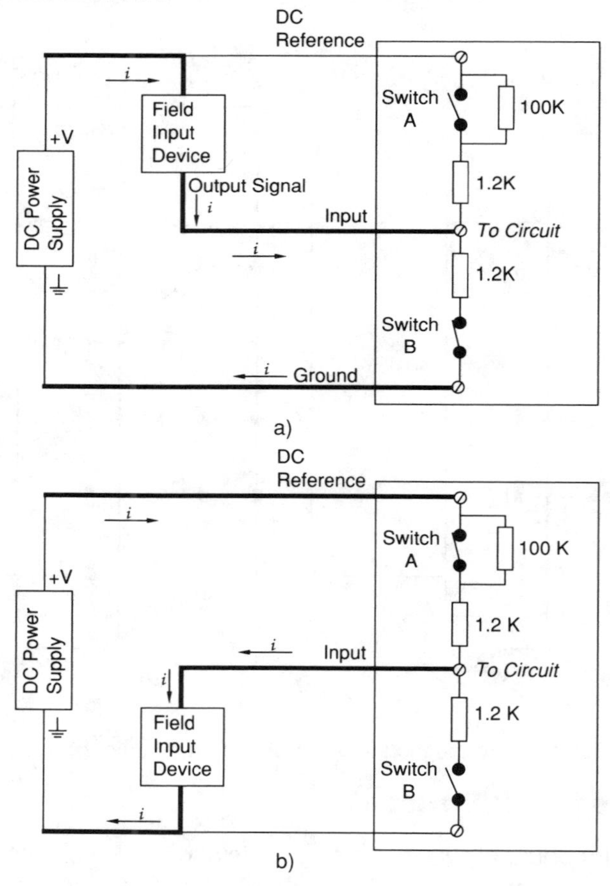

Figure 6-10. Current for a) Sinking input module/sourcing device and b) Sourcing input module/sinking input device.

The general operation of the DC input module is very similar to that of the AC/DC input interface. The DC module's block diagram has an identical representation as the AC/DC with the exception of the bridge rectifier section. The user must keep in mind during the interfacing of this sinking and/or sourcing module, the minimum and maximum specified currents that the input devices and module are capable of sinking or sourcing as the case may be. Also, note that if the module provides for selection of a sink or source operation via selector switches, the user must assign them properly. Figure 6-11 illustrates three field device connections to the DC input module that has capabilities for both sinking and sourcing input devices.

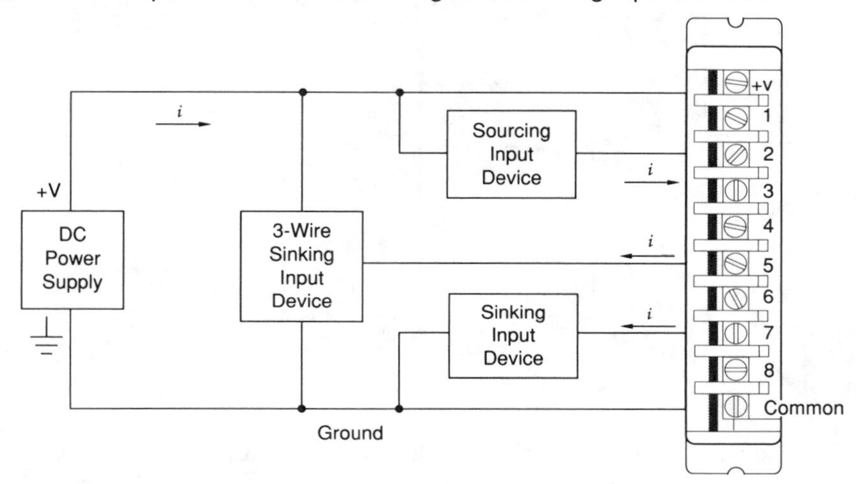

Figure 6-11. Field device connections for a sink/source DC input module.

A potential interface problem could arise, for instance, if an 8-input module is set for sink operation and all input devices except one are operating in a source configuration. The source input devices may be ON, but the module would not detect properly the ON signal, even though a voltage could be detected with a voltmeter across the modules' terminals.

Isolated AC/DC Inputs

Isolated input interfaces operate in the same manner as standard AC/DC modules. However, they differ from standard AC/DC input modules in that each input has a separate individual common or return line. Depending on the manufacturer, standard AC/DC input interfaces may have one return line per four, eight, or sixteen points. Although this single return line may be perfectly ideal for 95% of AC/DC input applications, it may not be suitable in applications requiring individual or isolated commons. Such applications include input devices that are connected to different phase circuits coming from different power distribution centers. Figure 6-12 illustrates a sample device connection for the AC/DC input isolation interface.

Isolated input interfaces generally provide less points per module than their standard counterparts. This modularity simply exists because of the extra terminal connections required to connect each of the return lines. If isolation modules are not available, the user can select the standard interfaces; however, several standard inputs will be lost because only one input line would be allowed per return line to keep

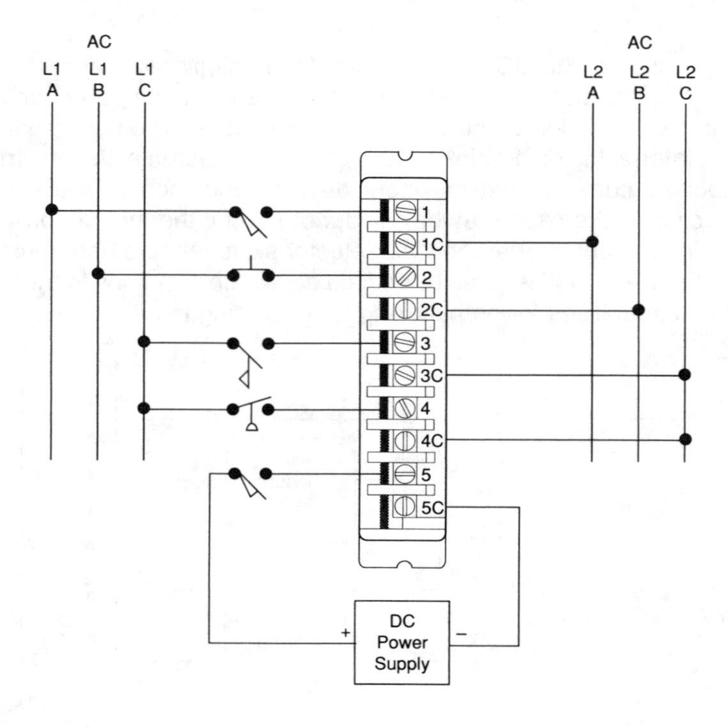

Figure 6-12. Device connections for an AC/DC isolated input interface.

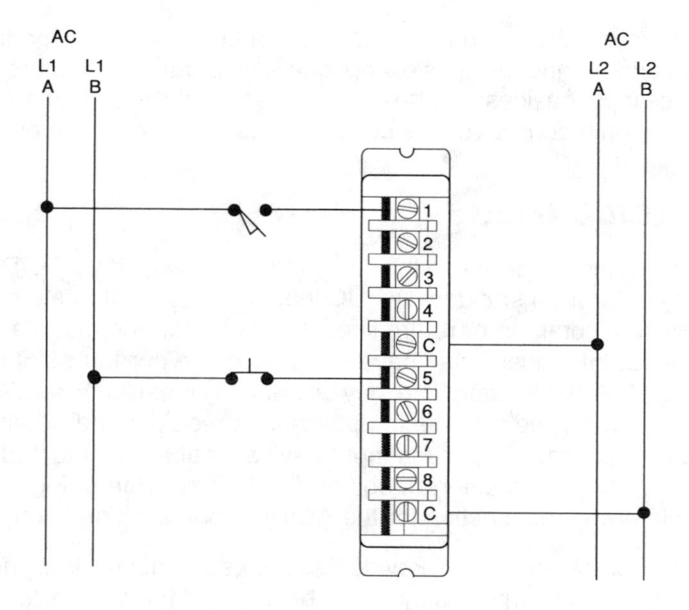

Figure 6-13. Use of an 8-point standard input module as an isolated module.

isolation among inputs. For instance, a 16-point standard module, having one common line per four points would be able to accommodate four distinct isolated field input devices (each one from a *different* source); twelve points however, would be lost in this module. Figure 6-13 illustrates an 8-point module with different commons for every four inputs, thus allowing for two possible isolated inputs.

TTL Inputs

The Transistor-Transistor Logic, or TTL, input interfaces allow the controller to accept signals from TTL compatible devices, including solid-state controls and sensing instruments. TTL inputs are also used for interfacing with some 5 VDC-level control devices and several types of photoelectric sensors. The TTL interface has a configuration similar to the AC/DC inputs; however, the input delay time caused by filtering is generally much shorter. TTL input modules generally receive their power from within the rack enclosure; however, some interfaces may require an external +5 VDC power supply (rack or panel mounted).

The TTL module may also be used in applications which uses BCD thumbwheel switches (TWS) that operate at TTL levels. These interfaces generally provide up to eight inputs per module and may have as many as sixteen inputs (high density input module). The TTL input module can also interface thumbwheel switches if these input devices are TTL compatible. The only difference between a high density TTL input module and the BCD register input module (covered next) is that the latter can receive inputs between 5 VDC and 24 VDC. The register module may also have the capability of multiplexing inputs which allows more than one input signal to be connected to each input module's terminal. Figure 6-14 shows a typical TTL input connection diagram with an external power supply.

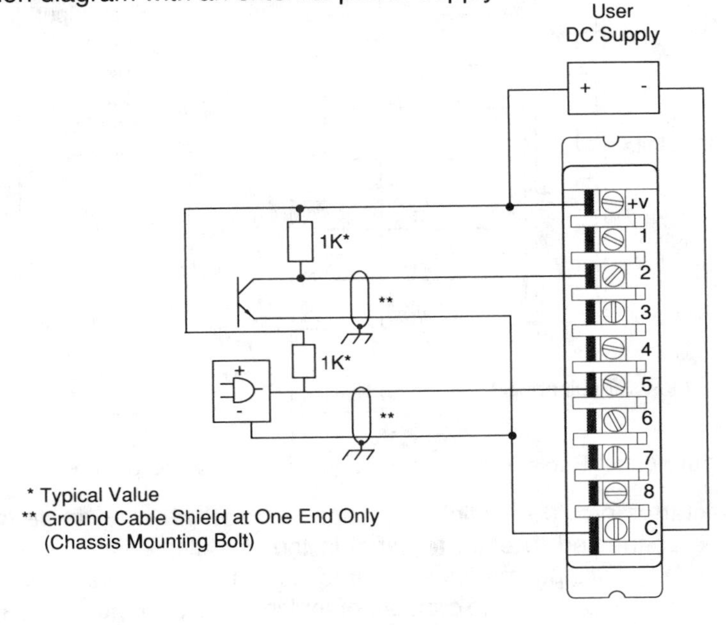

Figure 6-14. TTL input connection diagram.

Register or BCD Inputs

These multi-bit input modules came about due to the need to enhance operator's input interfacing methods to the programmable controller using standard thumbwheel switches. This register or BCD configuration allows groups of bits to be input as a unit to accommodate devices that require that the bits be handled in parallel form.

This interface is generally used to input parameters into specific register or word locations in memory to be used by the control program. Typical parameters are timer and counter presets, and set point values. The operation of each input is almost identical to that of the TTL or DC input module.

These interfaces generally accept voltages in the range of 5 VDC (TTL) to 24 VDC and are grouped in a module containing 16 or 32 inputs, which correspond to one or two I/O registers (mapped in the I/O table). Data manipulation instructions, such as GET or Block Transfer In, are used to access the data from the register input interface. Figure 6-15 illustrates a typical device connection for a register input.

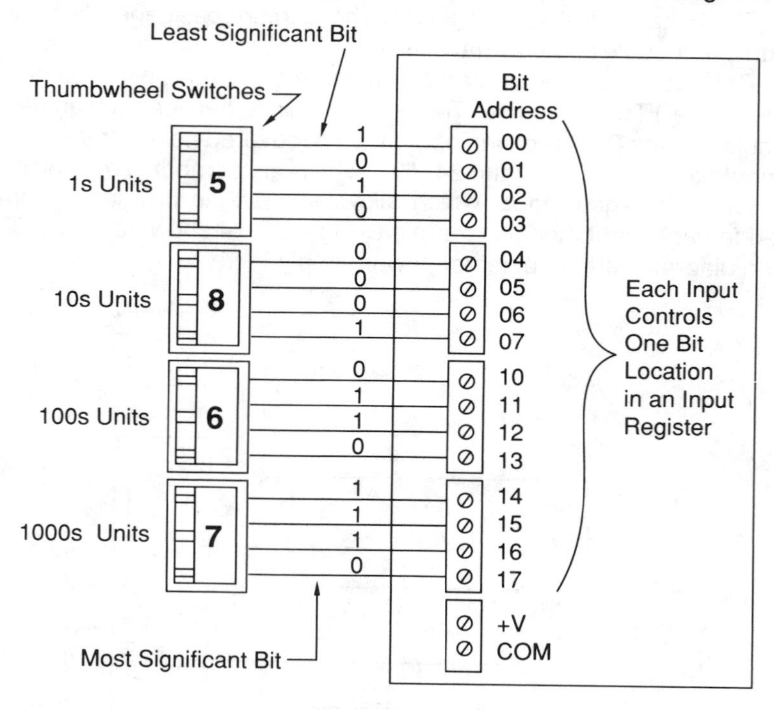

Figure 6-15. Register or BCD input module connection diagram.

Some manufacturers provide multiplexing capabilities which allow more than one input line to be connected to each terminal in the module. Using this technique minimizes the number of input modules required to read several sets of 4-digit TWS. For instance, a 16-bit input module capable of multiplexing six inputs for each input terminal (total of 96 inputs—6 x 16) would be able to receive information from six 4-digit thumbwheel switches. The user or programmer does not need to decode each of the sixteen input group (six sets) since the module automatically enables each group of sixteen inputs to be read in one scan at a time using additional *enable* connection lines. However, depending on the PLC, the programmer is responsible for specifying the register or word addresses where the sixteen bit data will be stored. The location where the multiplexed data is placed in the storage area is specified by the instruction used. A length or number of registers to be stored is generally entered with the instruction. Figure 6-16 illustrates a block diagram connection for a module with the capability of multiplexing four 4-digit TWS (four 16-bit input lines).

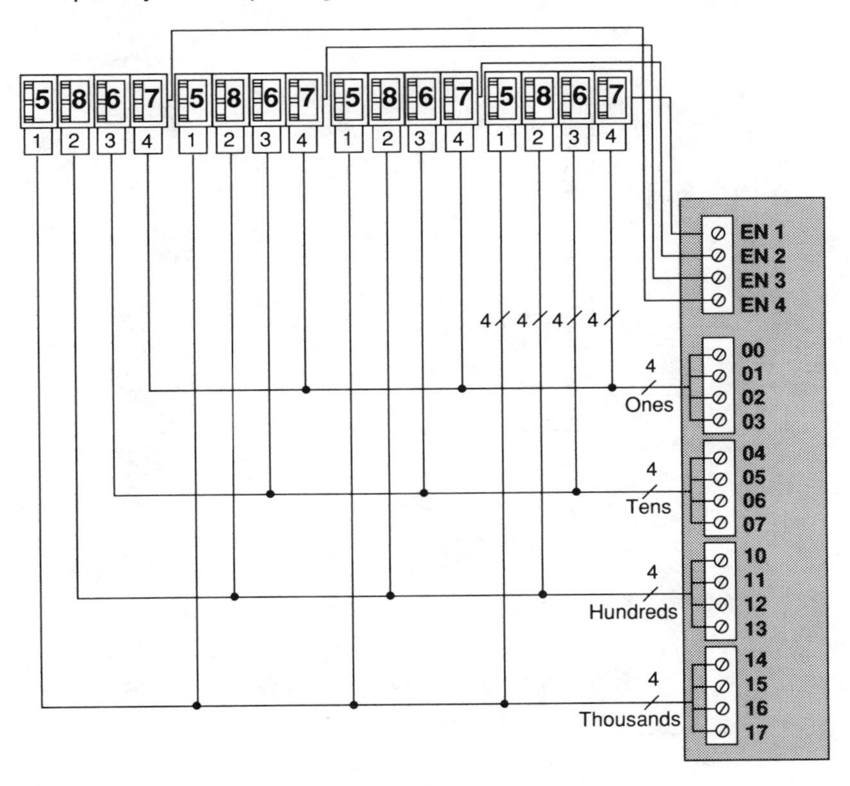

Figure 6-16. Multiplexing input module connection diagram. Note that each TWS provides four wires (BCD data) to each four groups of input bits in the multiplexer module.

Non-Voltage Input

The non-voltage input module accepts the same type of discrete input devices as those mentioned in Table 6-2. The only difference that exists is that these input modules do not require that the field devices provide power when energized. In other words, the field devices may not be powered by an external power source. Typical field devices include dry contacts from standard relays, some types of proximity or photoelectric switches, and solid state relay or instrumentation devices which provide an open collector output.

The operation of the non-voltage input is similar to the functions provided by the standard input modules. The non-voltage input can detect *contact closure* type inputs by detecting ground or common levels. These modules can also provide DC voltages, usually 12-24 VDC at 10 mA, to input devices, thus providing the necessary power for sensing the device in the ON state. Figure 6-17 illustrates several types of non-voltage field connections.

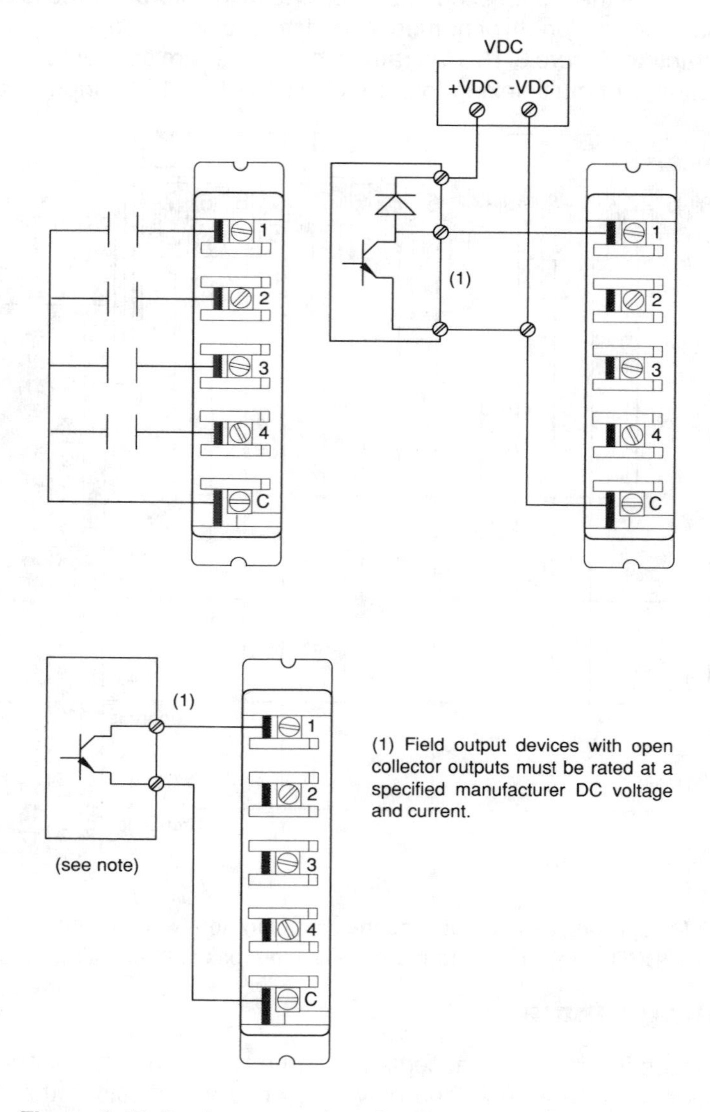

Figure 6-17. Device connections for non-voltage input module.

6-5 DISCRETE OUTPUTS

Like discrete input interfaces, discrete output interfaces are the most commonly used type of PLC output modules. These outputs provide connections between the programmable controller and output field devices. The devices controlled are discrete or digital in nature, which exhibits one of two states such as ON/OFF or OPEN/CLOSE. Table 6-4 illustrates some types of devices that fall into this category.

Output Devices
Alarms
Control Relays
Fans
Lights
Horns
Valves
Motor Starters
Solenoids

Table 6-4

Discrete output modules receive their necessary voltage and current from the enclosure's back plane where they are plugged-in (see Chapter 4 for Loading Considerations). The field devices to which the modules interface may differ; therefore, several types and/or magnitude of voltages are provided to control them (e.g. 120 VAC, 12 VDC). Table 6-5 lists the standard output ratings generally found in discrete output applications.

Output Interfaces
12-48 Volts AC
120 Volts AC
230 Volts AC
12-48 Volts DC
120 Volts DC
230 Volts DC
Contact (relay)
Isolated Output
TTL level
5-50 Volts DC (Sink/Source)

Table 6-5

The output interface circuitry switches the supplied voltage ON or OFF according to the status of its corresponding addressed bit in the output image table. This status (1 or 0) is set during the execution of the control program and sent to the output module at the end of scan (output update). If the signal received by the module from the processor is 1, the output module will switch the supplied voltage (e.g. 120 VAC) to the output field device. If the signal received from the processor is 0, the module will CLOSE and deactivate the field device by switching to 0 volts.

Typically, the output coil (——◯) instruction activates the output interface whenever the reference address is logic 1 (ON). Multibit outputs, such as BCD register outputs, generally utilize functional block instructions like *block transfer out* in order to output a word or register to the module. These instructions, in conjunction with the input instructions, will be heavily utilized during the programming and control of discrete I/O signals. More information about the use and operation of these instructions is provided in Chapter 9.

AC Outputs

AC output circuits, like input circuits, vary widely among PLC manufacturers and in general can be depicted by the block diagram configuration shown in Figure 6-18. This block configuration describes the main sections that form an output module and are used here to illustrate the module's operation. The circuit consists primarily of the logic and power sections, coupled by an isolation circuit. The output interface can be thought of as a simple switch through which power can be provided to control the output device.

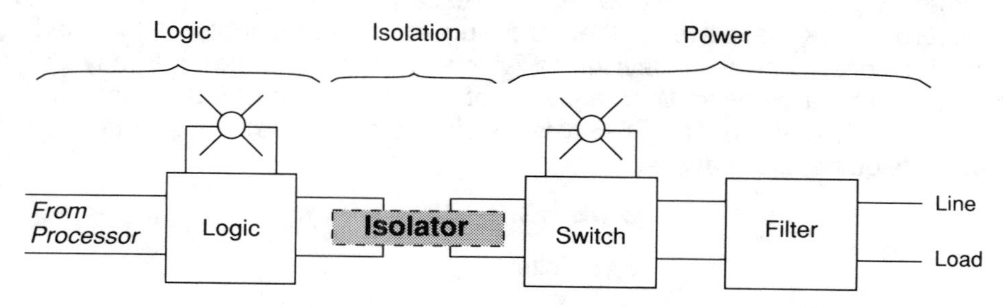

Figure 6-18. AC output circuit block diagram.

During normal operation, the processor sends to the module's logic circuit the output status according to the logic program. If the output is to be energized, reflecting the presence of a "1" in the output table, the logic section of the module will latch a "1" and the ON signal will be passed through the isolation circuit, which in turn will switch the voltage to the field device through the power section of the module. This condition will remain ON as long as the output table's corresponding image bit remains a "1." When the signal turns OFF, the 1 that was latched in the logic section is unlatched and the signal passed will provide no voltage to the power section, thus de-energizing the output device. Figure 6-19 illustrates a typical AC output circuit.

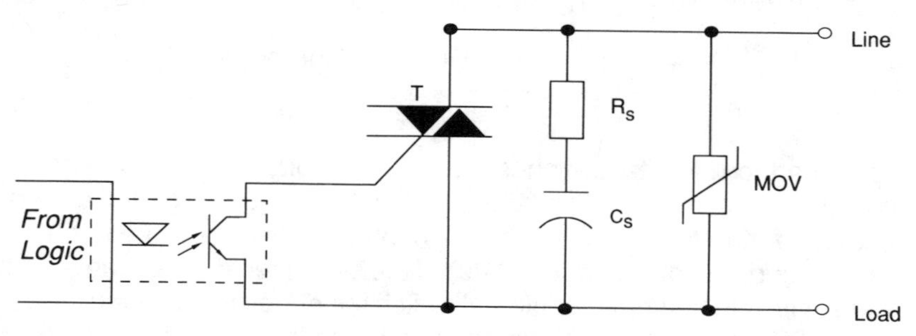

Figure 6-19. Typical AC output circuit.

The switching circuit in the power section generally uses a *Triac* or a *Silicon Controlled Rectifier (SCR)* to switch the power. The AC switch is normally protected by an *RC snubber* and often a *Metal Oxide Varistor (MOV)*, which is used to limit the peak voltage to some value below the maximum rating and also to prevent electrical noise from affecting the circuit operation. A fuse may be provided in the output circuit to prevent excessive current from damaging the AC switch. If the fuse is not provided in the circuit, it should be supplied by the user and should conform to the manufacturer's specifications.

As with input circuits, the output interface may provide LED to indicate operating logic and power circuits. If the circuit contains a fuse, a fuse status indicator may also be incorporated. An AC output connection diagram is illustrated in Figure 6-20. Note that the switching voltage which is used to turn ON the output device is supplied to the module from the field (L1). Several factors that must be considered when connecting AC outputs are discussed in Chapter 16.

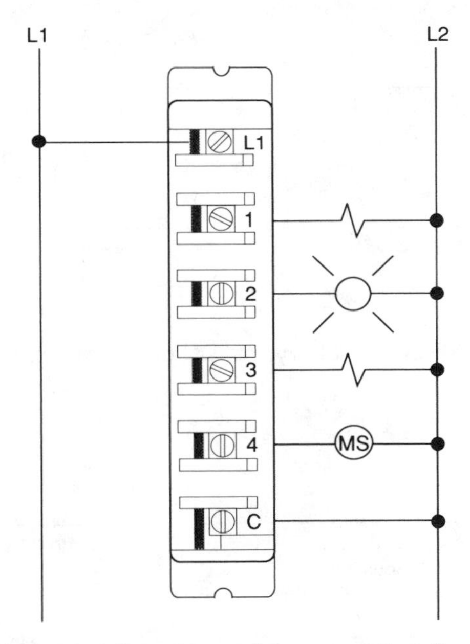

Figure 6-20. AC output module connection diagram.

DC Outputs (Sink/Source)

The DC output interface is used to control discrete DC loads by switching ON and OFF. Functional operation of the DC output is similar to the AC output; however, the power circuit generally employs a *power transistor* to switch the load. Like triacs, transistors are also susceptible to excessive applied voltages and large surge currents, which could result in overdissipation and a short circuit condition. To prevent this condition from occuring, the power transistor will normally be protected by a *free wheeling* diode placed across the load (field output device). Other DC outputs may also incorporate a fuse which is used to protect the transistor during moderate overloads. These fuses however, must be capable of opening or breaking continuity very quickly before excessive heat due to overcurrents can build up.

As in DC inputs, the DC output modules may be supplied by the manufacturer in *sinking* or *sourcing* configurations. If the sinking module configuration is used, current *from the load* flows into the module's terminal therefore switching the negative (return or common) to the load. The current flows from the positive voltage through the load and to the common via the module's power transistor.

The sourcing module configuration is used when current flows *from the module* into the load, therefore switching the positive voltage to the load. Figure 6-21 illustrates a typical DC output circuit (sourcing type) and device connections for sourcing and sinking configurations (Figure 6-22). Note that the sinking output devices have current flow into the device's terminal from the module (module must source current). Conversely, sourcing output devices have current flow out of the device's terminal into the module (module must sink current).

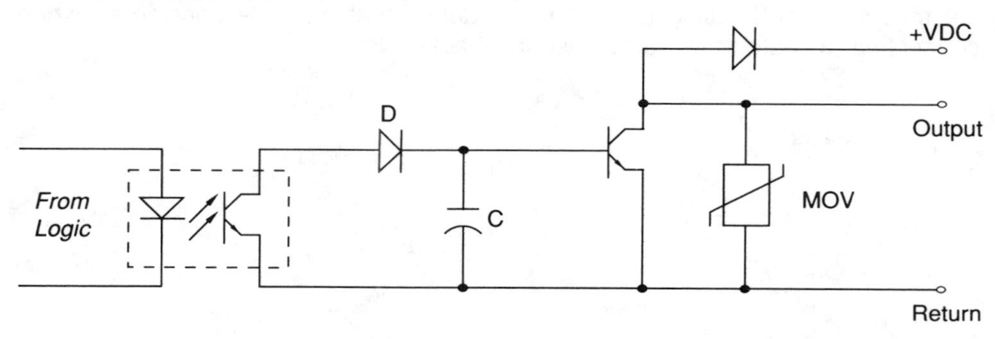

Figure 6-21. Typical DC output circuit.

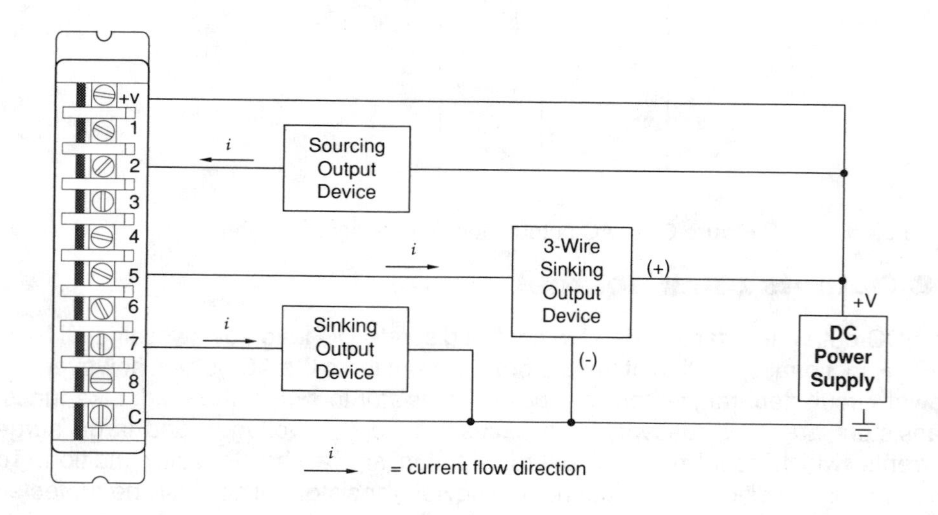

Figure 6-22. DC output module connections with sinking and sourcing capability.

Isolated AC and DC Outputs

Isolated outputs operate in the same manner as the standard AC and DC output interfaces. The only difference is that each output has its own return line circuit (common) independent or isolated from the other outputs. This configuration allows the control of output devices powered by different sources which may also be at different ground (common) levels.

The standard modules have one return connection common to all outputs; however, some standard modules may provide one return line per four outputs if the interface has eight or more outputs available. In general, the isolated interfaces provide less modularity (i.e. less points per module) than their standard counterpart because extra terminal connections are necessary for the independent return lines. Connections for this type of output interfaceare illustrated in figure 6-23 .

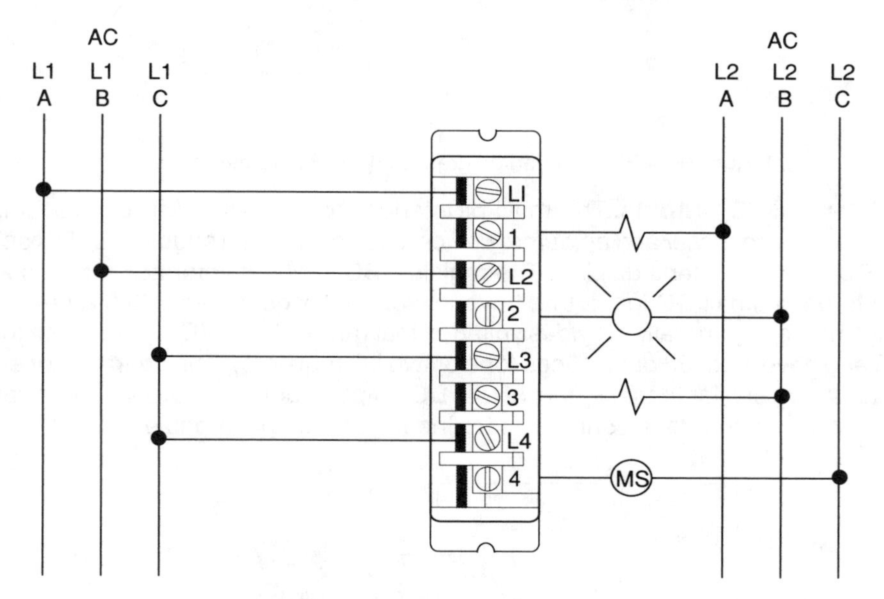

Figure 6-23. Connection diagram for an isolated AC output.

TTL Outputs

The TTL output interface allows the controller to drive output devices that are TTL compatible, such as seven-segment LED displays, integrated circuits, and various 5 VDC devices. These modules generally require an external +5 VDC power supply with specific current requirements. TTL modules are generally found with eight available output terminals to interface with TTL field devices. Some TTL modules may be connected to as many as sixteen devices at a time (high density TTL modules). Typical output devices that utilize the high density TTL modules include 5 Volt 7 segment indicators. The main difference between the high density TTL modules and the register (BCD) output module (covered in the next section) is that the latter usually handles DC voltages between 5 VDC and 24 VDC and could provide perhaps more current to the loads. Figure 6-24 illustrates typical output connections to TTL compatible devices.

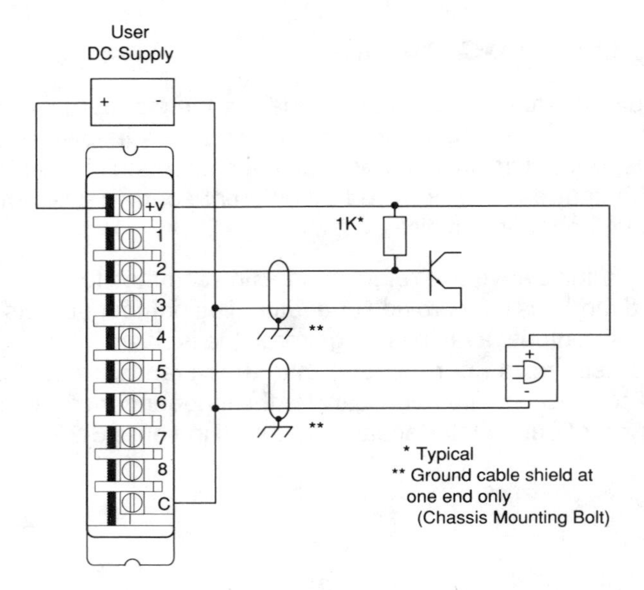

Figure 6-24. Connection diagram for the TTL output module.

Register or BCD Outputs. This multi-bit interface provides *parallel communication* between the *processor* and an *output device*, such as a seven segment LED display or a BCD alphanumeric display. The register (BCD) output interface may also be used to drive small DC loads that have low current requirements (0.5 amp). This output interface generally provides voltages that range from 5 VDC (TTL level) to 30 VDC and have 16 or 32 output lines (one or two I/O registers). The operation of each output is very similar to that of the TTL or DC output module. Figure 6-25 illustrates a typical device interface connection for the register output module.

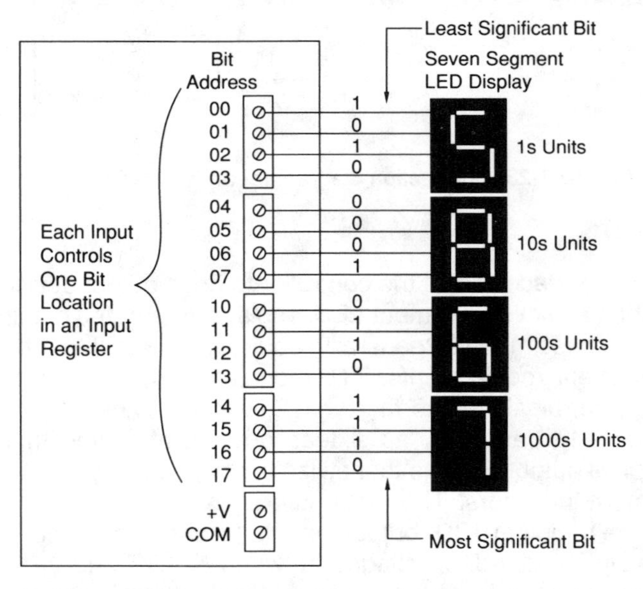

Figure 6-25. Register (BCD) output connection diagram to 7 segment indicators.

The information sent to the module originates in the output data table, where a 16-bit word or register is sent to the specified module address according to the data transfer or I/O register instruction (e.g. block transfer out). Once the data arrives at the module, it is latched and made available at the output circuits.

This output module may also have multiplexing capabilities. Under this option, several groups of 16 or 32 outputs, depending on the modularity, can be controlled with one interface. For example, if the multiplexing output has the capability of handling four sets of 16-bit outputs, then up to four sets of 4-digit 7-segment indicators could be driven. Register data from the output table is sent to the module generally once a scan, thus updating each multiplexed set of output devices.

Using the multiplexing outputs generally does not require special programming since there are instructions that specify the multiplexing operation. The only requirement is that the output devices (e.g. LED displays) must have enable circuits that allow the module to connect the enable lines to each set of loads controlled by each set of sixteen bits. Figure 6-26 shows a block diagram of a multiplexing output module for four sets of 7-segment LED indicators.

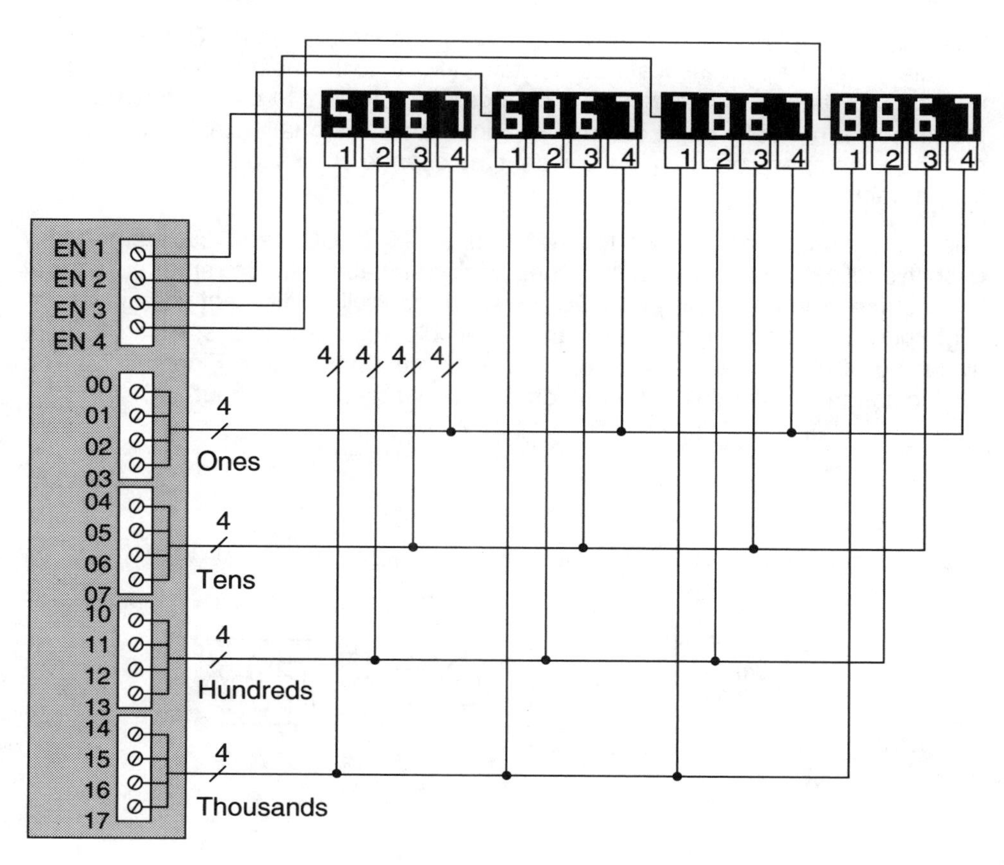

Figure 6-26. Connection diagram for a multiplexing input module. Note that each line connected to each 7-segment display is composed of four wires (BCD data).

It is important to note that if output modules with enable lines are multiplexed, only passive type output devices can be controlled (i.e. 7-segment indicators, displays) as opposed to controlled elements such as low current solenoids. The reason being that while the multiplexing outputs are very useful, its output data does not remain "static" for one channel or set of sixteen bits or 32 bits, but changes for each circuit that is being multiplexed. The only way to use this module and still be able to have the output devices operate correctly is if additional latching/enabling circuits are incorporated in the output devices' hardware. Such a situation may be encountered in the transmission of parallel data to instrumentation or computing devices which have enable and latching lines for incoming data.

Contact Outputs

The contact output interface allows output devices to be switched by a N.O. or N.C. relay contact. Electrical isolation between the power output signal and the logic signal is provided by the separation not only between the contacts, but also between the coil and contacts. Filtering, suppression, and fuses are also generally incorporated.

The basic operation of the module is the same as the standard AC or DC output module. When the processor sends the status (1 or 0) to the module during the output update, the state of the contacts will change. If a 1 is sent to the module from the processor, a normally open contact will close and a normally closed contact will change to an open condition. If a 0 is sent, no change occurs to the normal state of the contacts.

The contact output can be used to switch either AC or DC loads, but are normally used in applications such as multiplexing analog signals, switching small currents at low voltage, and interfacing with DC drives for controlling different voltage levels. High power contact outputs are also available for applications which require switching of high currents. A contact output circuit is shown in Figure 6-27. The device connection for this output module is similar to the AC output.

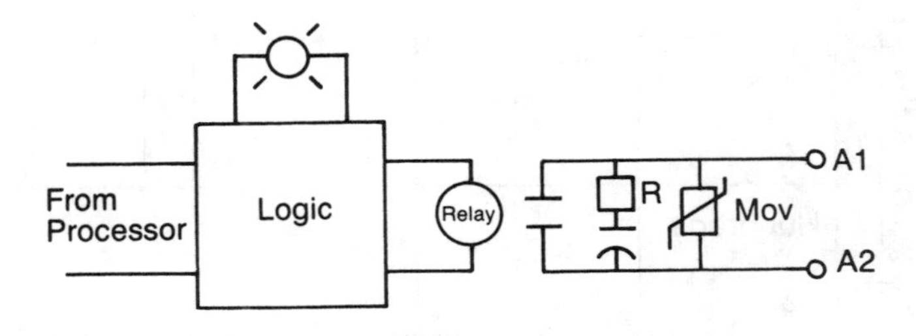

Figure 6-27. Typical contact output circuit.

Figure 6- 28 illustrates an interfacing example, where four analog voltage references are connected to the contact output module. These references may represent preset speed values which, if connected to a speed drive controller, could be used to switch different motor velocities (e.g. two forward, two reverse). Note that each contact in this interfacing must be mutually exclusive—that is, only one contact can be closed at a time. This is performed in the control program by interlocking logic which will prevent other output coils from being energized at the same time.

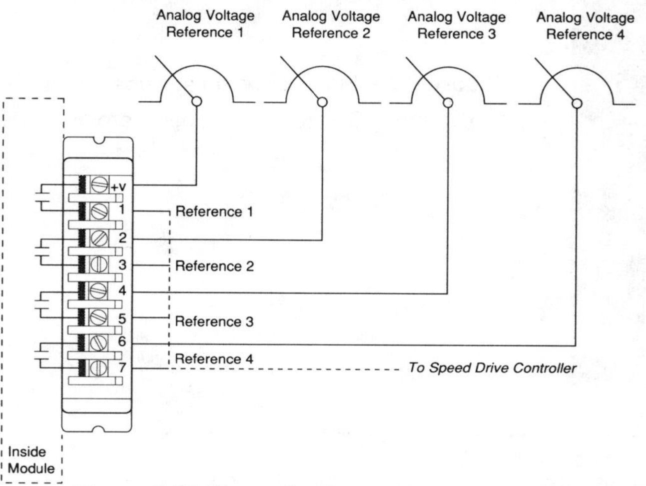

Figure 6-28. Example of contact output interface connection.

6-6 DISCRETE BY-PASS/CONTROL DEVICES

By-pass or manual back-up devices can also be incorporated into a programmable controller system to allow for more flexibility during start-up or during an output failure. These devices can be used to override the PLC output signal by adding a selector switch which allows the field output device to be switched ON regardless of the state of the output module. The by-pass device can also be configured to allow the field output to be under PLC output control or to simply have the field output in an OFF condition.

One example of available by-pass devices is the manual control station or MCS-7408 back-up from Control Technology Corporation (shown in Figure 6-29). The MCS-7408 discrete manual back-up station provides eight isolated and circuit breaker points for use with any PLC's discrete output modules. This control station is used between the PLC's output interface and the digitally controlled element. The primary use of the MCS-7408 is to provide a means of controlling a field device with or without PLC control. Using this station, the device can be disabled (OFF), or it can be activated (ON) without the PLC being in control. Figure 6-30 shows a simplified circuit of the MCS-7408 and typical load connection.

The uses of this control station prove to be very helpful during emergency disconnect of particular field devices, maintenance situations, and especially during the system start-up. Indicators are also incorporated into the control station to show the ON or OFF state of the field device.

Figure 6-29. Manual control stations for discrete outputs (Courtesy Control Technology Corp. Knoxville, TN).

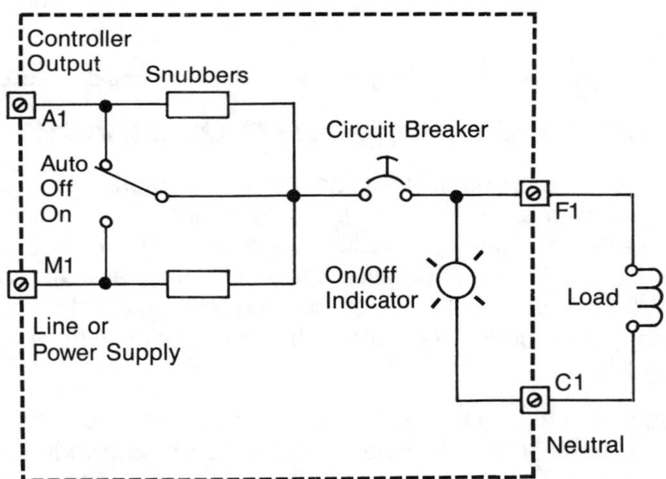

Figure 6-30. Typical control station for the MCS 7408 model (Courtesy of Control Technology Corp. Knoxville, TN).

6-7 SUMMARY OF DISCRETE I/O

The discrete or digital type of I/O interfaces covered in this chapter will most likely be required in all PLC system applications. Other applications may also include, in addition to the discrete interfaces, the application of analog and special I/O modules (covered in the next two chapters) to implement the required control.

An important topic that is common to all I/O interfaces is the fact that they accept input status data that will be placed into the input table and accept processed data from the output table whether the interfaces are discrete, analog or special function. The information is placed in or written from the I/O table (in word locations) according to the location or address of the modules. This address is generally dependent on the module's placement in the I/O rack enclosure, therefore the placement of I/O interfaces is an important detail to keep in mind.

The software instructions that are generally used with discrete type interfaces are the basic relay instructions (ladder type). However, multi-bit modules generally use functional block instructions as well as some advanced ladder functions. These software instructions are covered in Chapter 9. Several programmable controller input and output modules and enclosures are shown in Figure 6-31.

a)

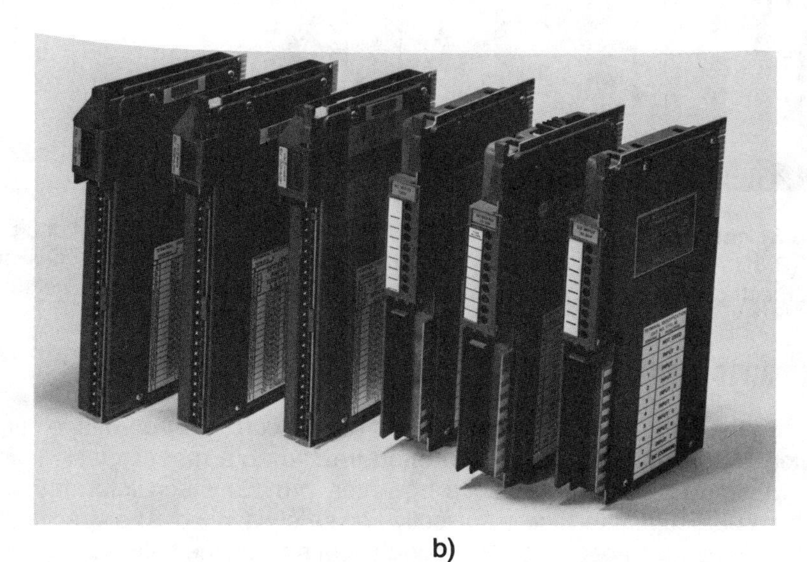

b)

Figure 6-31. a) PLC family shares the use of I/O modules (Gould, Andover, MA); **b)** Discrete I/O modules (Allen Bradley, Highland Heights, OH).

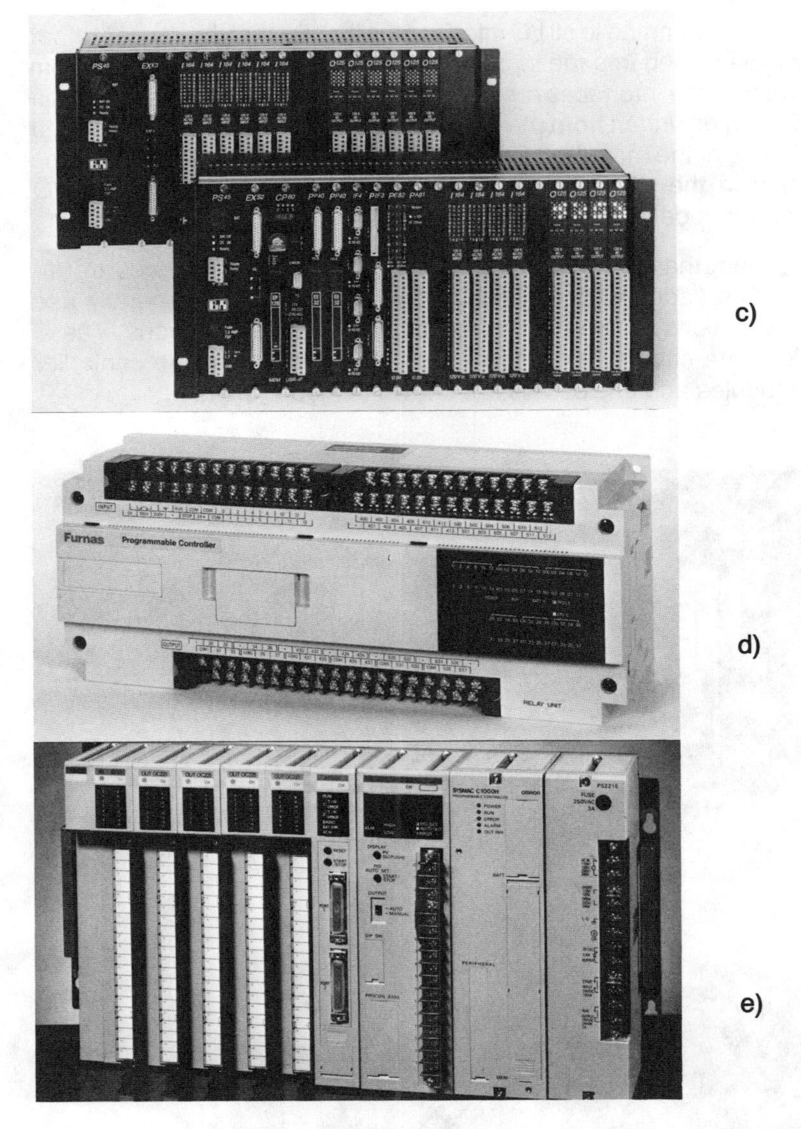

c)

d)

e)

Figure 6-31 continued. c) I/O modules placed in PLC racks along with CPU and power supply (B &R Automation, Stone Mountain, GA); **d)** Small PLC with built-in I/O system (Furnas Electric, Batavia, IL); **e)** PLC with high density I/O modules (Omron Electronics, Schaumburg, IL).

6-8 INTERPRETING I/O SPECIFICATIONS

Perhaps with the exception of standard input/output current and voltage ratings, specifications for I/O circuits are all too often treated as a meaningless listing of numbers. Manufacturers specifications however, provide much information that defines how a particular interface is correctly and safely applied. The specifications place certain limitations not only on the module, but also on the field equipment that it can operate. Failure to adhere to these specifications could result in misapplication of the hardware, leading to faulty operation or equipment damage. The following specifications should be evaluated for each PLC application. This summary of specifications also applies for the interfaces covered in the next two chapters (analog and special function).

Electrical

Input Voltage Rating. This nominal AC or DC value defines the magnitude and type of signal that will be accepted. Deviation from this nominal value will usually be accepted between plus or minus 10% and 15%. This specification may also be stated as the input voltage range. In this case, the minimum and maximum acceptable working voltages for continuous operation are given. An input circuit rated at 120 VAC may accept a signal of 108 to 132 VAC.

Input Current Rating. This nominal value defines the minimum input current at the rated voltage that the input device must be capable of driving to operate the input circuit. This specification may also appear indirectly as the minimum power requirement.

Input Threshold Voltage. This value specifies the voltage at which the input signal is recognized as being absolutely ON. This specification is also called the ON threshold voltage. Some manufacturers also specify an OFF voltage, defining the voltage level at which the input circuit is absolutely OFF.

Input Delay. The input delay will be specified as a minimum or maximum value. It defines the duration for which the input signal must exceed the ON threshold before being recognized as a valid input. This delay is a result of filtering circuitry provided to protect against contact bounce and voltage transients. The input delay is typically 9-25 msec for standard AC/DC inputs and 1-3 msec for TTL or electronic inputs.

Output Voltage Rating. This nominal AC or DC value defines the magnitude and type of voltage source that can be controlled. Deviation from this nominal value is typically plus or minus 10%-15%. For some output interfaces, the output voltage is also the maximum continuous voltage. The output voltage specification may also be stated as the output voltage range, in which case both the minimum and maximum operating voltages are given. An output circuit rated at 48 VDC, for example, may have an absolute working range of 42 to 56 VDC.

Output Current Rating. This specification is also known as the *ON-state continuous current rating*, a value that defines the maximum current that a single output circuit can safely carry under load. The output current rating is a function of component electrical and heat dissipation characteristics. This rating is generally specified at an ambient temperature (typically 0-60° C). As the ambient increases, the output current is typically derated. To exceed the output current rating or oversize the manufacturer's fuse rating could result in permanent short circuit failure or other damage.

Output Power Rating. This maximum value defines the total power that an output module can dissipate with all circuits energized. The output power rating for a single energized output is the result of multiplying the output voltage rating by the output current rating (e.g. 120V x 2A = 240VA). This value (for a given I/O Module) in units of volt-amps (or watts) may or may not be the same, if all outputs on the module are energized simultaneously. The rating for an individual output when all other outputs are energized should be verified with the manufacturer.

Current Requirements. The current requirement specification defines the current demand that a particular input/output module's logic circuitry places on the system power supply. Totaling the current requirements of all the installed modules that one supply supports and comparing the value with the maximum current that can be supplied will indicate if the power supply is adequate. The specification will normally provide a typical rating and a maximum rating (all I/O activated). An undercurrent condition, resulting from insufficient power supply current, can result in intermittent operation of field input and output interfaces.

Surge Current (max). The surge current, also called *inrush current*, defines the maximum current and duration (e.g. 20 amps. @ .1 sec.) for which an output circuit can exceed its maximum ON-state continuous current rating. Heavy surge currents are usually a result of transients on the output load line or on the power supply line and the switching of inductive loads. Output circuits are normally provided with internal protection by a free wheeling diode, Zener diode, or an RC network across the load terminals; if not, the protection should be provided externally.

Off State Leakage Current. Typically this is a maximum value that defines the small leakage current that flows through the triac/transistor in its OFF state. This value is normally in the order of a few microamperes to a few milliamperes and presents little problem in most cases. It could perhaps present problems in the case of switching very low currents or could give false indication when using a sensitive instrument, such as a volt-ohm meter, to check contact continuity.

Output ON-Delay. This specification defines the response time for the output to go from OFF to ON once the logic circuitry has received the command to turn ON. The turn-ON response time of the output circuit will affect the total time required to activate an output device. The actual worst-case time required to turn the output device ON after the control logic goes TRUE, will be two program scans plus the I/O update, the output ON-delay, and the device ON-response times.

Output OFF-Delay. The OFF-delay specification defines the response time for the output to go from ON to OFF once the logic circuitry has received the command to turn OFF. The turn-OFF response time of the output circuit will affect the total time required to deactivate an output device. The actual worst-case time required to turn the output device OFF after the control logic goes FALSE, will be two program scans plus the I/O update, the output OFF-delay, and the device OFF-response times.

Electrical Isolation. This maximum value in volts defines the isolation between the input/output circuit and the controller logic circuitry. Although this isolation protects the logic side of the module from excessive input/output voltages or currents, the power circuitry of the module may be damaged.

Output Voltage/Current Ranges. In reference to analog outputs, this specification is typically a nominal expression of the voltage/current swing of the digital-to-analog converter. The output will always be a proportional current or voltage within the output range. A given analog output module may have several hardware or software selectable unipolar or bipolar ranges (e.g. 0 to 10V, -10V to +10V, 4 to 20mA).

Input Voltage/Current Ranges. In reference to analog inputs, this specification is typically a nominal expression that defines the voltage/current swing of the analog-to-digital converter. The input will always be a proportional current or voltage within the input range. A given analog input module may have several hardware or software selectable unipolar or bipolar ranges (e.g. 0 to 10V, -10V to + 10V, 4 to 20mA).

Digital Resolution. This specification defines how closely the converted analog input/output current or voltage signal approximates the actual analog value within the specified voltage or current range (Chapter 6). Resolution is a function of the number of bits used by the A/D or D/A converter. An 8 bit converter has a resolution of 1 part in 2^8 or 1 part in 256. If the range is 0 to 10V, then the resolution is 10 divided by 256 or 40mV/bit.

Output Fuse Rating. Fuses are often supplied as a part of the output circuit, but only to protect the semiconductor output device (triac or transistor). The particular fuse that is employed by the interface or recommended (if not incorporated) by the manufacturer has been carefully selected based on the fusing current rating of the output switching device. Fuse rating incorporates a fuse opening time along with a current overload rating, which allows opening within a time-frame that will avoid damage to the triac or transistor. The recommended specifications should be followed when replacing fuses or in the case where fuses are not incorporated into the output.

Mechanical

Points Per Module. This specification simply defines the number of input or output circuits that are on a single encasement (module). Typically, a module will have 1,2,4,8, or 16 points per module depending on the manufacturer. The number of points per module has two implications that may or may not be of importance to the user. First, in general, the less dense (number of points) a module is, the greater the space requirements are; second, the higher the density, the lower the likelihood that the I/O count requirements can be closely matched with the hardware. For example, if the module contains 16 points, and the user requires 17, two modules must be purchased. Thus the user will have purchased 15 extra inputs or outputs.

Wire Size. A wire size specification is not always given, but may be of importance to the user and should be verified. It simply defines the number of conductors and the largest wire gauge that the I/O termination points will accept (e.g. 2 #14 AWG).

Environmental

Ambient Temperature Rating. This value defines what the maximum temperature of the air surrounding the input/output system should be for best operating conditions. This specification takes into consideration the heat dissipation characteristics of the circuit components, which are considerably higher than the ambient temperature rating itself. The ambient temperature is rated much less, so that the surrounding temperature does not contribute excess heat to that already generated by internal power dissipation. The ambient temperature rating should never be exceeded.

Humidity Rating. The humidity rating for PLCs is typically 0-95% non-condensing. Special consideration should be given to insure that the humidity is properly controlled in the area where the input/output system is installed. Humidity is a major atmospheric contaminant, which can cause circuit failure if moisture is allowed to condense on printed circuit boards.

Proper observance of the specifications provided by the manufacturer's data sheets will help to insure good, safe operation of the control equipment. In Chapter 15, other considerations for properly installing and maintaining the input/output system are discussed in greater detail.

Chapter

—7—

THE ANALOG I/O SYSTEM

"One line alone has no meaning; a second one is needed to give it expression."

—Delacroix

7-1 INTRODUCTION

With new technological advances and the growth of application areas for programmable controllers came the need for the implementation of systems which were beyond the discrete I/O world. Although discrete interfaces encompass most of the I/O modules used in PLC applications, analog and special I/O interfaces are also extensively used to accomplish specific control tasks.

Analog measurement and control is generally used in applications that are concerned with continuous processes such as batching, temperature control, and the monitoring of sensors and process instruments.

This chapter will cover how these interfaces operate and how they can be used with certain software instructions described in Chapter 9. Bear in mind that the applications of these interfaces are so numerous that covering all of them in detail would require a book in its own right.

7-2 ANALOG INPUTS

Analog input modules are used in applications where the signal being furnished by a field device is in a continuous form (see Figure 7-1); as opposed to discrete signals (ON or OFF), analog signals are present at various levels. For instance, temperature is a type of signal that continuously changes with time by infinitesimal amounts; the change from 70°F to 71°F does not comprise a change of 1°F instantaneously, but several smaller changes that may number in the hundreds to thousands of in fact infinite fractions of a degree.

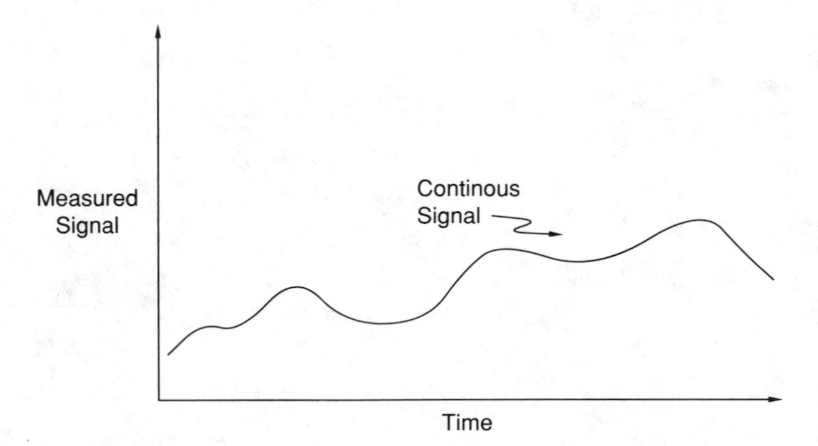

Figure 7-1. Representation of a continuous analog signal.

Even though these signals are continuous in nature, a programmable controller cannot interpret them in its analog form. A PLC is, like other digital computers, a discrete system that only understands ones and zeros. Therefore it is the duty of the analog input interface to translate the continuous analog signal into a discrete value that can be interpreted by the PLC processor and ultimately utilized by the user in a meaningful manner in the control program. Table 7-1 illustrates typical devices that are interfaced with analog input modules.

Analog Inputs
Temperature Transducer
Pressure Transducers
Load Cell Transducers
Humidity Transducers
Flow Transducers
Potentiometers

Table 7-1. Devices used with analog input interfaces.

7-3 ANALOG INPUT DATA REPRESENTATION

When we talk about a field device that provides an analog output as its signal, we most likely will find that the field device is really connected to a transducer or transmitter which in turn will provide the analog input signal to the module. These transducers convert the field device's variable (i.e. pressure, temperature, etc.) into an electrical signal (current or voltage) that can be input to the analog interface (see Figure 7-2).

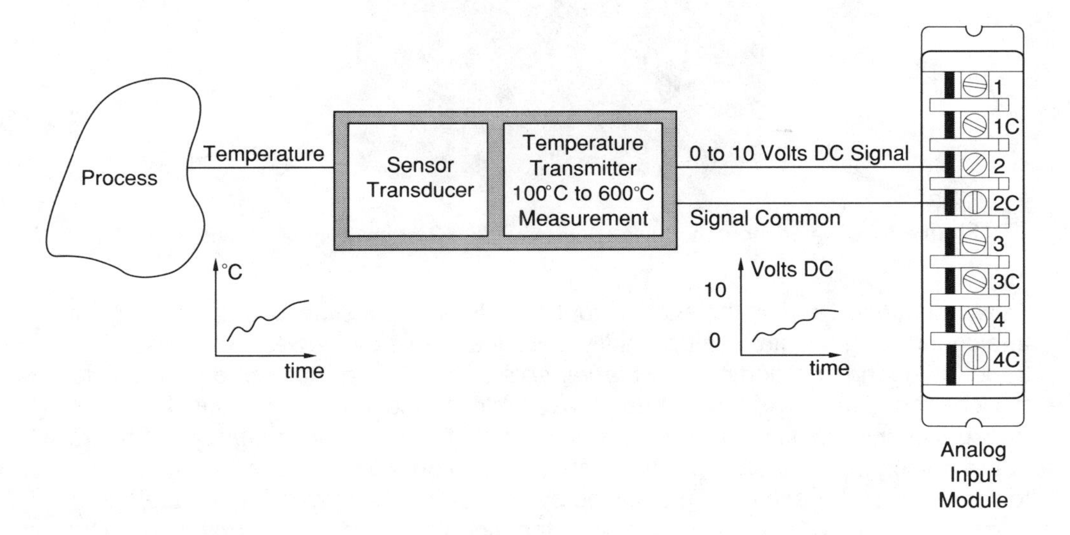

Figure 7-2. Representation example of an analog input signal.

Due to the many types of transducers available to convert the variable being measured into an electrical signal, several standard electrical input ratings are available for analog input modules. Table 7-2 lists the standard ratings mostly found in PLC applications (current and voltage). These analog interfaces are available in *unipolar* (positive voltage only, i.e. 0-5 VDC) and *bipolar* (negative and positive voltage, i.e. -5 to +5 VDC) ratings. Figure 7-3 shows an analog input module.

Input Interfaces
4-20 mA
0 to +1 Volts DC
0 to +5 Volts DC
0 to +10 Volts DC
1 to +5 Volts DC
± 5 Volts DC
± 10 Volts DC

Table 7-2. Typical analog input interface ratings.

Figure 7-3. Analog input modules (Courtesy Allen Bradley, Highland Heights, OH).

As we mentioned earlier, the analog input module is responsible for transforming the analog input signal into a value readily understandable by man and machine. This transformation is proportional to the *variable signal* being measured by the field device (e.g. pressure in psi) and is available at the module's input, from the transducer or transmitter, as a current or voltage. The current or voltage that appears at the analog input module is proportional to the signal being measured. The input to the analog interface is then *digitized* by converting the current or voltage into a number in binary which is proportional to the module's input current or voltage. The binary number converted by the module is therefore proportional to the variable being measured.

The number conversion performed by the module is accomplished utilizing an *analog-to-digital converter (A/D or ADC).* The input signal is divided or digitized in many digital counts which represent the magnitude of the current or voltage. This division of the parts of the input signal is generally referred to as *resolution.* The resolution of the module indicates into how many partitions the module's ADC will divide the input signal and is given as a function of how many bits the ADC uses during conversion. For instance, an input signal can be broken down using 12 bits in the ADC (see Figure 7-4). Therefore, the signal can be represented by a binary number using 12 bits (2^{12}) with values in the range of 0000 to 4095 decimal. The remaining bits could be employed by the manufacturer to represent module's conditions such as *active, OK, channel operating,* etc. Referring to Figure 7-4, these status monitoring bits would be 17, 16, 15, and 14.

The values provided by the ADC are used or transferred to the processor which in turn are available for use in register or word locations. Depending on the manufacturer, the values provided by the analog input module can vary based on the format used. The most common formats used are the binary and BCD. Table 7-3 shows several signal representations using the BCD and decimal equivalent of the binary information from the ADC. If the BCD format is provided, an extra linearity computation is performed by the module (or the processor) to provide valid BCD numbers. Other PLCs may offer direct scale conversion to engineering units (0 to 9999) proportional to the input signal.

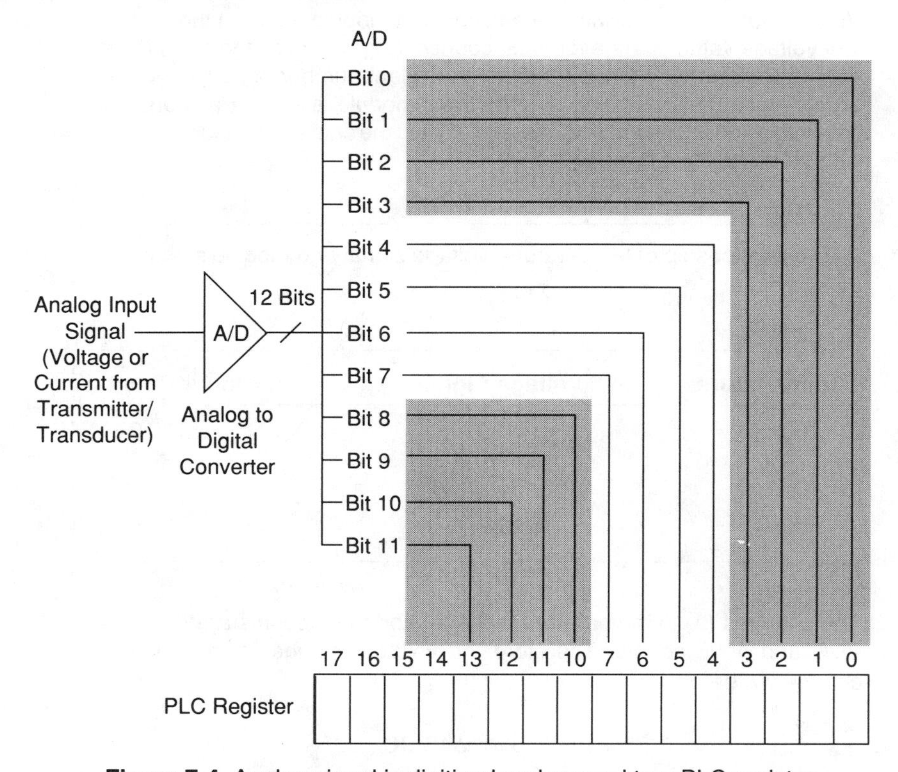

Figure 7-4. Analog signal is digitized and passed to a PLC register.

Analog Voltage Input	Digital Representation BCD Scale 000-999	Digital Representation Decimal Scale 0-4095
0-V	000	0
1-V	100	410
2-V	200	819
3-V	300	1229
4-V	400	1638
5-V	500	2047
6-V	600	2457
7-V	700	2866
8-V	800	3276
9-V	900	3685
10-V	999	4095

Table 7-3. PLC representation of analog input voltage.

Example 7-1

An input module has an ADC with a 12-bit resolution and is connected to a temperature transducer. The transducer receives the necessary signal from a temperature sensing device (100° to 600°C) and provides a 1 to 5 VDC signal compatible with the analog input module. Find **a)** the equivalent voltage value change for each count change (the voltage change per degree Centigrade change) and the equivalent number of counts per degree Centigrade, assuming that the input module is set to transform the data into a linear 0 to 4095 counts and, **b)** find the same values for a module with 10-bit resolution.

Solution

a) The relationship of temperature, voltage signal and module's counts is as follows:

Temperature	Voltage Signal	Input Counts
100°C	1VDC	0
•	•	•
•	•	•
•	•	•
600°C	5VDC	4095

The changes (Δ) in temperature, voltage, and input counts are 500°C, 4 VDC and 4095 counts. Therefore the voltage change for one degree change will be:

$$\Delta 500°C = \Delta 4 \text{ VDC}$$

$$1°C = \frac{4 \text{ VDC}}{500} = 8.0 \text{ mVDC}$$

The changes in voltage for each module's count will be:

$$\Delta 4095 \text{ counts} = \Delta 4\text{VDC}$$

$$1 \text{ count} = \frac{4\text{VDC}}{4095} = 0.9768 \text{ mVDC}$$

The corresponding number of counts per degree centigrade will be:

$$\Delta 500°\text{C} = \Delta 4095 \text{ counts}$$

$$1°\text{C} = \frac{4095 \text{ counts}}{500} = 8.19 \text{ counts}$$

b) A 10-bit resolution ADC will digitize the unipolar input signal into 1024 counts (one in 2^{10} parts); the relationship of temperature, voltage signal and counts will be:

Temperature	Voltage Signal	Input Counts
100°C	1VDC	0
•	•	•
•	•	•
•	•	•
600°C	5 VDC	1024

The changes in temperature, voltage and counts are 500°C, 4 VDC and 1024 counts. The voltage change per degree will be the same as **(a)** and is:

$$\Delta 500°\text{C} = \Delta 4 \text{ VDC}$$

$$1 \text{ count} = \frac{4 \text{ VDC}}{500} = 8.0 \text{ mVDC}$$

The changes in voltage per input count will be:

$$\Delta 1024 \text{ counts} = \Delta\ 4 \text{ VDC}$$

$$1°\text{C} = \frac{4 \text{ VDC}}{1024} = 3.906 \text{ mVDC}$$

The corresponding number of counts per degree will be:

$$\Delta 500°\text{C} = \Delta 1024 \text{ counts}$$

$$1°\text{C} = \frac{1024 \text{ counts}}{500} = 2.048 \text{ counts}$$

(0 to 10 VDC) will be converted into values ranging from 0 to 4095 which are proportional to the 0 to 500 psi signal variable being measured. The value of 0 counts will be associated with 0 psi and the value 4095 will be related to 500 psi. The following examples illustrate how the equivalent analog counts are computed for an analog input field signal.

Example 7-2

A temperature transducer provides a voltage signal of 0 to 10 VDC which is proportional to the temperature variable being measured. The temperature measurement ranges between 0°C and 1000°C. The analog input module that is going to be used accepts the 0 to 10 VDC unipolar signal range and converts it into a count range of 0 counts to 4095 counts. The process application where this signal is being used specifies that a low and high alarm shall be detected at 100°C and 500°C respectively. Find a) the relationship (equation) between the input variable signal (temperature) and the counts being measured by the PLC module, and b) the equivalent number of counts for each of the alarms temperatures specified.

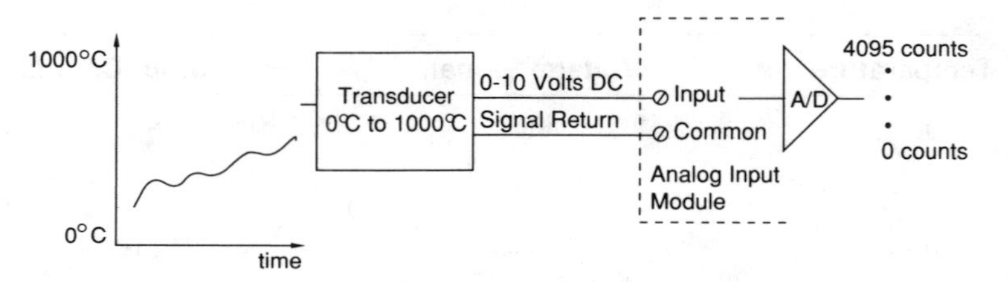

Figure 7-5. Description of analog input signal.

Solution

a) The illustration shown in Figure 7-5 shows the relationship between counts and the input signal in volts and degrees centigrade. The numerical representation of the input signal and the number of counts (assuming a linear relationship) is described by the line *Y*. There are two ways to achieve the required answer. First, we can find the numerical representation of the equation for line *Y*. This equation takes the form $Y=mX+b$ (see Appendix H), where *m* is simply the slope of the line which is described by:

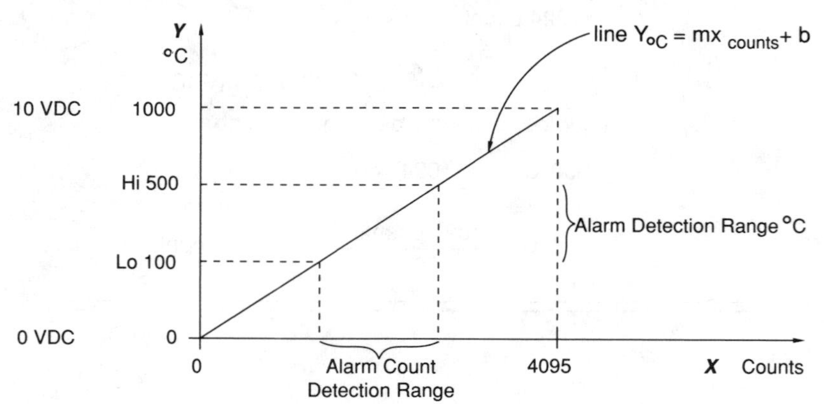

$$m = \frac{Y_2 - Y_1}{X_2 - X_1} = \frac{°C_2 - °C_1}{\text{Count 2 - Count 1}} = \frac{1000 - 0}{4095 - 0} = \frac{1000}{4095}$$

where Y_2, Y_1,X_2 and X_1 are known points. The term b is described as the intersection of Yor °C whenX or counts is equal to 0. This value can be computed as:

$$b = Y_{°C} - mX_{Counts}$$

where Y and X are values at known points (i.e. at 0°C and 0 counts) when X is at 0 counts we know that y is at 0°C and we have

$$b = 0 - \left(\frac{1000}{4095}\right) 0$$

$$b = 0$$

therefore b is equal to 0 and $Y_{°C} = \frac{1000}{4095} X_{Counts}$

if we had taken the X(counts) and y (°C) values at 4095 counts and 1000°C, we would have arrived at the same conclusion (try it as an exercise). Based on the equation, we can easily find the number of counts for each alarm range:

$$Y_{°C} = \frac{1000}{4095} X_{Counts}$$

$$X_{Counts} = \frac{4095\ Y_{°C}}{1000}$$

so for $Y_{°C}$ values of 100°C and 500°C we have

$$X_{Counts} = \frac{4095 \times 100}{1000} = 409.5$$

$$X_{Counts} = \frac{4095 \times 500}{1000} = 2047.5$$

The count value for 100°C would be 409.5 counts and for 500°C would be 2047.5 counts. Since there is no half count, we can round off the number of counts to 410 and 2048 respectively. Therefore at a count of 410 the low level temperature alarm would be enabled and at a count of 20 48 the high level temperature alarm would be enabled.

The second method that can be used to implement the solution is to obtain the number of counts that are equivalent to 1°C. In this example, we have a change of 1000 degrees per 4095 counts expressed as:

$$\frac{\text{change in counts}}{\text{change in degree}} = \frac{\text{max counts - min counts}}{\text{max degrees - min degrees}} = \frac{4095 - 0}{1000 - 0} = 4.095$$

Therefore, each degree is equivalent to 4.095 counts. The count value for the 500°C and 100°C would be 500 x 4.095 = 2047.5 and 100 x 4.095 = 409.5, rounding off gives us 2048 and 410 for 500°C and 100°C respectively. If the counts had not started at 0, an offset count addition would have been necessary for the counts computed per degree.

7-4 ANALOG INPUT DATA HANDLING

The previous section showed how a signal from the field is transformed into a discrete signal at the analog input module. Once the signal has been digitized or transformed into binary counts the value is available to the processor.

During the input reading section of the scan, the processor reads the value from the module and transfers the information to a location specified by the user. This location is generally a word or register storage or an input register. Depending on the controller, the count value is entered by the processor into memory using instructions that differ from those used by standard discrete input modules.

In general, analog modules provide more than one channel or input per interface, and therefore can connect to several input signals as long as they are compatible with the module (voltage levels, current). The instructions used in PLCs take advantage of these multiple channels and generally bring or input several values into registers or words (see Figure 7-6). The instructions used are generally called block transfer in, analog in, block in, or location in (see Chapter 9). However, it is possible to find PLC manufacturers that allow arithmetic or other instructions to access the analog module's address to obtain the count values.

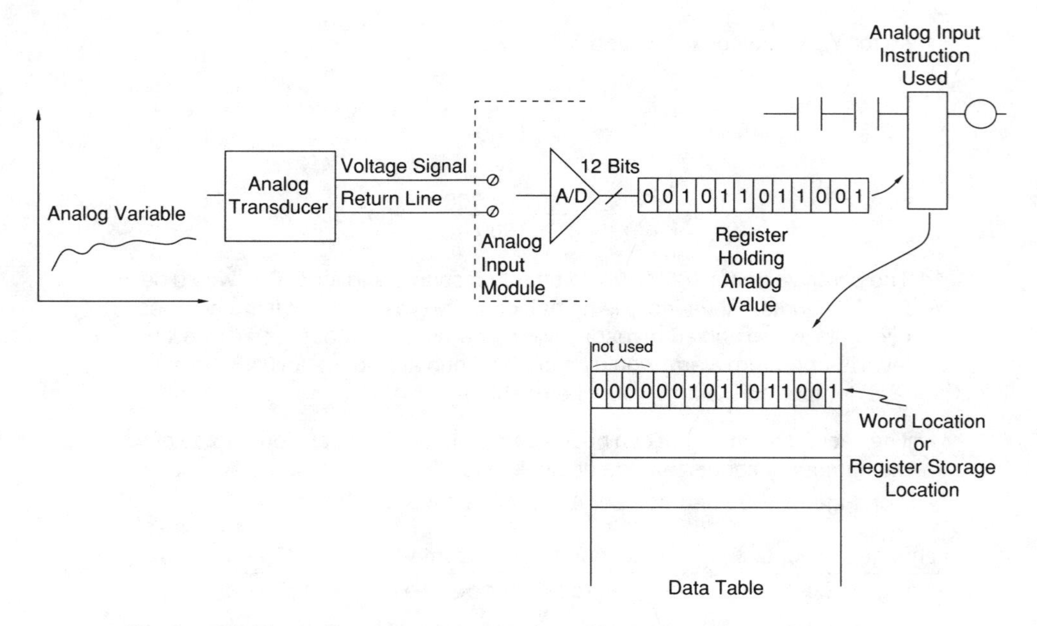

Figure 7-6. Block diagram representation of how an analog value is stored.

When the processor executes the instruction that reads an analog input, it will obtain the module's data during the next I/O scan and place the value in the destination register used or specified in the instruction. If multiple channels are being read, the processor will most likely read and store one channel on every scan; this would not cause any delay in signal processing since the scan is so fast and the signals being read are generally slow in nature (e.g. temperature).

Depending on the the manufacturer, the address location of the module is generally defined by the physical location of the module within a rack or enclosure (see Chapter 5 for I/O enclosures). Since the processor can recognize whether the module inserted in the enclosure is an analog module or not, it would read the available data in groups of 16 bits of which, depending on the resolution, 12 bits would contain the analog value in binary or BCD. Some controllers may provide diagnostic information about each channel of the module and about the module in general by allowing an extra word or register to be read after all channels are input. Figure 7-7 illustrates an example of an address for a possible analog module location. A typical instruction will reference the address location of the module by specifying the rack number, the slot where the module is placed, the number of channels or analog inputs used, and the *starting* register destination address. If eight channels are used and the destination storage register starts at address 200_8, the last register storage would be 207_8. Generally, the processor takes care of the register range automatically according to the number of channels; however, the programmer must remember not to overlap the usage of already assigned registers.

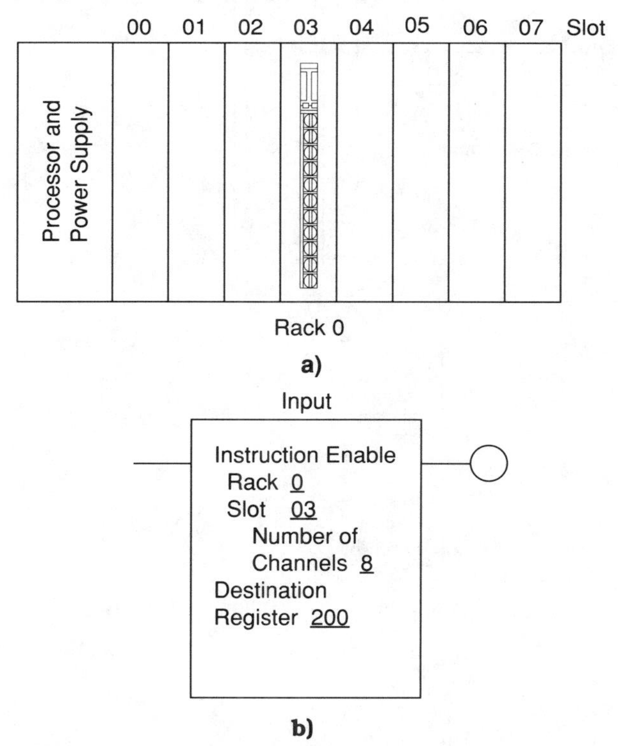

Figure 7-7. **a)** Illustration of an analog module in a rack, and **b)**Typical instruction used to read the analog data.

7-5 ANALOG INPUT CONNECTIONS

Analog input modules usually provide a high input impedance (in the Megaohm range) for voltage type input signals which allows them to interface to high source-resistance output from input sensing devices (transmitters or transducers). The current type input modules provide a low input (between 250 and 500 ohms) impedance which is necessary to operate properly with their compatible field sensing devices.

Input interfaces may be obtained with the capability of handling *single-ended* or *differential inputs.* They differ in that the single-ended input has all the input's commons electrically tied together, whereas the differential input mode accepts individual return or common lines for each channel. Undoubtedly, the single-ended module most likely will offer more points per module than its differential counterpart. Depending on the manufacturer, the choice between single-ended or differential modes may be selected by setting up the interface using rocker switches during the software set up of the module (generally in the instruction). Figure 7-8 illustrates a typical analog connection for the single-ended and differential inputs.

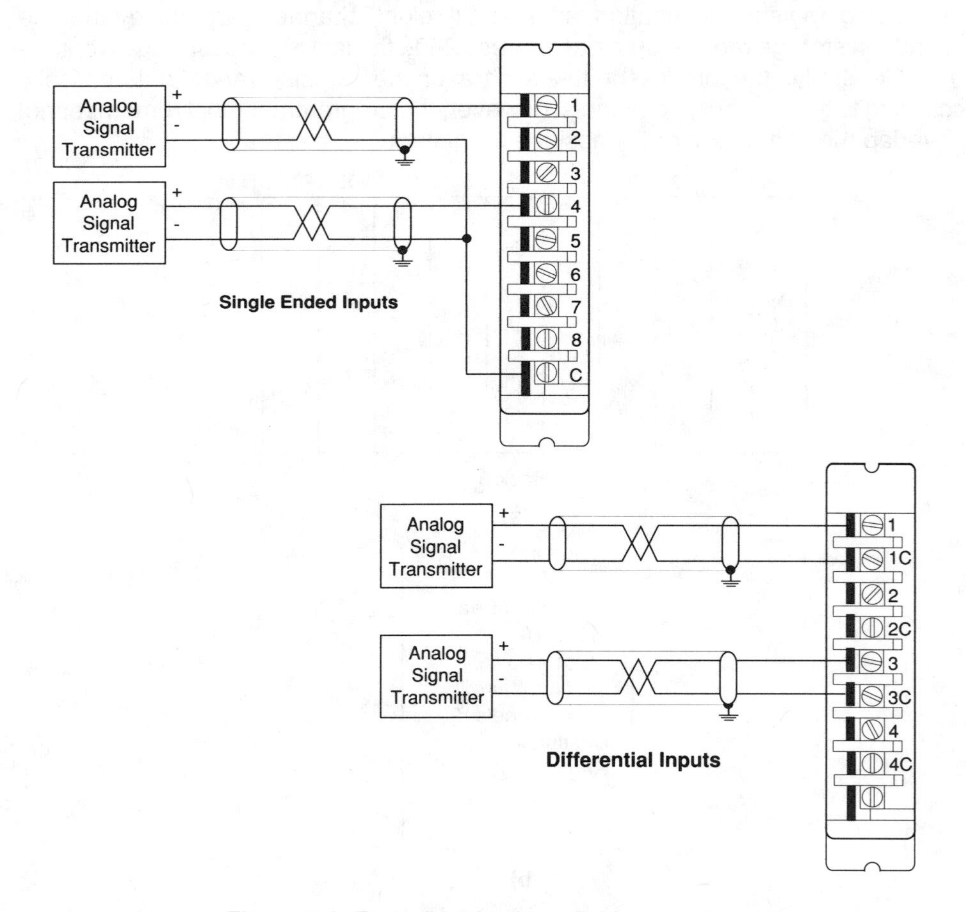

Figure 7-8. Typical analog input field connections.

Each channel in the interface provides signal filtering and isolation circuits to protect the module from field noise. In addition to the noise precautions that are included in the module, the user should take into consideration other electrical noise during the installation of the module (see Chapter 15). Typically, the connection between the input module and the transducer is made using shielded conductor cables which provide a better interface. This allows the line impedance imbalances to be kept low and to maintain a good common mode rejection ratio of noise levels, such as power line frequencies.

Analog input interfaces seldom require external power supply sources; they get their required power to operate properly from the backplane of the rack or enclosure. These interfaces, however, draw more current than the discrete type module; therefore, loading considerations should be kept in mind during the configuration of the PLC system and the computation of modules' currents for power supply selection.

7-6 ANALOG OUTPUTS

Analog output interfaces are used in applications requiring control capability of field devices which respond to continuous voltage or current levels. For example, a volume adjust valve used in hydraulic based punch presses could require a 0 to 10 VDC signal to operate the valve and vary the volume of oil being pumped to the press cylinders, thereby changing the speed of the ram or platen. The number of output field devices that respond to analog signals is so large and varied that to list them would take page after page. The more the user confronts process control applications, the more analog output devices he or she will encounter. The important thing is to know what to use and how to control those devices. Table 7-4 lists some output devices that can be categorized as most common.

Analog Outputs
Analog Valves
Actuators
Chart Recorders
Electric Motor Drives
Analog Meters
Pressure Transducers

Table 7-4. Typical analog output field devices.

7-7 ANALOG OUTPUT DATA REPRESENTATION

Like analog inputs, the output interfaces generally are connected to the controlling devices via the use of transducers. These transducers take the voltage signal and may amplify, reduce, or change it into another signal which in turn would control the output device (Figure 7-9). Since there are many types of controlling devices, transducers are designed with standardization in mind and are available in several voltages or current ratings. Table 7-5 lists some of the standard ratings that are usually used in programmable controllers which provide analog output capabilities.

Output Interfaces
4-20 mA
10-50 mA
0 to +5 Volts DC
0 to +10 Volts DC
+ 2.5 Volts DC
± 5 Volts DC
± 10 Volts DC

Table 7-5. Analog output ratings.

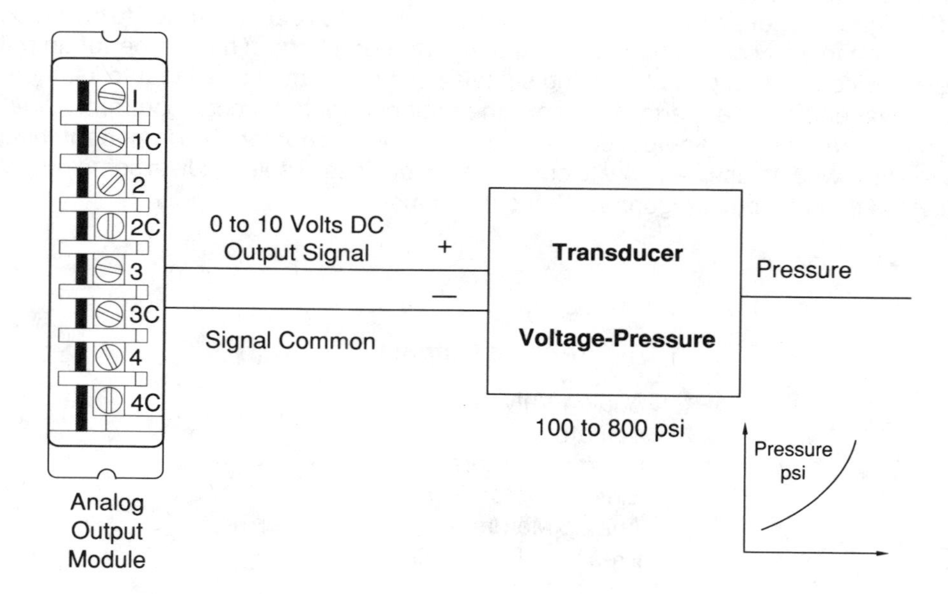

Figure 7-9. Representation example of an analog output signal.

The analog output interface operates in a similar manner as the input module, except that the data direction is reversed. As mentioned throughout the book, the PLC processor can only interpret digital binary numbers, and as such, it assumes that the outside world also operates in the same manner. Therefore, it is the responsibility of the analog output module to change the data from binary into an analog real-world signal which can be understood by field devices.

The transformation of data, which occurs at the output interface, is exactly the reverse of the analog input (see Figure 7-10). Numerical data received at the module from the processor, whether in BCD or binary, is converted into an analog signal using a *digital-to-analog converter (D/A or DAC).* The analog output value is proportional to the numerical value received by the module. Instead of digitizing the signal, the D/A converter creates a continuous analog signal whose magnitude is proportional to the minimum and maximum capable voltage or current (i.e. 0 to 10 VDC).

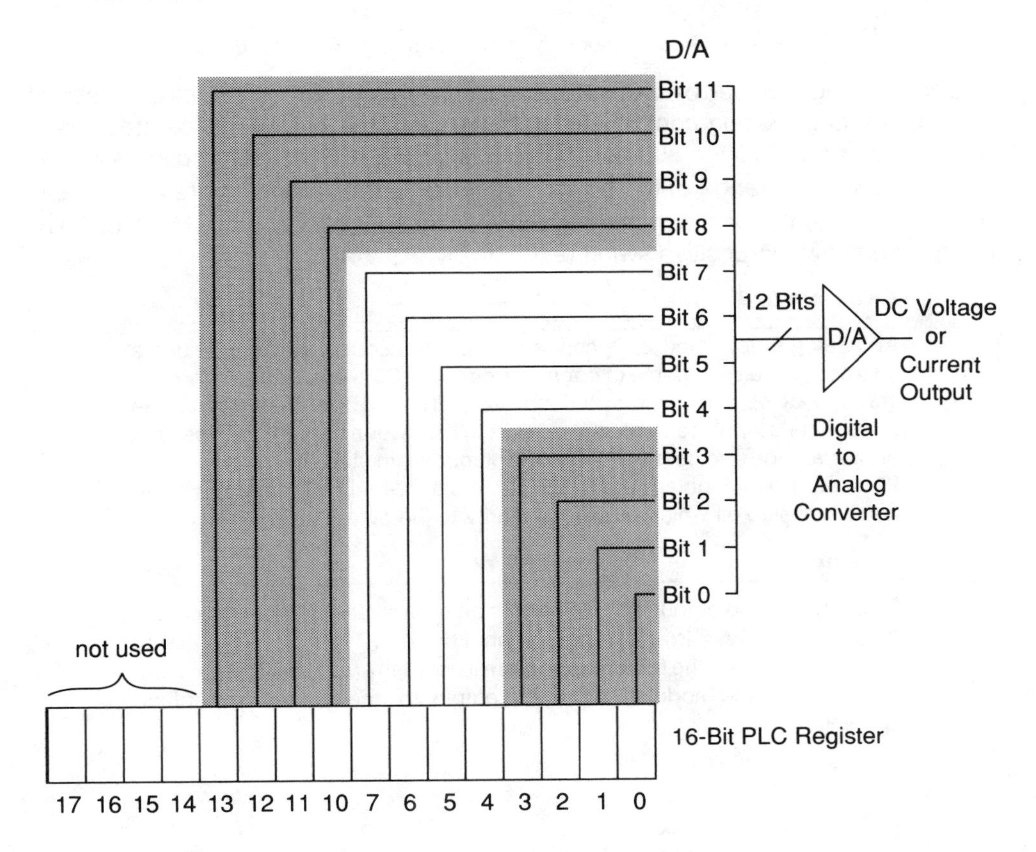

Figure 7-10. Analog value from a register is passed to the D/A converter in the output module.

The *resolution* of the DAC is defined by the number of bits that it uses for the analog conversion. For example, a resolution of 12 bits is said to exist if the module (DAC) can create an analog signal utilizing analog increments of one bit value in twelve bits (1 part in 4096) of the full analog value. A value of 2048 in a 12 bit resolution will provide a half analog value of the full range (e.g. in 10 VDC it would equate to 5 VDC). Table 7-6 shows the register value corresponding to a voltage and current module that has a 12-bit resolution.

PLC REGISTER		OUTPUT	
Decimal	Binary	0-10 VDC	4-20 mA
0	0000 0000 0000 0000	0 VDC	4 mA
2047	0000 0111 1111 1111	5 VDC	12 mA
4095	0000 1111 1111 1111	10 VDC	20 mA

Table 7-6. Output values for a 12-bit analog output module.

The value being provided by the processor will ultimately be proportional to the signal or variable that is being controlled. For instance, if the output device provides a pressure control of 100 to 800 psi, the values from the processor in counts will be proportional to the range of the pressure device. Output modules are available in unipolar and bipolar configurations which provide control voltages with all positive swing or negative to positive swing respectively.

Example 7-3

An analog output module is connected to a transducer which provides a flow control valve capable of opening from 0 to 100% of total flow. The percentage of opening is proportional to a -10 to 10 VDC signal at the transducer's input. Tabulate the relationship between percentage opening, output voltage and counts for the output module in steps of 10% (i.e. 10%, 20%, etc.). The bipolar output module has a 12-bit DAC (binary) with an additional sign bit which provides polarity to the output swing.

Solution

Since the analog output module has a sign bit provision, it receives counts ranging from -4095 to +4095 proportional to the -10 to +10 VDC required by the transducer. The following diagram illustrates graphically the relationship between the modules' count, the output voltage and the percentage opening.

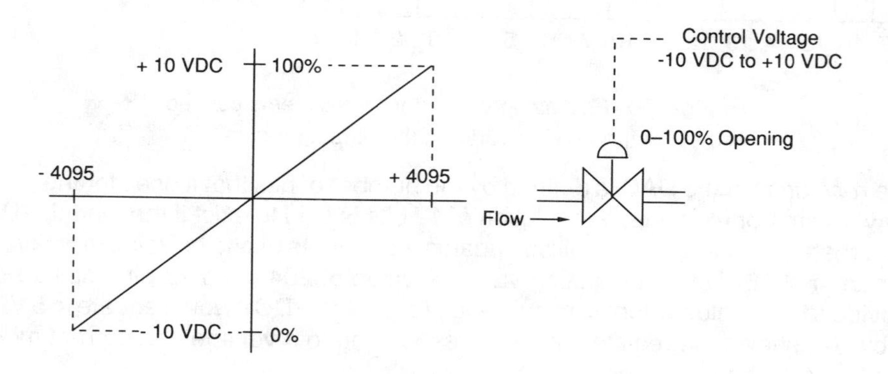

To formulate the desired table, let's first obtain the equivalent values corresponding to each variable. Since we are required to find the solution in increments as a function of percentages, we find that each percentage change is calculated as follows.

Δ*Percentage*	Δ*Voltage (-10 to +10)*	Δ*Counts (-4095 to +4095)*
100	20	8190

$$1\% \text{ as function of voltage } = \frac{20 \text{ VDC}}{100} = 0.2 \text{ VDC}$$

$$1\% \text{ as function of counts } = \frac{8190}{100} = 81.90 \text{ counts}$$

Note that the computations performed are *magnitude* changes. To implement the table, we need to add the offset values for the voltage and counts taking into consideration the bipolar effect of the module and the negative to positive changes in counts. Therefore, to obtain the voltage and count equivalence per percentage change we have to add the values of voltage and counts when the percentage is at 0%. We then have:

Percentage as function of voltage = (0.2 x P) - 10 VDC

Percentage as function of counts = (81.9 x P) - 4095 counts

where *P* is the percentage number to be used in the table. Therefore, the required table (Table 7-7) is implemented by multiplying each voltage and count relationship by the desired percentage of opening. These types of computations are generally useful when installing the system during system start-up and test. The calculation of output counts is performed in the software program according to a predetermined algorithm. Sometimes, the output computations are obtained using engineering units which indicate a 0000 to 9999 (binary value or BCD) change in output value which must be ultimately converted to, in this case, a -4095 to +4095 count conversion (covered later).

Percentage Opening	Output Voltage	Counts
0 %	-10 VDC	-4095
10	-8	-3276
20	-6	-2457
30	-4	-1638
40	-2	-819
50	0	0
60	+2	+819
70	+4	+1638
80	+6	+2457
90	+8	+3276
100	+10	+4095

Table 7-7.

7-8 ANALOG OUTPUT DATA HANDLING

In the previous section we discussed how a signal is transferred from the module to the transducer which signals the controlling output device. Here, we will discuss how the data is handled by the processor and some common methods of linearizing or scaling output data to reflect engineering units.

The data to be sent to an analog output module can be located in the storage section or I/O table section of the data table area of the PLC's memory system (see Figure 7-11). This data is generally a result of program computations which, when sent to the module, will control an analog output device.

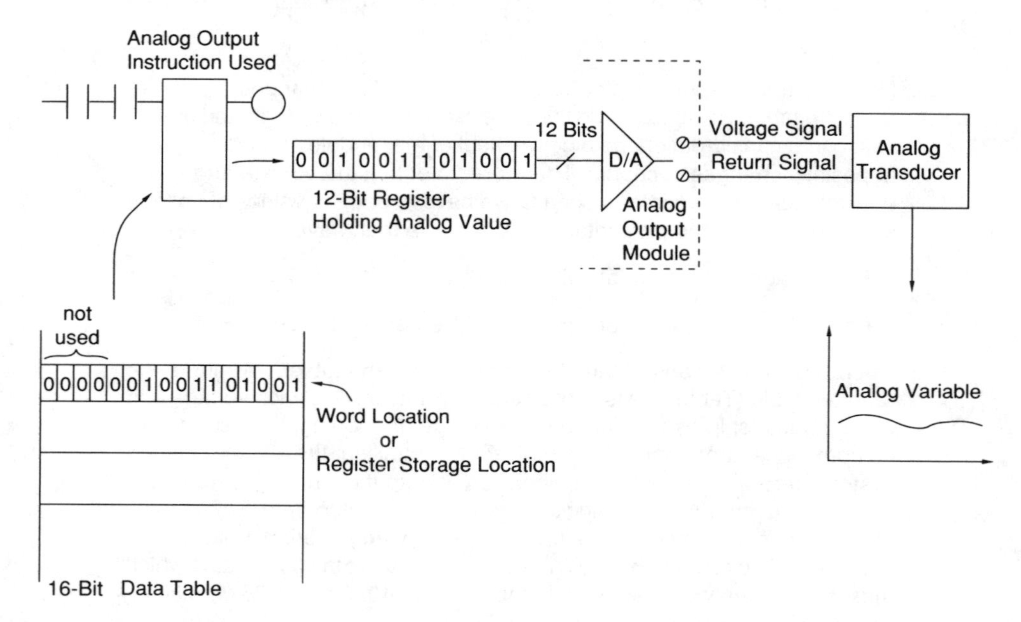

Figure 7-11. Block diagram of a PLC word transfer to an analog outut module.

During the execution of the output update, the processor sends the register or word contents to the analog module as specified by an address in the instruction executed during the program scan. The contents of the register or word is the binary or BCD value that will be used by the module to transform it into an analog output voltage or current. Since the value is calculated by the user (in the program), care must be exercised during programming to avoid computing or sending non-valid ranges for the module. For instance, if a word location containing a binary value of +5173 is sent to a 12-bit resolution module without checking for range validity, the module will be unable to associate the data, and therefore will emit an incorrect analog output signal (5173 in binary uses more than 12 bits).

Like the input counterpart, analog output modules generally handle more than one channel at a time so that more devices can be controlled with one module. The instructions that are used with these output interfaces provide the capability to transfer several words or register locations to the module. These instructions may be called block transfer out, analog out, block out, or location out (see Chapter 9).

It is possible however, to find PLCs that utilize arithmetic or other instructions to send data to the analog module address using the destination register of the instruction. Some PLC manufacturers may offer software instructions which allow scaling to be performed at the module or during the execution of the analog output instruction. This scaling will take a value, let's say from 0000 to 9999 BCD (16 bits), that relates engineering units (e.g. gallons per minute—gpm) and send it to the module linearized for count values. For instance, if the BCD value is 5000, the module will receive a linearized 12-bit binary value 0111 1111 1111, or 2047 counts, which represents the half way mark of the 0 to 4095 range.

Data transfers to analog modules with multiple output channels are performed or updated generally one channel per scan. Similarly, as in the analog input case, this update does not create any noticeable delay since the devices that respond to analog signals are slow in nature. The address location of the module is usually defined by the physical location of the module within the enclosure (see Chapter 6 for I/O enclosures).

Figure 7-12 illustrates an example of an analog output module in an enclosure with its corresponding address location. A typical output instruction may reference the module by the slot and rack location of the interface and the number of channels available or in use. The data to be transferred is stored in a register which is generally described as the source register. The starting source register address is specified in the instruction and the specified number of channel data is sent from the starting source register. For example, if the starting register is 300_8 and the number of channels is 4, the processor will send data from register 300_8 to register 303_8.

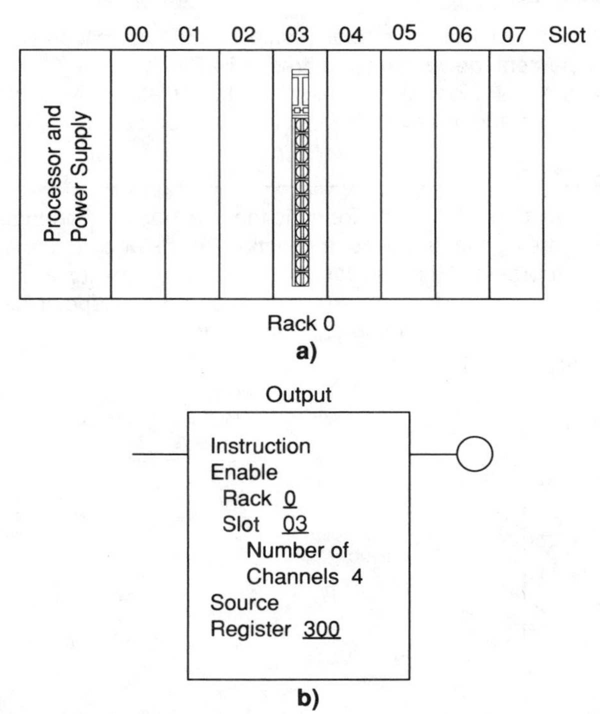

Figure 7-12. a) Illustration of an analog output module in a rack; and **b)** Typical instruction used to output data to the module.

It is important to remember that the analog output signal that was converted in the module is dependent on the register or word value that was transferred by the processor to the module. This value is generally the result of an operation(s) which determines the necessary value to be output by the module to implement the control requirements. There may be situations in which the value computed for a control action is based on a 0000 to 9999 range (engineering units). This value in turn must be converted (if linearity is not available in the output instruction) to the necessary output module count range (i.e. 0 to 4095 or -2048 to +2048) before the transfer to the module. Example 7-4 addresses this type of conversion.

Example 7-4

A programmable controller system uses a bipolar -10 to +10 VDC signal to control the flow of material being pumped into a reactor vessel. The flow control valve has a range of opening from 0% to 100% to allow the chemical ingredient to flow into the reactor tank. The required flow computation which obtains the percentage of valve opening is performed in the processor by executing a predefined algorithm. Feedback information about other chemicals being mixed is received via analog flow meters. The computed value for percentage opening is stored in a register and ranges from 0000 to 9999 BCD (0% to 99.99%).

a) Find the equation of the line representing the required analog output signal (in counts) for a 12-bit resolution module and analog output transformation from -4095 to +4095 counts which includes a sign bit as a function of voltage output and as a function of percentage opening.

b) Graphically illustrate the relationship of outputs in counts and the computed percentage opening as stored in the PLC register (0000 to 9999). Also find the equation which describes the relationship between the required counts and the available calculated value stored in the register.

Solution

a) The equation that has to be found is the line labelled *Y* in Figure 7-13. This equation is computed as a function of voltage and percentage opening. Y represents the equivalent number of counts for each of these cases. The line has the form $Y = mX + b$ where *m* is the slope of the line and *b* (see Appendix H) is the value of *Y* when *X* is 0.

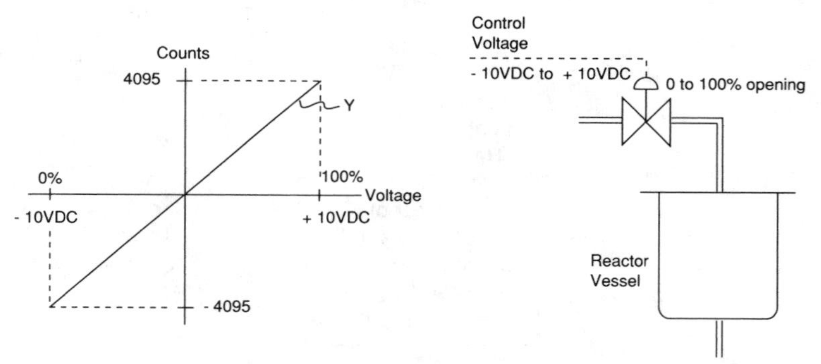

Figure 7-13. Graphic representation of percentage opening and analog output counts.

The X axis can represent the output voltage or percentage opening depending upon the equation that is derived. The Y axis represents the number of counts output by the module for each X value (% or VDC). The equation as a function of voltage is expressed by:

$$Y = mX + b$$

$$m = \frac{\Delta Y}{\Delta X} = \frac{4095 - (-4095)}{10 \text{ VDC} - (-10 \text{ VDC})} = \frac{8190 \text{ counts}}{20 \text{ VDC}}$$

$$Y = \frac{8190}{20} X + b$$

b is calculated by replacing a known value for y when x is 0 counts which is also 0.

$$b = \frac{8190}{20} \quad 0 = 0$$

Therefore $$Y = \frac{8190}{20} X + 0$$

$$Y = \frac{8190}{20} X$$

This equation gives the value Y in counts for any voltage X. The other Yequation as a function of percentage is computed in a similar manner:

$$Y = mX + b$$

$$m = \frac{\Delta Y}{\Delta X} = \frac{8190 \text{ counts}}{100\%}$$

$$Y = \frac{8190}{100} X + b$$

b is computed by replacing the count value Y when X is equal to 0 percent; this value of Y is -4095 (see Figure 7-13). Therefore we have:

$$b = -4095 - \frac{8190(X)}{100} = 4095 - \frac{8190(0)}{100} = -4095$$

Therefore $$Y = \frac{8190}{100} X - 4095$$

This equation for Y gives the number of output counts for any percentage value X.

b) The relationship between the output in counts and the value stored in the register which provides a 0000 to 9999 is shown in Figure 7-14. This graph is very similar to the one shown in part *a*; the only difference is that the output "equation" will have to be recomputed if the register value is used. This equation will be

$$Y = mX + b$$

$$m = \frac{\Delta \text{ Counts}}{\Delta \text{ register value}} = \frac{8190}{9999}$$

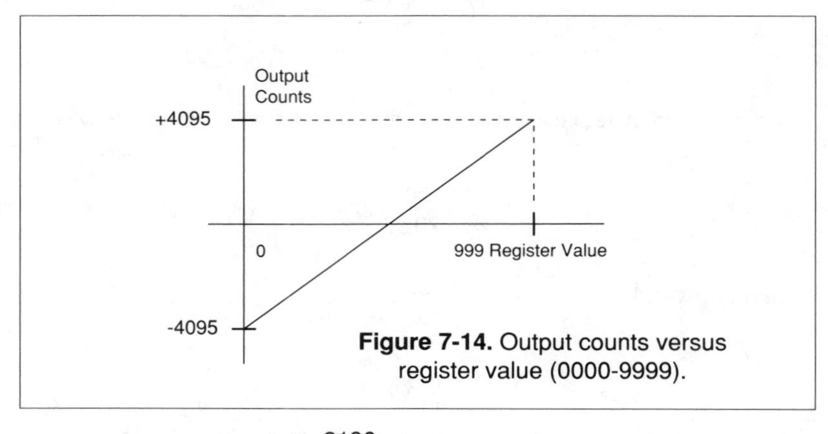

Figure 7-14. Output counts versus register value (0000-9999).

$$Y = \frac{8190}{9999} X + b$$

$$b = Y - mX \quad \text{at } X = 0$$

The value of Y when X is 0 is -4095

$$b = -4095 - \frac{8190}{9999} X = -4095 - \frac{8190}{9999} (0) = -4095$$

Therefore $$Y = \frac{8190}{9999} X - 4095$$

The value of Y will be the output count for any value X (percentage) ranging from 0000 to 9999. If this type of equation is implemented in the PLC using standard decimal arithmetic instructions and the 0000 to 9999 register value is in BDC code, conversions from BCD to decimal will be required. These conversions are generally implemented using software instructions available in the PLC system.

7-9 ANALOG OUTPUT CONNECTIONS

Analog output interfaces are available in configurations that range anywhere from 2 to 8 outputs per module; however, on the average, modules are found with four analog output channels. These channels can be configured as single-ended or differential outputs. Differential is most commonly used when individually isolated outputs are required.

Each analog output is electrically isolated from other channels and the PLC side itself (isolation between field outputs and the PLC), which protects the system from damage due to overvoltage damage at the module's outputs. Depending on the manufacturer, these interfaces may or may not require external panel-mounted power supplies. However, today most of the modules get their power from the PLC's power supply system. Current requirements for analog modules are higher in general than discrete outputs and must be taken into consideration during the computation of current loading (see power supply loading considerations in Chapter 4). Figure 7-15 illustrates a typical connection for an analog output module for single-ended and differential output modules.

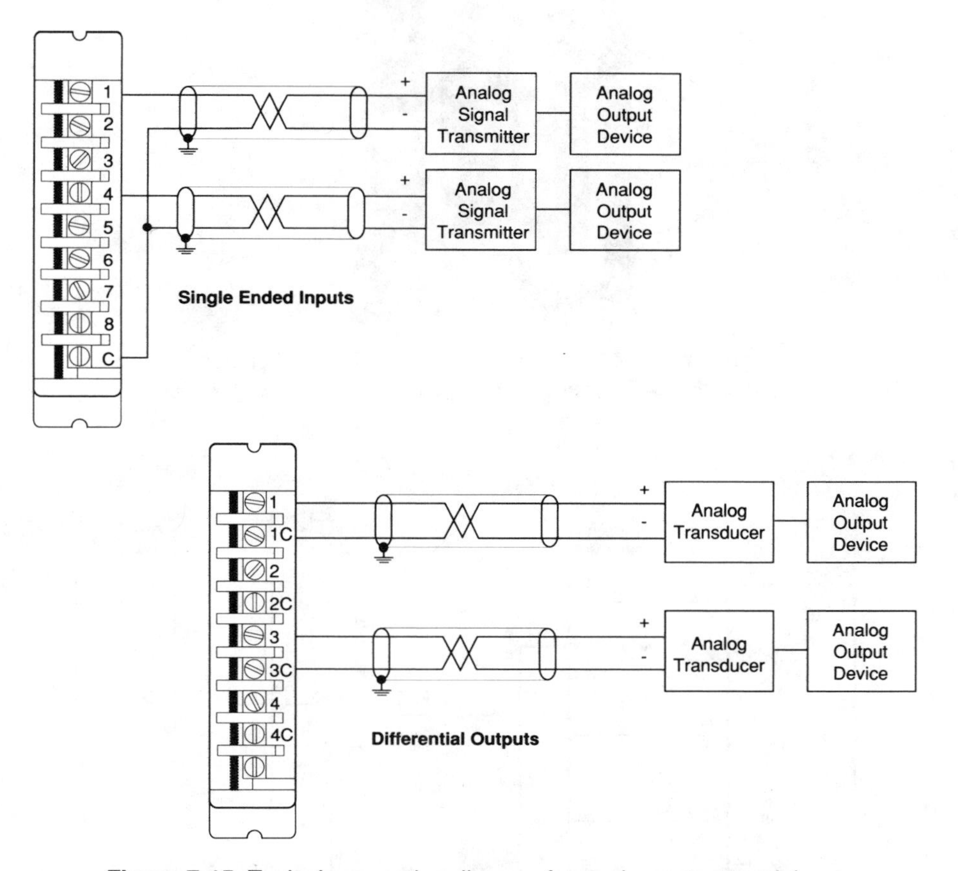

Figure 7-15. Typical connection diagram for analog output modules.

7-10 ANALOG OUTPUT BY-PASS/CONTROL STATIONS

Similar to the application of discrete outputs, the PLC system itself may also require the addition of a by-pass/control back-up unit which can be used with an analog output module. A typical example of these devices is the Manual Control Station (MCS-7200) from Control Technology Corporation (shown in Figure 7-16).

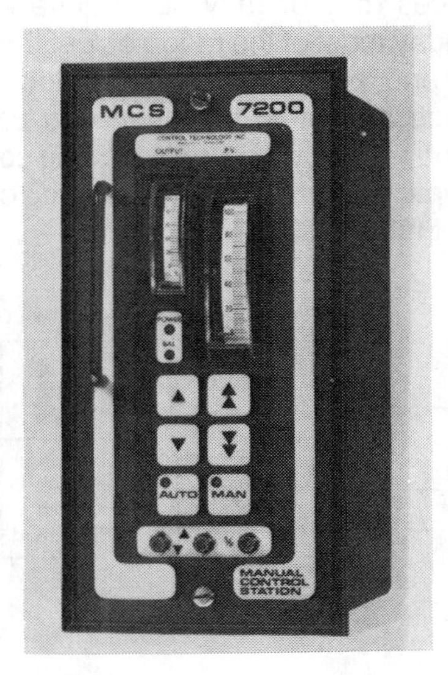

Figure 7-16. MCS-7200 analog manual control station (Courtesy of Control Technology Corp., Knoxville, TN).

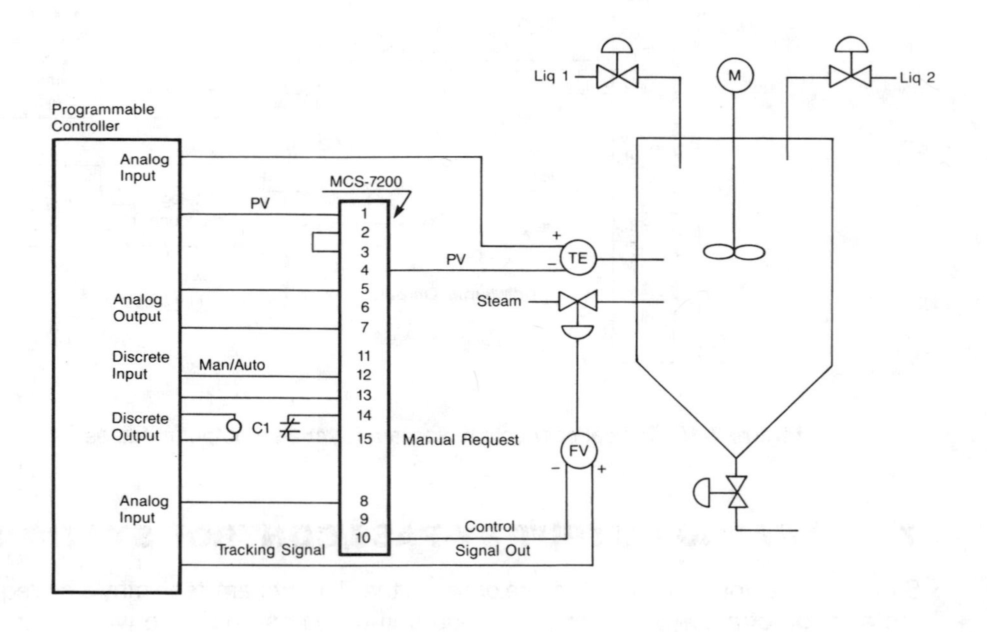

Figure 7-17. Block diagram configuration of the MCS-7200 control station (Courtesy of Control Technology Corp., Knoxville, TN).

The MCS-7200 is used in most cases between the PLC's analog interface and some controlled element to provide a means of assuring continued production or control in a variety of abnormal process situations. This control station becomes very useful during start-up, override of analog outputs, and back-up of analog outputs in case of failures.

During start-up, the operator can manually position the final control element by using the MCS-7200 to manipulate initial control parameters, such as valve position, speed control, hydraulic servos, and pneumatic converters. This task can be accomplished without the PLC or prior to its checkout. When the final elements are working properly, final check with the PLC can be accomplished, and the control station can be switched to automatic mode for direct control of the process from the PLC.

The MCS-7200 can also be used in applications that require signal level modifications of some combination of analog outputs, such as the addition of more or less ingredients in a batching process. This control station is very helpful in the event of any PLC component failure. After the process is stabilized with the MCS-7200 in the manual mode, the PLC failure can be diagnosed and corrective measures can be taken to put the system back in automatic mode. Figure 7-17 illustrates a typical configuration of the use of the MCS-7200 as a back-up unit.

Chapter

—8—

SPECIAL FUNCTION I/O AND SERIAL COMMUNICATION INTERFACING

"Still glides the stream, and shall for ever glide;
The form remains, the function never dies."

—William Wordsworth

8-1 INTRODUCTION

In the previous chapters we have discussed input and output interfaces which are used in perhaps 90 to 95% of PLC applications. Although these I/O systems (discrete and analog) allow control implementation for most applications, there are processes and PLC implementations which require special types of signals. The handling of these special signals by the processor, without special interface pre-conditioning, would be impossible to implement.

Some special I/O interfaces, called *preprocessing* modules, condition low-level signals and fast speed input signals that cannot be interfaced with standard I/O modules. Other special function I/O modules incorporate on-board microprocessors to add intelligence to the interface. These intelligent modules can perform complete processing tasks independent of the PLC's processor (and program scan). The method of allocating various control tasks to I/O interfaces is known as *distributed I/O processing*.

Special I/O modules are available along the whole spectrum of PLC sizes from small controllers to very large PLCs and in general are compatible throughout a PLC family of controllers. In this section we will discuss the most commonly found special interfaces.

8-2 SPECIAL DISCRETE INTERFACES

Fast Input

The fast response input interface is used to detect input pulses of very short duration. Certain devices generate signals that are much faster than the PLC scan time and cannot be detected through regular I/O modules. The fast response input interface operates as a *pulse stretcher* to enable the input signal to remain valid for one scan. If the controller has immediate input capabilities, it can be used to respond to a fast input that could initiate an interrupt routine in the control program.

The input voltage range is normally between 10 and 24 VDC for a valid ON (1) signal, and the logic data signal can be activated by the leading or trailing edge of the triggering input (see Figure 8-1). When the input is triggered, the interface stretches the input signal and makes it available to the processor. Filtering and isolation are also provided; however, the filtering causes a very short input delay since the normal input devices connected to this interface do not have contact bounce. Typical devices include proximity switches, photoelectric cells, or instrumentation equipment to provide pulse signals with durations ranging between 50 to 100 microseconds. Connections to the fast input module are the same as for those of standard DC input modules. Depending on the module, the field device must meet the sourcing or sinking requirements necessary to the interface for proper operation. In general, field devices must source a required amount of current to the input module at the rated DC voltages.

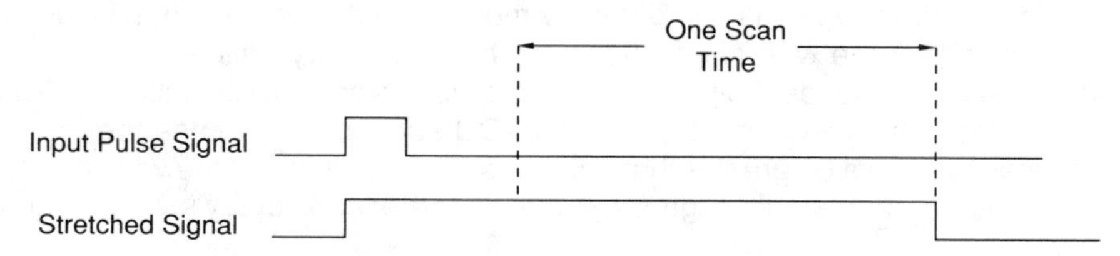

Figure 8-1. Pulse stretching for the fast input module.

Wire Input Fault

The wire input fault module is a special input interface designed to detect short circuit or open circuit connections between the module and the input devices. This module operates similarly to the standard DC input module in that a signal is detected and passed to the processor for storage in the input table. However, the module contains special provisions to detect any malfunction associated with the connections. Figure 8-2 illustrates a simplified block diagram of Allen-Bradley's wire input fault module. Typical applications of this module include critical input connections that must be monitored for correct wiring and field device operation.

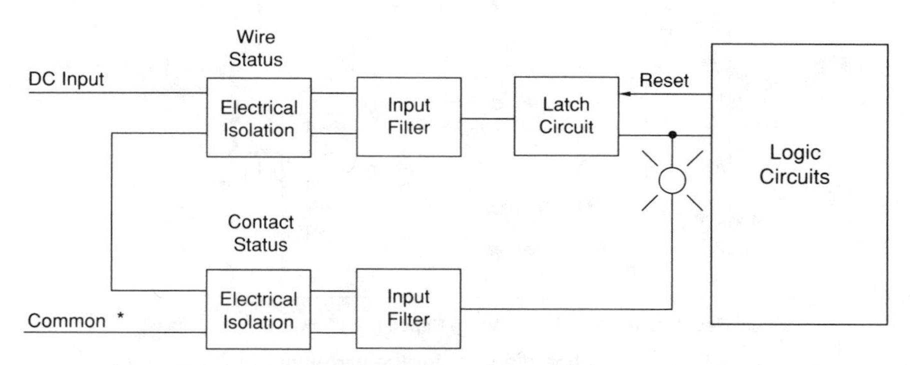

Figure 8-2. Block diagram of the wire input fault module.

The wire fault interface detects a short circuit or open circuit wire by sensing a change in the current. A 6 mA current is provided for each input to pass through a shunt resistor (placed across the input device) when the input is OFF (logic 0). A 20 mA current is provided when the input is ON (logic 1) and there is no problem in the wiring. An opened or shorted input will disrupt the monitoring current and a wire fault is detected by the module. A fault is annunciated by the module by flashing the corresponding status LED. The fault can also be detected by the control program where appropriate preprogrammed action can take place.

Figure 8-3 illustrates a typical connection diagram for the wire input fault interface. Note that shunt resistors must also be connected even though an input device is not wired to the module. The value of the shunt resistor is dependent on the DC power supply voltage level used. This supply may range from 15 to 30 VDC. An important consideration that must be kept in mind is that the total wire resistance of the connecting wire must not exceed a specified ohm rating for a DC supply voltage level.

This wire resistance value is computed by multiplying the per foot ohm value times the total length of the wire connection. It is however, very unlikely to exceed the maximum wire resistance. For instance, a wire size 14 has a resistance of 0.002525 ohms per foot which if connected to a 15 VDC supply, in this case the total wire resistance should not be greater than 25 ohms. This implies that the wire should not exceed 9,900 feet in length. Figure 8-4 shows the Allen Bradley's wire input fault module.

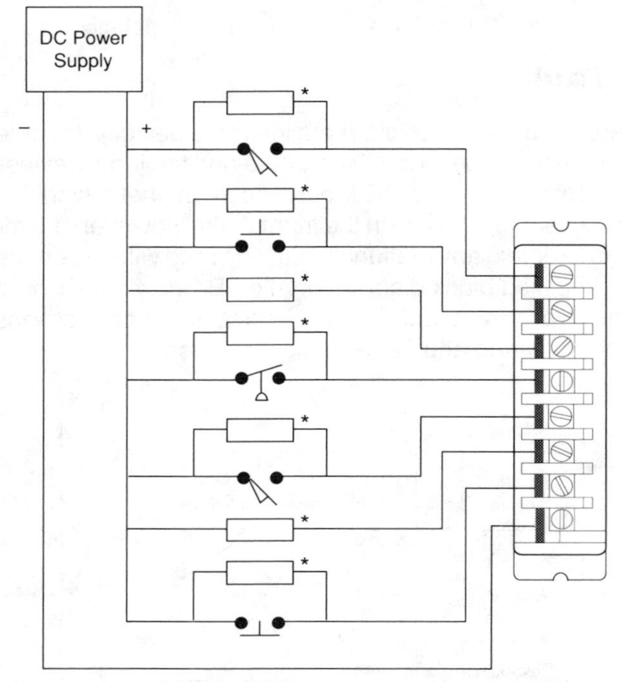

* Shunt Registers 1/2 Watt Rating. Value Depends on Power Supply Voltage.

Figure 8-3. Connection diagram for the wire input fault module.

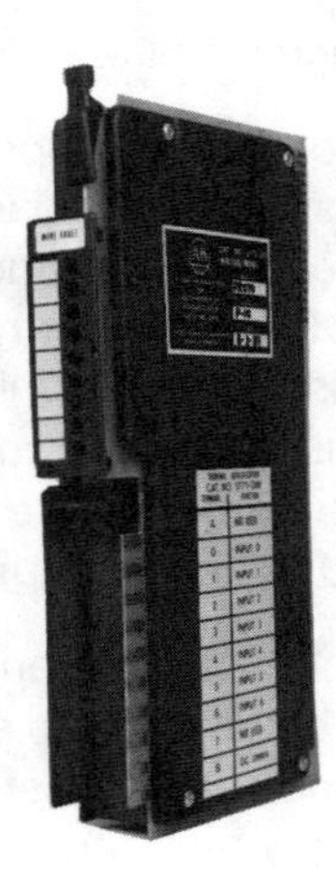

Figure 8-4. Wire input fault module (Courtesy Allen Bradley, Highland Heights, OH.)

Fast Reponse

The fast response interface can be described as an extension of the fast input module. This interface detects a fast input and responds with an output. The speed at which this occurs could be as short as 1 msec from the sensing of the input to responding with an output. The output response is independent of the PLC processor and the scan time.

This module has advantages that include the ability to respond to very fast input events which require an almost immediate output response. For instance, the detection of a jam of a feeder in a high speed assembling or transporting line may require a disengage signal of the product feed, thus reducing the amount of product being jammed and maybe lost.

The module, during operation, usually receives an enable signal from the processor (through the control program) which sets the module ready for "catching" the fast input. Once the module is active and a signal is received, the output is activated and remains ON until the processor (via the ladder program) disables the module, thereby resetting the output. Figure 8-5 illustrates a block diagram operation of this interface and the logic and timing it implements.

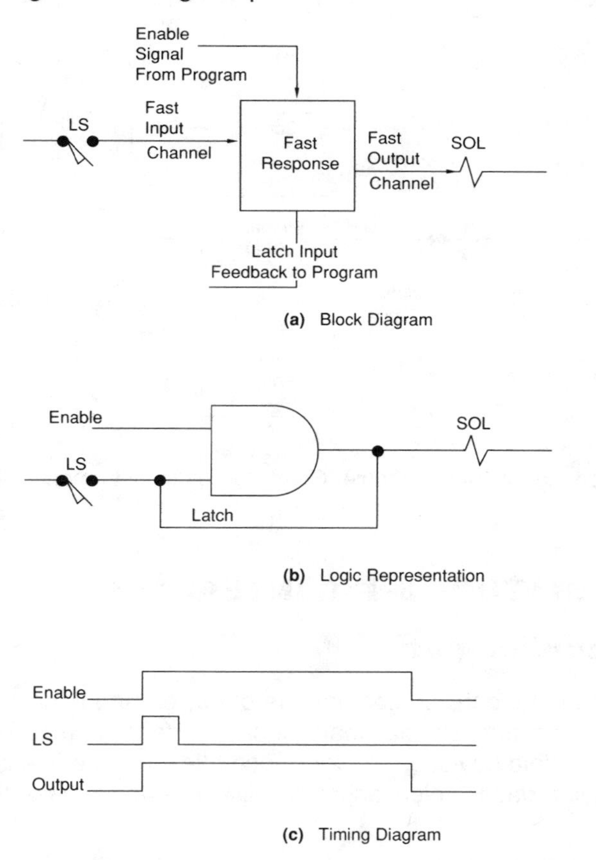

Figure 8-5. Block diagram operation of the fast response interface.

Redundant Output

The redundant output module, although seldom provided by most manufacturers, offers a tremendous advantage in applications in which an output module's malfunction is intolerable. The module upon detection of a failure caused by an open or shorted triac, a blown fuse, or any other failure would switch automatically all the primary circuits to the back-up or secondary circuits.

Figure 8-6 illustrates a block diagram operation of the Texas Instrument 110VAC redundant output interface. The operation of the redundant module is similar to that of the standard output module. The processor however, is capable of obtaining the status of the output module via the ladder control program. In case of a second failure the processor can, through the program and feedback information, take precautionary steps about the process or machine being controlled.

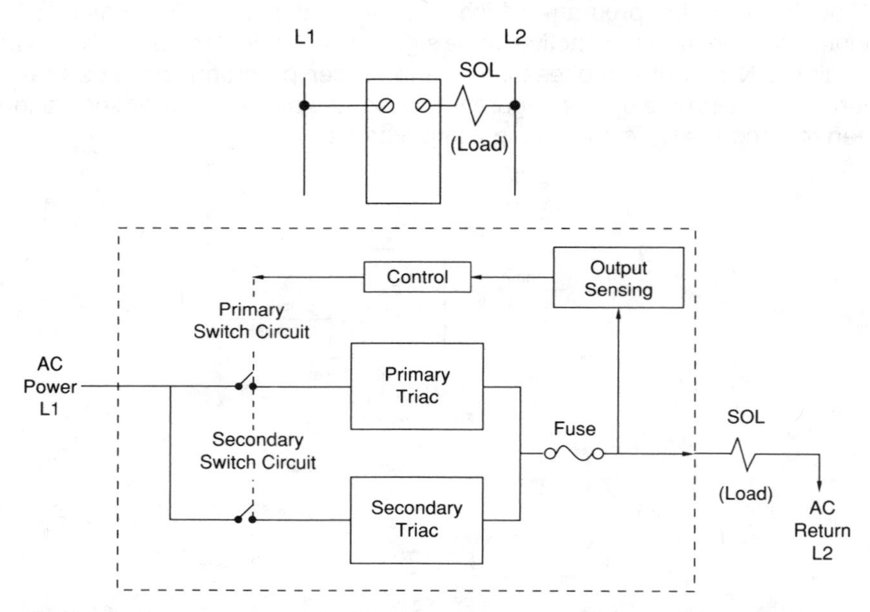

Figure 8-6. Redundant output module (Courtesy Texas Instruments, Johnson City, TN).

8-3 TEMPERATURE & PID INTERFACES

Thermocouple Input

In addition to the standard analog voltage or current input interfaces that receive signals from transmitters, special analog input interfaces can accept signals directly from the sensing field devices. The thermocouple input module is an example of a preprocessing interface which accepts millivolt signals from the thermocouple transducers.

Depending on the thermocouple used, different types of thermocouple interface modules are available. This module may be able to interface to several types of thermocouples by selecting jumpers or rocker switches in the module. For example, a module may be offered with capability to interface to thermocouples (ISA standard) type E, J, and K, where the type of thermocouple is used for a particular measurement range. In Chapter 12 we list some of the ranges, types and application comments for the most commonly found thermocouples.

The operation of the themocouple module is very similar to the standard analog input interface. The input signal (in millivolts) is amplified, digitized and converted into a digital signal which may be in binary or BCD format. Depending on the manufacturer, the number converted will represent, in binary or BCD, the degrees Celsius or Fahrenheit being measured by the thermocouple selected.

The thermocouple module does not provide a range of counts proportional to the temperature because thermocouples exhibit nonlinearities along its range usually between 0°C and their upper temperature limits. The module itself generally contains an on-board microprocessor which calculates the appropriate degrees (°C or °F) that correspond to the millivoltage read. This is accomplished by referencing a chosen thermocouple table (millivolts versus °C or °F) and performing linear interpolations (see thermocouples in Chapter 12).

Thermocouple interfaces usually provide *cold junction* compensation to the thermocouple (device) readings. This compensation available in the module allows the thermocouple to operate as though there were an ice-point-reference (0 °C) since all thermocouple' s tables depicting the generation of emf are referenced at this point. In addition to this cold junction compensation, the module generally provides what is called *lead resistance compensation* for a determined resistance value.

Lead resistance deals with the loss of signal due to the resistance present in the wires. This resistance value can be obtained from the thermocouple manufacturer for a given wire size length at a known temperature. Thermocouple interfaces may provide different lead resistance compensations depending on the PLC manufacturer. One manufacturer may provide 200 ohms of compensation while another may provide 100 ohms. If the lead resistance is greater than the available compensation, a calculation can be made to add °C in the control program to compensate for the resistance. When possible, it is a good practice to use the same type of lead wire material of which the thermocouple is made. Smaller gauge wire provides a slightly faster response, but heavier gauge wire tends to last last longer and resists contamination or deterioration at high temperatures. Figure 8-7 shows typical thermocouple interface connections.

The following example illustrates a case where compensation is performed. It might come in handy in applications where very long lead wires are employed or when several thermocouples are connected in parallel thus creating long lead wires to measure average readings.

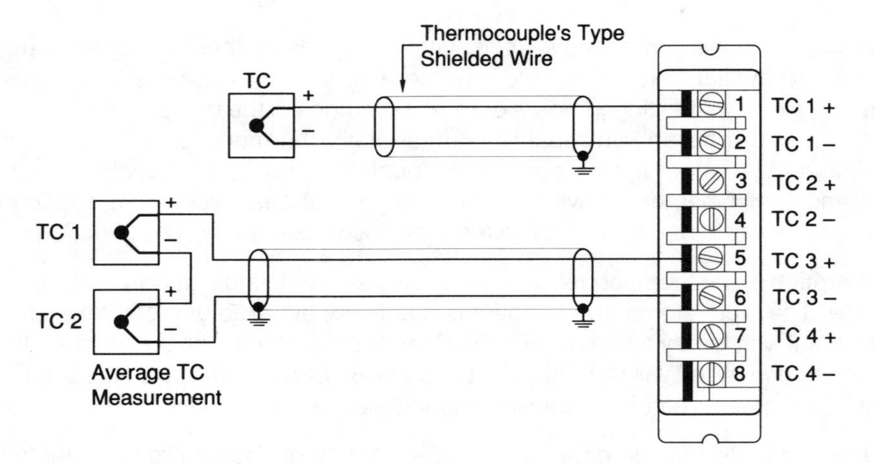

Figure 8-7. Thermocouple interface connections.

Example 8-1

A type J thermocouple is to be connected to a thermocouple module placed in an I/O rack located 500 feet away. This thermocouple is connected to a heat trace circuit which measures temperature ranges throughout the length of a process pipe. Find the total lead resistance and the necessary compensation in degrees Celsius to be added in the program to the value measured. The thermocouple has 18 AWG lead wires which have a resistance of 0.222 ohms for each foot of double wire (positive and negative wire conductors) at 25°C. The thermocouple module used has a lead resistance compensation of 50 ohms and the manufacturer has a 0.05 °C per ohm compensation error factor.

Solution

The total lead resistance is computed as:

Total Lead Resistance

= Thermocouple lead resistance x Length of lead wire

= 0.222 x 500

= 111 ohms

The compensation requirement will be the difference between the module's compensation and the total resistance times the compensation error factor:

compensation in °C

= (111 - 50 ohms) x 0.05°C/ohm

= 3.05°C

The required compensation that would be added to the thermocouple reading would have the value of 3.05 in °C.

RTD Input

The Resistance Temperature Detector (RTD) interface receives temperature information from RTD devices. RTDs are temperature sensors that generally have a wire-wound element whose resistance changes with temperature in a known and repeatable manner. RTDs in their most common form consist of a small coil of platinum, nickel, or copper protected by a sheath of stainless steel. These devices are frequently specified for temperature sensing because of their accuracy, repeatability and long term stability.

The operation of this module is similar to other analog input interfaces. This module is designed to send a small (mA) current through the RTD and read back the resistance to the current flow. In this manner, the module can measure the changes in temperature since the RTD changes resistance with temperature changes.

The module converts the resistance changes into temperature and has this value available for the processor in either °C or °F. Other interfaces may, in addition to the temperature measurement, be able to provide resistance value in ohms to the processor. Depending on the manufacturer, the module may be able to sense more than one type of RTD. Table 8-1 lists some of the most common RTD devices and their resistance used in this type of interface.

Type RTD	Resistance Rating (ohms)	Temperature Range	
Platinum	100	-200 to 850°C	-328 to 1562°F
Nickel	120	-80 to 300°C	-112 to 572°F
Copper	10	-200 to 260°C	-328 to 500°F

Table 8-1. Commonly found RTD types.

RTD sensing devices are available in 2,3, or 4-wire connections. Devices with a 2-wire scheme do not provide a means for compensating for lead resistance, while the 3 and 4-wire RTDs allow for lead resistance compensation. The most commonly used RTD device is the 3-wire which is used in applications requiring long lead wires where wire resistance is significant in comparison with the ohms/°C sensitivity of the RTD element. It is good practice to try to match the lead resistance of the wires by using quality cabling and heavy gauge wires (16-18 gauge). Figure 8-8 illustrates a typical connection representation for an RTD module (2,3 and 4 wire RTDs).

PID Module

The Proportional-Integral-Derivative interface is used in process applications in which closed-loop control employing the PID algorithm is required. This module provides proportional, integral, and derivative control action from sensed parameters, such as pressure and temperature, which are the input variables to the system. The PID interface is typically applied to any process control operation which requires continuous closed-loop control. PID control is often referred to as a "three-mode closed-loop feedback control."

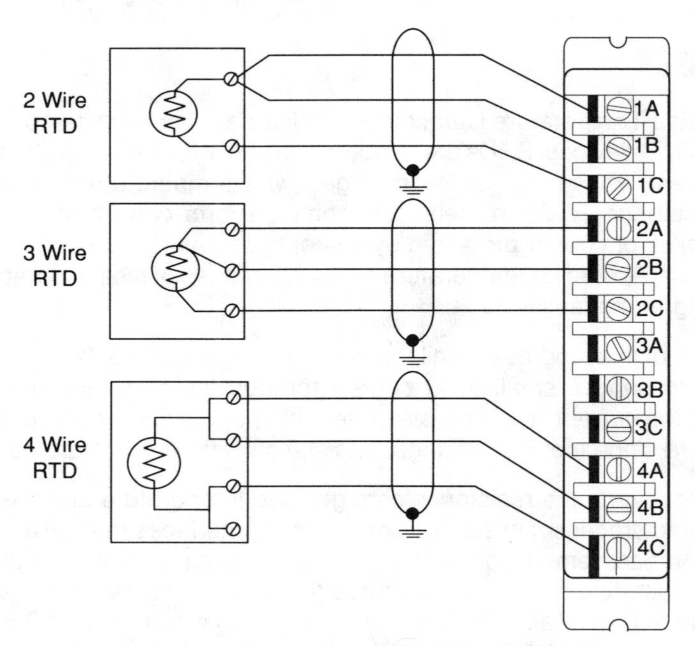

Figure 8-8. Connection diagram for the RTD module.

The basic function of closed-loop process control is to maintain certain process characteristics at desired set points. Generally, the process deviates from the desired set point reference as a result of load material changes and interaction with other processes. During control, the actual condition of the process characteristics (liquid level, flow rate, temperature, etc.) is measured as a process variable (PV) and compared with the target set point. When deviations occur, an error is generated by the difference between the process variable (actual value) and the set point (desired value). Once an error is detected, the function of the control loop is to modify the control variable (CV) output in order to force the error to zero.

The control algorithm implemented in the module is represented by the following equation:

$$\text{Vout} = K_P E + K_I \int E\, dt + K_D \frac{dE}{dt}$$

where

K_P is the proportional gain

$K_I = \frac{K_P}{Ti}$ is the integral gain (Ti = reset time)

$K_D = K_P$ Td is the derivative gain (Td = rate time)

E = PV - SP is the error

V_{out} is the control variable output

The module receives the process variable in analog form and computes the error difference. The error difference is used in the algorithm computation to provide corrective action at the control variable output. The functionality of the control action is based on an output control, which is proportional to the instantaneous error value. The integral control action (reset action) provides additional compensation to the control output, which causes a change in proportion to the value of the error over a period of time. The derivative control action (rate action) adds compensation to the control output, which causes a change in proportion to the rate of change of error. These three modes are used to provide the desired control action in Proportional (P), Proportional-Integral (PI), or Proportional-Integral-Derivative (PID) control fashion.

The information that is sent to the module from the main processor is primarily the control parameters and set points. Depending on the module used, data can be sent to describe the *update time*, which is the rate or period in which the output variable is updated, and the *error deadband*, which is a quantity that is compared to the error signal. If the error deadband is less than or equal to the signal error, no update takes place. Some modules provide square root calculations of the process variable, which is used to perform a square root extraction to obtain a linearized scaled output for use by the PID loop. An example application in which the square root extractor can be used is in the PID control of flow. Other parameters, such as maximum error and maximum/minimum control variable output for high and low alarms, can also be transmitted to the module if these signals were provided. During operation, the PID interface maintains status communication with the main CPU, exchanging module and process information. Figure 8-9 illustrates a block example of the PID algorithm for this interface and a typical module connection arrangement.

8-4 POSITIONING INTERFACES

Positioning interfaces are intelligent modules that are used in applications requiring position related feedback information and control output to perform the control of machine axes. Much can be said about axis positioning (motion control) which could serve to fill an entire book. This section covers basic aspects of motion control as they relate to PLC applications that will prove to be very beneficial.

Today, the capabilities of these modules have allowed some programmable controllers to perform functions such as point-to-point control and axis positioning, using servo mechanisms that once required computer numerical control (CNC) machines.

Encoder/Counter

The encoder/counter module provides interface between encoders or high speed counter devices and the programmable controller. This type of module generally operates independently of the processor and I/O scan. The encoder/counter module is an integral part of a programmable controller system when it is used in applications requiring position information. Such applications include closed-loop positioning of machine tool axes, hoists, and conveyors, as well as cycle monitoring of high speed machines such as can making equipment, stackers, and forming equipment.

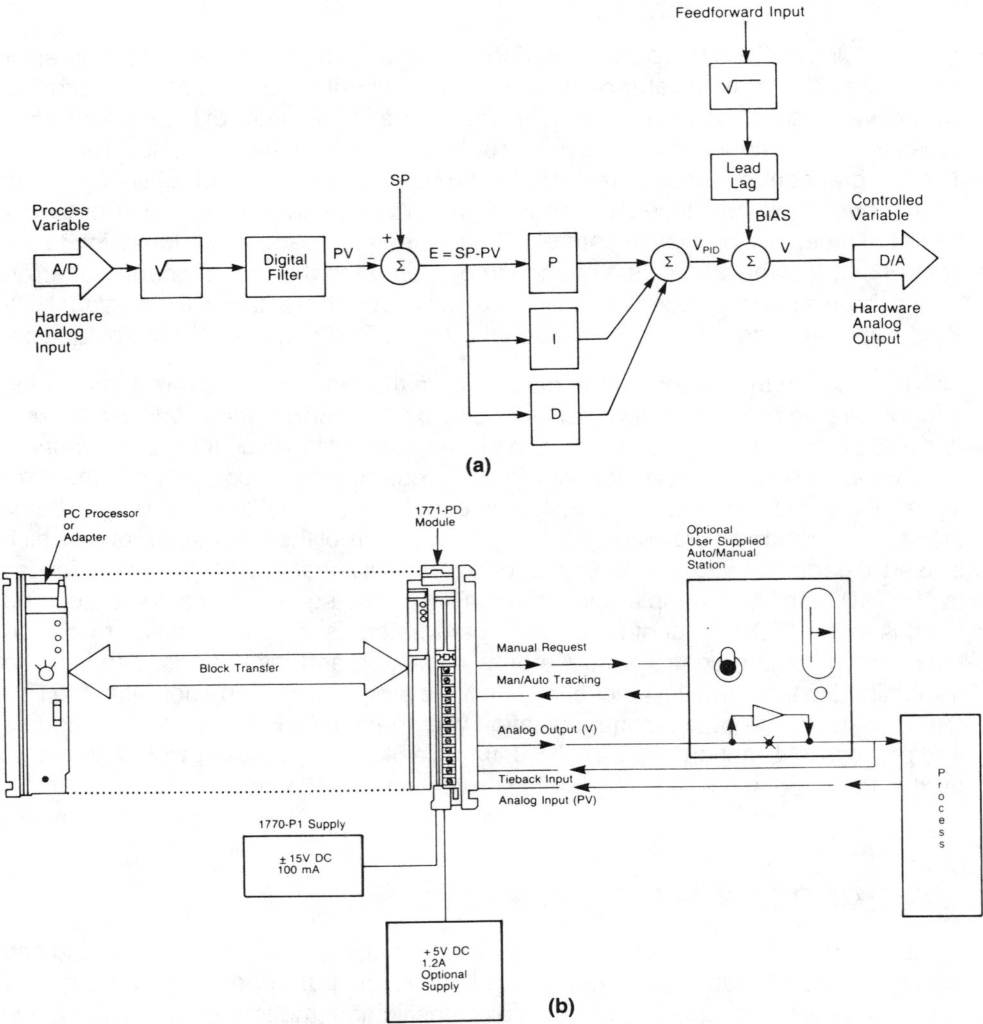

Figure 8-9. a) Block diagram of the PID algorithm, and **b)** connection arrangement for Allen Bradley's 1771-PID module (Courtesy of Allen Bradley, Highland Heights, OH).

The encoder/counter interface may be available in an *incremental* or *absolute* type of encoder. Incremental encoders provide measurement of shaft rotation over distance by outputting a fixed number of pulses per shaft rotation. There are two pulse signals which have a 90° phase difference (quadrature); the direction of rotation is determined by which of the two pulse channels is the leading waveform. Absolute encoders provide angular measurement of the shaft. This angular position is provided in parallel to the encoder interface module by the encoder in BCD, binary, or gray code. Incremental encoders provide a *marker* or *index* channel which sends a pulse for every revolution. This marker, which is an input to the module, can be used in conjunction with the module's *limit switch* channel input to establish a home position along the encoder's measurements. When the encoder interface is used in a counter configuration, only one of the input channels is connected to the device that provides the pulse count.

During operation, the module receives two pulse channel inputs (incremental encoder mode) which are counted and compared with a user specified preset value. The interface can have one or two output lines available which may be energized once the incoming pulses are equal to, greater than or less than the preset values. The input channels and outputs available are generally rated for TTL or 12-48 VDC ratings. The maximum input pulse frequency that can be properly counted ranges between 50Khz and 60Khz.

The communication between the encoder/counter interface and the processor is bi-directional. The module accepts the preset value and other control data from the processor and transmits values and status data to the PLC memory. The output controls are enabled from the control program, which tells the module to operate the outputs according to the count value received. The processor, also through the program, enables and resets the counter operation. The interface can let the PLC know when the marker and limit switch are both energized, indicating a home position.

Typically, the maximum length between the module and the encoder should not exceed 50 feet, and shielded cables are generally used. Since the module has both inputs and outputs, isolation is provided between the input and output circuits as well as between the control logic and both I/O circuits. The isolation is enhanced by the use of separate power supplies that must be provided by the user. Figure 8-10 shows a typical connections for an incremental encoder configuration.

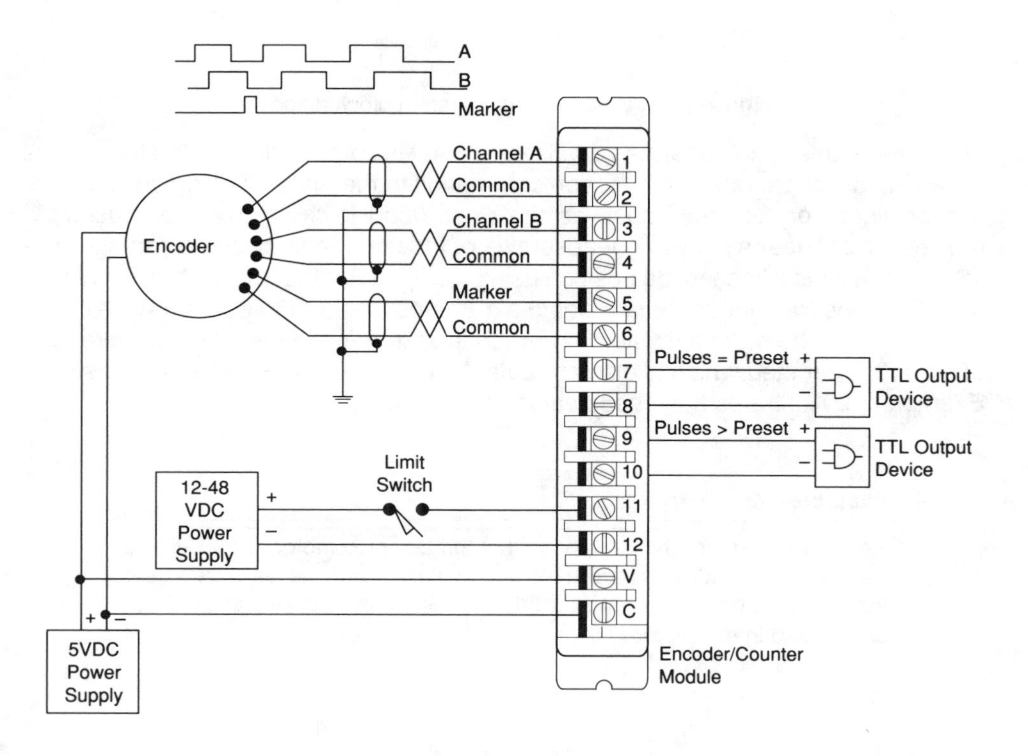

Figure 8-10. Encoder Connection Diagram.

Stepper Motor

The stepper motor interface, as its name implies, is used in applications requiring control of stepper motors. These steppers are permanent type magnet motors and are used to translate incoming pulses, through a stepper translator, into mechanical motion. The desired motion of a stepper can be accelerated, decelerated, or maintained constantly by controlling the pulse rate output from the stepper module. "Stepper" is a generic term which describes this member of a brushless motor capable of making fixed angular motion in response to a step input. The ability to respond to an input voltage (in the form of DC pulses) makes the stepper motor well suited for incremental motor programmable control systems. The stepper motor follows under controlled conditions the number of input pulses. This ability to respond to a fixed input enables the system to operate in an *open loop* mode, obviously representing cost savings in the total system. However, in high response applications, closed-loop operation is generally necessary (using encoder feedback). Figure 8-11 illustrates a simplified block diagram of a stepper motor system.

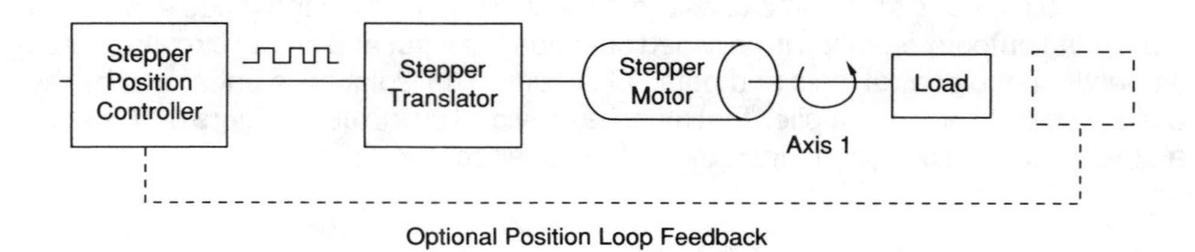

Figure 8-11. Stepper motor system block diagram.

The stepper interface generates a pulse train that is compatible with the translators indicating distance, rate and direction commands to the motor. The motion induced can be linear or rotational such as the ones used in the forward or backward movement of linear slide using leadscrews or indexing table positioning. Figure 8-12 shows a typical linear slide position using a step motor that makes one revolution per 200 steps (resolution) thus yielding a 1.8° step angle (360/200 or one 200th of a revolution). The stepper system shown in the figure provides a linear movement of 0.00125" per step which is also attributable to the 4 threads per inch leadscrew. Example 8-2 illustrates how these values are obtained.

Example 8-2

Referencing Figure 8-12, suppose that the 200 step motor is operating at *half stepping* conditions (400 steps per revolution) and the leadscrew used has 5 threads per inch; what is the linear displacement per step and the step angle used in the system?

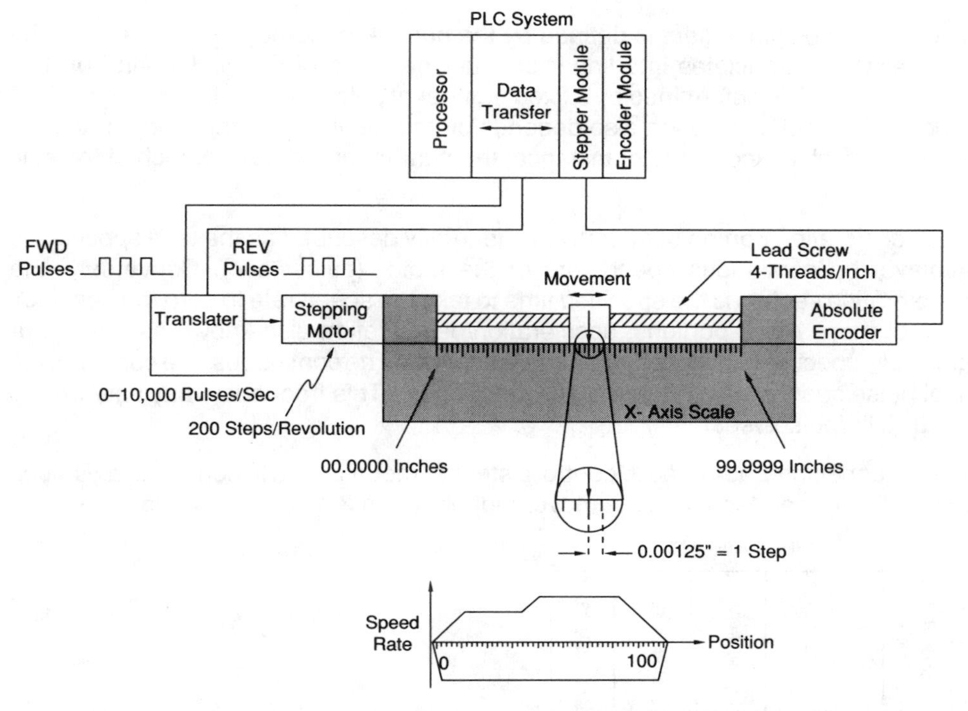

Figure 8-12. Typical linear slide block diagramming a PLC system.

Solution

The step angle is computed by dividing the number of degrees in one revolution (360°) by the number of steps required to turn the motor; therefore the step angle is:

$$\textit{step angle} = \frac{360^\circ}{400} = 0.9^\circ$$

with a resolution of one 400th of a revolution. The linear displacement is computed by taking the number of threads it takes to move one inch; each thread requires one revolution (rotational to linear displacement). Therefore to move one inch it requires five revolutions; each revolution requires 400 steps.

$$\textit{1'' travel} = 5 \text{ rev} \times 400 \text{ steps/rev} = 2000 \text{ steps}$$

Therefore

$$\textit{1 step} = \frac{1}{2000} = 0.0005 \text{ inch}$$

The position displacement is defined by the number of outburst pulses sent to the stepper which translates into linear or rotational units of travel. The end or final position is usually determined by a fixed number of pulses sent to the motor from the module. The actual location also depends on the resolution of the stepper and the application which includes, for instance, the number of threads per inch of travel in a leadscrew.

The acceleration portion of the move is generally described as the time specified to achieve the continuous speed rate of the motor (pulses/sec). Conversely, the deceleration section is the specified time to take the speed rate to zero (pulses/sec). Both of these move portions, acceleration/deceleration also known as *ramps* are generally specified as a function of time (seconds). The continuous rate concerns the final pulse/sec rate sent to the motor (frequency). This frequency may vary from 1 to 20 KHz (pulses/sec).

Each stepper interface used to control a stepper motor is said to control an *axis* since the motion generated causes a movement about an X,Y, or Z axis (Figure 8-13).

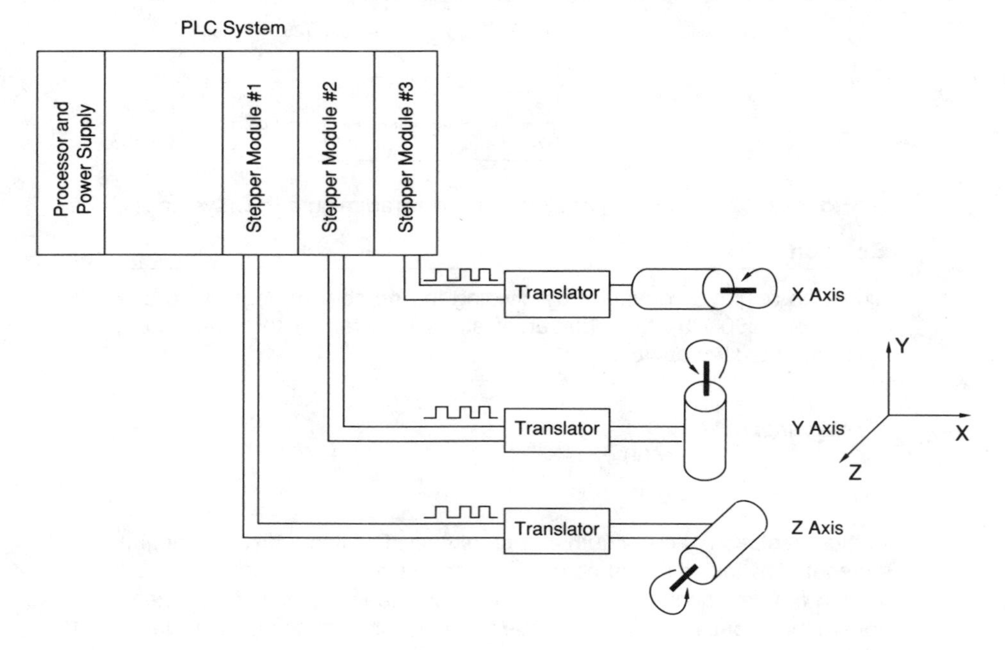

Figure 8-13. PLC system using steppers to control 3 axes.

These stepper motor interfaces generally operate in two modes which are the *single-step profile mode* and *continuous profile mode.* Under the single-step mode, the PLC processor sends individual move sequences to the interface. These sequences include the acceleration, final or continuous speed rate, and deceleration rates of the move (see Figure 8-14). Once this move sequence is terminated, another one may be started by the processor by transferring the next move's profile information and commands. Several single-step modes may be stored in the processor and sent to the module under PLC program control.

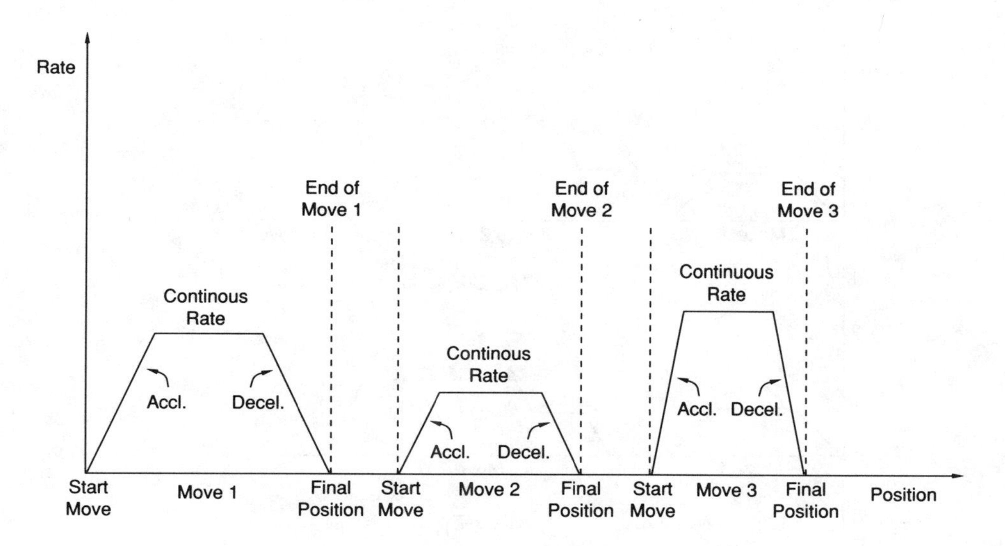

Figure 8-14. Step profile mode.

The continuous mode is used when a motion profile is cycled through various accelerations, decelerations and continuous speed rates to form a blended motion profile (see Figure 8-15). Rather than require additional commands for the changes in motion speed, the interface receives the whole profile in a single block of instructions. The interface then performs the step motor control duty independently until the motion is completed and the next profile is received from the processor. As in the single-step, the processor can also store several continuous mode profiles in its memory and send them to the interface during the program execution.

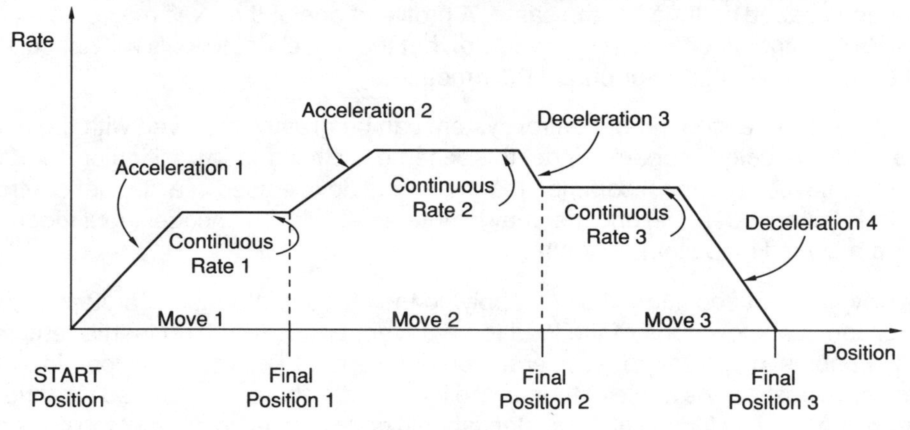

Figure 8-15. Continuous profile mode.

Depending on the PLC manufacturer, more than one axis may be controlled using several stepper module interfaces. When multiple axis motions are being implemented, they can be controlled independently or synchronized (see Figure 8-16a and b respectively). What this really means is that, under the independent mode, each axis is independent of the other, each executing its own profile mode whether it is single-step or continuous. The beginning and end of each axis motion may be different.

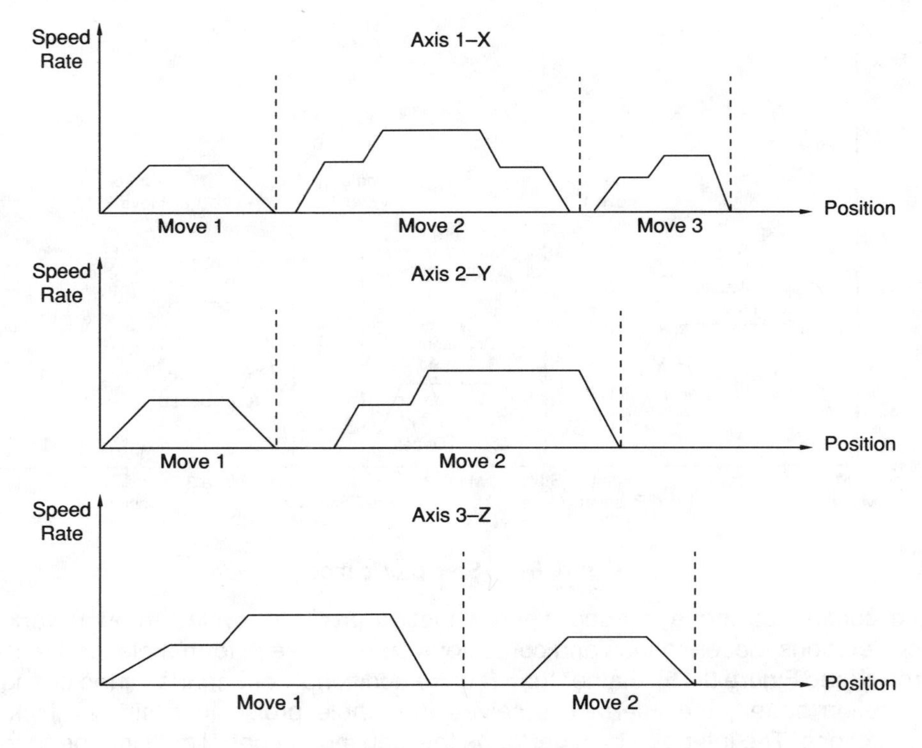

Figure 8-16. a) Independent mode.

Under the synchronized control of several axes, the beginning and end motion commands based on time are the same. A profile of one of the axes may start later or end before the other axes (Figure 8-16b), but the move that follows will not occur until all axes have started or ended their motions.

The operation of a step motor control system can be greatly improved with the use of a position/velocity feedback. Under this scheme, a closed-loop positioning control can be achieved. The most common feedback field device used in a stepper control system is the encoder. The encoder may be interfaced with an encoder input module to form a closed loop stepper control.

A worthy note to keep in mind when applying the stepper interface in a step motor application is a knowledge of the load to be driven. Loads with high inertia require large amounts of power to accelerate or decelerate. Therefore, proper inertia matching is desired. As a rule of thumb, the load inertia should not exceed ten times the rotor inertia. The friction of the system should be examined to prevent the system from being underdamped (not enough friction) or from losing position accuracy (too much friction).

Coupling mechanisms used to connect the stepper to its load include metal bands, pulley and cable, direct drive, and the leadscrew, which is applied mostly for linear actuation. Figure 8-17 illustrates a typical stepper interface connection with Jog Forward and Jog Reverse capabilities.

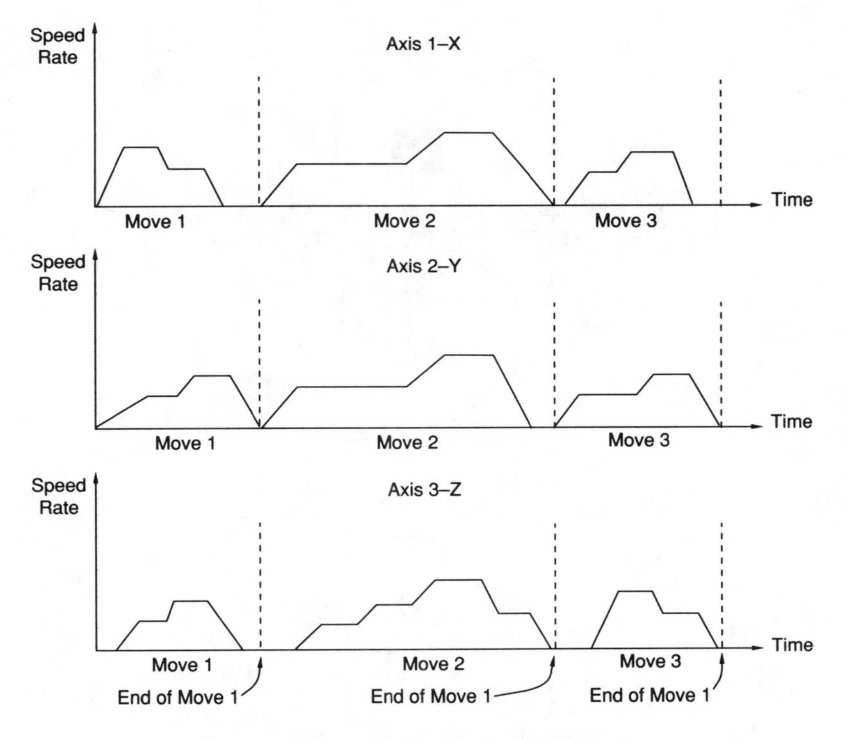

Figure 8-16. b) Synchronized mode.

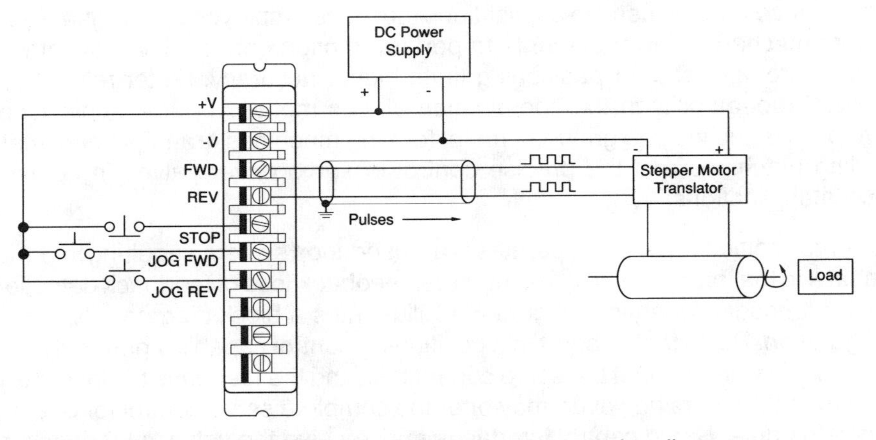

Figure 8-17. Typical stepper interface connection diagram.

Servo Motor

Servo interfaces are used in applications requiring control of servo motors via a servo drive controller. The servo module provides a ±10 VDC to the drive which defines the forward and reverse speed of the servo motor. This type of equipment is generally utilized when axis motion control is required. The motion along the axis may be linear or rotational. A common linear motion example is a leadscrew assembly which translates rotational movements from a servo motor into linear displacement (see Figure 8-18).

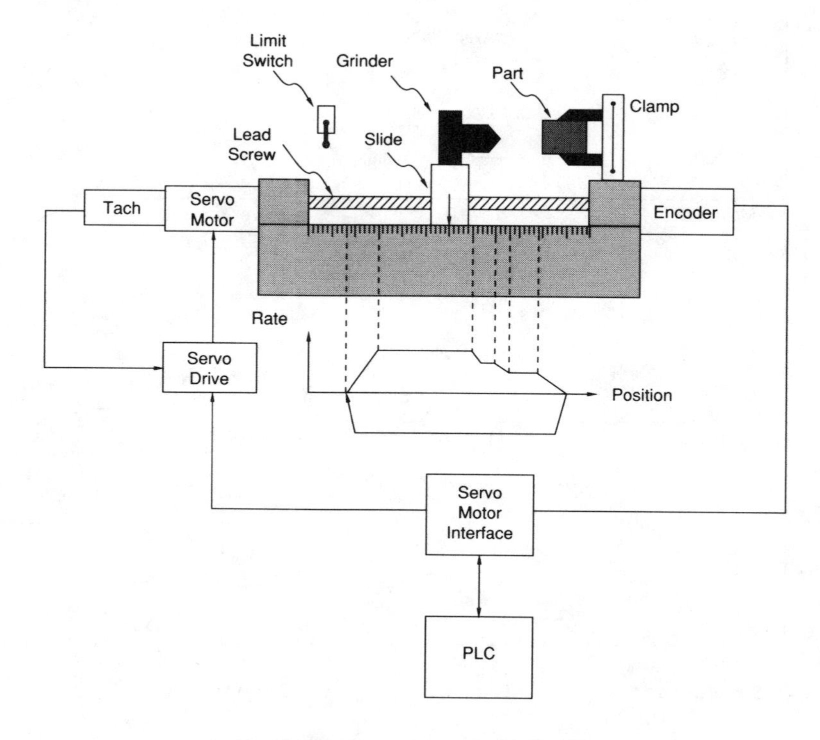

Figure 8-18. Servo control system block diagram.

This interface is being used in applications that once employed clutch-gear systems or other mechanical arrangements to perform motion control. The advantages of servo control are: shorter positioning time, higher accuracy, better reliability, and improved repeatability in the coordination of axis motion. Typical applications of servo positioning include grinders, metal forming machines, transfer lines, material handling machines and the precise control of servo driver valves in continuous process applications.

The servo positioning control operates in a closed-loop system, requiring information feedback in the form of velocity (tachometer feedback/loss of feedback detection) or position (encoder feedback). Figure 8-19 illustrates a block diagram servo control configuration. PLCs that support this positioning control capability generally require two modules to implement the servo control task and the feedback to close the loop. Some manufacturers however, may offer the complete servo control for one axis in a single module. Servo control interfaces may receive the velocity feedback in the form of a tachometer input, or positioning feedback in the form of an encoder input, or the module may provide both. The feedback signal provides the module with information concerning the actual speed of the motor and position of the axis so that it can be compared with the desired velocity and the desired position on the axis. If a difference is present, the module will correct its output until the error between the feedback and the set point velocity and position is zero.

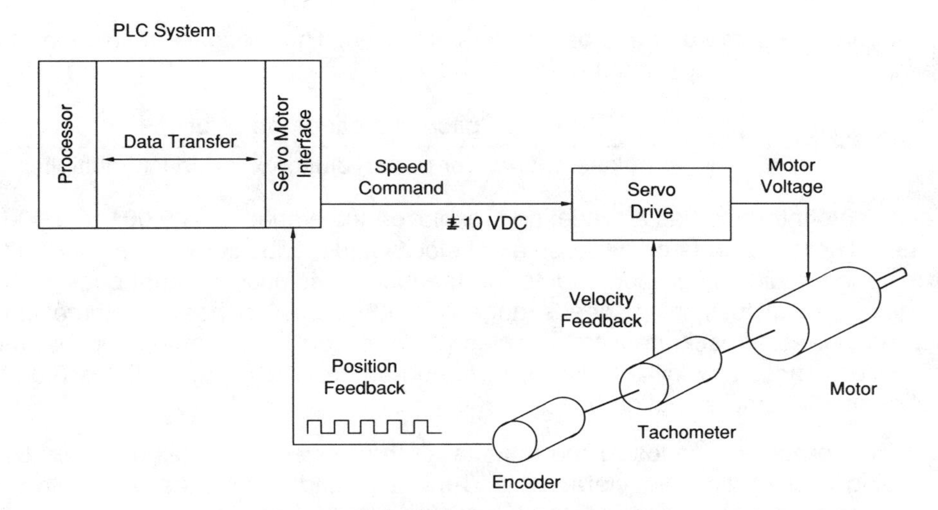

Figure 8-19. Servo control configuration.

Servo control, like the stepper motor, can be achieved in *single-step* positioning mode or *continuous* positioning mode (see Figure 8-20). The control of several axes (depending on the manufacturer) can also be accomplished for synchronization of the axes in either single-step or continuous modes.

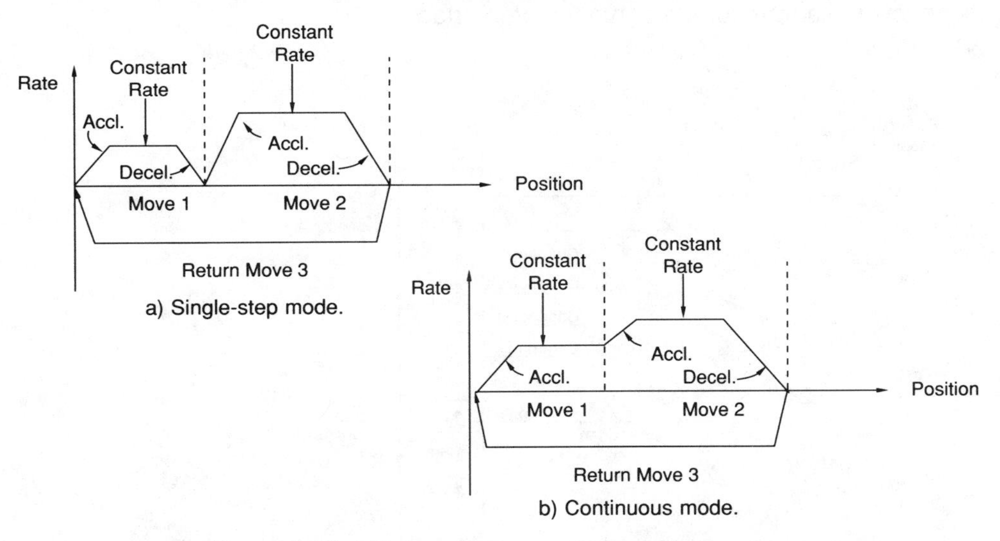

Figure 8-20. 2 Types of servo control position modes.

An important thing to keep in mind when using servo interfaces for positioning control is to understand the feedback resolution provided by the system. For instance, if a leadscrew (motion translator from rotational to linear) is used for the axis displacement and an encoder is employed to feed a signal back to the servo module, a knowledge of leadscrew pitch, the number of encoder pulses per revolution, and the multiplier value in the encoder section of the interface must be required. The multiplier selection is available in some interfaces and it allows better feedback resolution without changing the encoder. An example at the end of this section will show you how some of these parameters are used. The feedback resolution of a servo positioning (linear) can be defined as:

$$\text{feedback resolution} = \frac{\text{pitch of motion translator}}{\text{pulses of encoder per revolution x feedback multiplier}}$$

Each servo interface has a predefined resolution that varies from 0.001 to 0.0001 inches. There is a trade-off between axis velocity and feedback resolution since the axis speed is directly proportional to the feedback resolution brought back to the module. Typical positioning speeds range from 500 to 1000 inches per minute (ipm) and encoder feedback input frequencies of up to 250 KHz. Remember that the resolution or accuracy will diminish as the speed is increased i.e. resolution of 0.0001 inches at 450 ipm, resolution 0.001 at 900 ipm.

The PLC processor sends to the module all the move and position information including acceleration, deceleration, and the final and continuous feed velocity. When the module is operating, the processor monitors the status of the module without interfering with the module's complex and rapid calculations. A new move for an axis is updated from the processor whenever the previous move has been completed and the module is ready for the new profile. The acceleration and deceleration parameters are generally specified as speed in inches per minute per second (ipm/sec) at a specific resolution. Figure 8-21 illustrates a typical field connection diagram for the servo motor interface.

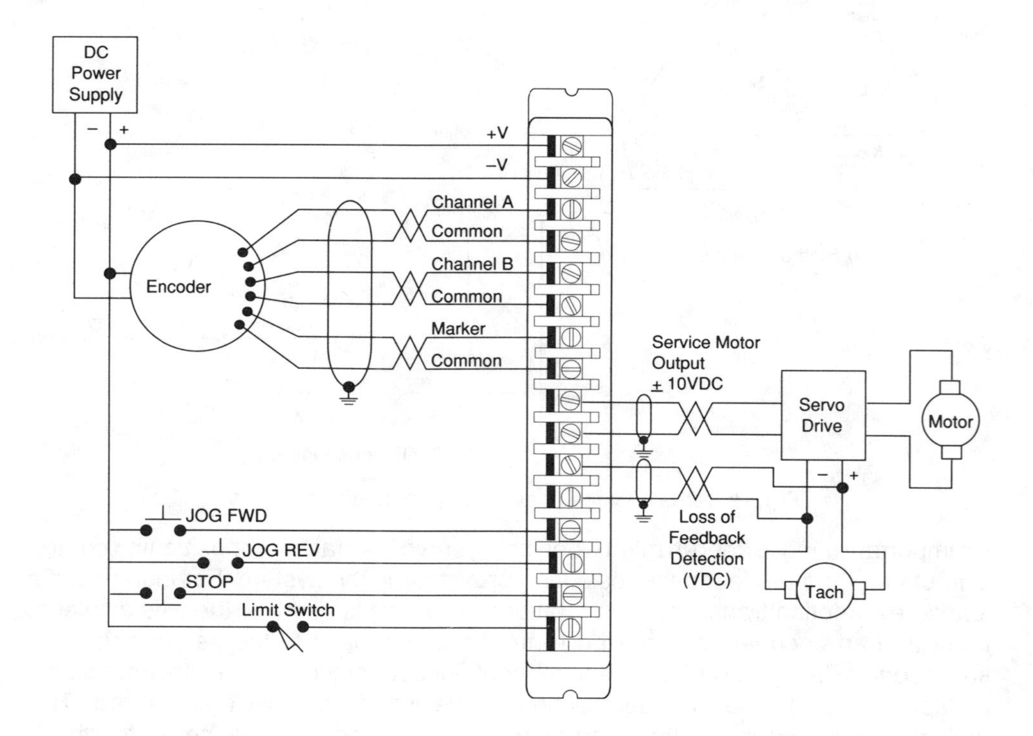

Figure 8-21. Field connections for a servo motor interface.

Example 8-3

A servo interface is being used in a PLC system to perform a one axis positioning of metal parts. This part will be machined at a defined profile which will be stored in the processor's memory. The part is moved along an x axis by means of a leadscrew which allows the axis to travel 1/8 inch per revolution. Position feedback is provided by a quadrature incremental encoder with a 200 KHz pulse frequency that provides 250 pulses per revolution. The encoder is connected to the encoder feedback terminal in the servo interface which provides a software programmable multiplier of (X = times) X1,X2, and X4 increments per pulse.

Find the feedback resolution that will be obtained and the number of pulses that will be received if the part travels 12.5 inches. What would be a way to double the feedback resolution without changing the encoder?

Solution

The feedback resolution is a function of the leadscrew pitch and the number of pulses per revolution generated by the encoder times the feedback multiplier. The leadscrew has a 1/8 inch pitch which means that the part will travel 0.125 inches for every rotation (see Figure 8-22).

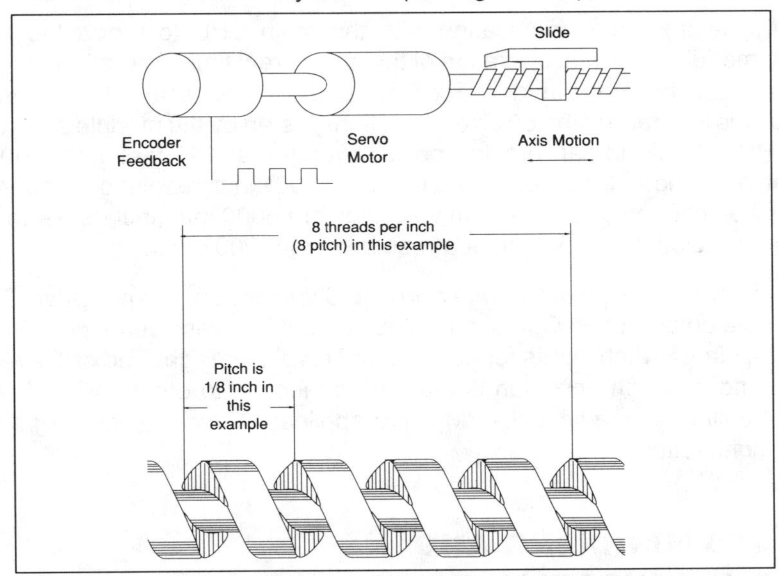

Figure 8-22. Lead screw (linear) displacement system.

The feedback resolution is therefore:

$$= \frac{0.125 \text{ inch/rev}}{250 \text{ pulses/rev} \times 1}$$

$$= 0.0005 \text{ inches}$$

A metal part moving 12.5 inches will generate a position feedback of:

$$= \frac{12.5 \text{ inches}}{0.0005 \text{ inches/pulse}}$$

$$= 25{,}000 \text{ pulses}$$

The resolution of 0.0005 inches (movement per encoder pulse) could be improved by using a multiplier of "2" which will make the revolution 0.00025 inches instead. This times 2 multiplier option allows the quadrature pulses (A and B) to both be counted, therefore yielding twice as many pulses in one rotation.

Axis Positioning

The Axis Positioning Module (APM), developed by General Electric for their Series Six programmable controllers, is another special interface that is used for servo position control, in addition to the standard sequencing and control capabilities. The APM interface utilizes an on-board microprocessor, with a 2.34 msec scan time, which controls axis positioning independent of the main CPU. Contrary to other servo control modules, the APM is a single module capable of receiving feedback information from resolvers and controlling the velocity of a servo-motor (through a servo drive) for each axis. This interface receives the set-up data parameters, such as velocity feedforward, position loop gain, and reversal compensation, during initialization at power up.

The APM maintains communication with the main CPU to report the status of machine, module, and total operation of the loop in real time. Alterations of the APM commands can be made under program control if necessary. One outstanding feature of this interface is the 60 error conditions tested by the module and sent back to the CPU. The APM can provide position ranges of 140 feet with 0.0001 inch resolution or a longer distance of 1400 feet with 0.001 inch resolution. Depending on the resolution chosen, the maximum velocities are 6000 ipm (inches per minute) at 0.001 inch resolution or 600 ipm at a resolution of 0.0001 inch.

This servo interface is provided with a ±10VDC velocity command, Drive OK input, Drive Enable output, Loop Contactor output, up to three resolver excitation outputs and inputs, Limit Switch inputs for Left End of Travel and Right End of Travel, Home Position, and a Synchronization output for connection to other APMs if required. Figure 8-23 illustrates a typical APM servo application and a block diagram of the APM I/O connections.

8-5 DATA HANDLING AND COMMUNICATION INTERFACES

In this section we discuss the most common intelligent modules which are used to accept and transmit data to other field devices. This data is handled in the form of ASCII characters, through another computing language, or through a proprietary media such as in the case of a network. Local and remote I/O processors fall into this category of communication interfaces since they communicate information to the PLC's subsystems. However, they are covered under processors in the remote I/O section since these modules also fall under the PLC processors category. This section will give you a look at how information can be transmitted or communicated to the real world.

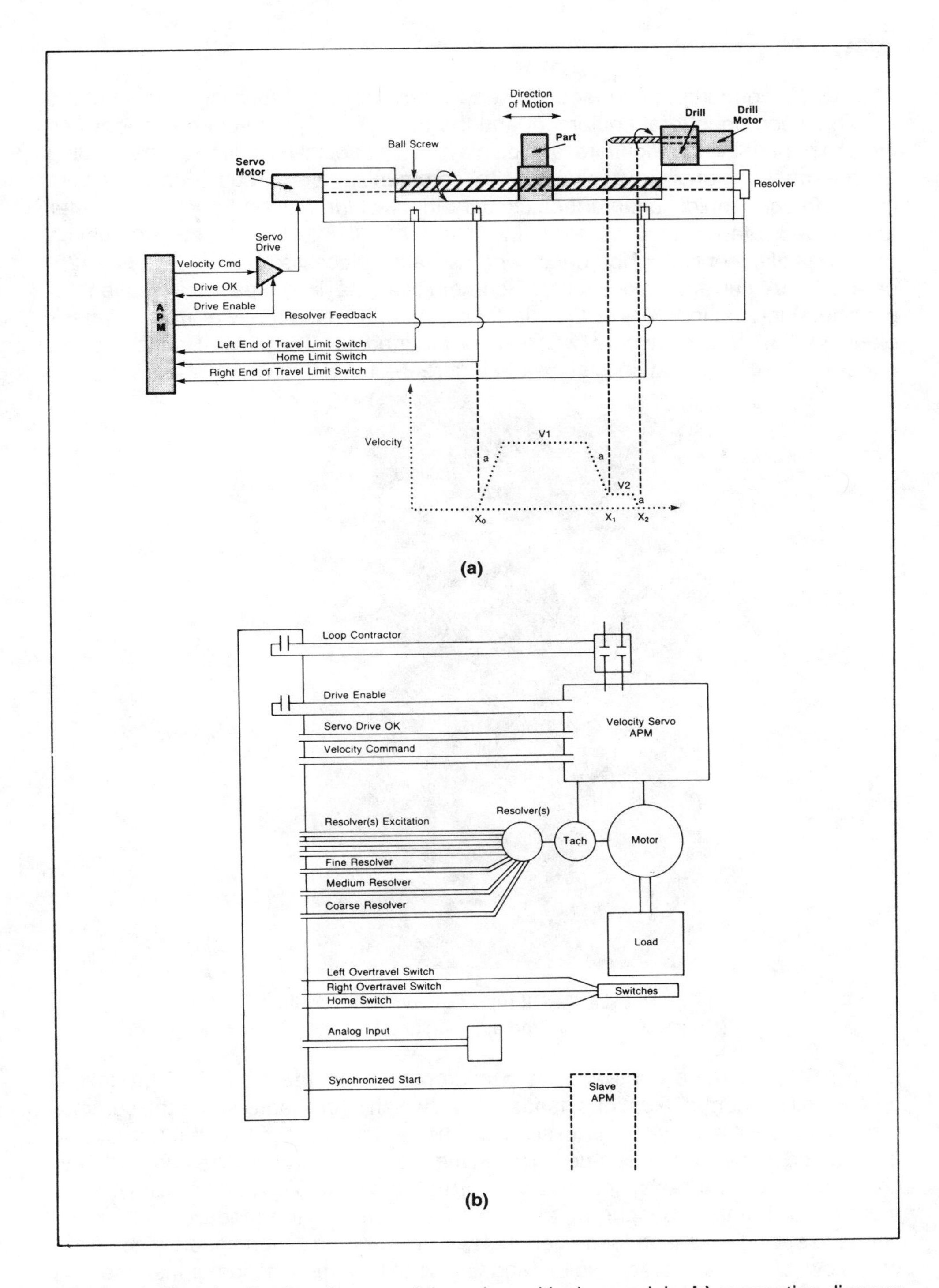

Figure 8-23 a) Application diagram of the axis positioning module; **b)** connection diagram for the APM (Courtesy of General Electric Co., Charlottesville, VA).

ASCII

The ASCII input/output interface is used for sending and receiving alphanumeric data between peripheral equipment and the controller. Typical peripheral devices include printers, video monitors, and displays. This special I/O interface, depending on the manufacturer, is available with communications circuitry only, or with complete communication interface circuitry that includes an on-board RAM buffer and a dedicated microprocessor (intelligent ASCII interface). The information exchange of either type of interface generally takes place via an RS-232C, RS-422, or a 20 mA current loop standard communications link (see next section for peripheral interfacing). An ASCII interface generally gets its power from the back plane of the rack enclosure where it is plugged in. The Allen-Bradley report generation (ASCII) module is shown in Figure 8-24.

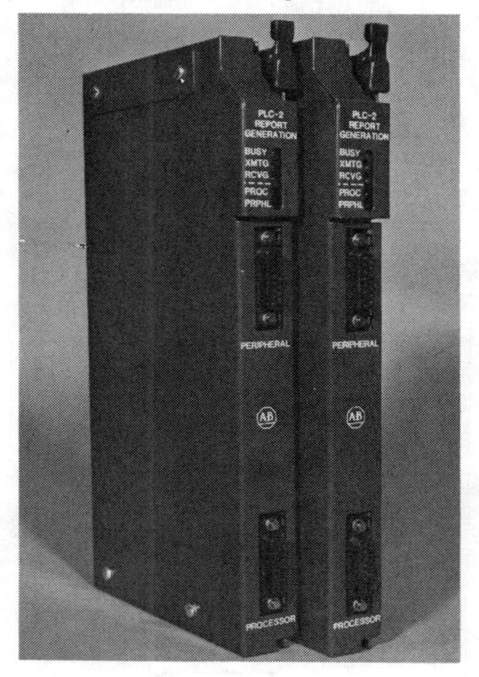

Figure 8-24. ASCII report generation module
(Courtesy of Allen Bradley, Highland Heights, OH).

If the ASCII interface does not use a microprocessor, the main PLC processor handles all the communications handshaking, which significantly slows the communication process and the program scan. Each character or string of characters to be transmitted to the module or received from the module is handled on a character-by-character (interrupt) basis. The module interrupts the main CPU each time it receives a character from the peripheral, and the CPU accesses the module each time it needs to send a message to the peripheral. The communication speed is generally very slow, and for a character to be read, the scan time must be faster than the time required to accept one character. For example, if a scan time is 20 msec, and the baud rate is 300 (30 characters per second), a character will be received every 33.3 msec. Conversely, if the baud rate is 1200 (120 characters per second), more than one character per scan will be transmitted from the peripheral (one character every 8.33 msec). If this is the case, it is obvious that several characters will be lost. This non-smart module can be used in applications which do not require too many characters to be output at a relatively slow speed.

If a smart ASCII interface is used, the transmission is accomplished between the peripheral and the module also on an interrupt basis, but at a faster transmission speed. This is possible since the on-board microprocessor is dedicated to performing the I/O communication. The on-board microprocessor contains its own RAM memory, which can store blocks of data that are to be transmitted. When the input data from the peripheral is received at the module, it is transferred to the PLC memory through a data transfer instruction at the I/O bus speed. All the initial communication parameters, such as number of stop bits, parity (even or odd) or non-parity, and baud rate, are hardware selectable using rocker switches or jumpers or selectable through control software depending on the interface. This method significantly speeds up the communication process and increases data throughput. The smart module is generally used in applications requiring lengthy reports or fast information exchange with alphanumeric devices.

Example 8-4

A standard non-intelligent ASCII module is to be used in a PLC system which has a scan time of approximately 15 msec. This interface will be used to read and write information to and from a remote alphanumeric keyboard/display user interface. What would be the maximum baud rate (bits per second) that can be used for proper transmission?

Solution

The scan time of 15 msec implies that a character cannot be received in less time than the 15 msec. Each ASCII character has 10 bits (7 for the code plus start and stop and parity) which are used during each character transmission.

The inverse of the scan time will give you the minimum time that would be required by the processor to read an incoming character of 10 bits, therefore time for one character (10 bits):

$$= \frac{1}{\text{scan}} = \frac{1}{15 \times 10^{-3}} = 66.67 \quad \text{characters/scan}$$

and the baud rate would be:

$$= 66.67 \times 10 = 666.7$$

The maximum baud rate would be 666.7 (or 667) which transmits 66 characters per second. However, since this baud rate is not a standard baud rate, the user would have to use a more standard one, perhaps a 600 baud.

BASIC—Data Processing

The BASIC module, also referred to as a data processing module, is an intelligent I/O capable of performing computational tasks without burdening the PLC's processor's computing time. In contrast to other intelligent I/O interfaces, such as the servo control, the BASIC module does not actually command or control specific field devices, but complements the performance of the PLC system.

The data processing module is in reality a personal computer packaged in an industrial I/O module which allows user-written BASIC programs to be input and run independently of the PLC's processor. The BASIC language instructions available in this type of interface follow the same instructions available in what may be encountered in a personal computer's BASIC. However, the PLC manufacturer generally incorporates additional instructions which allow the access of the PLC's memory (i.e. I/O data table, or storage area). These added instructions are very useful when information from the process is required by the module to perform BASIC run calculations.

We may also encounter data processing modules which will be able to run other types of languages such as PASCAL, C or other high-level languages. Similar added functions would be found in these languages to allow direct internal communication (data transfers) between the module and the PLC processor.

Communication between the module and the processor generally occurs by performing MOVE instructions which transfer blocks of data *to* and *from* the module. Typical instructions may be a MOVE BLOCK READ, MOVE BLOCK WRITE or similar. Running the BASIC program may be initiated directly by the user through the module's programming port (using the terminal), upon recognition of a user-defined data decoding transferred from the PLC, or after power-up initialization of the PLC system.

The programming port of the BASIC interface is generally compatible with RS-232C or RS-422 communication standards (see next section) intended to support ASCII terminals or the manufacturer's PLC programming terminal. At least one serial peripheral port is available to provide interfacing to printers, asynchronous modems or other serial peripherals. The serial port is generally employed for report generation under BASIC program control, for operator interfaces, and even perhaps a local area network of other personal computers gathering process data for storage purposes.

Other typical applications of the module include complex number crunching calculations that would require awkward PLC programming. Uses and applications of this data processing module are only limited by the user's imagination and in some cases the use of the module would seem to be unequivocally necessary. The utilization of this interface in a PLC system may be convincing proof of the integration of the standard computing power of a personal computer with the powerful I/O handling and control capability of a programmable controller.

Network Interface

Network interface modules are designed to allow a number of PLCs and other intelligent devices to communicate and pass PLC data over a high-speed local area communication network (see Chapter 14). Normally, PLCs that can interface to the network are restricted to products designed by the network manufacturer. The other types of devices that can be interfaced and the total number of devices that can be connected are dependent on the network interface.

In general, when a message is sent by a processor or other network device, its network interface retransmits the message over the network at whatever the network's baud rate. The receiving network interface accepts the transmission and sends it to the intended device. The protocol for the communication link varies depending on the network.

Figure 8-25 shows the SY/NET (Square D Company) network interface module. The module allows up to 200 Square D controllers and other devices to be connected using twin-axial cable, at distances up to 9,000 feet. The interface will allow the CRT programmer to program any PLC on the network. The baud rates for the communication ports on each interface module are adjustable, from 300 to 9600 in order to accommodate printers, modems, and data terminals. The network interface module can also be used to connect two or more networks together. Note in the figure that this network interface module has two communication ports to allow two devices to communicate with each other, as well as with the network.

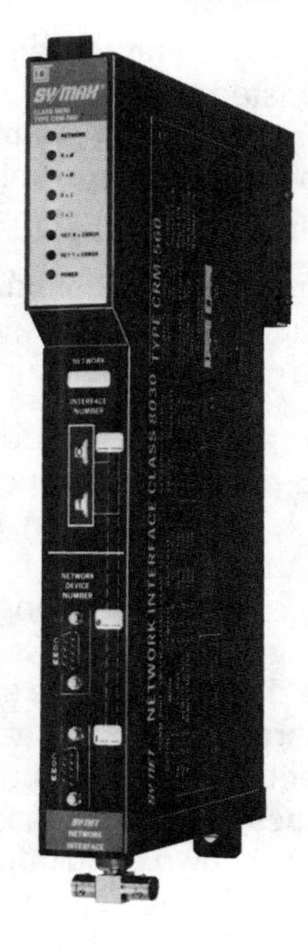

Figure 8-25. Local area network interface module (Courtesy of Square D Co., Milwaukee, WI).

8-6 PERIPHERAL INTERFACING

Regardless of the peripheral used, the user must properly connect the device to the PLC, or smart module, to achieve correct communication. Typical peripherals communicate in serial form at speeds ranging from 110 to 19,200 bits per second (baud), with parity or non-parity, asynchronously, and by using different communication interface standards.

Communication Standards

Communication standards fall into two categories: proclaimed and de facto. Proclaimed standards have been officially established by various electronic organizations, such as the Institute of Electrical and Electronics Engineers (IEEE) or the Electronic Industries Association (EIA). These institutions attempt to define public specifications by which manufacturers can properly establish communication schemes that allow compatibility among different manufacturers' products. Proclaimed standards, such as the IEEE 488 instrument bus, the EIA RS-232C, and the EIA RS-422, are examples of well-defined standards.

When we refer to de facto standards, we are referring to interface methods that have been adopted and have gained popularity through widespread use without official definition. Some de facto standards have caused interfacing problems in the past; however, other standards, such as the PDP-11 Unibus and 20 mA current loop, are good examples of well-defined de facto standards.

Serial Communication. This type of communication, as the name implies, is done in a serial form through simple twisted pair cables. ASCII information is usually sent to peripheral equipment such as terminals, modems, and line printers. Serial data transmission is used for most peripheral communication, since these devices are slow-speed in nature and require long cable connections.

Two of the most popular standards for serial communications are the RS-232C and the 20 ma current loop. Another upoming PLC standard is the RS-422, which will tend to improve performance and give greater flexibility in data communication interfaces.

Data communication links generally used with peripheral equipment can be unidirectional or bi-directional. If a peripheral is strictly either an input or an output device, data needs to be sent only in one direction. In either case, a serial signal line is all that is required to complete the link. Devices that serve as both input and output devices (e.g. video terminal) require bi-directional links. There are two ways to achieve this bi-directional communication. First, a single data line can be provided as a shared communication line. The data can be sent in either direction, but only in one direction at a time. This operation is known as half-duplex. If simultaneous bi-directional communication is required, two lines can be sent from the PLC to the peripheral. One line would be assigned permanently to an input, while the other would be an output. This mode is known as full-duplex. Figure 8-26 illustrates the uni-directional, full-duplex, and half-duplex communication methods.

EIA RS-232C. The EIA RS-232C is a proclaimed standard that defines the interfacing between data equipment and communication equipment employing serial binary data interchange. Electrical signals, as well as mechanical details of the interface,

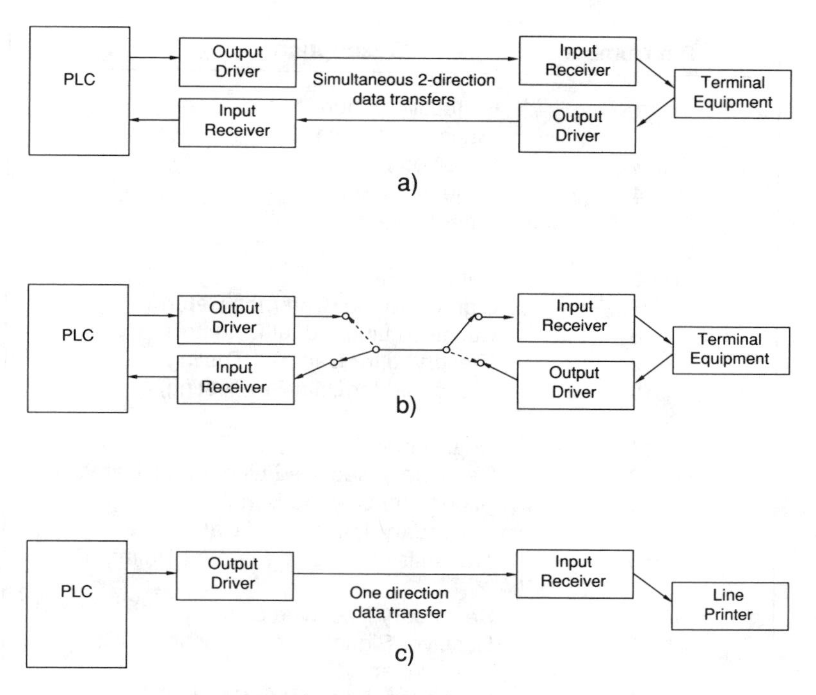

Figure 8-26. a)Full duplex, b) Half duplex and c) Unidirectional data communications format.

are well-defined by this standard. The complete RS-232C interface consists of 25 data lines, which have all possible signals for simple to complex communication interface. Although several of these lines are specialized and few are left undefined, most peripherals require only from 3 to 5 lines to operate properly. Table 8-2 describes the 25 lines as specified by EIA.

Figure 8-27a illustrates an RS-232C data communication system using a telephone modem (modulator/demodulator); 8-27b shows the RS-232C wiring connections from a computer to a smart EIA PLC interface module; and 8-27c illustrates a typical interfacing to a teletype printer.

Note that the communication between a computer and a PLC has few lines swapped, if no modem or other data communication equipment is used. This cable is called a NULL MODEM cable. When connecting a PLC to an RS-232C peripheral (printer, etc.), four wires are normally required; however, it is recommended that users refer to the connection specifications for both devices for any special details.

The RS-232C standard calls for certain specific electrical characteristics. A listing of some of these specifications follows:

a) The signal voltages at interface point are a minimum of +5V and maximum of +15V for logic 0, and for logic 1, a minimum of -5V and maximum of -15V.

b) The maximum recommended distance is 50 feet or 15 meters; however, longer distances can be permissible, provided that the resulting load capacitance, measured at the interface point and including the signal terminator, does not exceed 2500 picofarads.

Pin Number	Description
1	Protective Ground
2	Transmitted Data
3	Received Data
4	Request to Send
5	Clear to Send
6	Data Set Ready
7	Signal Ground (Common Return)
8	Received Line Signal Detector
9	(Reserved for Data Set Testing)
10	(Reserved for Data Set Testing)
11	Unassigned
12	Secondary Received Line Signal Detector
13	Secondary Clear to Send
14	Secondary Transmitted Data
15	Transmission Signal Element Timing (DCE)
16	Secondary Received Data
17	Receiver Signal Element Timing (DCE)
18	Unassigned
19	Secondary Request to Send
20	Data Terminal Ready
21	Signal Quality Detector
22	Ring Indicator
23	Data Signal Rate Selector (DTE/DCE)
24	Transmit Signal Element Timing (DTE)
25	Unassigned

Table 8-2. EIA RS-232C Interface Signals

c) The drivers used must be able to withstand open or short circuits between pins in the interface.

d) The load impedance at the terminator side must be between 3000 and 7000 ohms, with no more than 2500 picofarads capacitance.

e) Voltages under -3V (logic 1) are called MARK potentials (signal condition); voltages above +3V (logic 0) are called SPACE voltages. The area between -3V and +3V is not defined.

Figure 8-28 illustrates a typical RS-232C serial ASCII pulse train. The transmission begins with a START bit (0) and ends with either one or two STOP (1s) bits. Parity is also included and can be even or odd (see Chapter 4 for parity).

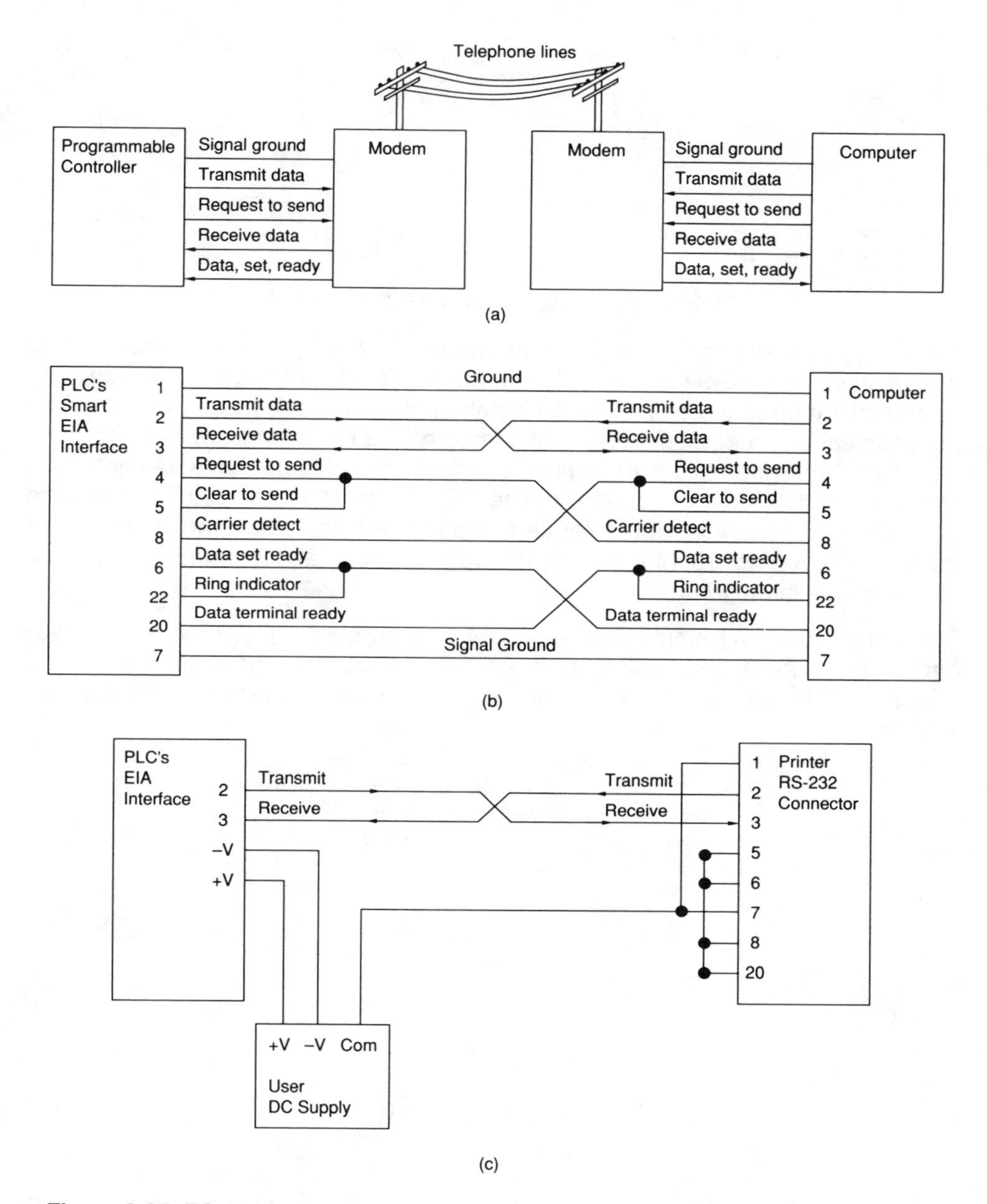

Figure 8-27. RS-232C communication connections for a) modem, b) PLC to computer and c) PLC to printer.

EIA RS-422. The RS-422 standard has been devised to overcome some of RS-232C shortcomings, including an upper data-rate of 20 K baud, maximum cable distance of 50 feet, and an insufficient capacity to control additional loop-test functions for fault isolation. The RS-422 standard still deals with the traditional serial-binary switch-signals of two voltage levels across the interface. Mechanical specifications for this electrical interface standard (RS-422) are defined by the RS-449 standard to meet new operational requirements.

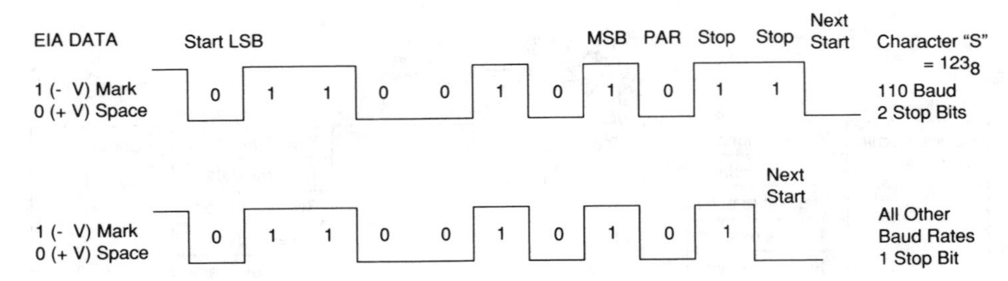

Figure 8-28. RS-232C serial ASCII pulse train.

The RS-232C is an unbalanced link communication which specifies a primary station always in control (MASTER/SLAVE) and responsible for setting logical states and operational modes of each secondary station, thereby controlling the entire data communication process. The RS-422 is a balanced link in which either party can configure itself and initiate transmission, when both stations have identical data transfer and link control capability. The RS-422 specifies electrically balanced receivers and generators that tolerate and produce less noise to provide superior performance up to 10 Megabaud. (10,000 K baud) and to meet even more requirements in the future.

The balanced circuit employs differential signaling over a pair of wires for each circuit. The unbalanced configuration signals use one wire for each circuit and a common return circuit. Figure 8-29 illustrates both configurations for RS-422 and RS-232C.

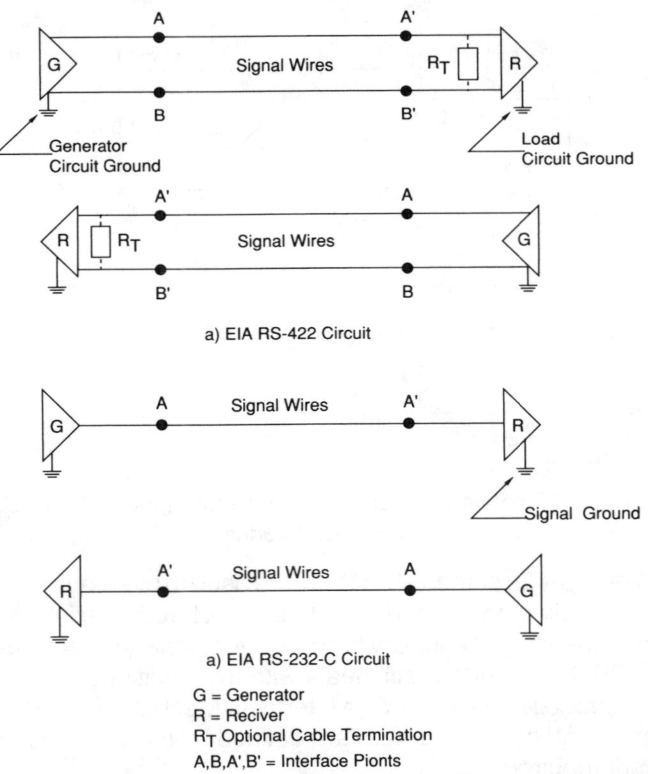

Figure 8-29. Circuit configuration for a) RS-422 and b) RS-232C.

In general, the RS-422 may be required when interconnecting cables are too long for effective unbalanced operation, and noise in excess of 1 volt can be measured across the signal conductors. The driver circuits for RS-422 have the capability to furnish the DC signal necessary to drive up to 10 parallel connected RS-422 receivers; however, this capability involves considerations such as stub line lengths, data rate, grounding, fail-safe networks, etc. Cable characteristics are not specified, but to insure proper operation, paired cables with metallic conductors should be employed and, if necessary, shielded.

The maximum distance applicable for RS-422 is a function of the data transmission rate. A relationship between distance and data rate is illustrated in Figure 8-30. The balanced electrical characteristics of RS-422 provide an even better performance with an optimal cable termination of approximately 120 ohms in the receiver load. These curves, however, are conservative for RS-422 balanced operation. Actually, several miles can be realized at lower data rates with good engineering practice. The graph here describes empirical measures using a 24 AWG, copper conductor, twisted pair cable with a shunt capacitance of 52.5 pF/meter (16 pF/foot), terminated in a 100 ohm resistive load. If longer distances are required, analysis on absolute loop resistance and capacitance of the cable should be performed. Longer distances, in general, would be possible when using 19 AWG cable. The type and length cable used must be capable of maintaining the necessary signal quality needed for the particular application.

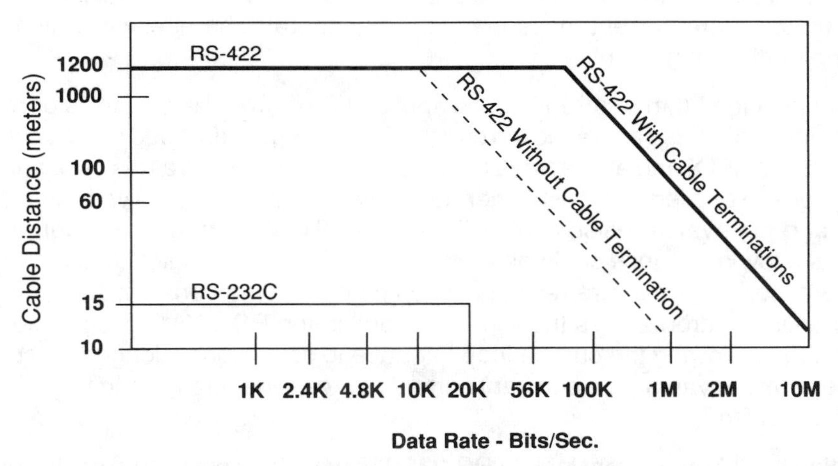

Figure 8-30. Cable distance versus data rate relationship for RS-422 and RS-232C communications standard.

The RS-449 mechanical standard (that supports the electrical RS-422) offers several extra circuits (signals) that were added to provide greater flexibility to the interface and to accommodate the new common return circuits. These additional functions and wires were beyond the capacity of the RS-232C 25-pin connector; therefore, EIA selected a 37-pin connector that will satisfy the needs of the interface channels. If secondary channel operation is to be used as a low-speed TTY or acknowledgements channel, a separate nine-pin connector is also needed.

20 mA Current Loop. The 20 mA current loop de facto standard consists of four basic wires: transmit plus, transmit minus, receive plus, and receive minus. Figure 8-31 illustrates the four lines used to form the 20 mA current loop.

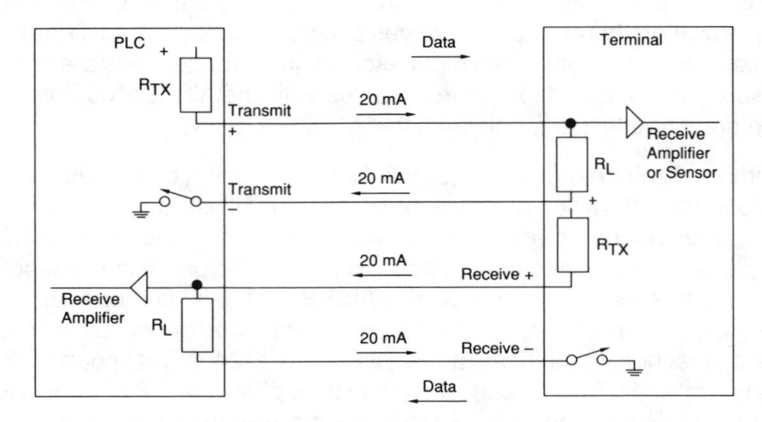

Figure 8-31. Illustration of 20 mA current loop operation.

Recognition of ones and zeroes is achieved by opening and closing the current loop. When the current loop standard was first used in teletypewriters, the loop was connected and broken by rotating switch contacts within the teletypewriter sending the data, and the 20 mA signal drove a print magnet in the receiving teletypewriter. Today, most 20 mA current loops electronically operate the opening switch and printer magnet arrangement.

The voltage in the 20 mA current loop is applied to a current limiting resistor at the data sending end, thus generating the current. The voltage is dropped across current limiting resistor RTX and also across the load resistor RL . The values of R and the positive voltage applied to it must generate a flow current of 20 mA. Typically, a high voltage and high value resistance (RTX) are usually chosen, even though a low voltage and low resistance could be used. Current loop communications present a big advantage, since the wire resistance has no effect on a constant current loop. Voltage does not drop across the wire as it does in the RS-232C voltage-oriented interface, thus allowing the current loop interface to drive signals longer distances. To achieve this advantage, a constant current source is generally used to generate the 20 mA current.

Converting the 20 mA current loop to RS-232C can be done simply by employing an RS-232C level receiver to drive a switching transistor on the transmission end and an optical-isolator and load resistor to drive the RS-232C driver on the receiving end.

Chapter

—9—

PROGRAMMING LANGUAGES

"Most of the fundamental ideas of science are essentially simple, and may, as a rule, be expressed in a language comprehensible to everyone."

—Albert Einstein

9-1 INTRODUCTION

Programming languages used in programmable controllers have been evolving since the inception of the PLC back in the late 1960's. However, several basic instructions have found firm footing as a PLC instruction for a long time to come. Among these instructions are included some of the original basic relay ladder formats which came along with the birth of the programmable controller.

With the proliferation of new products came new, more versatile instructions. These new instructions provided more computing power in the sense of the single operations performed by the instruction itself. For instance, blocks of data could be transferred, transparent to the user, from one memory block location to another and at the same time perform a logic or arithmetic operation with another block. As a result of these instructions, data can be handled more easily by the control program.

In addition to the new programming instructions, developments of powerful I/O modules introduced changes to the existing new instructions. These changes include the capability of sending and obtaining data to and from the module by addressing the modules' locations. For example, it is possible to read and write data to and from analog modules. All these advances, in conjunction with the future needs of industry, will create a demand for more powerful instructions which will allow for easier, more compact, function-oriented types of PLC programs.

In this chapter we present the various instructions used to program a PLC. These instructions include the basic PLC instructions as well as the more advanced functions. The functional descriptions presented here will give you an understanding of how the instructions operate and their relationship to the data table. Although the instructions covered are generic in nature, you will find them in the same shape, form, and function in most programmable controllers.

9-2 TYPES OF PLC INSTRUCTIONS

The programmable controller was conceptualized with the ease of programming using effective representation of the program logic needed to control the machine or process. This effective representation involved the use of the existing relay ladder symbols and expressions to define the control logic. The result was a *programming language* which used the original basic relay ladder elements of symbols—the name given was *ladder language.* Figure 9-1 illustrates a relay ladder logic and the PLC ladder language.

Progression and evolution of the original language turned ladder programming into a more powerful instruction set. New functions were added to the basic relay, timing and counting operations. The name *function* was used to describe instructions that perform, as its name implies, a handling and transferring function of data within the programmable controller. These instructions were also based on the simple principle of the basic relay logic, although complex operations could be implemented and

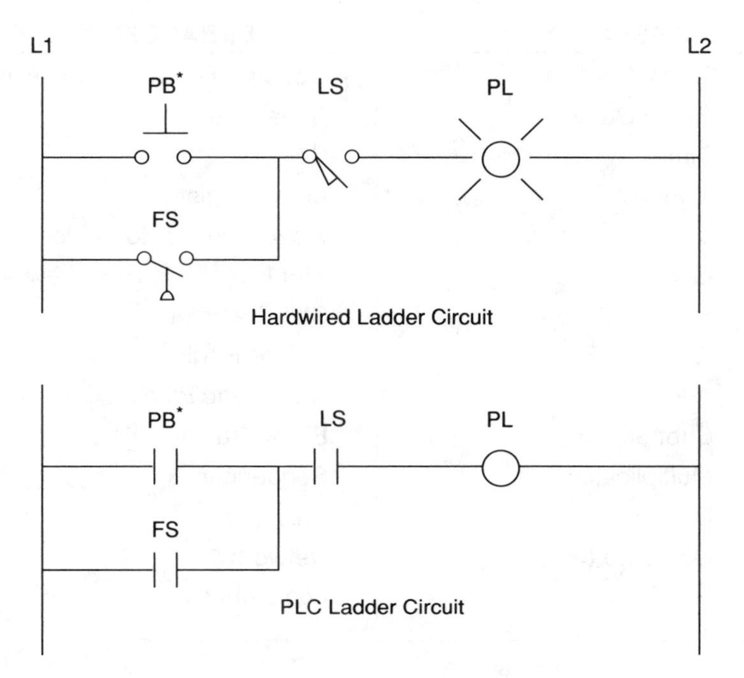

Figure 9-1. Electrical ladder circuit and PLC ladder circuit.

performed. The new addition to the basic ladder logic was known as *functional blocks*, and can be considered an *enhanced ladder language*. Figure 9-2 shows enhanced functions driven by the basic elements of relay ladder instructions. As can be seen in the figure, the enhanced functional block may be represented, literally, by a block or by a functional instruction between two contact symbols. The enhanced ladder function format depends on the PLC maker; however, regardless of representation the function performed is the same in nature. Throughout this chapter, we will make reference to the block type of format enhanced ladder instructions. Instructions available in PLCs can be grouped in two groups:

- Basic Ladder Instructions
- Enhanced Ladder Instructions

Each of these two groups have several PLC instructions which form that category. Classifications of what instruction falls in which group may differ among manufacturers and users since a definite classification does not exist. However, a *de-facto* standard may have been created throughout the years placing certain instructions under a group. Table 9-1 shows a typical classification under a specific instruction group.

BASIC	ENHANCED
Relay Contact	Double Precision Arithmetic
Relay Output	Square Root
Timer	Sort
Counter	Move Register
Latch	Move Register to Table
GO TO	FIFO
MCR	Shift Register
END (MCR)	Rotate Register
Addition	Diagnostic Block
Subtraction	Block Transfer (IN/OUT)
Multiplication	Sequencer
Division	PID
Compare (=,>,<)	Network
GO SUB	Logic Matrix

Table 9-1. PLC instruction set classifications.

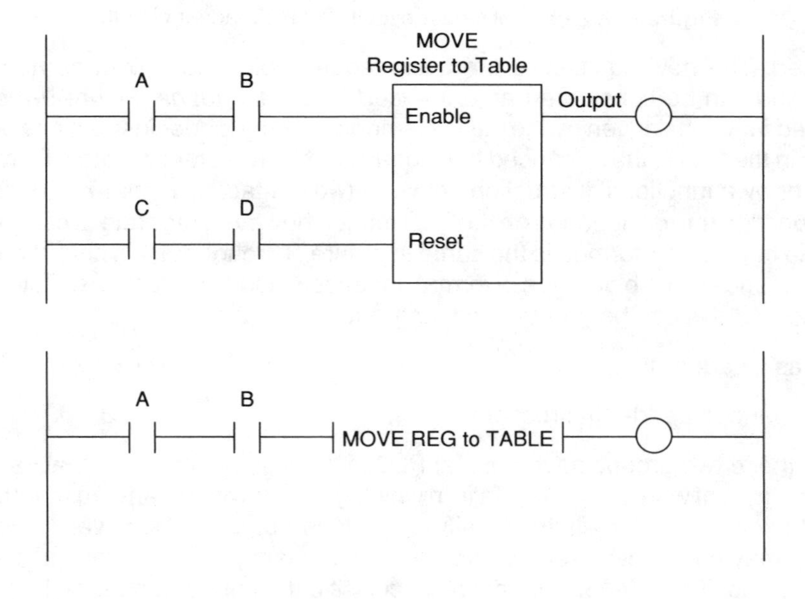

Figure 9-2. Enhanced ladder instructions format.

Some classifications may also be defined by naming the basic ladder instructions as a low-level language and the enhanced ladder functions as a high-level language. In the end, both groups are saying the same thing: a low-level language performs basic functions while a high-level language performs high-level functions! The true definition of the type of PLC instructions really comes when functional categories are defined. These instruction categories are:

- Basic Relay
- Timing and Counting
- Program/Flow Control
- Arithmetic
- Data Manipulation/Handling
- Data Transfer
- Special Function

Here is where the grouping of instructions becomes subjective. Some people may say for instance that the basic ladder instructions include basic relay, timing and counting, program/flow control, arithmetic, and *some* data manipulation. While others may say that only basic relay and timing and counting should be considered basic ladder instructions.

The point to make is that the more of these instruction categories a PLC has, the more powerful its control capability becomes. Small PLCs tend to have the basic instruction requirements and perhaps some enhanced instructions. As a PLC product family progresses in hierarchy, more instructions are included along with the I/O capacity and size of memory.

Other Instruction Types/Languages

Some programmable controllers may offer the ability to program in other languages beside the conventional ladder language. An example is the BASIC programming language. Other manufacturers use what is called *Boolean Mnemonics* to program a controller. The Boolean language is really a method used to enter and explain the control logic which follows Boolean algebra (Chapter 3). Towards the end of the chapter we discuss these two typical languages or instruction sets as well as other types of symbolic program representation.

9-3 LADDER DIAGRAM FORMAT

The ladder diagram language is a symbolic instruction set that is used to create a programmable controller program. The ladder instruction symbols can be formatted to obtain the desired control logic that is to be entered into memory. Since this type of instruction set is composed of contact symbols, it is also referred to as *contact symbology*.

The main function of the ladder diagram program is to control outputs and perform functional operations based on input conditions. This control is accomplished through the use of what is referred to as a ladder "rung". Figure 9-3 shows the basic structure of a *ladder rung*. In general, a rung consists of a set of input conditions, represented by contact instructions, and an output instruction at the end of the rung, represented by the coil symbol. Throughout this chapter, the contact instructions for a rung may be referred to as input conditions, rung conditions, or control logic.

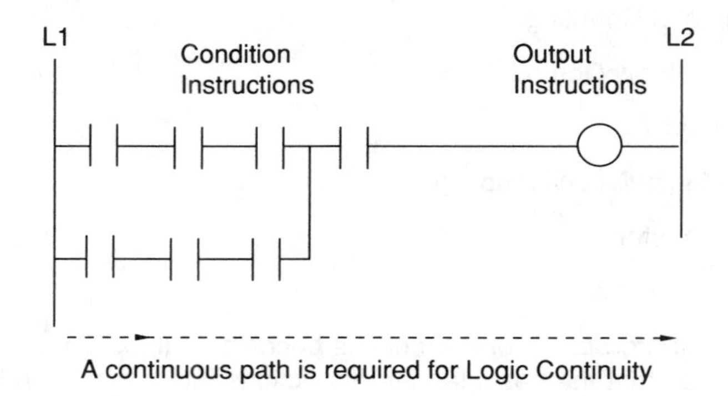

Figure 9-3. Ladder rung.

A ladder rung is said to be true or to be energizing an output or functional instruction (block) when logic continuity exists in the rung. Logic continuity exists when power is provided in the rung from left to right. This continuity is formed by the desired logic of events that take place in order to enable the output. The left-most side (left power line) simulates the L1 line of a relay ladder diagram while the right-most side (right power line) simulates the L2 line of the electromechanical representation. Continuity is achieved whenever a path contains contact elements in a closed condition so that power is flowing from left to right. These contact elements will either close or remain closed according to the status of its reference inputs. Figure 9-4 illustrates several continuous paths that may take place to achieve continuity and energize the output of the rung. How these contact symbols are interpreted to be ON or OFF is covered in detail in the basic relay instruction section.

When functional blocks are implemented, input conditions are also represented by contact instructions driving the logic for the block. The block format can also have one or more output coils which signify the status of the function being performed. The functional block may have one or more enable inputs controlling the functional block operation. For example, the block shown in Figure 9-5 has an enable block line which, when energized (continuity exists), will activate the block to perform the function or instruction. It is like saying: if the enable is ON because the desired logic has continuity, then execute the block instruction. Depending on the functional block instruction, other enable lines may indicate *reset* or other control functions. If you want the block to be active at all times without any driving logic, simply omit any contact logic and place a continuity line during programming (see Figure 9-6).

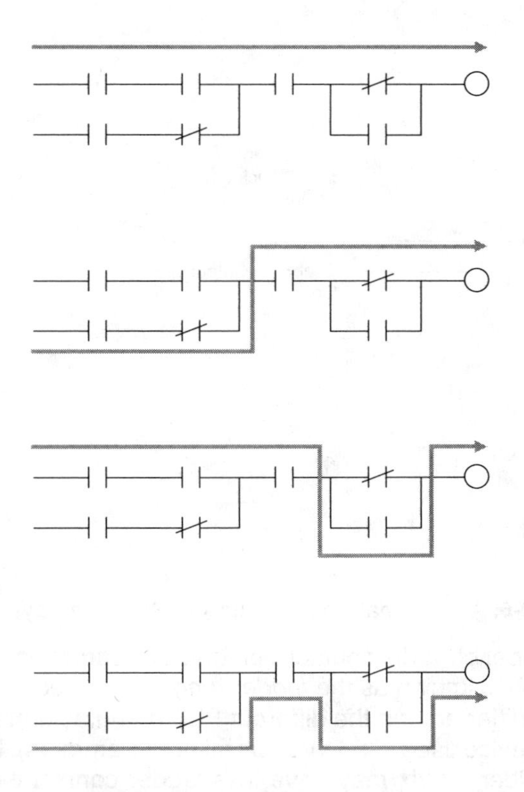

Figure 9-4. Possible continuous path that may turn an output ON.

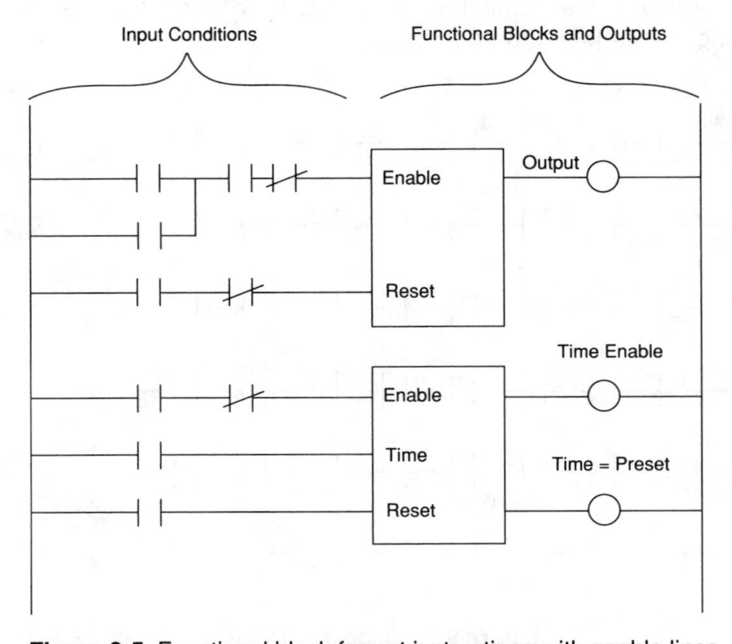

Figure 9-5. Functional block format instructions with enable lines.

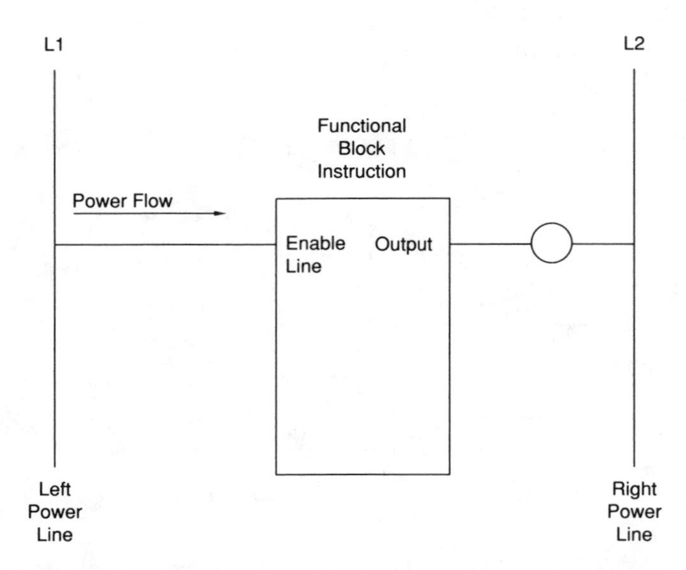

Figure 9-6. A functional block instruction which is always enabled.

The maximum number of ladder contact elements that can be used to program a rung is restricted by what is known as the *ladder rung matrix* (see Figure 9-7). The size of this matrix may differ among the different PLC manufacturers and of course, by the programming device used (CRTs versus miniprogrammers). For functional block operations, the ladder matrix may have less ladder contact elements due to the actual block size display (see Figure 9-8a). If enhanced functional instructions are available in a particular PLC instead of the block type of instruction, the instruction may take one or more contact symbol spaces to show the instruction in the programming device (see Figure 9-8b).

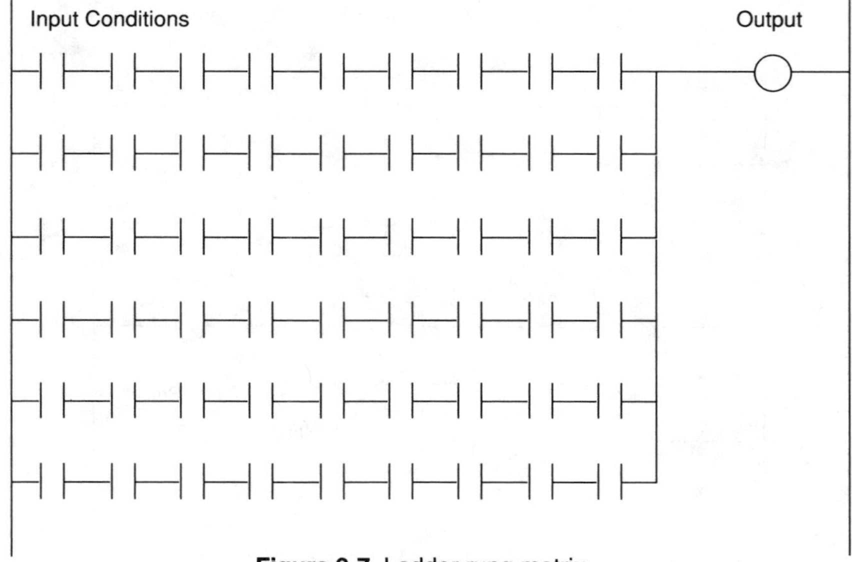

Figure 9-7. Ladder rung matrix.

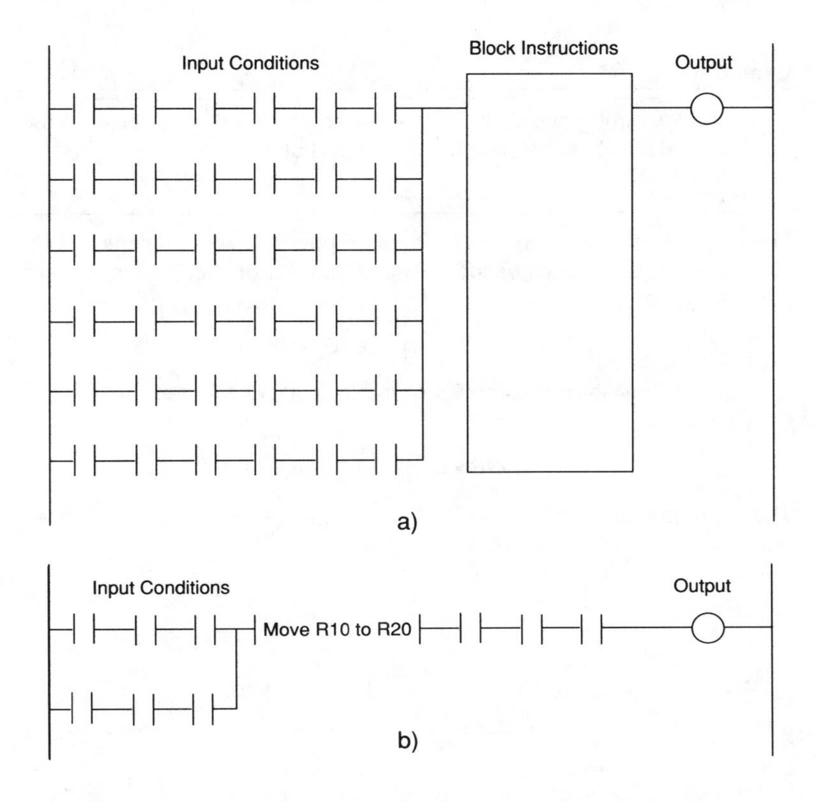

Figure 9-8. Ladder rung matrix for enhanced instructions.

The ladder matrix represents all the possible locations where a contact symbol instruction can be placed. The programming device generally displays all the possible locations on the screen and allows you to fill the contact symbol in the desired location. However, according to the maker of the PLC certain rules must be followed. One rule that would be in almost all PLCs is that of *reverse power flow*. Reverse power, as shown in Figure 9-9, is not allowed in PLC logic to prevent possible *sneak paths* that could occur in hardwired electromechanical relay systems. If the logic calls for the implementation of reverse flow, the user must reprogram the rung with a forward power flow to all contact elements. The solution to the reverse power flow rung of Figure 9-9 is illustrated in the following example.

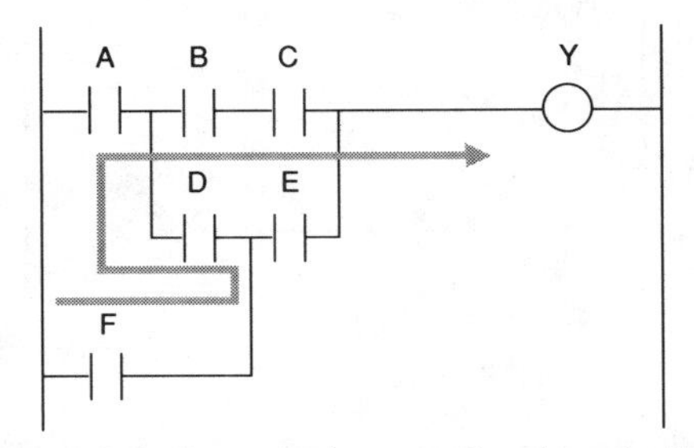

Figure 9-9. Reversed power flow is not allowed.

Example 9-1

Solve the logic rung shown in Figure 9-9 so that no reverse power flow condition exists. Reverse condition is not part of the logic.

Solution

The output Y is determined by the forward power flow of the logic. Let's implement it using logic concepts. The output Y is defined, using forward paths only, as:

$$Y = (A \cdot B \cdot C) + (A \cdot D \cdot E) + (F \cdot E)$$

which can be minimized using Boolean algebra's distributed rule to (see Chapter 3):

$$Y = A \cdot (B \cdot C + D \cdot E) + (F \cdot E)$$

The logic gate implementation is shown in Figure 9-10 while the ladder equivalent solution is given in Figure 9-11.

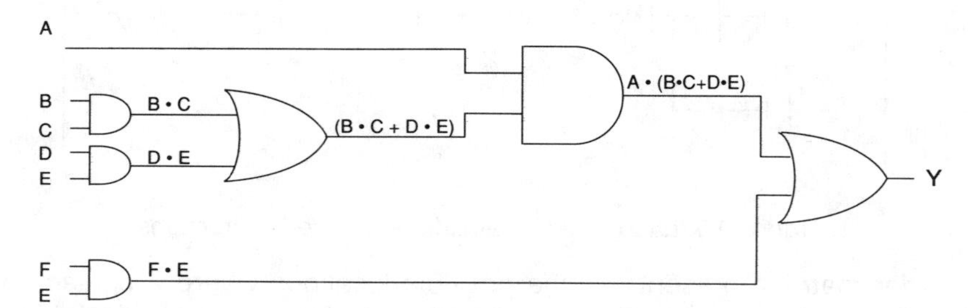

Figure 9-10. Logic Solution.

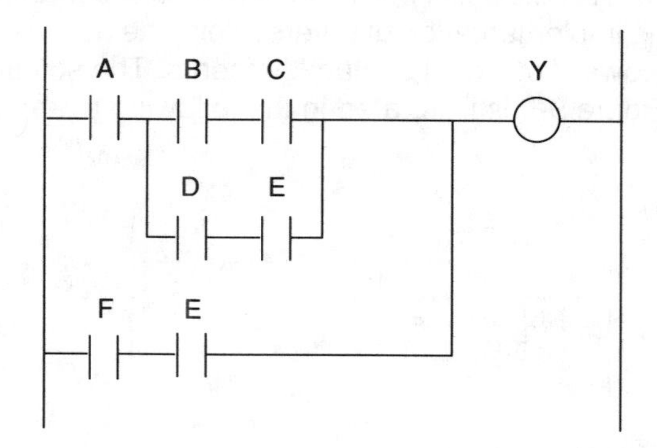

Figure 9-11. Ladder diagram implementation for Example 9-1.

Example 9-2

Solve the ladder logic shown in Figure 9-9 so that no reverse power flow exists. The required logic solution must implement the reverse logic flow.

Solution

Following the same procedure as in Example 9-1, we can get the desired logic for output Y by obtaining the Boolean logic expressions. Therefore, the output Y, including the reverse power flow logic, is represented by:

$$Y = A \cdot B \cdot C + A \cdot D \cdot E + F \cdot E + F \cdot D \cdot B \cdot C$$

$$= A \cdot (B \cdot C + D \cdot E) + F\,(E + D \cdot B \cdot C)$$

(using distributed law)

The term $F \cdot D \cdot B \cdot C$ implements the sequence of the reverse flow which is required for output Y. The ladder solution is shown in Figure 9-12.

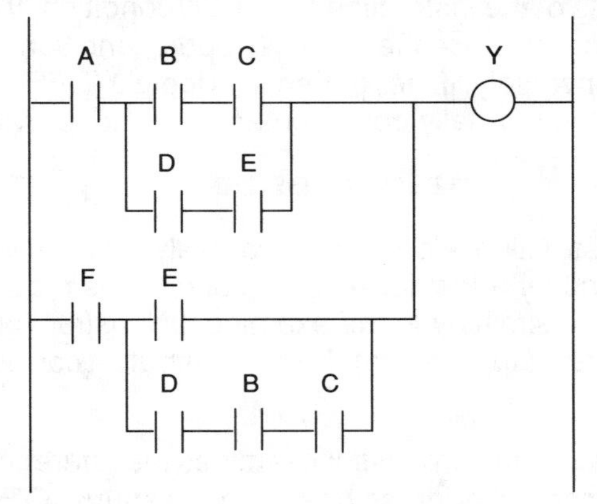

Figure 9-12. Ladder implementation for example 9-2.

9-4 BASIC RELAY INSTRUCTIONS

Coils and contacts are the basic symbols of the ladder diagram instruction set. The contact symbols programmed in a given rung represent conditions to be evaluated in order to determine the control of the output. All outputs are represented by coil symbols.

When programmed, each contact and coil is referenced with an address number which identifies what is being evaluated and what is being controlled. Recall that these address numbers reference the I/O table location of a connected input or output or an internal or storage bit output. A contact, regardless of whether it represents an input/output connection or an internal output, can be used throughout the program whenever that condition needs to be evaluated.

The format of the rung contacts is dependent on the desired control logic. Contacts may be placed in whatever series, parallel, or series/parallel configuration is required to control a given output. When logic continuity exists in at least one left to right contact path, it is said that the rung condition is TRUE. The rung condition is FALSE if no path has continuity.

The relay-type instructions covered in this section are the most basic of programmable controllers instructions. They provide the same capabilities as hardwired relay logic, but with greater flexibility. These instructions primarily provide the ability to *examine* the ON/OFF status of specific bit addresses in memory and to control the state of an internal or external output. The symbol used for each instruction is shown at the far right of the instruction name.

Examine ON—Normally Open

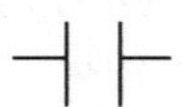

The examine ON instruction, generally referred to as a normally open (NO) instruction, is programmed to test for an ON condition from a reference address. The reference address could come from an input device (input table) or from an output bit in the internal bit storage section of the data table or from a bit corresponding to an output in the output table (see Chapter 3 and 5 for I/O addressing).

During the execution of a control program, (program scan), the processor *examines* the reference address of the instruction for an *ON* condition. If the reference is logic 1 (ON), the processor will *close* the normally open condition to provide power flow (logic continuity). Conversely, if the reference is logic 0 (OFF), the processor will not change the state of the normally open contact, thus not providing continuity.

Examine OFF—Normally Closed

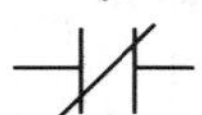

The examine OFF instruction, also called a normally closed (NC) contact instruction, is programmed whenever a test for an OFF condition is required from a reference address. The address, similarly to the examine ON instruction, can be referenced from the input table (real input device), from an output bit corresponding to an output in the output table.

During the program scan, the processor examines the reference address for an OFF condition. If the NC contact address has a logic 0 status (OFF) the instruction will continue to provide power (continuity) through the rung. If the reference address has a logic 1 status (ON), the instruction will open the NC contacts, thus not providing continuity to the rung.

Output Coil

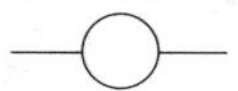

The output coil instruction is programmed to control either a real output connected to the PLC, via output interfaces, or to control an internal output (control relay). The reference address used for the output coil may be signified by a bit of the internal storage area (used as control relay). The output coil instruction may be represented also as a —( )— symbol.

During the program scan, the processor evaluates all the input conditions in the ladder rung; if continuity in any path is encountered, the processor will place a logic 1 in the output coil address (bit) referenced by the instruction. The logic 1 status indicates an ON condition to the output coil instruction. If the reference corresponds to an output bit in the output table, the processor will turn ON the output and therefore the field device connected to the module's terminal whose address maps the output coil address. Remember that the turning ON of the modules occurs once the ladder program has been completely solved (program scan) and an update of outputs is performed after the end of scan (EOS). If no continuity exists, the output coil instruction will be OFF (logic 0).

When the output coil address maps an internal bit storage address and the coil is turned ON, the corresponding bit in the storage area is set to logic 1. This case of using *internal outputs*, as they are also known, is generally employed when interlocking of sequences is required or when a real output is not necessary.

Reference contacts from the output coil (┤├ and ┤/├), i.e. having the same reference address, will open or close according to the status of the output. Figure 9-13 illustrates an example of a simple ladder with the NO and NC contacts driving an output rung. For output 20 to turn ON, two things must happen; first, PB1 must be pushed to turn ON the reference input 10 and second, the limit switch LS1 must not be activated in order to maintain the reference input 11 OFF. In this case, the processor examines for an ON condition at input 10 and an OFF condition at input 11; if both logic conditions are met, output 20 is energized. Since output 20 is ON, the NO contact 20 will close, turning output 100 ON. The NC contacts 20 will open because the examine for an OFF 20 condition is not true (reference 20 is ON), therefore turning output 101 OFF. At the EOS, the pilot light (PL1) will be lit because the processor sends a 1 to the module which latches the signal logic 1 until the continuity in rung (output) 20 is disrupted. Output 100 will not turn a real output device ON because it is an internal bit not mapped to the I/O table (see Note in Figure 9-13).

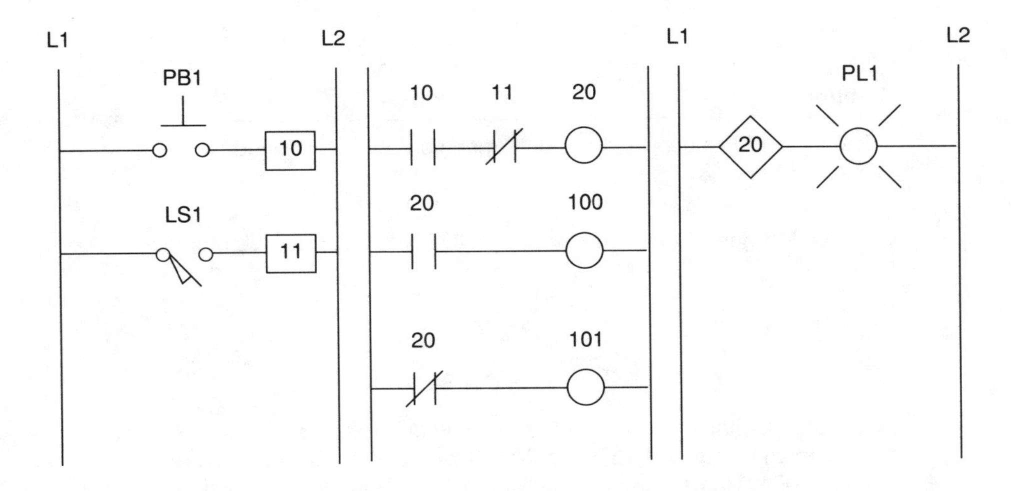

NOTE: 20 is a real output.
100 and 101 are internal output coils.

Figure 9-13. Normally open and normally closed contacts driving output coils.

NOT Output Coil

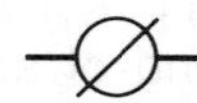

The NOT output coil instruction (recall the NOT logic function) is essentially the opposite of the output instruction. If continuity is not present, the reference output bit is turned ON. If continuity is present in the rung, the output is turned OFF. When the NOT output is ON, its reference contacts will change states (NO will close, NC will open); if the NOT output is OFF, the reference NO contacts will be open and the NC contacts will be closed.

The NOT output coil may be represented by the —(/)— symbol in some PLCs. The use of this instruction may be a little tricky to implement; however, it is easier to have the logic expression of the output rung and apply the Boolean logic rules (Chapter 3) to obtain the NOT output coil ladder rung (see following example).

Example 9-3

a) Implement the equivalent ladder rung shown in Figure 9-14 using the NOT output coil instruction. b) Also, implement the NOT Y *logic* without using a NOT coil.

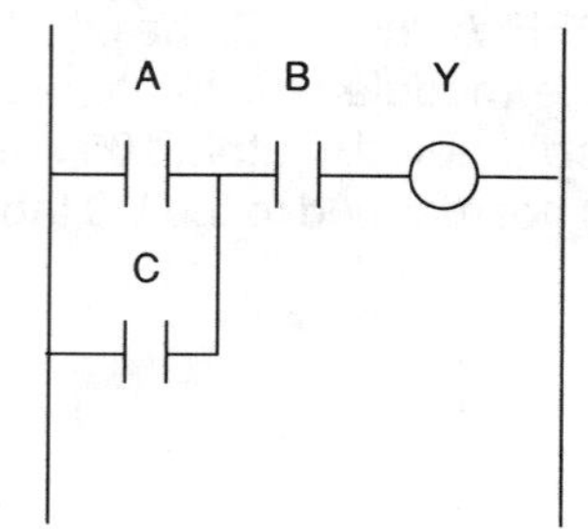

Figure 9-14.

Solution

a)The ladder logic expression representing output Y is defined by:

$$Y = (A + C) \cdot B$$

Using DeMorgan's Law (Chapter 3) the NOT Y function can be obtained :

$$\begin{aligned} \overline{Y} &= \overline{(A + C) \cdot B} \\ &= \overline{(A + C)} + \overline{B} \\ &= \overline{A} \cdot \overline{C} + \overline{B} \end{aligned}$$

Therefore, the implementation of the logic using a NOT output coil (Y) will be as shown in Figure 9-15. Output Y will be ON if A and B are ON or if C and B are ON. Remember that the NOT output is ON if continuity does not exist and OFF if continuity is present. The circuit shown in Figure 9-15 is, logically, indentical to that of Figure 9-14.

b) The easiest way to implement a logic NOT function of Figure 9-14 would be to use the same rung with the exception that the output Y would be a NOT coil. If we cannot use the NOT coil, then you can implement the NOT by adding another rung as shown in Figure 9-16. Here, output Z will be essentially the implementation of the NOT output Y.

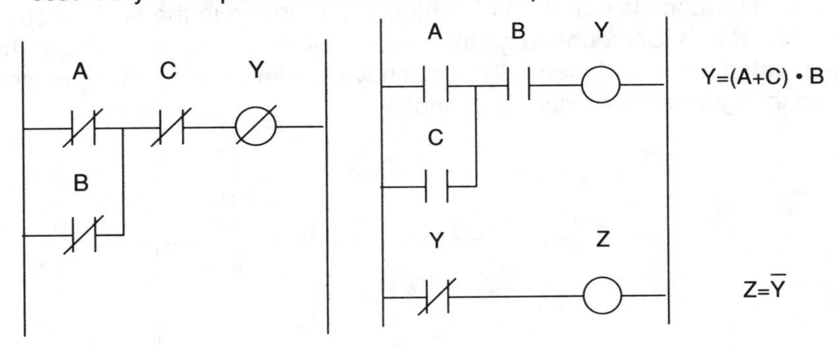

Figure 9-15. Implementation of the circuit of Figure 9-14 using a NOT coil

Figure 9-16. Implementation of a logic NOT function of figure 9-14.

Latch Output Coil

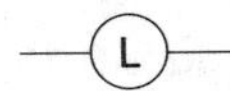

The latch coil instruction is programmed if an output is to remain energized, even though the status of the contacts which caused the output to energize may change. If any rung path has logic continuity, the output is turned ON and retained ON, even if logic continuity or system power is lost. The latched output will remain latched ON until it is unlatched by an unlatch output instruction of the same reference address. The unlatch instruction is the only automatic (programmed) means of resetting the latched output. Although most controllers allow latching of internal or external outputs, some are restricted to latching internal outputs only.

Unlatch Output Coil

The unlatch coil instruction is programmed to reset a latched output having the same reference address. If any rung path has logic continuity, the referenced address coil is turned OFF or unlatched to an OFF condition. The Unlatch output is the only automatic means of resetting a latched output. Figure 9-17 illustrates the use of the latch and unlatch coils.

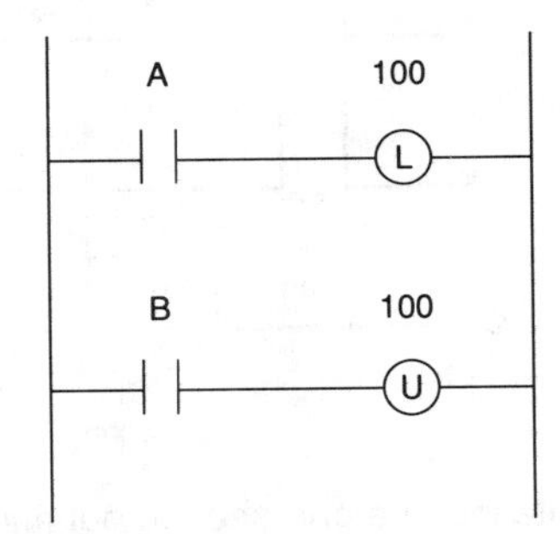

If A closes, coil 100 is latched until B closes to unlatch coil 100.

Figure 9-17. Example of latch and unlatch.

The latch and unlatch instructions may be found in block form as shown in Figure 9-18. The only difference is that the unlatching is performed in the same instruction. If the unlatch input is ON (continuity) the output coil will remain OFF. Note that the latch and unlatch outputs of Figure 9-17 can have ladder logic in between, while the one shown in Figure 9-18 obviously cannot.

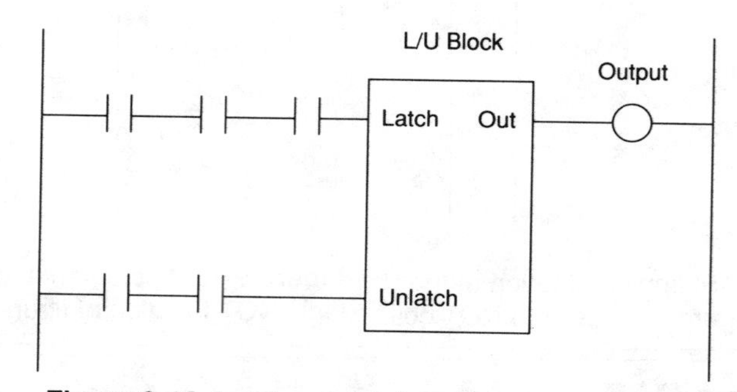

Figure 9-18. Latch-unlatch functional block instruction.

One Shot Output

The one shot output operates in a manner similar to the output coil. If a ladder rung has continuity, the one shot output will be energized (ON); however, the length of time the one shot is ON is *one scan or less*, depending where it is entered in the program.

One shot outputs are generally used for resetting conditions in one scan. Careful attention must be taken when using the one shot for resetting other output rungs or functional blocks. The logic to be reset must be programmed *after* the one shot rung is programmed. Figure 9-19 illustrates a one shot output and its corresponding timing diagram.

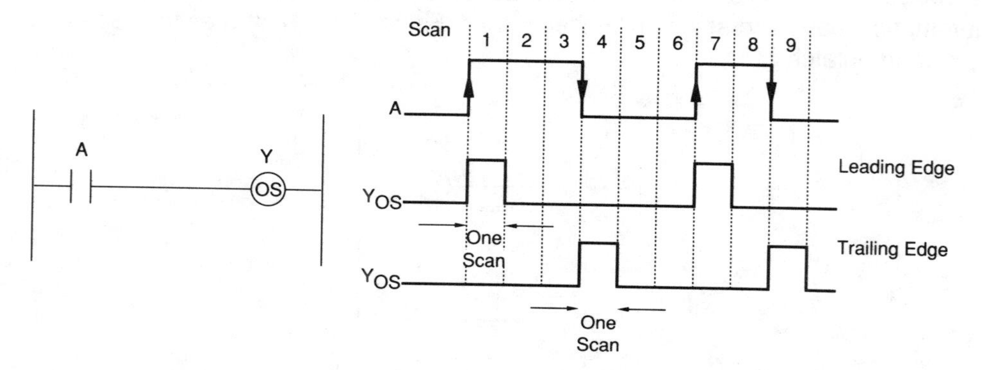

Figure 9-19. Implementation of a one-shot output and its timing diagram.

Depending on the PLC used, there may be a leading edge or a trailing edge triggered. The first one turns the output ON for one scan after the OFF to ON transition of the input. The second one turns the output ON for one scan after the ON to OFF transition of the input.

Transitional Contact

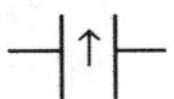

The transitional contact is programmed to provide a one-shot pulse when the referenced trigger signal makes a positive OFF-to-ON transition (leading edge) or an ON to OFF transition (trailing edge). This contact will close for exactly one program scan whenever the trigger signal goes from OFF-to-ON. The contact will allow logic continuity for one scan and then open, even though the triggering signal may stay ON. The triggering signal must go OFF and ON again for the transitional contact to close again. The contact address (trigger) may represent an external input/output or an internal output.

The transitional contact instruction is generally provided by PLCs which do not provide the one shot output. Like the one shot output, the transitional contact is generally used to reset conditions in one scan. For example, to reset a latch coil, i.e. unlatch. Figure 9-20 shows a circuit application of both transitionals and their respective timing diagram.

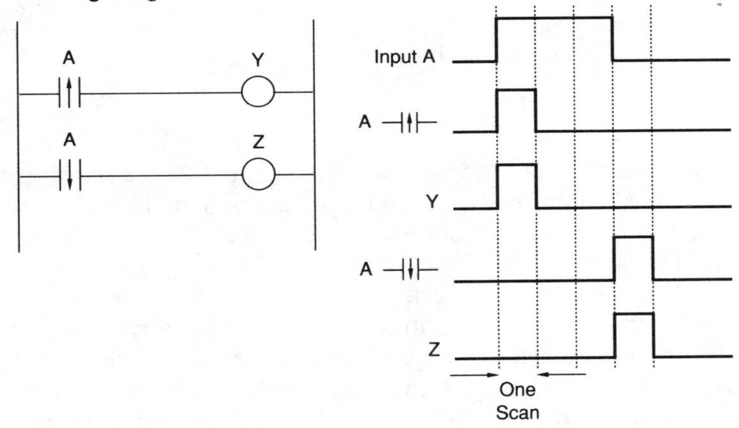

Figure 9-20. Transitional contacts and their timing diagram.

Ladder Scan Evaluation

Scan evaluation is an important concept since it defines the order in which the ladder diagram is executed. As we know, the processor starts solving the ladder program after it has read the status of all inputs and stored it in the input table.

The solution of the ladder program starts at the top of the program, beginning with the first rung. As the processor solves the control program it examines the reference of each programmed instruction so that it can determine logic continuity for the rung being solved. The processor does not go back to solve a previous rung even if output conditions that change in the present rung should affect previous rungs.

To make it more clear, let's take an example. Figure 9-21 illustrates four simple rungs. The first one is activated by a NO contact 10 which we will assume corresponds to a pushbutton; if 10 is ON, it will turn output 100 ON. In the second rung, the contact from 100 will turn output 101 ON; contact 101 will turn 102 ON and finally contact 102 will turn 103 ON. All these outputs turn ON *in the same scan*. When the processor finishes the program scan, it updates the real output devices connected to the modules. In this case if outputs 100, 101, 102 and 103 were connected to pilot lights, they would all turn ON at the same time.

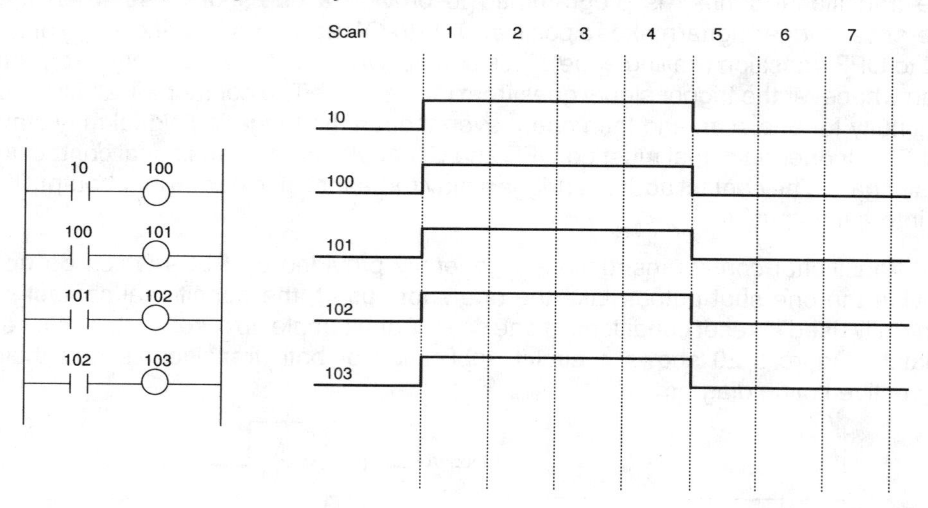

Figure 9-21.

Figure 9-22 illustrates the same ladder logic as in Figure 9-21 but with the placement of rungs reversed. What this shows is that to energize output 103 would take four scans. Assume that input 10 is pushed in the first scan. The processor examines reference 102 and finds it OFF (logic 0), therefore output 103 stays OFF. In the second rung, 101 is OFF and therefore 102 remains OFF. In the third rung, 100 is OFF so 101 remains OFF. In the fourth rung, 10 is ON because the pushbutton is pushed and turns 100 ON. In the next scan (second), if the pushbutton remains ON, the output 101 will turn ON because at the end of the first scan the reference address 100 was set to logic 1. This logic will continue until the fourth scan when all four outputs will be ON. The outputs will turn OFF in the same way once the pushbutton is released.

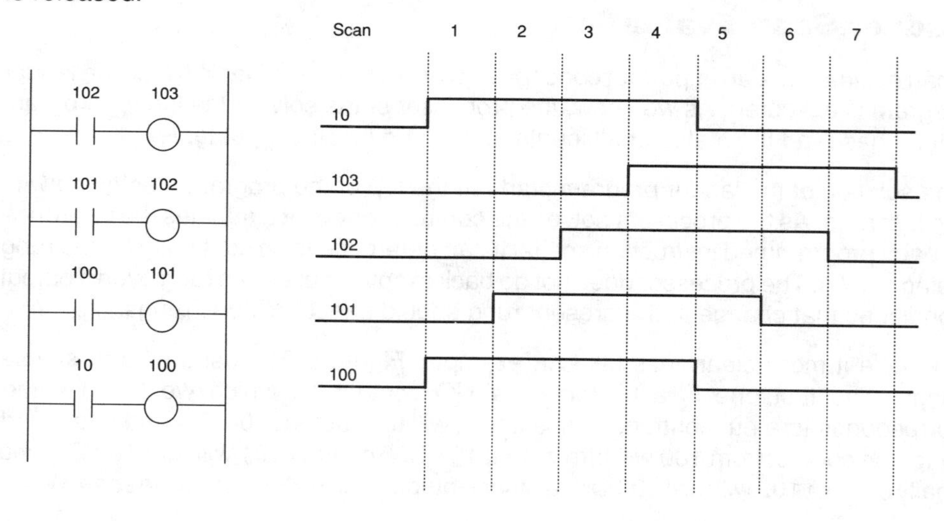

Figure 9-22.

The physical operation of the circuit in Figure 9-23 would be almost impossible to observe while the PLC is running the program because of the speed (scan in milliseconds) of the PLC. All the pilot lights would seem to come ON at the same time. The only sure way to observe the ladder outputs would be to use what is called a *single scan* operation of the PLC. Under a single scan, the processor reads the inputs, executes the logic and updates the outputs and stops until another single scan is executed. This single scan operation is generally used during testing of a control program.

The important thing to remember is that if you want an output to have a repercussion on another rung in the same scan, the first one must be programmed before the latter one. Order of execution problems can arise especially when using transitional contacts and one shot outputs to reset or unlatch other rungs. This is illustrated in Figure 9-23 where the output unlatch instruction will never take place.

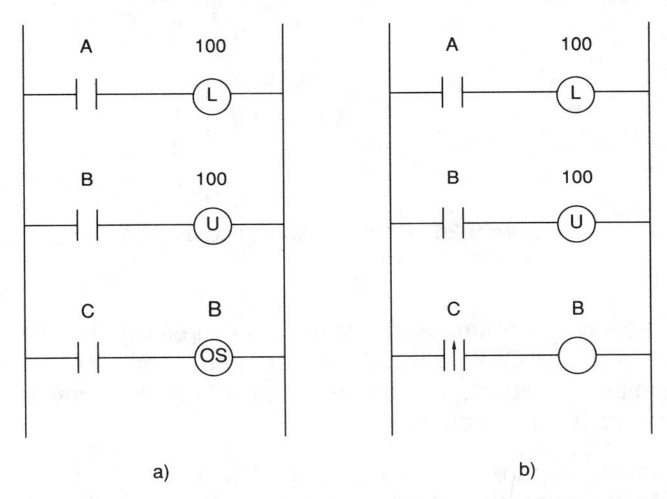

Figure 9-23. a) The one-shot and b) transitional logic rungs will never unlatch coil 100.

Programming a Normally Closed Input

As you may have noticed, we have tried to stay away from presenting input device connections which are in the normally closed condition. There is a simple reason for it: we did not want to confuse you. The understanding of how to program a NC input device may be trivial to some people, however to most of us, it can be one of the most difficult things to comprehend at first. Once you learn about it, try explaining it to someone else and watch their reaction.

Throughout the years, we have found that the best way yet to explain this topic is to use the following example. Suppose we want to implement the identical logic of the simple hardwire circuit shown in Figure 9-24. To implement the same logic we mean that the pilot light PL1 should behave in the same manner as in the hardwire circuit when implemented in the PLC. If PB1 is not pushed, PL1 should be ON; if PB1 is pushed, PL1 should be OFF. Figure 9-25 and 9-26 show the two possible methods of programming PB1 and implementing the logic. At first glance you may think that the solution in Figure 9-25 is the answer, but that is not true. Here are the reasons.

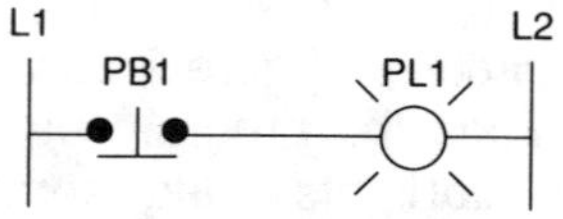

Figure 9-24. Hardwire circuit.

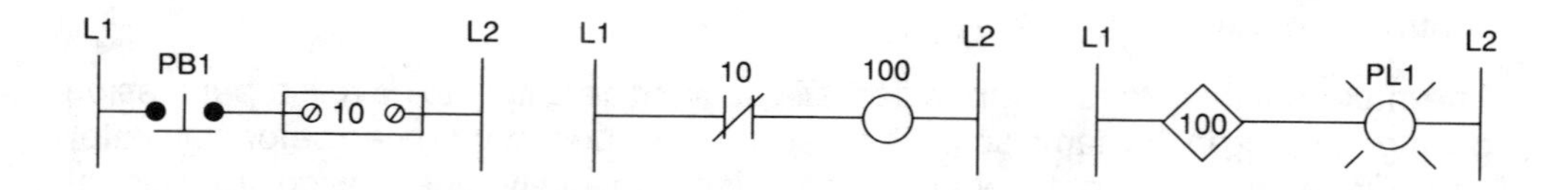

Figure 9-25. Programming using the NC.

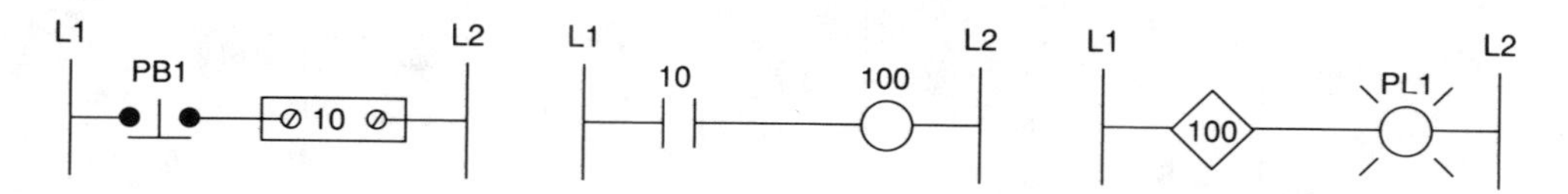

Figure 9-26. Programming using the NO.

In Figure 9-25 we have programmed the reference address of PB1 (10) as a normally closed contact (examine OFF) which drives the output coil 100 connected to pilot light PL1. When the PLC starts, it reads the status of the input device connected to input 10 and stores it in the input table.

If PB1 is not pushed (Figure 9-25), input 10 read as a logic 1 (power flowing to the module). During the execution of the ladder logic the examine OFF instruction is evaluated and will be since the reference (input 10) is ON, the NC contacts will be open, disrupting continuity. The output 100 will be OFF and the light PL1 would not turn ON. If PB1 is pushed, then the input module location 10 will read a 0 (power not flowing to the module). The examine for an OFF condition of reference 10 will be true, therefore the instruction will provide continuity and turn output 100 ON (and PL1).

In Figure 9-26, the input condition has been programmed as an examine ON instruction. During operation, if PB1 is not pushed the input module 10 will read an ON status. When the ladder rung is evaluated, the examine for an ON condition of reference 10 will be true, contact 10 will close to provide power, therefore turning output 100 ON and (PL1).

If PB1 is pushed (Figure 9-26), the module will have an OFF status, and the processor will store a logic 0 in the input table. During evaluation of the rung, the examine reference 10 for an ON condition will not be true (input 10 is OFF) and continuity will not occur because the contacts will be kept open. Therefore, output 100 and PL1 will be OFF.

The programming solution to the normally closed input connection, shown in Figure 9-26, exemplifies the following: *if you want a normally closed wired input device, when connected, to behave as a normally closed device, you must program it as an examine ON or normally open contact instruction.* Discrete inputs to a PLC can be made to act as NO or NC regardless of their original configuration. This ability to examine a single device for either an open or closed state is the key to the flexibility of a PLC; however the device is wired (NO or NC), the controller can be programmed to perform the desired action without changing the wiring. In most cases, an input device wired normally closed is programmed as a normally open PLC contact. Remember that the programming state of an input depends not only on how it is wired, but also on the desired control action. The following example shows the case in which one pushbutton, with two contacts, is brought to the PLC and is programmed differently depending on which contact is wired to the module.

Example 9-4

Implement the hardwired logic shown in Figure 9-27 in a PLC using only one pushbutton connection with **a)** the normally open connected to an input module and **b)** the normally closed connected to the input module. Use input address 10 for the pushbutton and addresses 30 and 31 for the pilot lights PL1 and PL2 respectively. Indicate the lights in the ON condition (without PB1 being pushed) by a filled PL indicator.

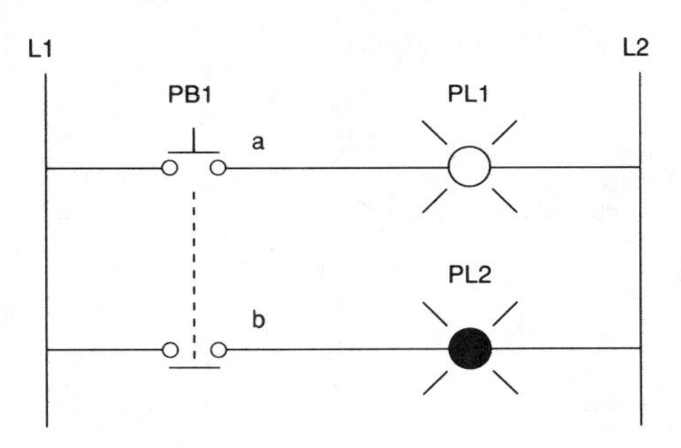

Hardwired Logic

Figure 9-27

Solution

Examining the circuit in Figure 9-27 shows that if PB1 is not pushed PL1 should be OFF. PL2 should be ON because the other contact of PB1 (NC) is providing power to PL2. We can wire any of the two connections of PB1 (a or b) to the input module and take care of the required logic. Remember that we can make any contact act in the program as we desire (as a NO or NC).

a) The circuit solution to the NO pushbutton connection is shown in Figure 9-28. PL1 is driven by an examine ON instruction and PL2 by an examine OFF instruction. When PB1 is not pushed PL1 is OFF and PL2 is ON. The first rung implements a wired NO pushbutton to act as a NO pushbutton while the second rung implements a wired NO pushbutton to act as a NC pushbutton.

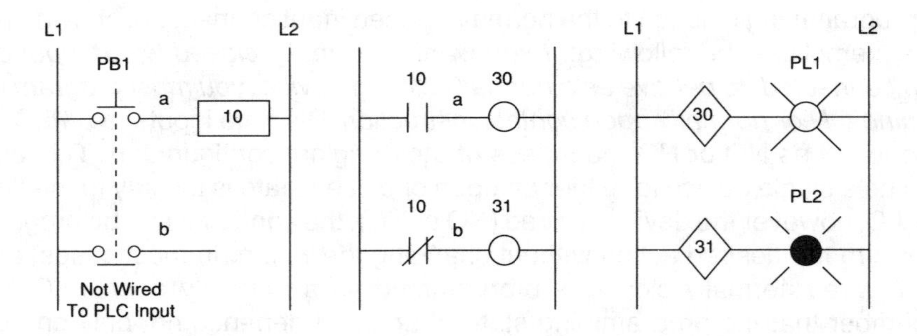

Figure 9-28. NO Connected to PLC.

b) The circuit solution to the NC pushbutton connection is shown in Figure 9-29. PL1 is driven by an examine OFF instruction. During operation, PB1 (b) provides power to the module if not pushed, therefore the reference address (10) is a 1. The NC contacts with address 10 will be open as long as PB1 is not depressed, keeping PL1 (output 30) OFF. In the second rung, the output for PL2 (31) is driven by an examine ON instruction which closes as long as PB1 is not pushed. The first rung implements a wired NC pushbutton to act as a NO pushbutton while the second rung implements a wired NC pushbutton to act as a NC pushbutton.

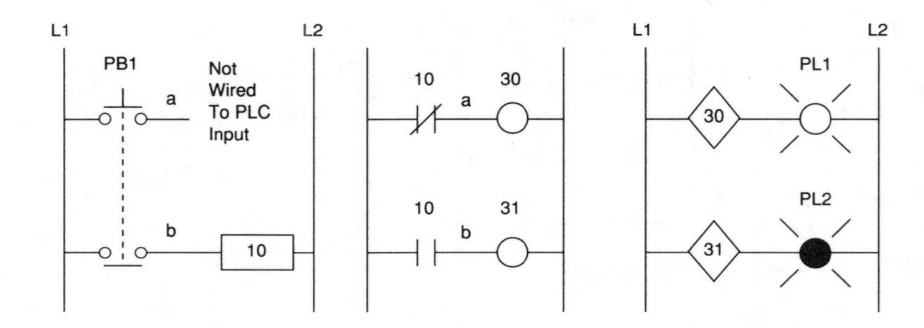

Figure 9-29. NC Connected to PLC.

9-5 TIMER AND COUNTER INSTRUCTIONS

Timers and counters are internal instructions that provide the same functions as would hardware timers and counters. They are used to activate or de-activate a device after a time interval has expired or a count has reached a preset value. The timer and counter instructions are generally considered internal outputs. Like the relay-type instructions, timer and counter instructions are fundamental to the ladder diagram instruction set.

Timers may have one or more time bases which are used to time an event. For instance, resolutions of one hundredth of a second may be available as a time base (0.01 sec). Typical time bases are 0.01 sec, 0.1 sec and 1.0 sec. Although it is unusual, you may find a PLC timer with a time base of 1 minute. Applications of timers are innumerable considering that PLC timers completely replace hardware timers. There are however, instances where timers need to be added to a control program to introduce delays in the hundredths of seconds. Such delays may be added because a PLC scan is turning its outputs ON very fast as compared to a hardwired relay system.

Counters are used in general to count events. Typical events include parts passing on a conveyor, the number of times a solenoid is turned ON, etc. As will be seen, timers and counters (T/C) must have two values, a *preset* and an *accumulated* value. These values are stored in register or word locations in the data table. Timers and counters may be implemented using the basic ladder format or the block instruction format.

Timers

There are several types of timer instructions available in PLCs. However, PLC manufacturers may provide different definitions for each type of timer function offered. The following list shows the types of timers available.

- Timer ON Delay Energize
- Timer ON Delay De-energize
- Timer OFF Delay Energize
- Timer OFF Delay De-energize
- Retentive Timer

The functionality of these timers will be essentially the same differing only in the type of output provided. Figure 9-30 illustrates two format representations generally found for timers.The block format timer may have one or two inputs depending on the PLC. These inputs are normally labeled *control* and *enable/reset*. If the control line is TRUE (i.e. has continuity), and the enable line is also true, the block function will start timing. The ladder format timer generally has only one input which is the control line. If the control line is ON, the timer will start timing.

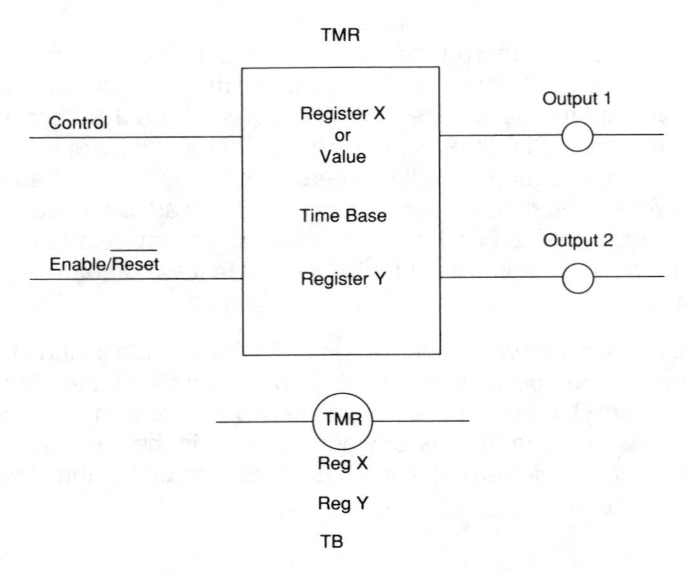

Figure 9-30. Timer formats: block (top) and ladder (bottom).

Common to both timer formats is the use of the Register X, where X implies any register address, to hold a preset value and register Y to store an accumulated value. The time base may be selectable depending on the PLC used (e.g. 0.01 sec, 0.1 sec, 1.0 sec, etc.) and it determines the resolution of the timer. When the accumulated *tick* count is equal to the preset count, the timing function is executed and output condition set depending on the type of timer (e.g. ON delay energize, etc.).

ON Delay Timer Energize

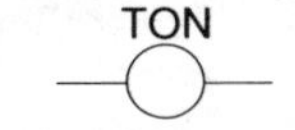

The delay energize timer output instruction is programmed to provide time delayed action or to measure the duration for which some event is occurring. Once the rung has continuity, the timer begins counting time-based intervals and times down until the accumulated time equals the preset value. When the accumulated time equals the preset time, the output is energized, and the timed out contact associated with the output is closed (see Figure 9-31). The timed contact can be used throughout the program as a NO or NC contact. If logic continuity is lost before the timer times out, the accumulator register is reset to zero.

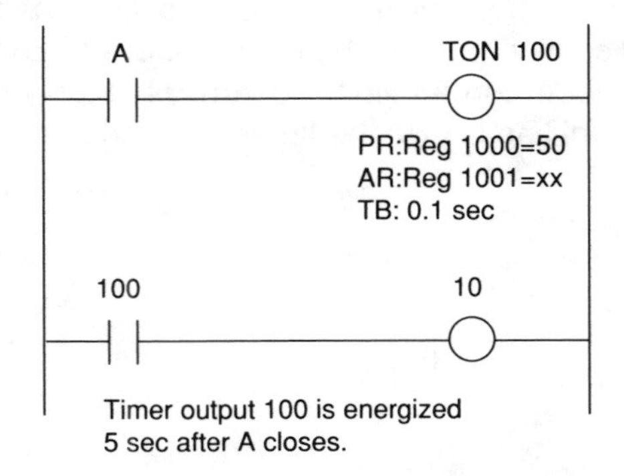

Figure 9-31. On-delay timer energize.

ON Delay Timer De-Energize

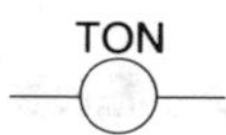

The ON delay timer de-energize operates in a similar manner as the energize type. The output of the timer however, is ON and will be de-energized once the rung has continuity and the time interval has elapsed (accumulated register = preset register). A PLC manufacturer generally provides either the ON delay energize or the ON delay de-energize since the remainder can be easily implemented (see Chapter 11—section on Programming Tip Circuits). Figure 9-32 illustrates a timing diagram for both types of ON delay timers.

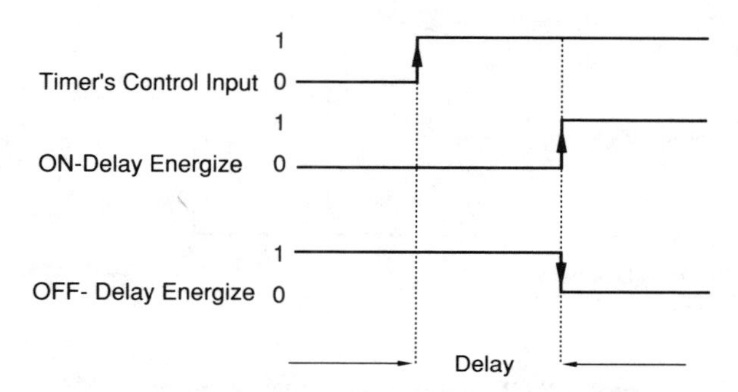

Figure 9-32. Timing diagram for the ON delay timers.

OFF Delay Timer Energize

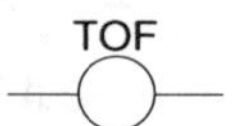

The delay de-energize timer output instruction is programmed to provide time delay action. If the control line rung does not have continuity, the timer begins counting time-based intervals until the accumulated time equals the programmed preset value. When the accumulated time equals the preset time, the output is energized and the timed-out contact associated with the output is closed (see Figure 9-33). The timed contact can be used throughout the program as a NO or NC contact. If logic continuity is gained before the timer is timed out, the accumulator is reset to zero. Reference Figure 9-34 for the timing diagram illustration of this timer.

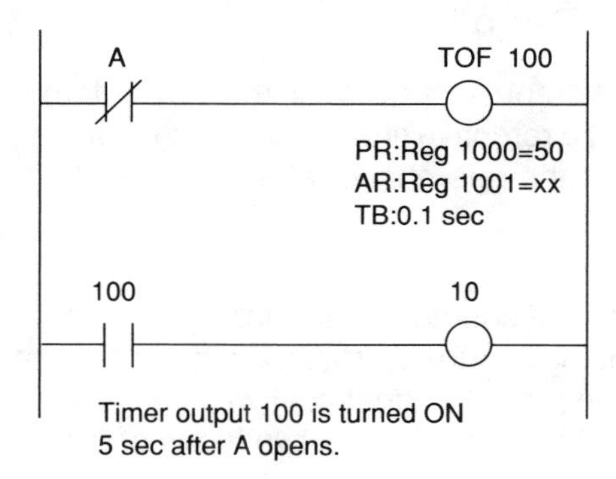

Figure 9-33. OFF delay energize timer.

OFF Delay De-Energize

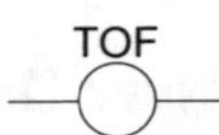

The OFF delay de-energize timer is similar in operation to its OFF delay energize counterpart. The output of the timer however, is ON or energized and will be de-energized once the rung does not have continuity in the control line and the time interval has elapsed or accumulated register = preset register. Just as with the ON delay timers, a PLC manufacturer will provide the OFF delay energize or de-energize. Figure 9-34 illustrates the timing diagram difference for both types of OFF delay timers.

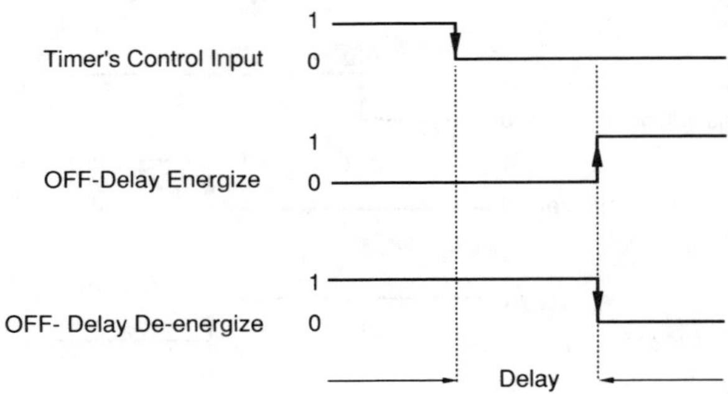

Figure 9-34. OFF delay timer timing diagram.

Retentive ON Delay Timer

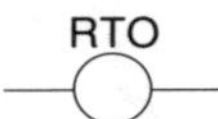

The retentive timer output instruction is programmed if it is necessary for the timer accumulated value to be retained, even if logic continuity or system power is lost. If any rung path has logic continuity, the timer begins counting time-based intervals until the accumulated time equals the preset value. The accumulator register retains the accumulated value, even if logic continuity is lost before the timer has timed out, or if power is lost. When the accumulated time equals the preset time, the output is energized, and the timed out contact associated with the output is turned ON. The timer contacts can be used throughout the program as a NO or NC contact. The retentive timer accumulator value must be reset by the retentive timer reset instruction.

Retentive Timer Reset

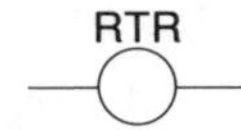

The retentive timer reset output instruction is the only automatic means of resetting the accumulated value of a retentive timer. If any rung path has logic continuity, then the accumulated value of the referenced retentive timer is reset to zero.

Counters

There are two basic types of counters, one that can count up and another one that can count down. Depending on the controller, the format may also vary. Some PLCs may have the ladder type format (output coil) while others may have the functional block format. Figure 9-35 illustrates these two types of formats.

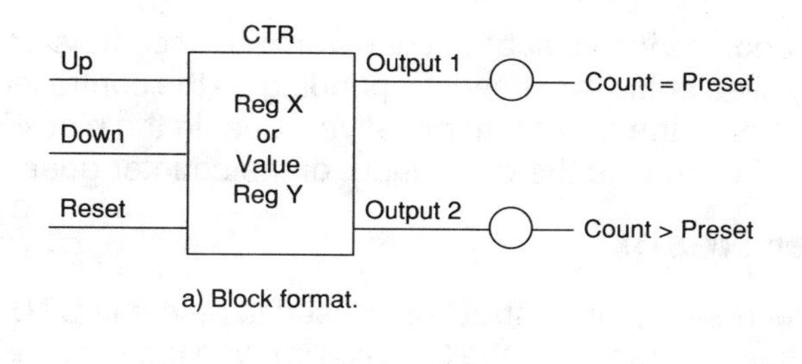

a) Block format.

CTU
Up Counter
Reg X
Reg Y
CTD
Down Counter
Reg X
Reg Y
CTR
Reset Counter

a) Basic ladder format.

Figure 9-35. Types of counter formats generally found.

Up-counter

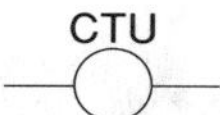

The up-counter output instruction will increment by one each time the counted event occurs. A control application of a counter is to turn a device ON or OFF after reaching a certain count (preset value in Register X). An application of a counter would be to keep track of the number of filled bottles that pass a certain point. The up-counter increments its accumulated value (Register Y) each time the up-count event makes an OFF-to-ON transition. When the accumulated value reaches the preset value, the output is turned ON and the count is finished and the contact associated with the referenced output is closed.

Depending on the controller, after the counter reaches the preset value it is either reset to zero or continues to increment for each OFF-to-ON transition. If the latter case is true, a reset instruction is used to clear the accumulator.

Down-counter

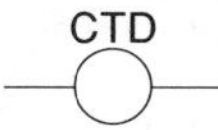

The down-counter output instruction will decrement by one each time a certain event occurs. Each time the down-count event occurs, the accumulated value is decreased. In normal use, the down-counter is used in conjunction with the up-counter to form an up/down counter given that both counters have the same reference registers. In this application, the down-counter provides a means for accounting for data correction. For example, while the CTU, or up counter, counts the number of filled bottles that pass a certain point, a CTD, or down counter, with the same reference address would subtract one from the accumulator each time an empty or improperly filled bottle is sensed. Depending on the controller, the down-counter will stop at zero or at the maximum negative value. In the block format, the down count is performed each time the down input of the counter goes from OFF to ON.

Counter Reset

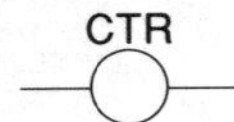

The counter reset output instruction is used to reset the CTU and CTD accumulated values. When programmed, the CTR coil is given the same reference address as the CTU and CTD coils. If the CTR rung condition is TRUE, the referenced address will be cleared. The reset line in the block format sets the accumulated count to zero (Register Y). Figure 9-36 illustrates a typical counter rung using the block formatted counter.

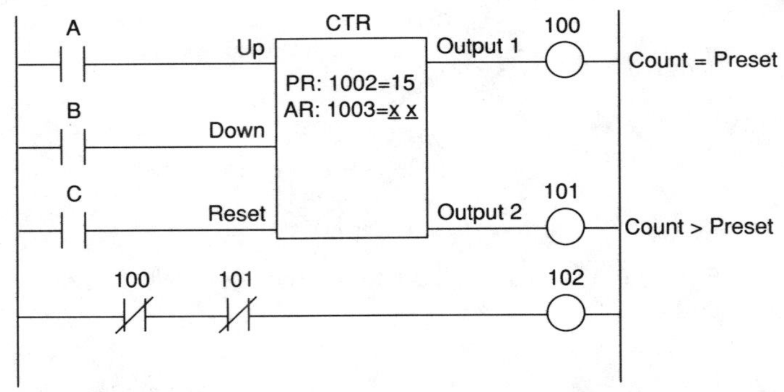

Counter will count up when A closes, count down when B closes and is reset to 0 (register 1003) when C closes. If count is equal to 15 as a result of an up or down count, output 100 will be ON. If Register 1003 contents are greater than15, output 101 will be ON. Output 102 is ON if the accumulated count is less than 15.

Figure 9-36. Functional block up/down counter instruction.

9-6 PROGRAM/FLOW CONTROL INSTRUCTIONS

Program control instructions are used to direct the flow of operations and execution of instructions within a ladder program. These operations are accomplished using branching and return instructions inside the program which are executed if certain control logic conditions already programmed are met. Typically, program control instructions form a fence within a program. This fence contains groups of other ladder instructions that are used to implement a desired function. Figure 9-37 illustrates the fence creation within a program.

Several types of flow control instructions may be encountered in a particular PLC depending on its designed capability and the scope of the application. These instructions provide the PLC with the ability to perform efficiently special user programmed routines that need be called only when required. The scan time is therefore reduced, optimizing total system response.

Table 9-2 shows the most commonly used flow control instructions. These instructions are generally used in pairs, that is, one instruction starts the change of flow control which means that it sends execution of instructions to another section of the

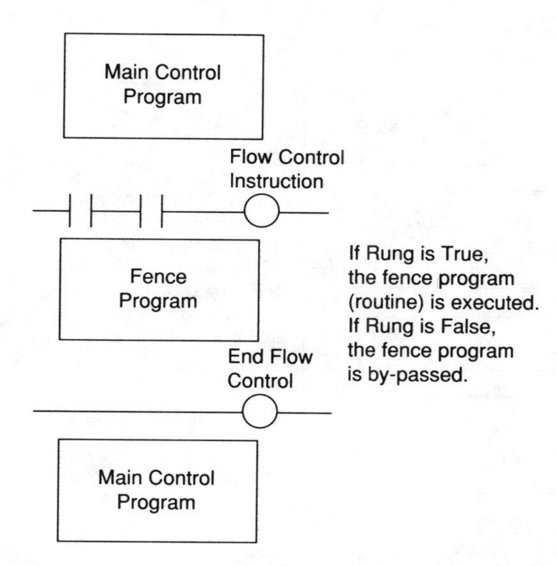

Figure 9-37. Illustration of a fence created by using a flow control instruction.

Instruction	Name
MCR	Master Control Relay
ZCL	Zone Control Last State
JMP (GO TO)	Jump To (Go To)
GOSUB (JSB)	Go Subroutine (Jump to Subroutine)
LBL	Label (To Jump to or To GoTo)
RET	Return (From Subroutine)
END	End (End MCR or ZCL)

Table 9-2. Types of flow control instructions.

program first while the other one returns it back to where it was when the change of control flow first took place.

Master Control Relay

MCR

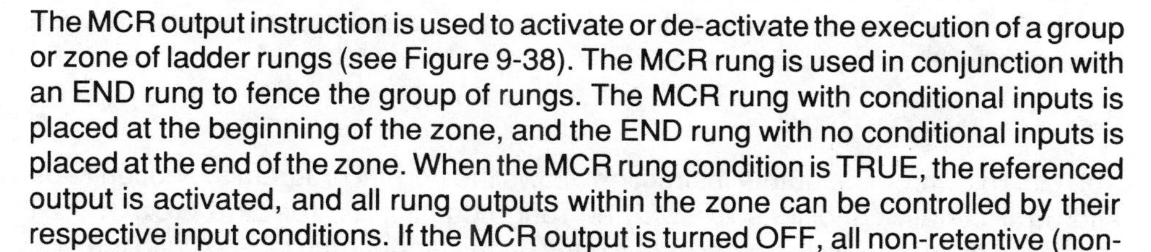

The MCR output instruction is used to activate or de-activate the execution of a group or zone of ladder rungs (see Figure 9-38). The MCR rung is used in conjunction with an END rung to fence the group of rungs. The MCR rung with conditional inputs is placed at the beginning of the zone, and the END rung with no conditional inputs is placed at the end of the zone. When the MCR rung condition is TRUE, the referenced output is activated, and all rung outputs within the zone can be controlled by their respective input conditions. If the MCR output is turned OFF, all non-retentive (non-latched) outputs within the zone will be de-energized.

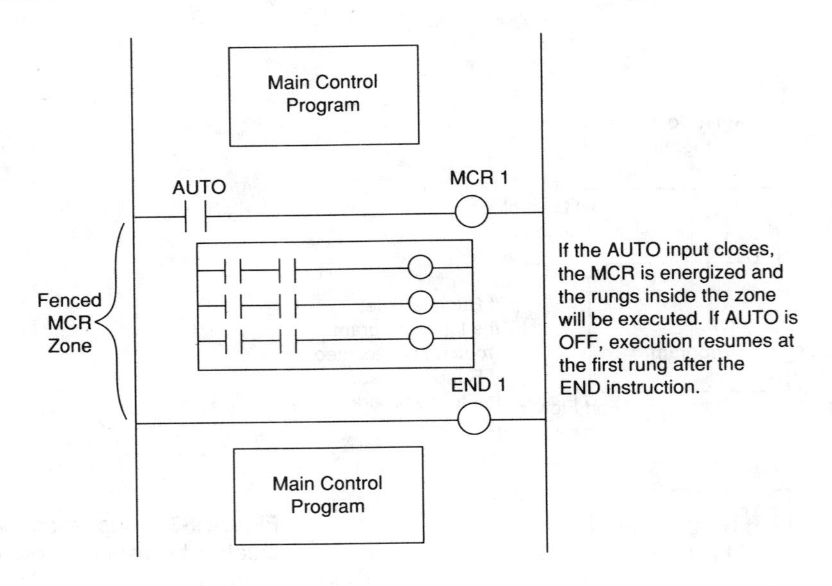

Figure 9-38. MCR instruction application.

Zone Control Last State

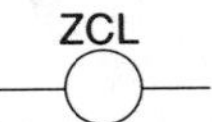

The zone control instruction is similar to the MCR instruction. It determines if a group of ladder rungs will be evaluated or not. The ZCL output with conditional inputs is placed at the start of the fenced zone, and an END ZCL output with no conditional inputs is placed at the end of the zone. If the referenced ZCL output is activated, the outputs within the zone are controlled by their respective rung input conditions. If the ZCL output is turned OFF, the outputs within the zone will be held in their last state.

Jump To

The jump to instruction allows the program sequence to be altered if certain conditions exist. If the rung condition is true, the JMP coil reference address tells the processor to jump forward and execute the target rung. Figure 9-39 illustrates the JMP instruction. The target rung to jump to is generally defined by a JMP address label which uses the —| LBL |— instruction. Other PLCs may use the jump to (coil) address reference as the logic rung to go to. Using this instruction, order of execution can be altered to execute a rung that needs immediate attention. This instruction may also be called a go to (coil address). Care should be exercised when jumping over timers and counters.

Go To Subroutine

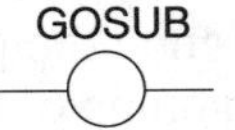

The Go To Subroutine output instruction allows the normal program execution to be altered if certain conditions exist. If the rung condition is true, the GOSUB coil reference address tells the processor to jump to the ladder rung labeled with the LBL instruction having the same reference number, and continue program execution until a RETurn coil is encountered. Each subroutine begins with a labeled rung and must end with an unconditional RETurn instruction. This instruction may also be called Jump to Subroutine (JSB).

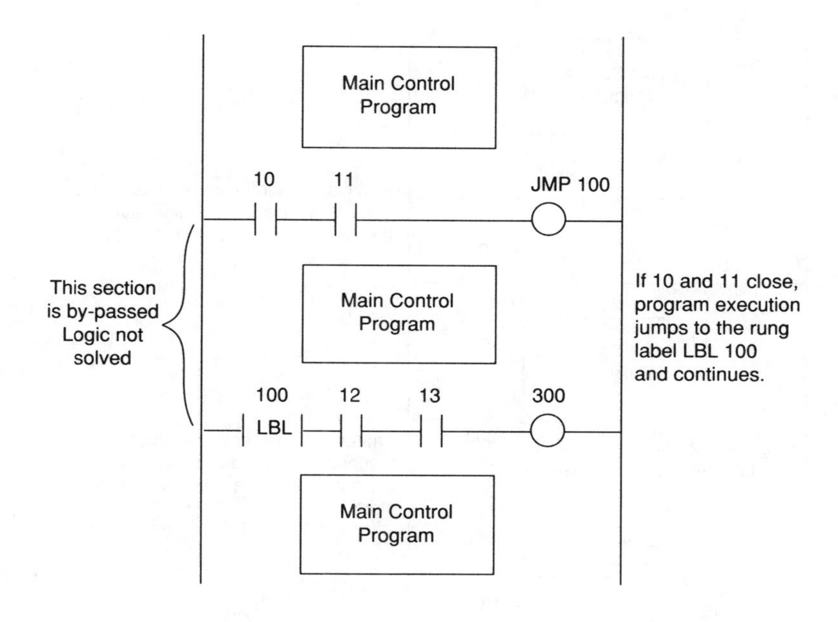

Figure 9-39. Illustration of Jump instruction.

This instruction is very useful whenever a routine in the program may be recalled by several sections of the main control program or on a timely basis (i.e. look-up analog interpretation table every 10 seconds). Subroutine sections are generally located at the end of the control program and are sometimes specified by the PLC maker (see Figure 9-40). It is recommended, for programming documentation order, that the subroutines be placed at the end of the program. If a particular PLC does not have a reserved subroutine area you can create the subroutine area by programming a dummy rung with direct control to another dummy rung at the end of *your* subroutine area (see Figure 9-41).

Label

—| LBL |—

The LBL instruction is used to identify a ladder rung which is the target destination of a JMP or GOSUB (Go to Subroutine) instruction. The LBL reference number must match that of the JMP and/or GOSUB instruction with which it is used. The LBL instruction does not contribute to logic continuity, and for all practical purposes is always logically true. This instruction is placed as the first condition instruction in the rung. A LBL instruction referenced by a unique address can be defined only once in a program.

Return Coil

RET
—( )—

The return instruction is used only to terminate a ladder subroutine. It is programmed with no conditional inputs. When encountered, program control is returned to the main program and begins at the ladder rung immediately following the JSB instruction that initiated the subroutine. Normal program execution continues from that point. A RET instruction must be programmed for each subroutine.

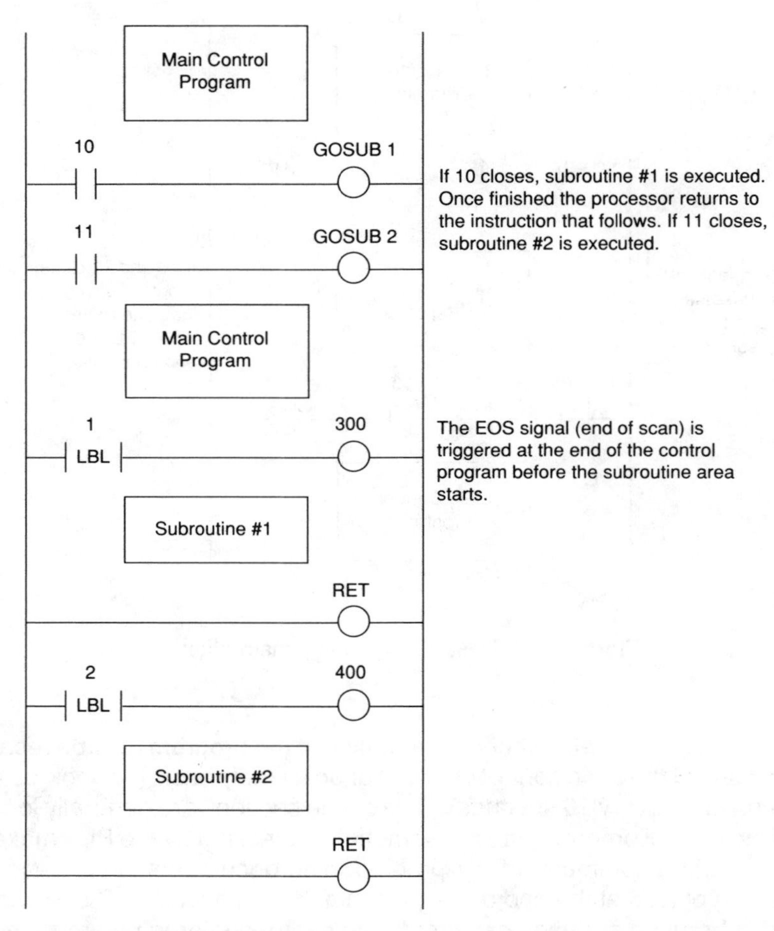

Figure 9-40. PLC with assigned subroutine at end of program.

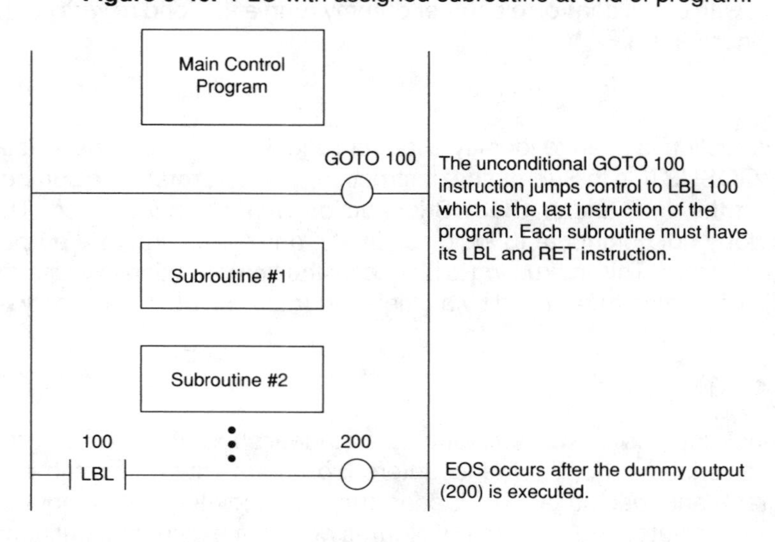

Figure 9-41. User created subroutine area.

END

The END instruction is used to signify the last rung of an MCR or ZCL instruction. The END function is generally programmed without any conditions to energize (unconditional). The reference to the MCR or ZCL, depending on the controller, may or may not be included in the END address. If it is included, it will determine the END of a particular MCR or ZCL. If it is not included, it will terminate the latest MCR or ZCL instruction.

9-7 ARITHMETIC INSTRUCTIONS

Arithmetic instructions in a PLC include the basic four operations of addition, subtraction, multiplication, and division. In addition to these four math functions, square root operations may also be found in large PLCs.

Like other instructions, the formats which may be encountered in a PLC include the basic ladder format and the functional block format. However, operation of either format is essentially the same. Figure 9-42 illustrates these possible formats. Most of the instructions require three reference registers which define the two operands and the destination register of the operation. Some instructions, such as multiplication and division may use four registers. When larger numbers are employed in a PLC system, double precision arithmetic may be used. Double precision implies that double the number of registers are used to hold the operands and results. For instance, a double precision addition would use a total of six registers, two for each operand and two for the result. The numerical format used in the math operations will vary depending on the PLC and are usually three, four, or five digits (BCD or binary).

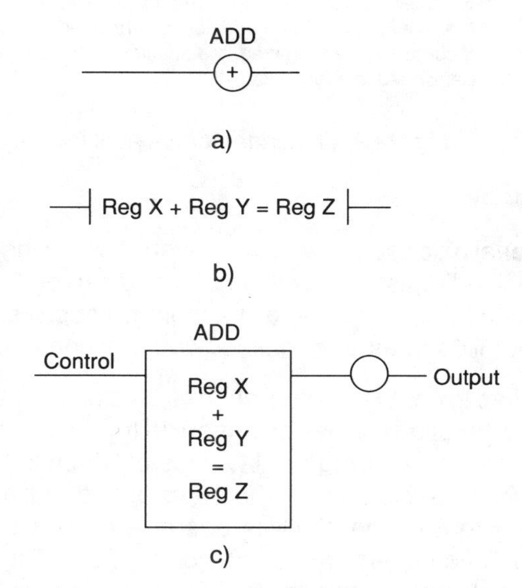

Figure 9-42. Types of format used for arithmetic instructions: a)Coil, b)Contact and c)Block.

The following instruction operations are presented for both types of formats to familiarize the reader with what may be expected. Note that the ladder type may require other ladder data transfer instructions to obtain the arithmetic operands. In the functional block format, some manufacturers offer the ability to cascade block functions. This cascading is very useful when dealing with multiple arithmetic operations. Other manufacturers allow you to perform arithmetic operations in block form; that is, using a block of several contiguous registers as either of the operands and storing the results in another block of registers.

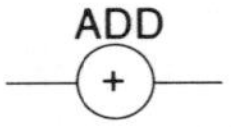

Addition—Ladder

ADD
—(+)—

The ADD instruction performs the addition of two values stored in the referenced memory locations. How these values are accessed is dependent on the controller. Some instruction sets use a GET data transfer instruction to access the two operand (registers) values as shown in Figure 9-43, while others simply reference the two registers using contact symbols (see Figure 9-42b) The result is stored in the register referenced by the add coil. If the addition operation is enabled only when certain rung conditions are true, then the input conditions should be programmed before the values are accessed in the addition rung. Overflow conditions are usually signalled by one of the bits of the addition result register.

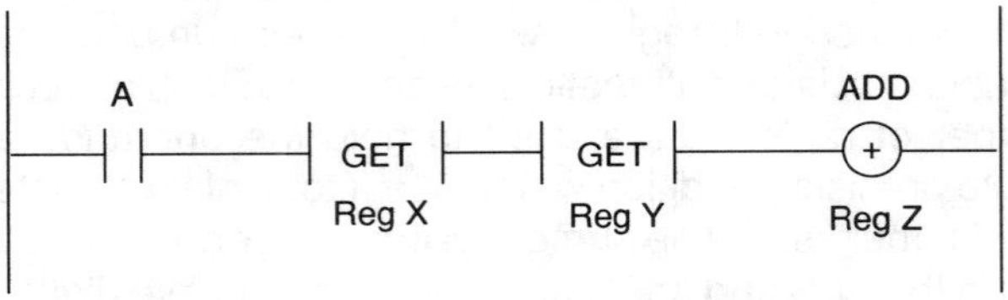

A
GET
Reg X
GET
Reg Y
ADD
(+)
Reg Z

If A closes, the contents of Register X and Register Y are added and stored in Register Z.
If A does not close, no addition is performed.
If contact A were omitted, the addition would be performed in every scan.

Figure 9-43. Ladder format addition.

Addition—Block

The addition functional block sums two values stored within the controller and places the results in a specified register. The values added can be fixed constants, values contained in I/O or holding registers, or variable numbers stored in any memory location. Figure 9-44 illustrates a typical addition functional block.

The control line is used to enable the operation of the block. When the rung conditions are true, the addition function will be performed. In this block, register X and register Y can be a preset value or a storage or I/O register. Each time the control signal is enabled (OFF-to-ON transition), the two numbers are added, and the result is placed in register Z. The output of the block is energized when the addition operation overflows. If the operation overflows, some controllers will clamp the results at the maximum value that the register can hold, while others will store the difference between the maximum count value and the actual overflow value.

Another method used by some controllers is the double precision addition using the block format. This operation is identical to the simple addition, but two registers are used to hold the numbers to be added, and two registers are used to store the result.

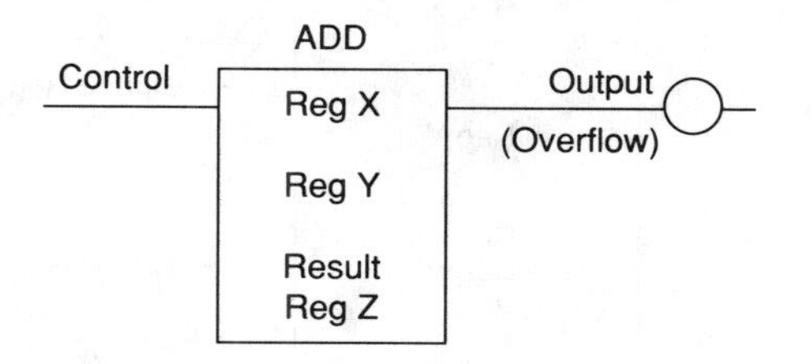

Figure 9-44. Block addition.

Subtraction—Ladder

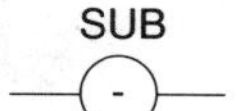

The SUB instruction performs the subtraction of two operands stored in two registers. As in addition, if the rung is enabled the subtraction operation takes place. The two registers used by the SUB instruction are generally accessed by a GET data transfer instruction. The subtraction result register will usually have an underflow bit to represent a negative result. Figure 9-45 shows a rung with the SUB instruction.

A
GET
Reg X
GET
Reg Y
SUB
Reg Z

If A closes, the contents of Register Y are
subtracted fom Register X (Reg. X - Reg. Y)
and the result is stored in Register Z.
If A does not close, no subtraction is performed.

Figure 9-45. Ladder format subtraction.

Subtraction—Block

The subtraction functional block, as in the ladder format, takes the difference of two values and stores the result in a register. A typical subtraction functional block is shown in Figure 9-46.

The control input operates in the same manner as in addition. When set to 1, the block operation is performed. The three operand registers are used to hold the data during the operation. The values these registers can assume also vary in format and may or may not include a sign. For example, register X could contain 9009 decimal, and register Y could hold -10,020. The result of this operation would be +19,029, stored in register Z [9009 - (-10,020)]. The formats for subtraction vary, and sometimes the result register may not include a result sign. In this case, the controller will normally provide three outputs: a positive result, register X greater than register Y; equal result, register X equal to register Y; or negative result, register X less than register Y. This type of block essentially performs a comparison function.

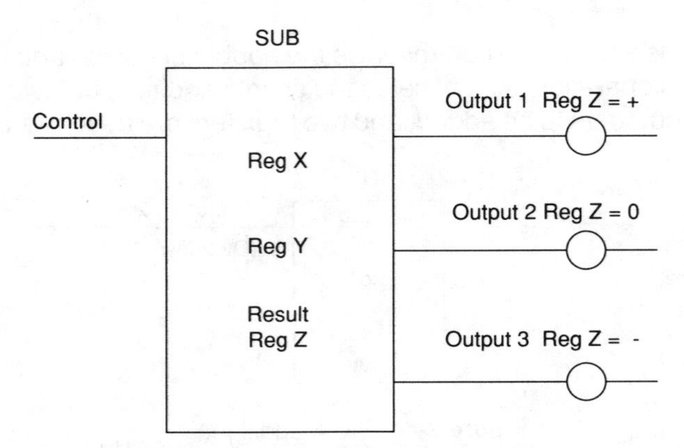

Figure 9-46. Subtraction block.

Some controllers allow you to add a constant, directly into the block function, to another register by having perhaps a **K** in front of the number where the register is inserted (e.g. K1035, constant 1035). Controllers that do not provide other I/O transfer instructions may use the subtraction block to read an analog or multibit I/O value through I/O registers and to compare it to a set parameter. The output coils are used to signal a greater than, equal to, or less than the set point comparison. If that value is only to be read, a constant of zero (K0) could be subtracted (Register Y = K0) from Register X (the input value).Figure 9-47 illustrates an example of a SUB block instruction to input an analog channel.

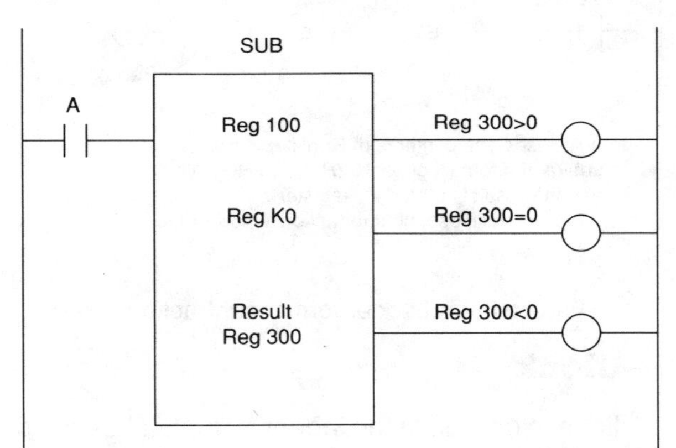

If A closes, the SUB operation is executed. Register 100 may represent or have stored the address where the input module is located (analog or multibit). After the execution, a constant of zero is subtracted from the input value and stored in Register 300 for use by the program.

Figure 9-47. Subtraction block used to read analog inputs.

Multiplication—Ladder

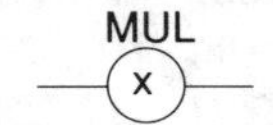

The multiplication instruction performs the multiplication operation. It uses two registers to hold the result of the operation performed by two operand registers. The two result registers are referenced by one or two output coils depending on the PLC.

These operand registers may be brought into the operation using GET instructions. If there is a condition to enable the operation, it should be programmed before the two operands are accessed in the multiplication rung.

Multiplication—Block

As with the previous math blocks, the multiplication functions have two registers to hold the operands and a register to store the result.The multiplication block is illustrated in Figure 9-48, where the control line is used to enable the operation.

Normally, the product of two 4-digit numbers will result in an eight digit number. In such a case, some controllers will provide two registers in which to store the result (Register location Z and Register location Z + 1). Other controllers use what is known as *scaling*, in which the result of the multiplication is held temporarily in two registers and then multiplied by the scale value. For example: if a PLC has a 4-digit BCD format, and register X and Y contain 9001 and 8172 respectively, with a scaling value of -5 (or 10^{-5}), the result 73556172 will be held in two result registers temporarily (7355 and 6172), and then multiplied by the scaling value (10^{-5}). The result will be 735.56172. The result register will have 736 (rounded-off). The programmer should know that the result has been scaled. The actual result then is 736 x 10^5 (73600000).

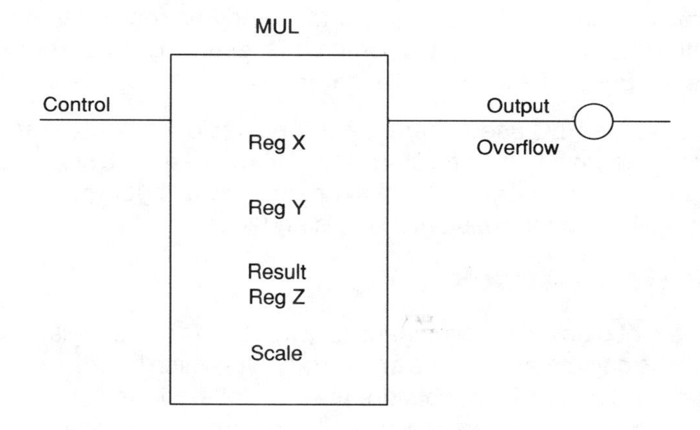

Figure 9-48. Multiplication Block.

Division—Ladder

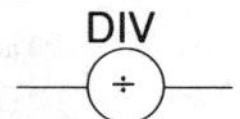

The division instruction performs the quotient calculation of two numbers. The result of the division is held in two result registers and referenced by the output coil. The first result register generally holds the integer, while the second result register holds the decimal fraction. Both operands used in the division operation may be obtained through the use of GET instructions.

Division—Block

The divide functional block performs the quotient calculation of two numbers that can be stored in one or more registers. This functional block is illustrated in Figure 9-49.

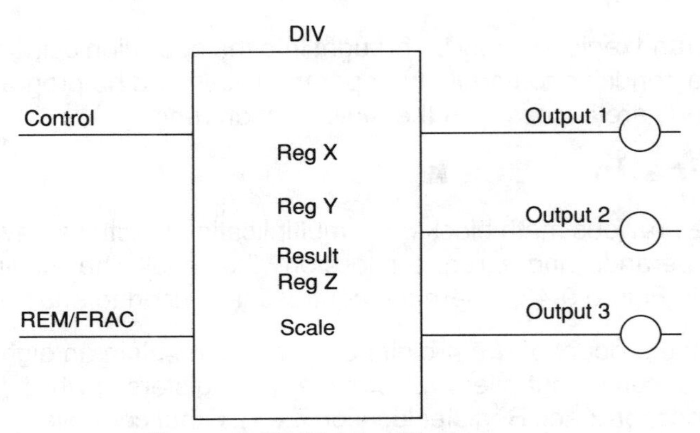

Figure 9-49. Division Block.

The division begins after the control rung has continuity with register X, the dividend, being divided by register Y, the divisor, and the result placed in register Z. The control line input will enable the block to perform the function. The REM/FRAC (REMainder/FRACtion) control line is used to determine whether one of the resultant register contents should be integer or fraction.

Some controllers also offer scaling factors to be specified in the block. This scaling permits fractional results to be scaled and stored in a register that otherwise would be lost. Other controllers have two contiguous result registers (Reg Z and Reg Z +1) where the integer and a fraction are stored.

Depending on the PLC used, there are three possible outputs that can be encountered. When energized, the top output generally represents a successful division, the middle output will represent an overflow or error (divide by zero), and the lower output will indicate whether the result has a remainder or not.

Square Root—Block

The square root block instruction generally has two to three registers, one that holds the value to be calculated and one or two registers which hold the result of the square root operation. These result registers may include one register for an integer and another one for a fraction ; scaling may also be provided. Once the control rung has continuity, the square root operation takes place. Of the possible outputs, the first one represents a successful or valid operation, and the second one may indicate a fractional value. Figure 9-50 illustrates a square root block instruction.

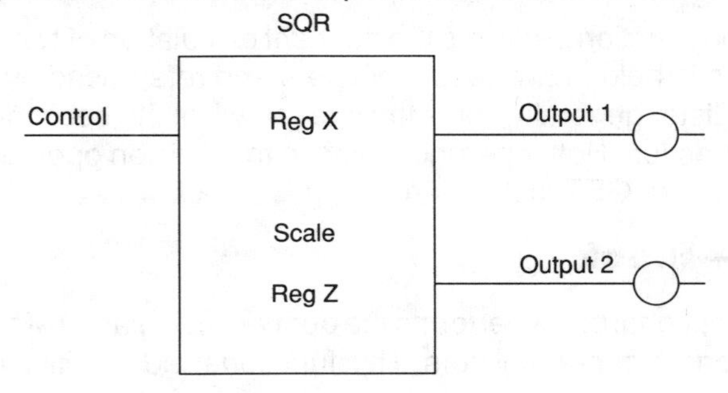

Figure 9-50. Square root block representation.

9-8 DATA MANIPULATION INSTRUCTIONS

The data manipulation instructions are an enhancement of the basic ladder diagram instruction set. Whereas the relay-type instructions were limited to the control of internal and external outputs based on the status of specific bit addresses; the data manipulation instructions allow multi-bit operations. Data manipulation instructions deal with operations that take place within one, two, or more registers. Typical functions encountered in data manipulation include *data comparisons, logic matrix, data conversion, set constant parameters, increment registers, logical shift* and *rotate operations,* and *examine bit* functions.

Data Comparisons

Data comparison instructions, as its name implies, are used to compare the values stored in two registers. These instructions are useful when checking for proper range of values used in a control or data entry section of the application program. The data comparison instruction may be found in some controllers in the basic ladder format while in other controllers it may be found as a block instruction. In both formats, there are three basic data comparisons which are: *compare equal to, compare greater than* and *compare less than.* Based on the results of these operations, outputs can be turned ON or OFF and other operations can be performed.

The comparison instructions that use the basic ladder format, shown in Figure 9-51, operate in a manner similar to the arithmetic instructions. If the rung has continuity, the comparison instruction is performed and if the comparison is true, continuity is passed to the output coil. The data transfer GET is used to obtain the first register to be compared to the CMP register. Note that all ladder conditions are programmed before the Get and Compare instructions.

The compare functional block, shown in Figure 9-52, is used to compare the contents of two registers. Register X and Y contain the two values that will be compared. Output 1 is energized when the comparison is performed, while Output 2 is energized if the comparison has been satisfied.

Typical comparison instructions are greater than (>), less than (<), equal to (=), and combinations such as less or equal to, greater to or equal to, and not equal to. Some controllers offer another comparison option which uses another register (Z) to perform a limit (LIM) function. This limit instruction compares the value of register Y for less or equal to register X, and register Y greater or equal to register Z ($X \geq Y \geq Z$). If the compare instruction is true, the output Z will be energized. Output 1 is ON whenever the instruction is enabled.

There are controllers that do not have a compare block and yet perform a similar comparison using a subtraction block (see arithmetic blocks). In this case, there are three outputs that signal whether the result of the subtraction is positive (greater than), equal (equal to), or negative (less than).

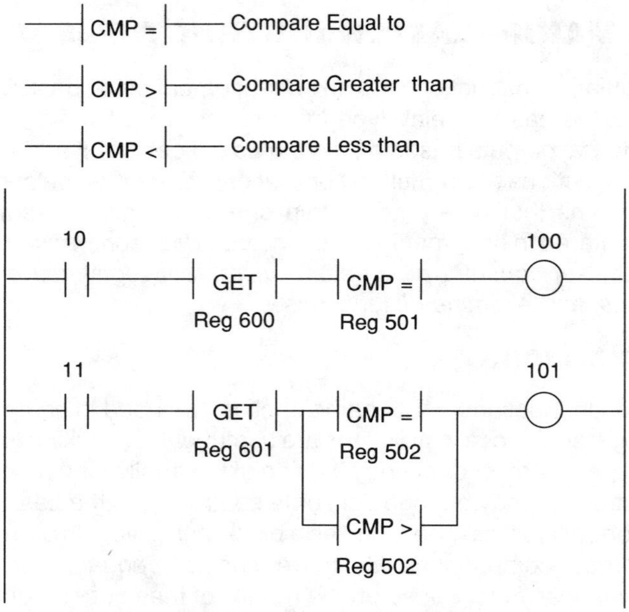

CMP =
Compare Equal to
CMP >
Compare Greater than
CMP <
Compare Less than
10
GET
Reg 600
CMP =
Reg 501
100
11
GET
Reg 601
CMP =
Reg 502
CMP >
Reg 502
101

If 10 closes, the contents of Register 600 are compared (=) to the contents of Register 501; if equal, coil 100 is turned on.
If 11 closes, the contents of Register 601 are compared for a greater than or equal to the contents of Register 502; if these conditions are met, output 101 is turned ON.

Figure 9-51. Ladder format comparisons.

LIM
Control
Output 1
Reg X
Reg Y
Reg Z
Output 2
CMP (>,=,<)
Control
Output 1
Reg X
Reg Y
Output 2

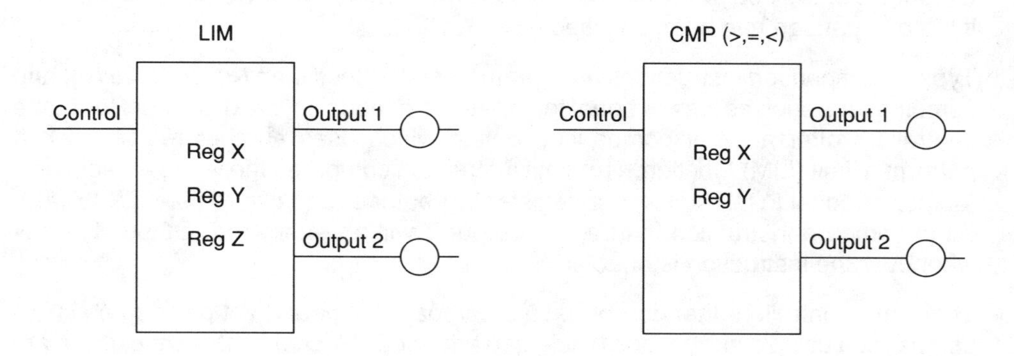

Figure 9-52. Block format comparisons.

Logic Matrix

The logic matrix functional block is used to perform AND, OR, EXCLUSIVE-OR, NAND, NOR, and NOT logic operations on two or more registers (see Chapter 3 for logic functions).

The performance of a logic function between two registers can be thought of as a matrix operation of length one, since there is one register for each operand. Figure 9-53 shows a typical block for the logic matrix function.

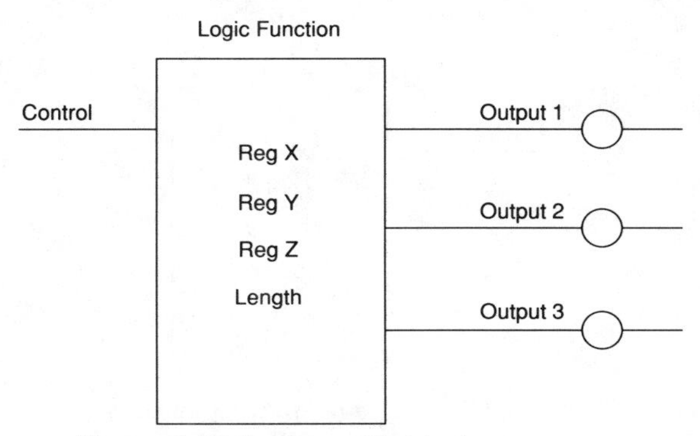

Figure 9-53. Logic matrix block.

Defined by the block to be executed, the logic function is performed when the control input is enabled. The registers inside the block are user specified and generally are holding or storage registers. The result register Z will hold the result of the operation. The length indicates the number of words or registers adjacent to each of the operands X, Y, and Z, where data is located in matrix form.

As an example, let us assume a length equal to 10 (0 through 9) and the logical function is an AND. When the block is enabled, the contents of register X to register X+9 will be ANDed with the contents of register Y to Y+9; the result of the operation is placed in registers Z to Z+9. Each register typically holds 16 bits of data; therefore, in this case, 160 bits in ten registers were ANDed with another 160 bits from registers Y to Y+9, and the result (160 bits) stored in another matrix (registers Z to Z+9).

There are three possible outputs in this block. The top output is energized once the control line is enabled, the middle output is energized when the operation is done, and the lower output is energized when there is an error.

Some controllers have only two operand registers (X and Y). When the logic operation is performed, the result is stored back in register Y, thus erasing the data previously stored in register Y. When this type of block is used, a data transfer to another register(s) prior to executing the logic matrix block will prevent loss of the data.

Data Conversions

These instructions are provided so that the contents of a given register can be changed from one format to another and certain operations can be performed.

Typical conversions include BCD to binary, binary to BCD, absolute, complement, and inversion.

BCD to binary code conversions are required when receiving BCD input data from field devices such as thumbwheel switches (TWS). A conversion in this case will allow the input data to be used in math operations (if required). The binary to BCD function operates in the same manner and is utilized when outputting data to field devices that operate in BCD, such as seven segment LED indicators. A typical functional block for this type of conversion is illustrated in Figure 9-54.

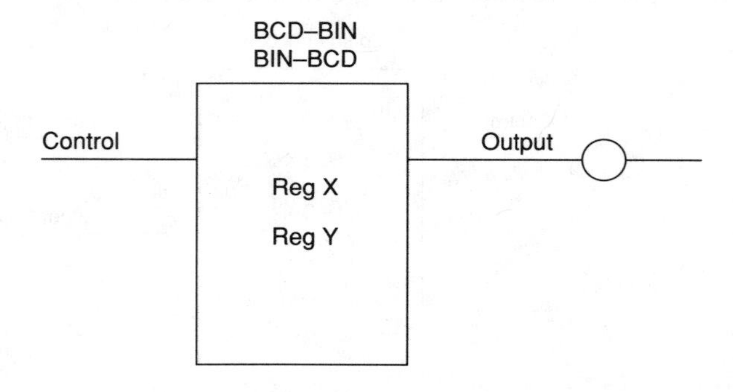

Figure 9-54.

The operation of this block, whether BCD-BIN or BIN-BCD, is basically the same. When the control input is enabled, the contents of register X (BCD or BIN) will be converted to binary or BCD depending on the conversion instruction. The result of the conversion is placed in register Y. The block output is energized when the instruction is completed.

The absolute, complement, and invert operations are generally done in a single register. In other words, the result of the operation will be stored back in the same location. A typical block is illustrated in Figure 9-55.

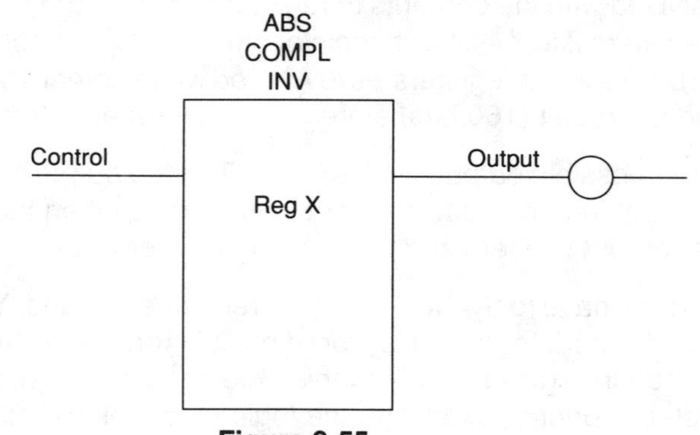

Figure 9-55.

The absolute block function computes the absolute value (always positive) of the contents of register X. If register X contains the value -5713, the result of X after the block instruction will be +5713.

The complement instruction when executed changes the sign of the contents of register X—positive to negative or negative to positive. For example, -6600 after execution will be +6600, and +3314 will be -3314.

The invert functional block inverts all the bits in register X. If the binary pattern in register X is 0000111100001111, after execution it will be 1111000011110000. The block output will be ON when the instruction is finished.

Set Constant Parameters

It is sometimes necessary to store a constant parameter in a register that will be used later in the program for comparisons or set points. For this reason, some PLCs provide a block instruction that allows a fixed value to be assigned to a register. A functional block for set constants is illustrated in Figure 9-56.

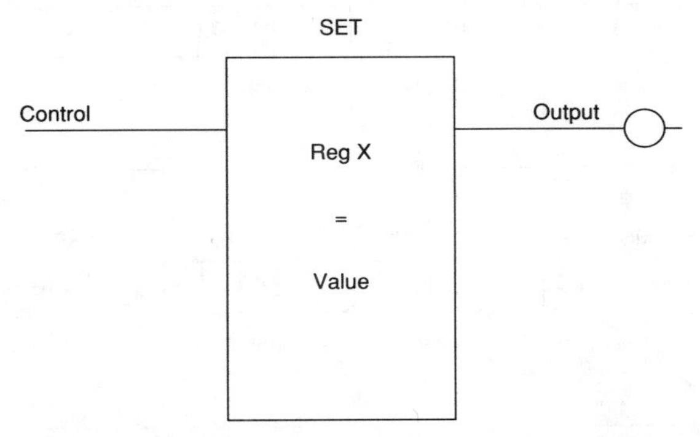

Figure 9-56. Set constant in register instruction.

When the block is enabled, register X is set equal to the value specified (in BCD, binary, etc.), and the output is turned ON once the operation is completed. This instruction is very useful when resetting to zero during initialization of several storage or I/O registers.

Logical Shifts and Rotates

The shift instruction is used to move bits of a register(s) to the right or to the left. Figure 9-57 illustrates the execution of a right shift instruction. The left shift is identical, except the bit is moved in the opposite direction (shifting out the most significant bit).

The rotate instruction, like the shift instruction, shifts data to the right or left, but instead of losing the shift-out bit, it becomes the shift-in bit at the other end (rotate the bit). Figure 9-58 illustrates the operation of the right rotate instruction. A functional block for the shift or rotate function is illustrated in Figure 9-59.

The control input enables the block operation for a rotate or shift execution. Some block instructions will have a right and left line to determine the shift or rotate direction. There are several variables that can be available inside the block, depending on the PLC model. Register X is generally the location where the data to

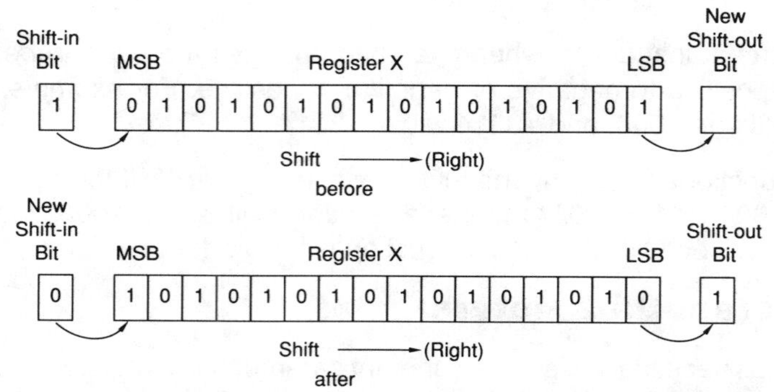

Figure 9-57. Shift (right) instruction contents of register before and after execution.

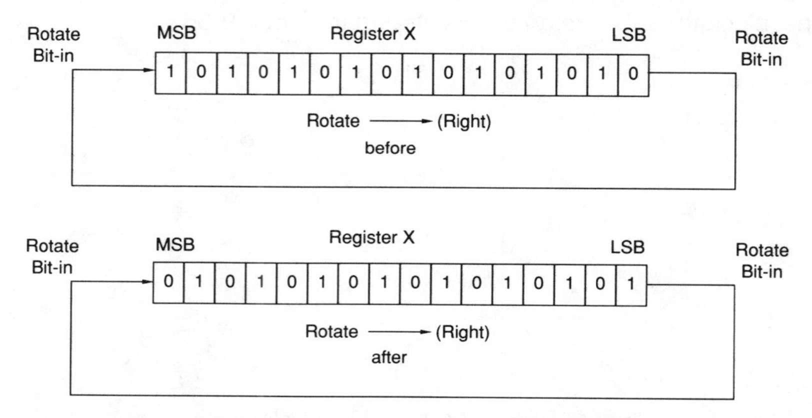

Figure 9-58. Rotate (right) instruction contents of a register before and after execution.

be shifted or rotated will be stored. If the length is to be specified, register X will be the starting location. For example, if the length is 10, then the block operation will be performed on 160 bits.

The bit-in and bit-out variables are used to specify the location of the bit whose value will be shifted in or out (available in shift block only). These bits can be real I/O locations that can be used to input or output data with the shift register operation. The number of bits indicate the amount of bit shifts or bit rotates that take place when the control input goes from OFF to ON. The shift and rotate instructions are very useful in applications where the status of inputs must be tracked along a path of travel (e.g. overhead conveyors in a part painting process).

Examine Bit

This functional block is used to examine the status of a single point, or bit, in a memory location. This type of instruction is generally used when *flags* are set during the program and then later tested and compared. A bit can be examined for an ON or OFF status. Figure 9-60 illustrates a typical block for this instruction.

The function is performed when the control input is ON. The bit position specified in register or location X will be examined for an ON or OFF condition. The output is energized if the instruction is ON.

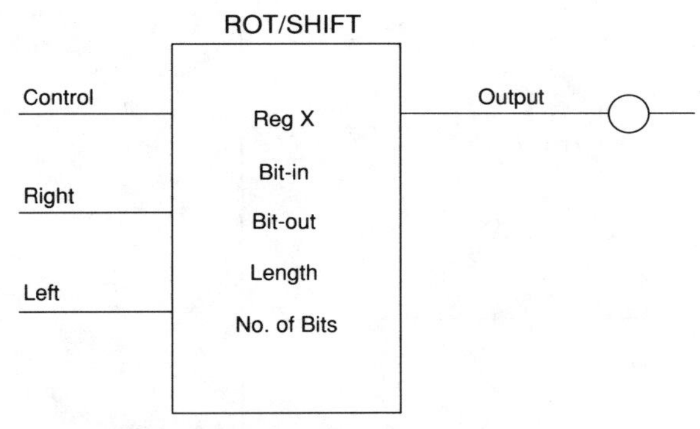

Figure 9-59. Example of a rotate or shift block.

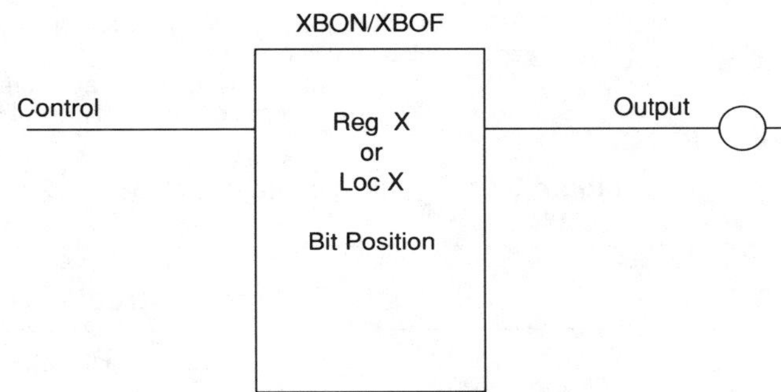

Figure 9-60. Example of an examine ON or examine OFF bit location.

9-9 DATA TRANSFER INSTRUCTIONS

Data transfer instructions involve the moving or transferring of numerical data within the PLC, either in single register units or in blocks (group of registers). Data transfers can generally address any location in the memory data table, with the exception of areas restricted by the system to the user. Typical uses involve the movement of constant and/or preset values to counters and timers, the reading of analog inputs as well as multibit input modules, and transferring of data to output modules.

As in other instructions, the format may be found in the basic ladder format or in the functional block format. However, the latter is most commonly found at this level of operations. The basic ladder format instructions included in this data transfer group are the GET and PUT instructions (shown in Figure 9-61). These two functions are used in general with PLCs which have basic ladder format implementation of arithmetic and data comparison instructions.

The functional block group of data transfer instructions forms perhaps the most useful set of functions available in an enhanced PLC after the basic relay instructions. Depending on the controller, some of the basic names may be different yet they may implement the same transfer function. Table 9-3 shows the different instructions available under data transfer operations.

A

Get Reg X

Get Reg Y

ADD

Reg Z

B

Get Reg Z

PUT

Reg W

If A closes, the contents of Register X and Register Y are added and stored in Register Z.
If B closes, the contents of Register Z are stored (Put) in Register W; the contents of Register Z are not altered.

Figure 9-61. Example of the use of GET and PUT instructions.

Data Transfer Instructions	
Move Point	Block Transfer IN
Move Register	Block Transfer OUT
Move Block	ASCII Transfer IN
Register to Table	ASCII Transfer OUT
Table to Register	FIFO STACK

Table 9-3.

Move

The move instruction is used to transfer information from one location to another. The destination location will be to a single bit or register. A move functional block is shown in Figure 9-62.

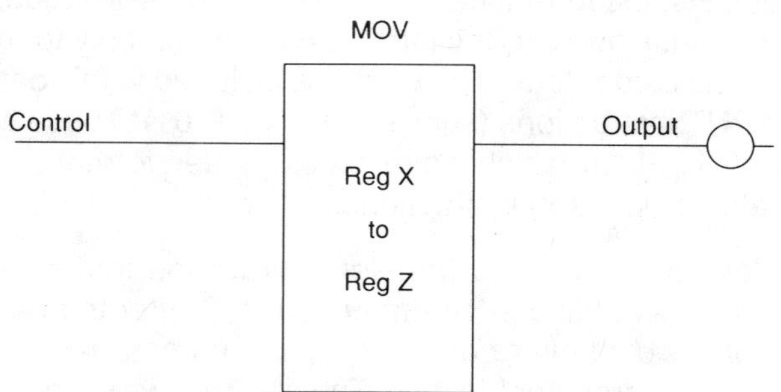

Figure 9-62. Move register block instruction.

When the control input is ON, the block operation is performed, and the contents of register X are copied into register Z. Register X, the source register, and register Z, the destination register, can be either a storage register or an I/O register. The output is energized when the instruction is completed.

In some PLCs, the move function is performed on special word table locations. In this case, the data copied is automatically converted to the proper numerical format of the destination location. For example, register X, or word X, might contain a BCD value that when transferred to register Z, or word Z, is stored as a binary value, thus executing essentially a BCD to binary conversion with the move instruction.

Another type of move instruction involves the masking of certain bits within the register. This block is shown in Figure 9-63. The data in register X is transferred to register Z, with the exception of the bits specified by a zero in the mask register Y. The bits "b" in register Z are set to zero due to the mask.

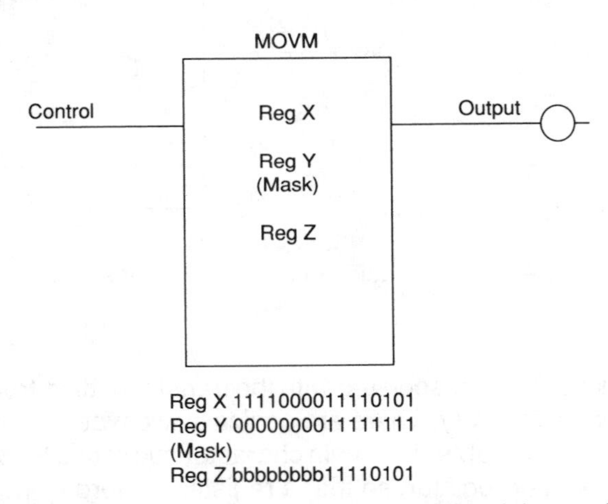

Figure 9-63. Example of a move with mask block instruction.

There is yet another move instruction that is found in several controllers. This is the MOVE STATUS instruction. With this block function, system status or I/O module status can be transferred to a storage register (register Z). The status information in the result register can then be masked, compared, or examined to determine the status of major or minor faults in the system or an I/O module and then used to take action under program control if necessary.

Move Block

The move block instruction causes a group of register or word locations to be copied from one place to another. The length of the block is generally user specified. Figure 9-64 illustrates a move block instruction.

When energized, the control input causes the execution of the block. The data starting at location X through location X+L (length=L) is transferred to locations Y through Y+L. The data in X to X+L is left unchanged. Some PLCs have the option of specifying how many locations can be transferred during one scan (rate per scan).

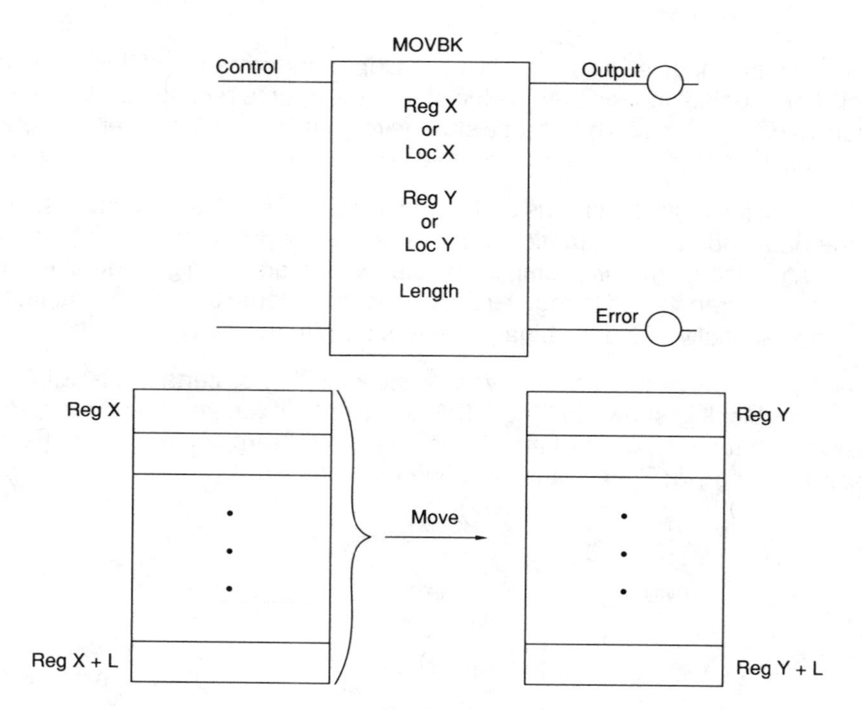

Figure 9-64. Illustration of a move block instruction.

Table Move

Table move instructions are associated with the transfer of data from a block or table to a register or word in memory. There are basically two types of table moves: table-to-register and register-to-table. The main characteristic of this functional block is the manipulation of a pointer register, so that a register or word value can be stored in a particular table location as specified by the pointer. The table move is shown in Figure 9-65.

When the control input goes from OFF-to-ON, the table move instruction is executed and the contents of the pointer register incremented. The middle input is used to disable the pointer increment. The bottom input of the table move block is used to reset the pointer to zero (initialize to top of table).

If it is necessary to store or retrieve data to or from a specific table location, the pointer register can be loaded with the appropriate value to point at the specified location. This loading is done prior to the table move through a set parameter or move register instruction.

The length L specifies the number of word locations in the table, counting from the starting location (X). The top output is energized when the block instruction is completed. The middle output of the table move block is energized when the pointer register has reached the end of the table.

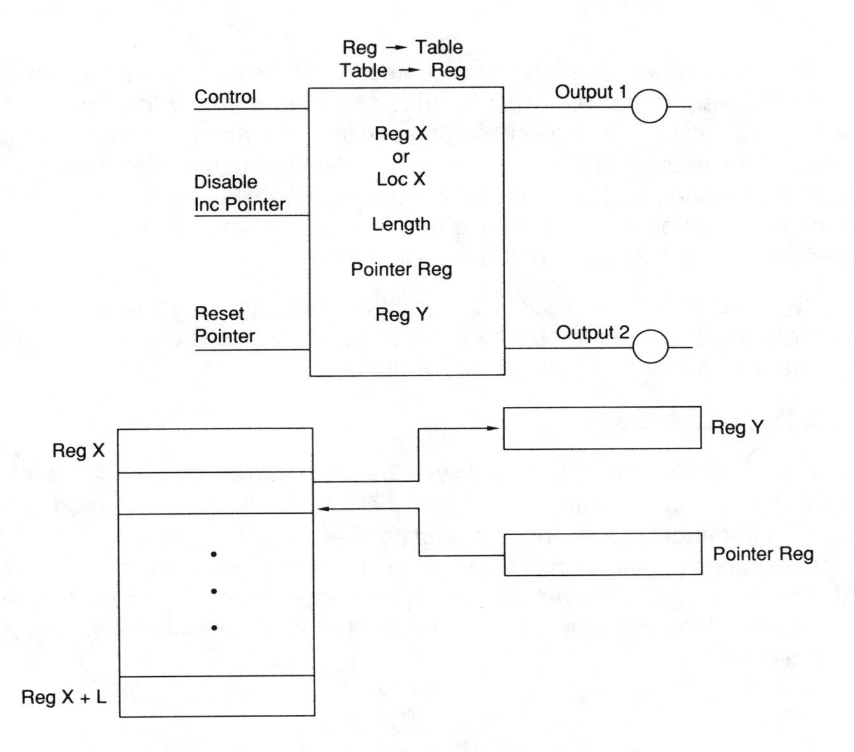

Figure 9-65. Illustration of a table move instruction.

Useful applications of the register to table transfer include the table loading of new data, storage of input information (e.g. analog) from special modules, or error information from the process being controlled. The table to register instruction is useful when changing preset parameters in timers and counters, and when driving a group of 16 outputs at one time through I/O registers. This instruction can also be used when looking up values in a table for comparison, linear interpolation, etc.

Block Transfer—In/Out

The block transfer instruction is available in some PLCs and primarily designed to be used with special I/O modules such as analog, encoder, BCD, and stepper motor. There are two basic types of block transfers that are used: input and output. Figure 9-66 shows a block transfer instruction. The Module Address Location may be explicitly marked as Rack and Slot where the interface is located.

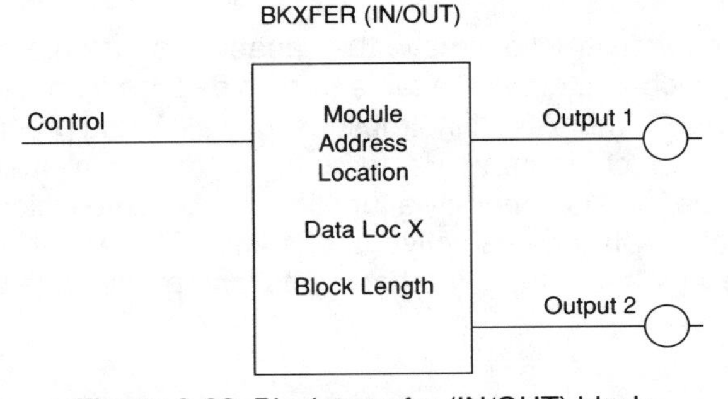

Figure 9-66. Block transfer (IN/OUT) block.

The control input, when enabled, executes the block transfer instruction. The data on the I/O module (for a BKXFER IN) is stored in memory locations or registers starting at location X. The block length specifies how many locations are needed to store the I/O module data. For example, the data from an analog input module with four input channels can be read all at once, if the length is specified as four. The block transfer output operates in a similar manner; the destination of the data transfer is determined by the address of the output module.

The top output of the block transfer instruction, when energized, signals the completion of the transfer operation. The bottom output is used to signal an error, such as bad I/O transmission or faulty module.

ASCII Transfers

The ASCII transfer instruction deals with the data transmission of ASCII characters from a PLC to a peripheral device. This functional block operates in conjunction with an ASCII communications module, and its execution is under program control. The communication usually occurs in two ways: reading data from a peripheral or writing data to a peripheral. This functional block is widely used when report generation is required in the application. Figure 9-67 illustrates a typical read/write ASCII functional block.

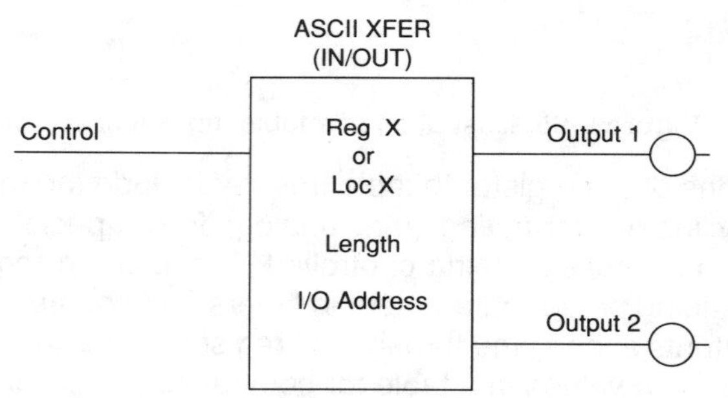

Figure 9-67. ASCII transfer block instruction.

The control input activates the ASCII transfer (In or Out). When reading data, the instruction will allow the special I/O module to perform a read function. The data is read from the module by the processor and then stored in special memory locations (X to length specified). The location of the module is indicated by the I/O address in the block. When writing data, the information is sent from the locations in which the message is stored to the address where the module is located.

Some ASCII transfer instructions have the capability of using a pointer register for accessing specific characters in the table (e.g. to decode a specific input character from the data table). Other ASCII instructions may allow the programmer to specify how many bytes or characters are transmitted during a scan. The speed of transmission (baud rate) is generally a function of scan time which can also depend on the number of ASCII devices active at one time. The ASCII transfer instruction assumes that proper baud rates, start/stop bits, and parity have been set-up in the I/O module.

FIFO Stack Transfers

The First-In-First-Out (FIFO) instruction is used to construct a table or queue where data is stored. The basic function of this operation is similar to an asynchronous shift register in which one word (16 bits) is shifted within the stack each time the instruction is executed. The data is shifted in the same order in which it is received—the first one in will be the first one out. The FIFO operation is analogous to the idea of first come (IN) first serve (OUT).

The FIFO operation generally consists of two instruction blocks: FIFO IN and FIFO OUT. FIFO IN is used to load the queue and FIFO OUT unloads the queue. The FIFO moves are useful for storing, and later retrieving, large groups of temporary data as it comes in. A typical application of the FIFO instruction is to store and retrieve data that is synchronized to the external movement of pieces or parts on a conveyor or transfer machine. Figure 9-68 shows a typical FIFO block instruction.

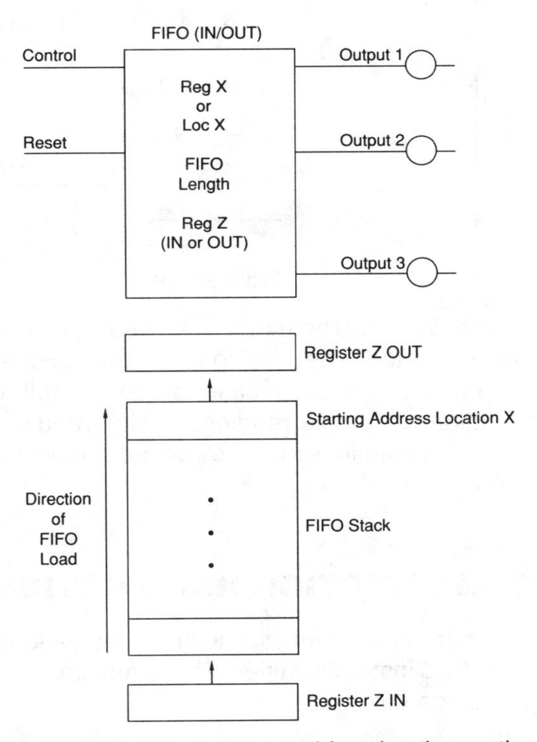

Figure 9-68. FIFO block instruction and functional operation.

The FIFO block is initiated by the OFF-to-ON transition of the control input logic. The reset signal may be available for resetting the FIFO stack (clear stack). For the FIFO IN instruction, register Z holds the data which will be transferred to the queue. This data is placed in the FIFO stack (bottom location) when the control input is ON. When using the FIFO OUT block, the data is output through register Z. The length of the stack is specified by FIFO length (register X to X + Length).

There are generally three outputs used in the FIFO blocks. The top output indicates that the functional block is operating, the middle output signals that the queue is empty, and the bottom output shows the queue is full. Most controllers have both the input and output FIFO blocks; however, some PLCs may offer both instructions (IN and OUT) in one FIFO functional block.

Sort Instruction

The sort block function (shown in Figure 9-69) sorts in ascending or descending order a block of registers according to their contents. Once the instruction is true the instruction examines each register for its contents and performs the sorting.

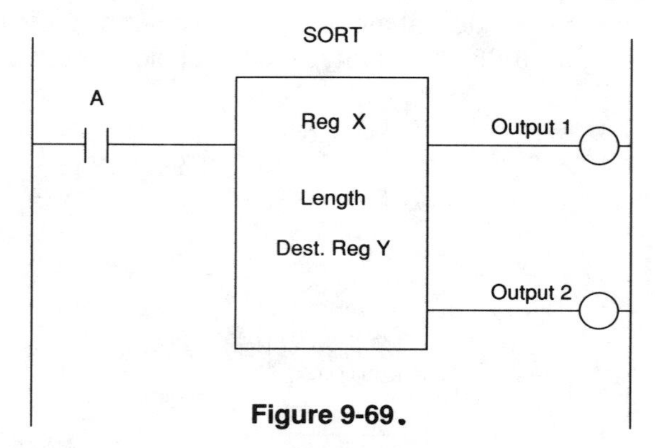

Figure 9-69.

Please refer to Figure 9-69 and note that, if A closes, the instruction is enabled and register X to X + L are sorted in ascending order and the sorted results are stored in register Y to Y + L. This type of function is very useful, for example, whenever computing the median of sample readings, which require that the sample be in numerical order. Some controllers may provide ascending, descending or both types of sorting storage.

9-10 SPECIAL FUNCTION INSTRUCTIONS

As the name implies, the special function instructions perform operations that do not fall under any other PLC instruction titles. These functions are generally available in medium to large size controllers.

Sequencers

The sequencer block is a powerful instruction that is used to simulate a drum timer. A sequencer is analogous to a music box mechanism in which each peg produces a tone as the cylinder rotates and strikes the resonators. In a sequencer, each peg (bit) can be interpreted as a logic 1, and no-peg as a logic 0. Sequencers are generally specified, using a table which is similar to the music box cylinder spread-out. Depending on the controller, the number of bits can vary from 8 to 64 or more. Figure 9-70 illustrates a cylinder and sequencer table comparison.

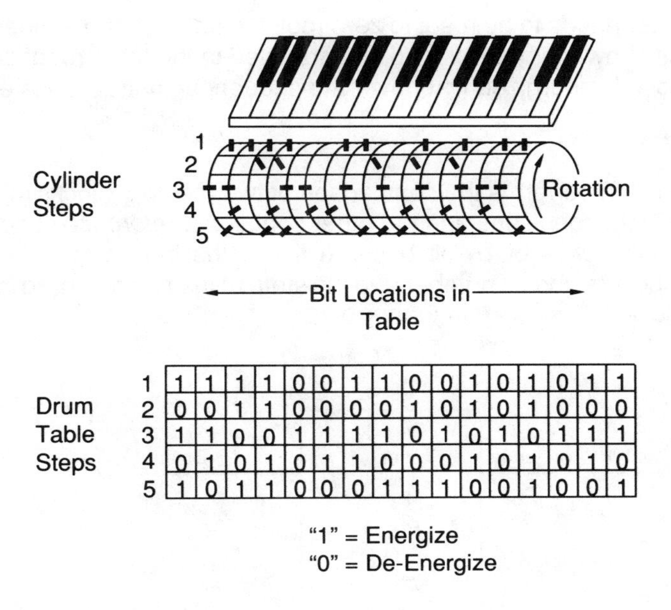

Step																
1	1	1	1	1	0	0	1	1	0	0	1	0	1	0	1	1
2	0	0	1	1	0	0	0	0	1	0	1	0	1	0	0	0
3	1	1	0	0	1	1	1	1	0	1	0	1	0	1	1	1
4	0	1	0	1	0	1	1	0	0	0	1	0	1	0	1	0
5	1	0	1	1	0	0	0	1	1	1	0	0	1	0	0	1

Figure 9-70. Sequencer table and cylinder equivalent.

The width of the table may vary, as does the size of the music cylinder. Each of the steps in the sequencer can be an output (through I/O registers) representing each of the pegs. Figure 9-71 shows a typical functional sequencer block.

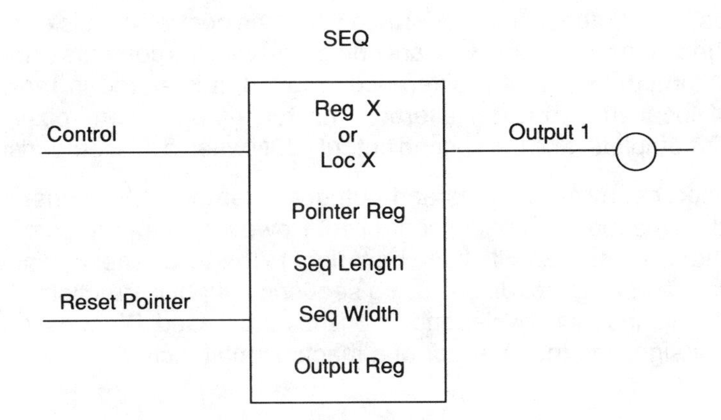

Figure 9-71. Sequencer functional block.

The block is initiated by the OFF-to-ON transition of the control input, and the contents of the sequencer table will be output in a sequential manner. Each step being output is pointed to by the contents of the pointer register. Every time the control input is energized, the pointer register is automatically incremented and points to the next table location. Depending on the PLC, the control input line may be driven by an *event* or by *time*, in which case the sequencers may be referred to as *event driven* or *time driven* sequencers. The reset pointer input is provided in case the pointer register needs to be reset to zero (point to step 1). The sequence length and width specify how many steps and bits are used in the table respectively. The output of the block is energized whenever the sequencer instruction is enabled.

Diagnostics

This block instruction performs the comparison of two memory blocks in which one contains actual input conditions and the other contains a *reference* condition. This comparison is done on a bit-by-bit basis to find if the blocks are identical. If a miscomparison occurs, the bit number and the state of the bit are stored in a holding register. This block is illustrated in Figure 9-72.

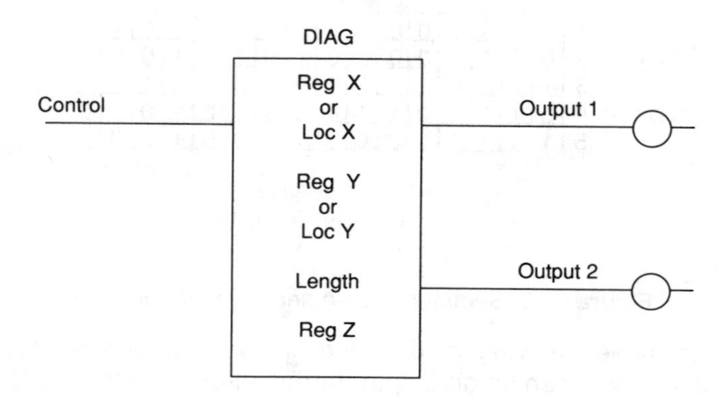

Figure 9-72. Diagnostic block.

The diagnostic block function is performed when the control input is energized. The contents of locations X through X + L are compared with the contents of the reference locations Y through Y + L. If a difference is found, it is stored in register Z. The contents of location Y are not altered. The top output is energized when the instruction is complete, and the second output is ON when a miscomparison is found.

Predetermined behavior of inputs and outputs (reference conditions) is generally determined by the machine being controlled. However, there are some controllers that allow the reference conditions to be taught to the PLC. These controllers input the reference *"teaching"* conditions using sequencer input instructions, block transfers IN, and other instructions depending on the model used. Diagnostic instructions are useful for signaling the operator of a machine malfunction.

PID

The PID functional block is used in PLCs that offer the capability of performing analog control using the Proportional-Integral-Derivative (PID) algorithm. The user specifies certain parameters associated with the algorithm to control the process correctly. Figure 9-73 illustrates a typical PID block.

The control input is generally used to enable the PID block to operate (auto). The bottom input *track*, when energized, will determine if the PID variables are being tracked, but not output. If the block is not enabled (in manual mode), the controller can still track the variables when track is enabled. The *input variable register (IVR)*

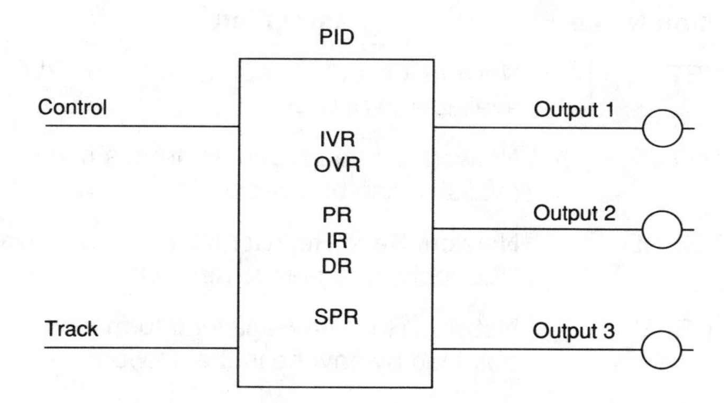

Figure 9-73. PID functional block.

and *output variable register (OVR)* are user specified and are associated with the location of the analog modules (input and output). The *proportional register (PR), integral register (IR),* and *derivative register (DR)* hold the gain values that need to be specified for the control of the process. The *set point register (SPR)* holds the target value for the process set point. Depending on the controller, other block variables can be specified, such as dead times, high and low limits, and rate of update. The top output of the PID block is used to indicate an active loop control, while the middle and bottom outputs usually indicate a low and high limit alarm respectively.

Some controllers provide PID capabilities without having the block instruction. Generally, in this case, a special PID module is used in which all the input/output parameters are in the module. The set point and gain parameters are transferred to the module during initialization of the program through an output instruction, such as block transfer out or move data to an output register. The module data can be altered under program control if any of the parameters need to be changed.

9-11 NETWORK COMMUNICATION

With the advantage of Local Area Networks (LANs) came the necessity to communicate and exchange information among PLCs in the network. The exchange of information takes place using network instructions specific to each PLC manufacturer. The operation involves status communication of contacts and/or output coils as well as registers.

Table 9-4 illustrates typical instructions that may be encountered in a PLC network environment. The implementation of these instructions are very simple to accomplish; however, the programmer must make sure that compliance with the PLC network rules are enforced. Also, programming organization and assignment of registers is a must if confusion is to be avoided.

Instruction Name	Description
NET ┤ ├	Network Output.Output activated by PLC logic; available to network.
NET —( )—	Network Contacts.Contact from a network output available can be used by anyone in the network.
NET SEND	Network Send Register information activated by PLC logic; available to network
NET RCV	Network Receive Register information can be obtained by anyone in the network
SEND NODE	Send to Node Send register information to a specific node (PLC) in the network; similar to copy information
GET NODE	Get from Node Get register information from a specific node in the network; similar to read and copy information

Table 9-4. Typical network instructions found in PLCs.

In general, once the instruction is executed and updated at the EOS, the information is passed on to the network hardware (modules or internal boards) for processing and transmission. Depending on the controller used, the format of the instruction may change. The instructions presented here are used as a guideline and illustrate the ease of implementation. Some controllers may use data transfer instructions to access the network while others have specific instructions to use.

The organization of a network depends on how it is configured. Some controllers have the network interface built in the main CPU while others may have it in the form of an interface module. Regardless, both will perform the same objective: network communications. If a network interface is installed in I/O racks, there are different ways the manufacturer will allow you to set up that particular PLC or *node* as it may be called. Some PLCs will allow configuration of the network during the configuration stage of the PLC, where the network module slot location may be specified. Other controllers may automatically recognize where the network interface is located. Yet another method may be the specification of the slot location in the network software instruction similar to how it is used in the block transfer IN and OUT.

The output coils and contacts in the network may be referred to as *network outputs* and *network contacts* while the registers may be called *network registers*. The network outputs are internal outputs that are located in a special area of the data table along with the network registers. These network elements may be part of the internal storage area with the additional LAN capability. For instance, if there are 512 possible internal outputs, 64 of them may be used as network outputs; likewise, if there are 128 storage registers, 32 of them may be used as network registers.

More on Local Area Networks operation and configurations is covered in Chapter 14. Now, let's take a look at the operational function of these network instructions. Here, we assume that the slot location of the network interface is specified during the total PLC system configuration. If this were not the case, then the slot entry specification would be most likely required for each instruction.

Network Output

A network output instruction, shown in Figure 9-74a, is used in conjunction with the network contact to pass one bit status information from a PLC to the network. If continuity exists in the logic path of the network output, the reference address will be turned ON and transmitted to the network interface for LAN transmission. The reference address must be a valid network coil depending on the controller used. The status of this output is available to all network stations or nodes (PLCs).

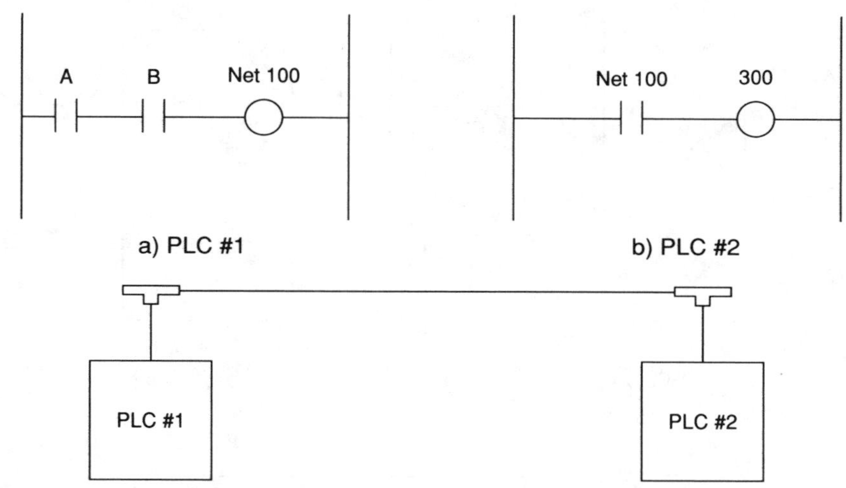

Figure 9-74. Illustration of the network output and network contact instructions.

Network Contacts

Network contact instructions are used to *capture* the status information from a network output. The reference address of the network contact must be that of an active network output; otherwise, that contact (examine ON or examine OFF) will never be evaluated. The reference must also be a valid reference address which may differ among PLC manufacturers. Figure 9-74b illustrates the operation of the network contact.

Information from the network is generally obtained during the *reading* of the inputs, at which time the processor will read the status buffer of the network module as though it were a mini data table. If the reference network contact is logic 1, the evaluation will take place and it will open or close its contacts to provide or remove continuity. The evaluation depends on how the network contact is programmed—(—| |— or —|/|—).

Network Send

The network send instruction is used to send register information to the local area network. This functional block is activated in the same manner as other blocks; if the rung is true, then the function is performed and the content of the register is sent onto the network line. The instruction may provide two outputs to indicate that the operation has been performed and that no error is detected (output1 and output2 respectively).

Figure 9-75a illustrates a typical network send instruction block. Several registers may be transmitted to the network if the length (L) is specified as more than one; the registers to be transmitted would start at register X and end at register X + L. The network send generally operates in conjunction with the network receive (NET RCV).

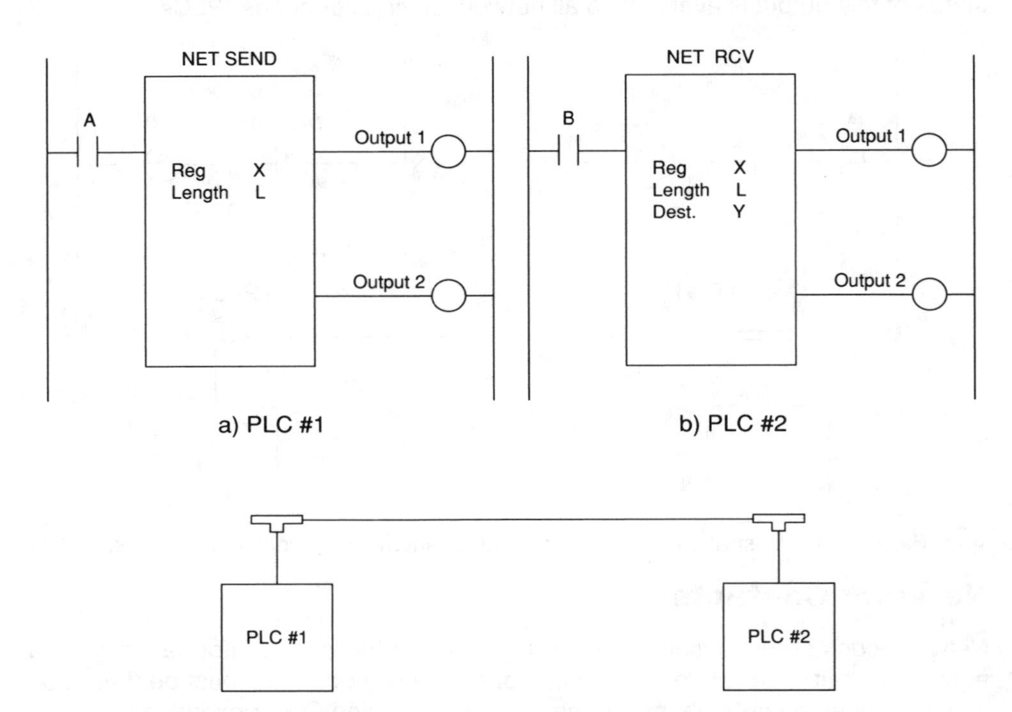

Figure 9-75. Illustration of the network send and receive instructions.

Network Receive

The network receive block function captures the registers available in the network lines and stores it into the receiving PLC's data table (register area). The user must make sure that the register information requested (i.e. register address number) matches the same addresses used by the NET SEND instructions. For instance, if a NET SEND instruction uses network registers 2000 to 2007 (length of 8), the PLC that is intended to retrieve or capture those network registers must reference the same network registers in its instructions. Figure 9-75b illustrates the use of a network receive instruction.

Once the instruction gets the register information, its contents are stored in the destination register Y to register Y + L if the length is greater than one. Of the two outputs available, the first one may represent the completion of the operation while the second one may indicate if an error has occurred.

Send Node

The send node instruction operates in a more direct fashion than the network send function. This send node instruction may be available to implement the sending of register information to specific locations (node) connected to the network. Essentially, the send node function implements a *copy to* function, where several registers from the sending node are written onto another node. Figure 9-76 illustrates a send node instruction.

When the instructions control line has continuity, the block is enabled and the contents of register X to register X + L are sent to the specified node (N); the information from registers X to X + L is stored in the destination register Y to Y + L. The first output is ON whenever the instruction is completed, while the second output may detect an error condition in the network.

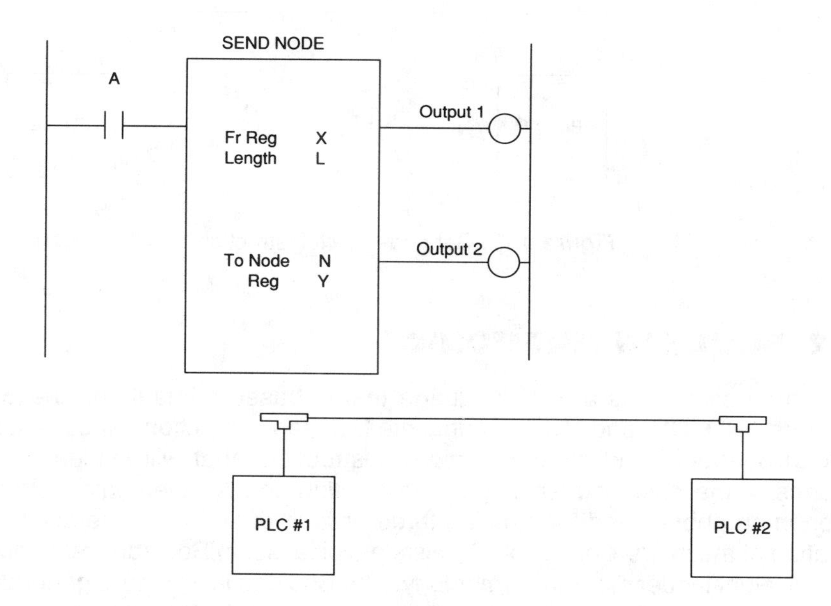

Figure 9-76. Send node block instruction.

Get Node

The GET NODE instruction is used to retrieve register information from another PLC node. It is as though a register copy is made from one node to the requesting node. Figure 9-77 illustrates the use of the GET NODE function.

When the block is enabled, the contents or registers X to X + L in the target node (N) are requested and stored in registers Y to Y + L of the PLC executing the GET NODE instruction. The first output is energized whenever the instruction is completed; output 2 is ON if there is a communication error in the network transmission.

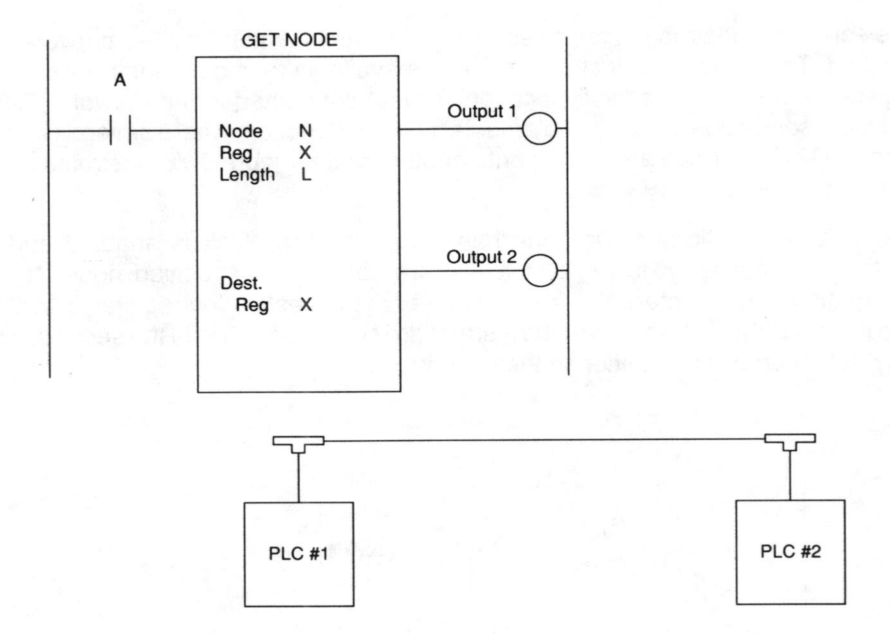

Figure 9-77. Get node block instruction.

9-12 BOOLEAN MNEMONICS

Boolean mnemonics is a PLC language that is based primarily on the Boolean operators AND, OR, and NOT. A complete Boolean instruction set consists of the Boolean operators and other *mnemonic* instructions that will implement all the functions of the basic ladder diagram instruction set. A mnemonic instruction is written in an abbreviated form, using three or four letters that generally imply the operation of the instruction. Table 9-5 lists a typical set of Boolean instructions and the equivalent ladder diagram symbology. The Boolean language is generally used to enter the required logic into the PLC's memory. The PLC however, may display the entered rung as a ladder diagram on the programming terminal.

Further enhancements of the Boolean instruction sets have resulted in enhanced Boolean output operators which perform additional functions. Figure 9-78 shows a short Boolean program and its equivalent ladder diagram representation. The principles of Boolean algebra, which are applied in the Boolean language, are discussed in Chapter 3.

MNEMONIC	FUNCTION	DESCRIPTION	LADDER EQUIVALENT
LD/STR	Load/Start	Starts a logic sequence with a NO contact.	
LD/STR NOT	Load/Start Not	Starts a logic sequence with a NC contact.	
AND	And Point	Makes a NO contact series connection.	
AND NOT	And Not Point	Makes a NC contact series connection.	
OR	Or Point	Makes a NO contact parallel connection.	
OR NOT	Or Not Point	Makes a NC contact parallel connection.	
OUT	Energize Coil	Terminates a sequence with an output coil.	
OUT NOT	De-Energize Coil	Terminates a sequence with a NOT output coil.	
OUT CR	Energize Internal Coil	Terminates a sequence with an internal output.	
OUT L	Latch Output Coil	Terminates a sequence with a latch output.	L
OUT U	Unlatch Output Coil	Terminates a sequence with an unlatch output.	U
TIM	Timer	Terminates a sequence with a timer.	TON
CNT	Up Counter	Terminates a sequence with an up counter.	CTU
ADD	Addition	Terminates a sequence with an addition function.	+
SUB	Subtraction	Terminates a sequence with a subtraction function.	-
MUL	Multiplication	Terminates a sequence with a multiplication function.	X
DIV	Division	Terminates a sequence with a division function.	÷
CMP	Compare =,<,>	Terminates a sequence with a compare function.	CMP
JMP	Jump	Terminates a sequence with a jump function.	JMP
MCR	Master Control Relay	Terminates a sequence with a MCR output.	MCR
END	End MCR, Jump, or Program	Terminates a sequence with an End of a control flow function.	END
ENT	Enter Value for Register	Used to enter preset values of registers.	Not Required

Table 9-5. Typical Boolean instructions and their equivalent PLC ladder instructions.

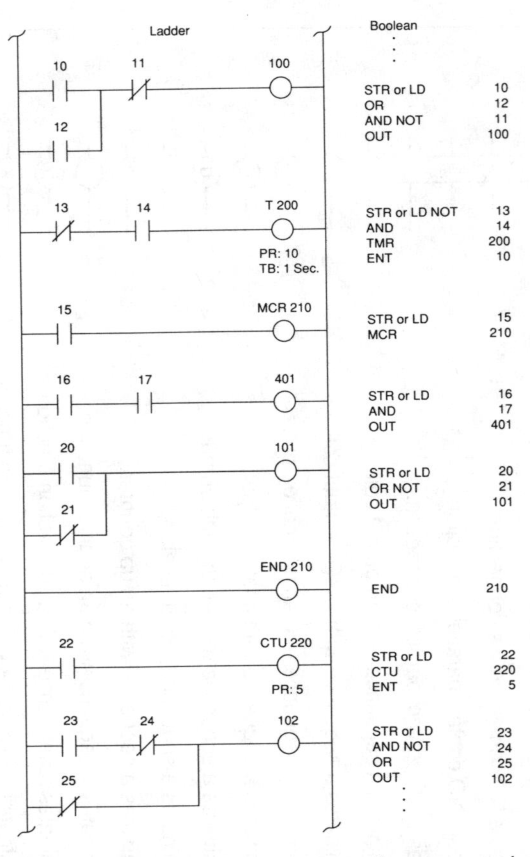

Figure 9-78. PLC and Boolean program codings comparison.

9-13 OTHER NON-LADDER PROGRAMMING LANGUAGES

In addition to the ladder language, there are some programmable controllers which provide instruction sets similar to the BASIC or Pascal (computer) language. Some manufacturers, like Texas Instruments, has added this type of language to the standard ladder instruction set which includes enhanced functional block instructions.

The capability of these non-ladder languages can be principally observed when the user requires complex math operations, statistical analysis, and advanced control strategy implementation. The number of instructions as well as the ease of implementation are greatly simplified.

The implementation of these languages in a PLC is very similar to having the functions of a dedicated BASIC or computer I/O module in the PLC rack. However, the major difference is that the whole executive for the processor is resident in the PLC memory system and easier programming interaction instructions and methods are provided. Direct instructions like PRINT *message* can be executed and documented as in any high level language. Figure 9-79 illustrates a sample portion of a program using high-level language.

```
     SF PROGRAM 1
        TITLE =                          BOILER 2
        CONTINUE ON ERROR (Y,N):         N
        PROGRAM TYPE (N,P,C,R) =         C
        ERROR STATUS ADDR (C,Y,U,WY):    V500
        CYCLE TIME (SEC) =               +0.5000

0001  *      THIS PROGRAM CALCULATES THE FLOW AND MASS FLOW RATE TO
             BOILER 2. GET SIGNAL, TAKE SQUARE ROOT, ADD CORRECTION,
             AND SAVE.
0002  MATH   V200 : = SQRT ( WX100 ) + V250
0003  *      THE CORRECT FLOWRATE IS NOW SAVED IN (CUBIC FEET/MIN). CALL
             SUBROUTINE 2 TO CALCULATE MASS FLOWRATE FROM TEMP AND
             FLOWRATE.
0004  CALL   SFSUB: 2       P1: V200       P2: V110
             P3: V202       P4:            P5:
0005  *      THE MASS FLOWRATE IS STORED IN LOCATION V202, IN (LBS/MIN).
0006  *      IF THERE IS A FLOWRATE MISMATCH WITH BOILER 1, SET FLAG
0007  IF     ABS ( V202 - V102 )>150
0008  IMATH  V100 : = 1
0009  ELSE
0010  *      ELSE RESET FLAG.
0011  IMATH  V100 := 0
****END****
```

Figure 9-79. Example of high-level programming (Courtesy of Texas Instruments, Johnson City, TN).

In addition to the high-level computer language, PLCs may also offer menu driven *fill-in-the-blanks* tables.These tables are displayed on the programming screen and allows the programmer to configure PID loops and other fill-in-the-blanks functions (e.g. servo motor and stepper control functions). Figure 9-80 illustrates an example of a fill-in-the-blanks configuration of an instruction.

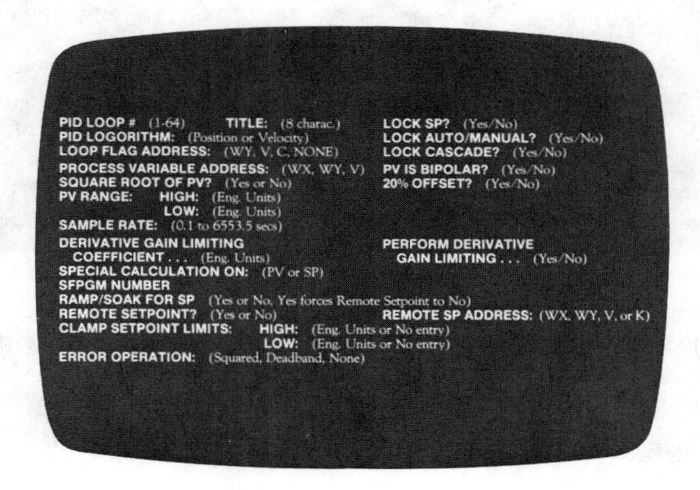

Figure 9-80. Example of fill-in-the-blanks menu instruction (Courtesy of Texas Instruments, Johnson City, TN).

Another version of non-ladder instructions is available from Bernecker & Ranier Industrial Automation Corp. (Stone Mountain, GA). B & R allows the programming of functional blocks in a customized manner. This custom block creation is accomplished by using a *functional block editor* which defines input and output specifications (i.e. discrete I/O, analog I/O, positioning I/O, etc.) and enters instructions that form the block.

The instructions used to implement the custom block are similar to assembly language mnemonics; however, these mnemonics also allow the implementation of high-level operations and data manipulation. Tailoring of block functions can enhance the PLC's control capability by allowing it to solve control tasks specific to an application with just a few custom blocks. Another advantage to the custom block is that it can be used with any of the members of the PLC family. Figure 9-81 shows a custom block.

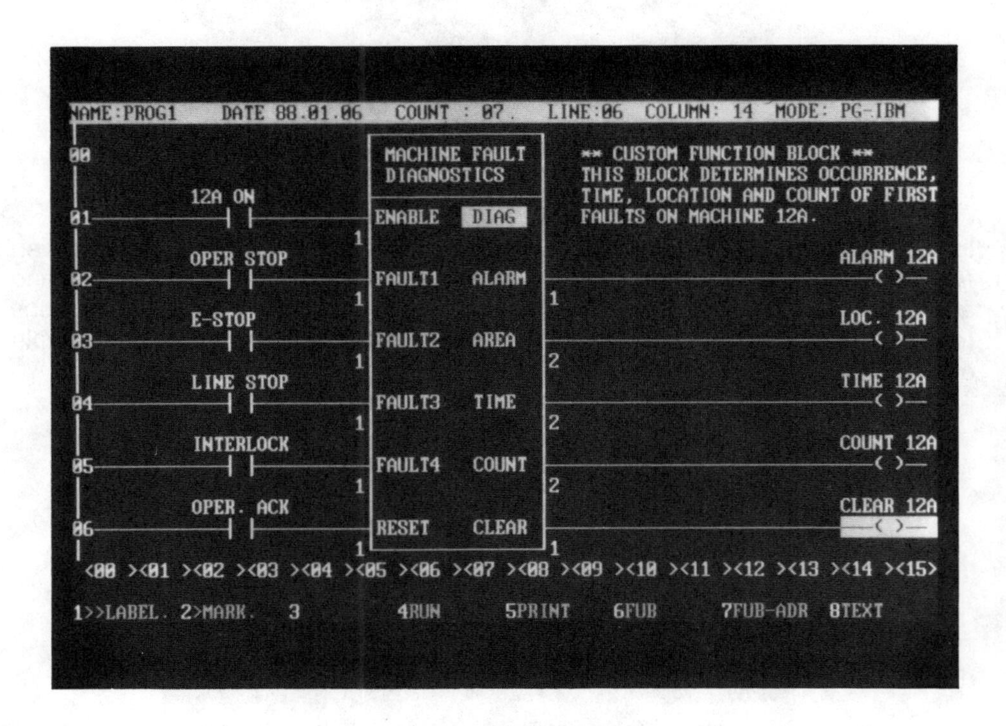

Figure 9-81. Custom functional block instruction formed by several instructions (Courtesy of B & R Industrial Automation, Stone Mountain, GA).

Language of the Future for PLCs

The continued growth of PLC applications and the necessity to implement more complex functions and operations will lead to a new type of PLC language. This language will not be a total change, but an addition to the existing ladder instructions.

The language that will be cast into programmable controllers will have the basic ladder instruction set as well as enhanced functional block ladders. However, a high level computer oriented language will be *embedded* in the ladder programming.

The new addition to ladders will take the form of BASIC, Pascal, or a hybrid instruction set that could be mixed in with the ladder program. These instructions will be capable of performing complex calculations, hard disk media storage and retrieval, modem communications, advanced mathematical functions and more. the implementation of these control functions would be very cumbersome and somewhat impossible to accomplish using the standard ladder set of instructions.

Figure 9-82 illustrates how this hybrid ladder language may take form. Using the hybrid ladder instructions the user will be able to program control logic sequences in an easy manner while more advanced computations can be implemented using the other instructions (BASIC, Pascal, etc.). All three formats, ladder, block and hybrid, will provide the programmer with the power and ease of understanding that definitely cast programmable controllers into a higher league of control equipment.

9-14 SUMMARY OF PLC LANGUAGES

In the previous sections we have covered the types of PLC instructions that can be found in most programmable controllers. A sharp definition of which instructions are advanced and which ones are basic is dependent on individual opinions.

We place emphasis on the ladder diagram order of execution which should be taken into consideration during the programming stages. In the last section, it was briefly mentioned that the next generation language in programmable controllers will probably be a hybrid ladder language that includes basic ladder instructions and functional blocks as well as a procedure oriented statement language similar to BASIC or Pascal.

As a final note, you should always keep in mind several factors during the selection of a PLC and its language characteristics. These factors include the following:

- Ease of use and implementation.
- The primary characteristics of the language.
- The type of problem to be solved.
- The execution time requirements that will be imposed on the program.

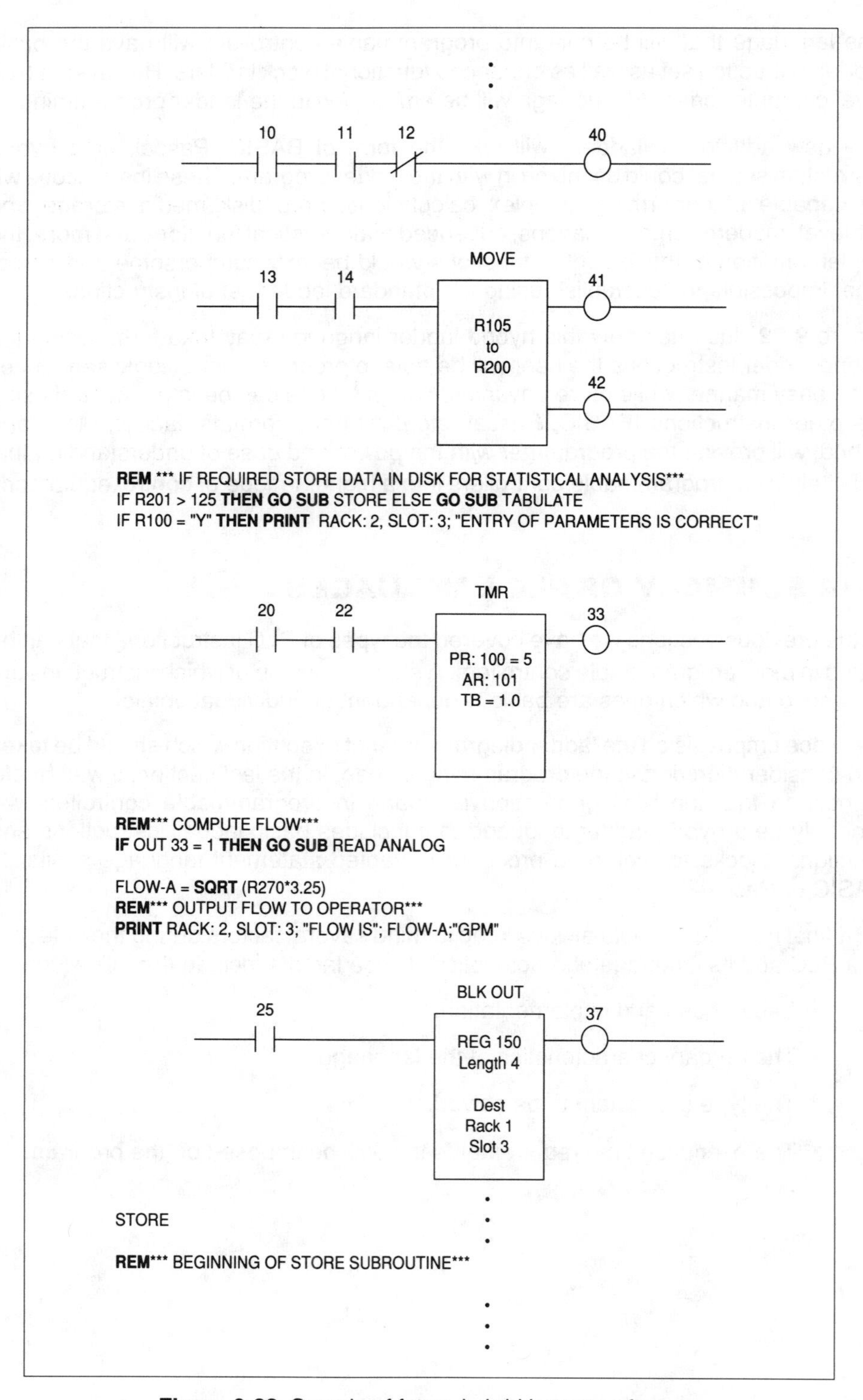

Figure 9-82. Sample of future hybrid language format.

Chapter

—10—

PLC SYSTEM DOCUMENTATION

"If you cannot—in the long run—tell everyone what you have been doing, your doing has been worthless."

—Erwin Schroedinger

10-1 INTRODUCTION

Documentation can be defined as an orderly collection of recorded information, concerning both the operation of a machine or process and the hardware and software components of the control system. These records provide valuable reference for use during system design, installation, start-up, debugging, and maintenance.

To the system designer, documentation should be a working tool that is used throughout the design phase. If the various documentation components are created and kept current during the design, they will not only serve as a reference to the designer, but will also (1) provide an easy means of communicating accurate information to all those involved; (2) put someone else, or the designer, in a position to later answer questions, diagnose possible problems, and modify the program if requirements change; (3) serve as training material for operators who must interface with the system and maintenance personnel who will maintain it; and (4) allow the system to be reproduced easily or altered to serve other purposes.

The achievement of proper documentation is realized through the gathering of hardware, as well as software information. This data is generally supplied from the engineering or electrical group that designs the system and is utilized by the end user. Although documentation is often thought of as something extra, it is a vital system component and should be a general engineering practice. In this chapter, we will cover important aspects needed to provide a "good" PLC documentation package that would help in the understanding of the control system.

10-2 STEPS OF DOCUMENTATION

System Abstract

As pointed out in the previous chapter, a good design starts with the understanding of the problem and a good description of the process to be controlled. This assessment is followed by a systematic approach that will lead to the implementation of the control system. Once the system is finished, the personnel involved in the design should provide a global description, or abstract, of the scheme and procedure used to control the process.

A system abstract should provide a clear statement of the control problem or task, a description of the design strategy or philosophy used to implement the solution of the problem, and a statement of the objectives that must be achieved. A description of the design strategy should define the function of the major hardware and software components of the system and why they were selected. For instance, a single CPU, located in the warehouse central area, will control two product conveyor lines. A remote subsystem, located in rooms four and five, will control the sorters for those areas. Data will be gathered on total production from both lines and reported in printed form at the end of each shift. Finally, the statement of the objectives will allow the user to measure the success of the control implementation.

The system abstract can always serve to transmit general design information to the end user or anyone who first needs to understand the original control task and, then, transmit the solution to the problem.

System Configuration

As the name implies, the system configuration is nothing more than a system arrangement diagram. In fact, it is a pictorial drawing of the hardware elements defined in the system abstract. It shows the location, simplified connection, and minimum details regarding major hardware components (i.e. CPU, subsystems, peripherals, etc.). Figure 10-1 illustrates a typical system arrangement diagram.

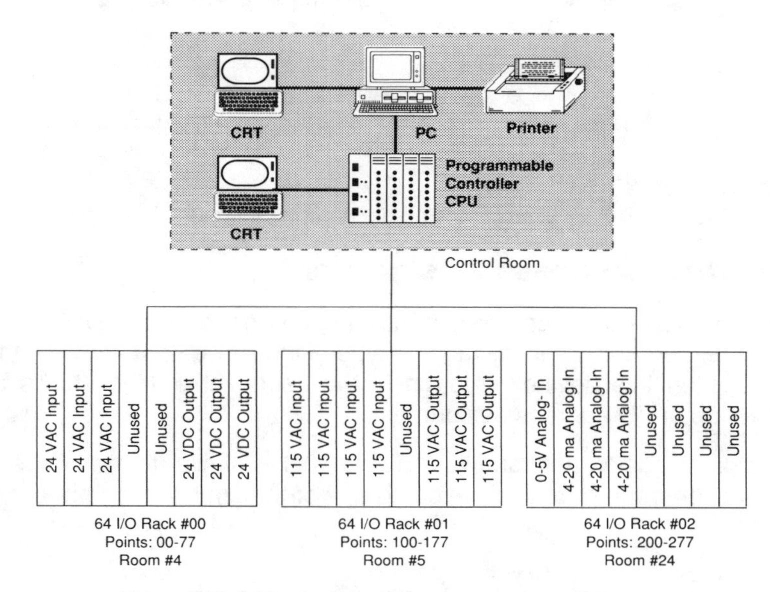

Figure 10-1. Typical system arrangement diagram.

This configuration not only indicates the physical location of subsystems, but also the designation of the I/O rack address assignments. Referencing of the rack address assignments will allow quick location of specific I/O. For example, during start-up the user can easily find out that I/O point 0200 (LS, PB, etc.), located in subsystem 02, is housed in room number four.

I/O Wiring Connection Diagram

The I/O wiring diagram shows actual connections of the field input and output devices to the PLC module. This drawing normally includes power supplies and subsystem connections to the CPU. Figure 10-2 illustrates an example of I/O wiring diagram documentation. The rack, group, and module locations are shown to illustrate explicitly the termination address of each I/O point. If the field devices are not wired directly to the I/O module, then terminal block numbers should be shown. In this example, the terminal blocks are represented by a dark circle with the TB number. Good I/O wiring documentation will be invaluable during installation and for later reference.

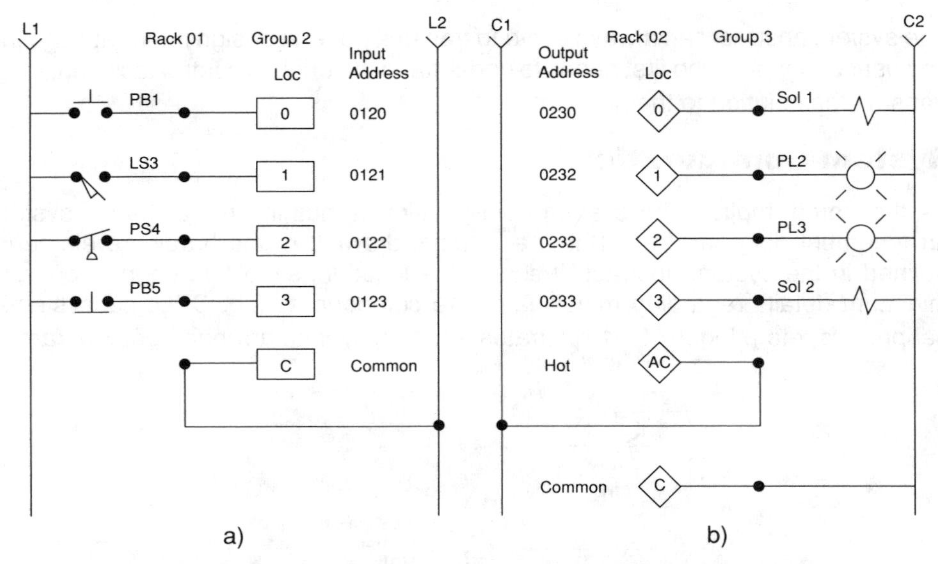

Figure 10-2. Typical I/O wiring documents; a)inputs, b) outputs.

Input/Output Address Assignment

The I/O Address assignment document identifies each field device by address (based on rack, group, and terminal), the type of input or output module (115 VAC), and the function this device performs in the field. Table 10-1 shows a typical documentation of I/O address assignment.

The I/O address assignment shown in Table 10-1 is similar to the I/O assignment table that will be done prior to developing any of the control program examples in Chapter 11.

Address	I/O Type	Device	Function
0120	115 VAC In	PB	Start Push Button PB 1
0121	115 VAC In	LS	Up Limit #2
0122	115 VAC In	PS	Hydraulic Pressure OK
0123	115 VAC In	PB (NC)	Reset PB 2
•	•	•	•
•	•	•	•
•	•	•	•
0230	24 VAC Out	Sol	Retract #1
0231	24 VAC Out	PL	#2 in position
0232	24 VAC Out	PL	Running
0233	24 VAC Out	Sol	Fast Up #3

Table 10-1. Real I/O address assignment.

Internal Storage Address Assignments

The documentation of internals is an important part of the total documentation and package. Because internals are used for programming timers, counters, and control relay replacement, and are associated with no field devices, there is a tendency to use them freely without much regard for accounting for their usage. However, just as with real I/O, misuse of internals could result in system misoperation.

Good documentation of internals will simplify field modifications during start-up. For example, imagine a start-up situation involving the modifications of one or more rungs by adding extra interlocking. The user will have to utilize internal coils that are not already being used. If the internal I/O address assignment is current and accurate, showing both used and unused addresses, then a useable address is quickly located, time is saved, and confusion is avoided. Table 10-2 illustrates a typical I/O address assignment documentation for internals.

Internal	Type	Description
1000	Coil	Used to Latch Position
1001	Coil	Set-up Instantaneous Timer Contact
1002	Compare	Used for CMP Equal
1003	Add	Addition Positive
•	•	•
•	•	•
•	•	•
T100	Timer	Time on delay—Motor 1
C400	Counter	Count pieces on Conv. #1
•	•	•
•	•	•
•	•	•

Table 10-2. Internal output address assignment.

Storage Register Assignments

Each available system register, whether a user storage register or I/O register, should be properly identified. Most applications use registers to store or hold information for timers, counters, or comparisons. Keeping an accurate record of use and changes to these registers is very critical. Just as with the I/O assignment documents, the register assignment table should show whether an address is being used or not. Table 10-3 shows typical documentation form for register assignments.

Register	Contents	Description
3036	Temperature In	I/O Register with Analog Module
3040	Temperature In	I/O Register with Analog Module
4000	1200	20 sec preset of TDR3
4001	2000	Count preset for CMP =
4002	5000	Count preset for CMP >
•	•	
•	•	
•	•	
4100	0	Not Used
to	•	•
4200	•	•
	0	•

Table 10-3. Register usage address assignment.

Control Program Printout

The program printout is a hard copy of the control logic program stored in the controller's memory. Whether stored in ladder form or some other language, the hardcopy should be an exact replica of what is in memory. Figure 10-3 shows a typical ladder printout.

Generally, a hardcopy printout shows each programmed instruction with the associated address of each input and output. However, information on what each instruction does or which field device is being evaluated or controlled is not readily apparent. For this very reason (lack of complete information), the program coding alone is not adequate for interpretation of the control system without the previously mentioned documentation. Some manufacturers may provide a documentation package that allows the programming device, generally a PC (personal computer),

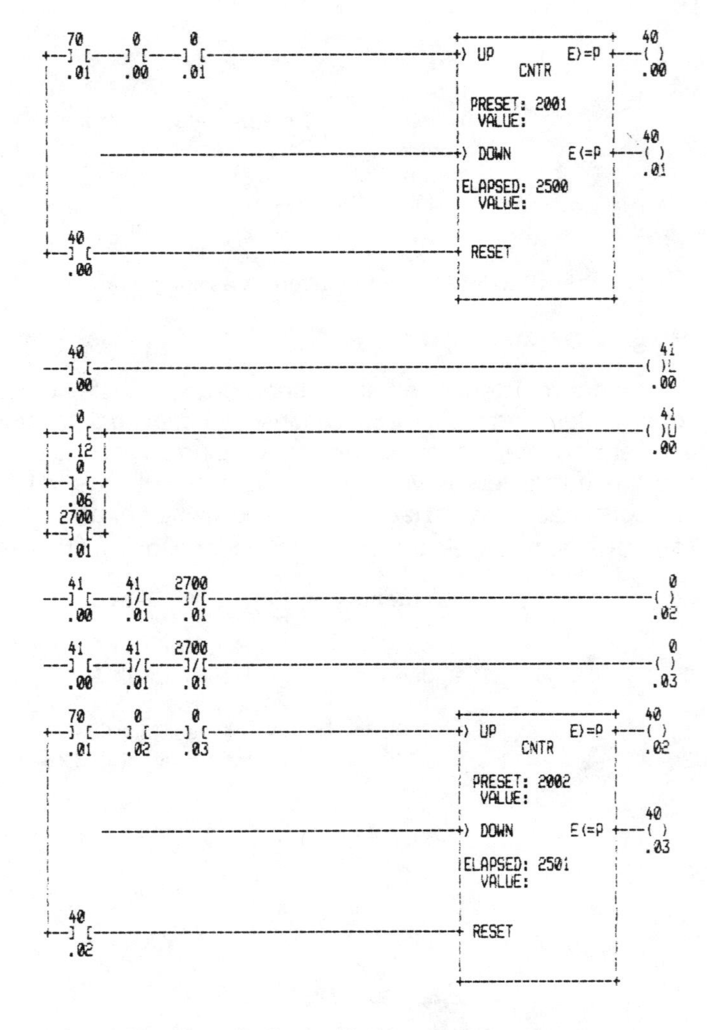

Figure 10-3. Typical ladder diagram printout.

to enter labels or mnemonic nomenclature to the program elements. The extent of the documentation may vary from one manufacturer to another and may or may not include input/output connection diagram documentation. Figure 10-4 illustrates a ladder control program that has generic documented elements in the ladder rung. PLC manufacturer's documentation may allow you to set global or generic mnemonic comments and then cross-reference the mnemonics with the I/O (real and internals) used in the system.

The controller will always have the latest software revision of the program stored in memory; therefore, it is wise to have the most recent hardcopy. During start-up, frequent changes are made to the program, which should be immediately documented. It is also a very good practice to obtain the latest hardcopy of the program at the earliest convenience.

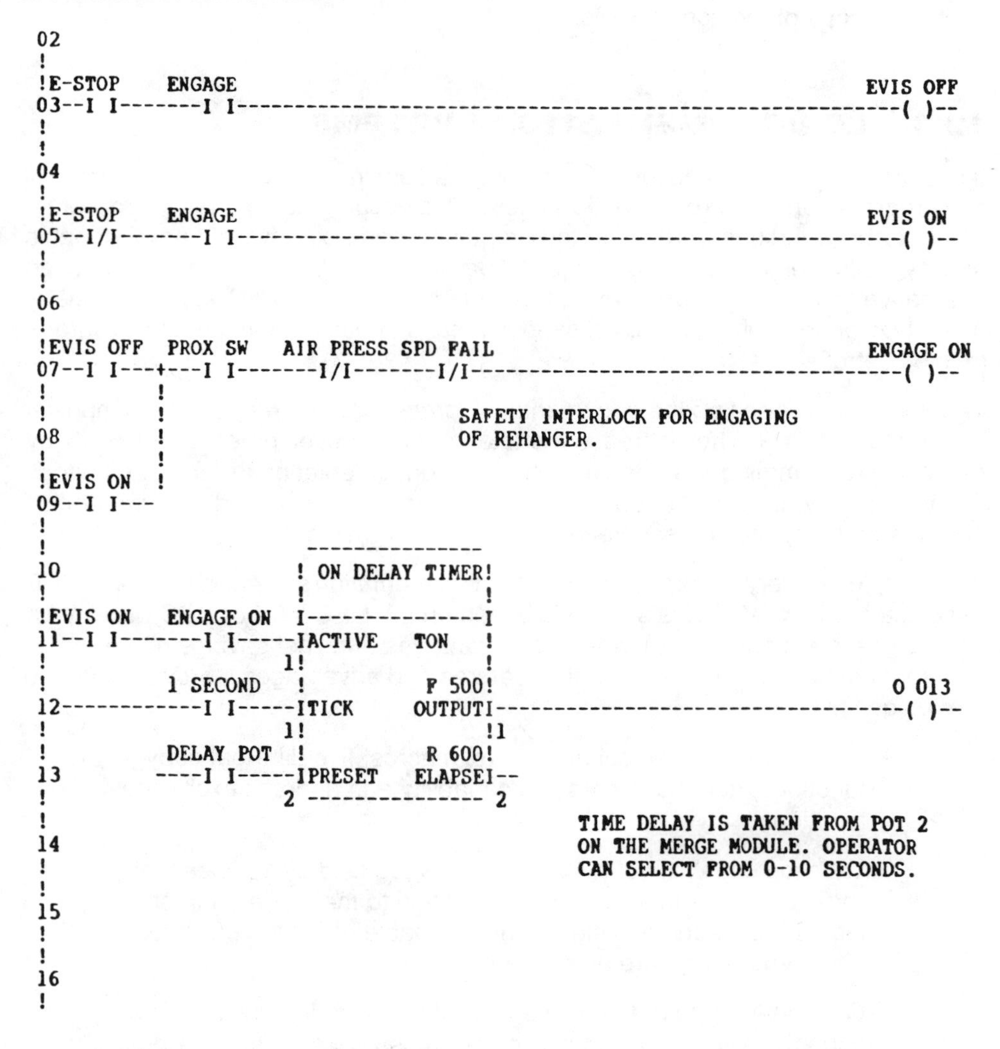

Figure 10-4. Ladder control program printout with manufacturer's documentation. (Courtesy of B &R Industrial Automation, Stone Mountain, GA)

Control Program Storage

For the most part, PLC programming takes place at a different location from where the controller will finally be installed. Saving the control program on a storage medium such as cassette tape, floppy disk, an electronic memory module, or some other means, is always recommended. This practice will allow you to send or carry the stored program to the installation site and reload the controller memory quickly. This approach is usually taken when the system uses a volatile type memory, and even when non-volatile memory is used.

The reproducible stored program, like any other form of documentation, should be kept accurate and current. A good safe practice is to keep two copies always, in case one is damaged or misplaced. Make sure that the stored program coincides with the latest hardcopy of the control logic.

10-3 PLC DOCUMENTATION SYSTEMS

It is obvious by now that documentation plays a very important role in the design of any programmable controller based system. This documentation may seem, sometimes, to be a tedious and costly task requiring perhaps several knowledgeable people to implement drafting, table preparations, or I/O assignments. As an alternative to this procedure, several manufacturers in the PLC support industry have developed sophisticated, yet simple, means of documenting a total programmable controller system.

These systems speed up the documentation procedure and reduce the manpower needed for the task. They increase the total program development productivity by reducing programming slip-ups and increasing documentation throughput. In addition to the standard types of documentation previously discussed, documentation systems normally provide several other useful documents.

An example of a very powerful and popular documentation system is *Ladders* (WRB Associates—Troy, MI). This system is based on the DEC VAX or DEC Micro VAX computer system and offers numerous advantages and cost savings over manual documentation methods. Some of the features and advantages it provides are the following:

- Provides electronic cut and paste, macros, five different copy functions, generic addressing capability and address exchange functions.
- Labeling capability of a 13-character wide field by 16 lines high (over contact and elements). In addition to the ladder element labels, the Ladders system is also capable of having unlimited comments anywhere else on the drawing.
- Complete range of I/O elements and hardwire I/O drawings capability with integrated automatic cross-reference into the logic program.

- Uploading and downloading program capability for most PLC systems.
- Changes in rung elements made in the program can be reported so that comparisons with previous logic can be made.
- Multi-user/Multi-tasking.

Besides the ladder printout, input and output usage (assignment) reports are also provided. These reports show a listing of the controller I/O addresses illustrating how each point is used. Construction mnemonics (e.g. contacts, limit switches, etc.) documentation is available as well as a complete report of all instructions available for the PLC used. Full cross reference reports provide direct information on all register contents and where in the program each element is used. An important advantage of program listings provided by documentation systems is that they show, on a single document, all the information regarding the control program. Figure 10-5 illustrates a typical printout of the I/O connection diagram from the Ladders documentation system. Figure 10-6 shows a sample printout section of a documented control program using Ladders.

In general, documentation systems are capable of uploading, verifying, and storing the PLC program directly from the controller or from a cassette, floppy disk, or other storage media.

10-4 CONCLUSIONS

Much could still be said for documentation and its relevance to the total system package. The importance of establishing complete, accurate documentation of the control problem and its solution, from the outset, cannot be overemphasized. If the system documentation is created during the system design phases, as it should be, it will not become an unwelcome burden imposed on designers as the project nears completion.

Documentation may seem trivial to some readers or too much work to others. Whether designing their own control system or sub-contracting the design, users should be sure that a good documentation package is delivered with the equipment. A well designed system is one that is not just put to work during start-up, but also one that can be maintained, expanded, modified, and kept running without difficulty. Good documentation will definitely help in this regard for both the designers and the end user. Remember, regardless of the application, a good design is not good unless its documentation is also good.

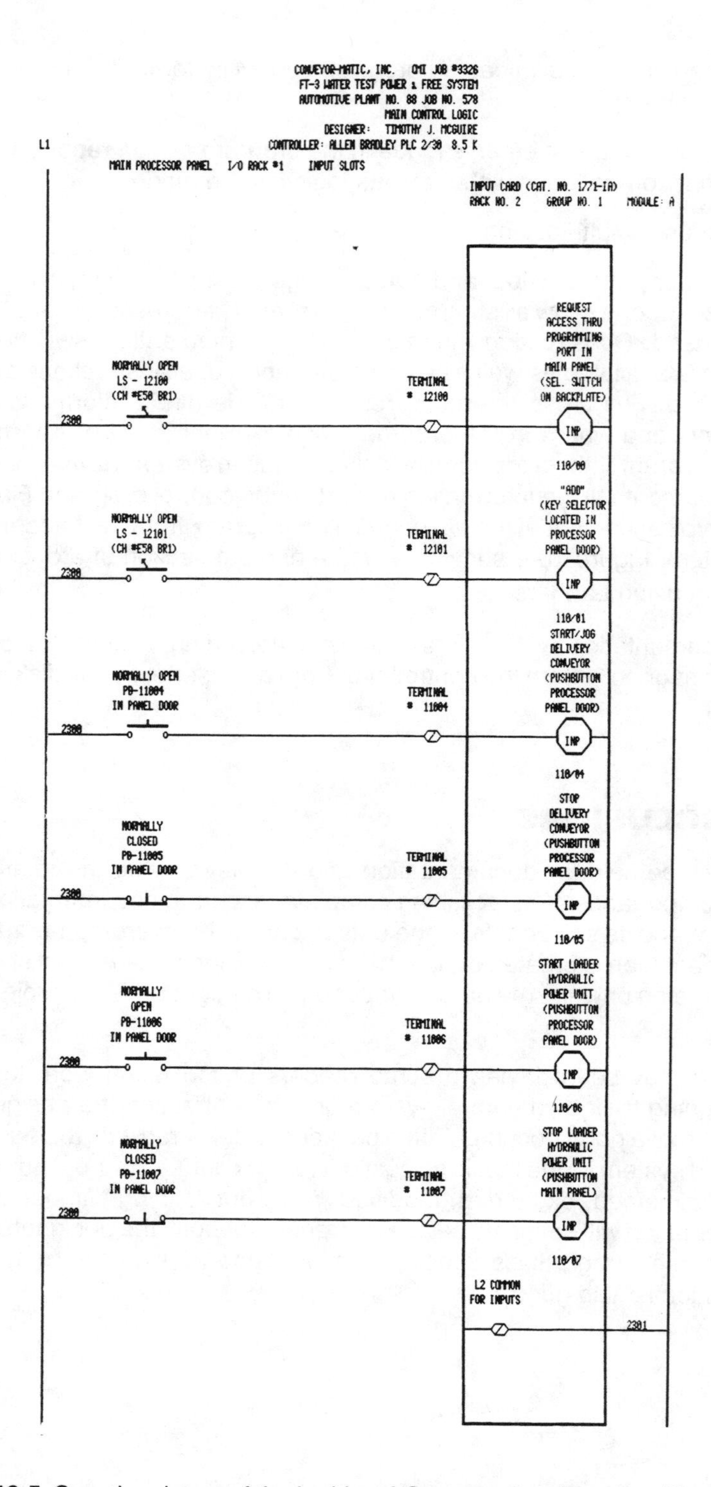

Figure 10-5. Sample printout of the Ladders I/O connection diagram: a)Input Connection. (Courtesy of WRB Associates, Troy, MI).

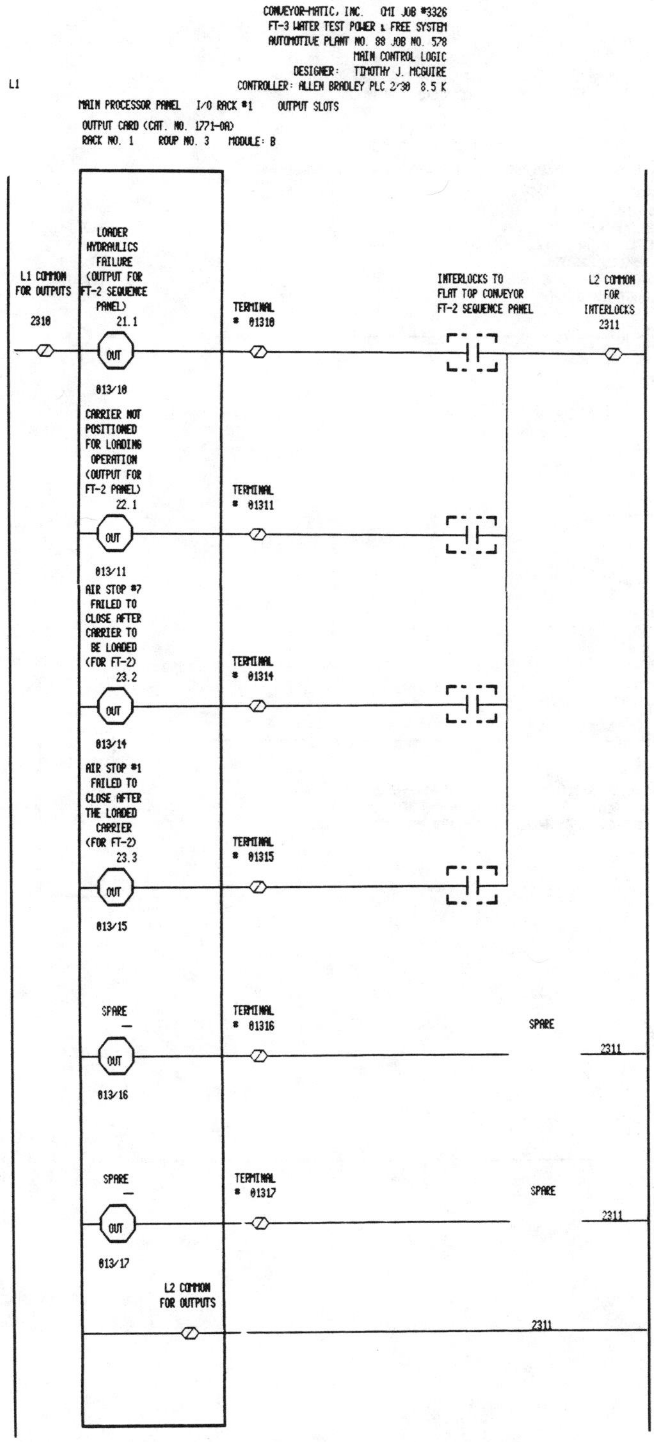

CONVEYOR-MATIC, INC. CMI JOB #3326
FT-3 WATER TEST POWER & FREE SYSTEM
AUTOMOTIVE PLANT NO. 88 JOB NO. 578
MAIN CONTROL LOGIC
DESIGNER: TIMOTHY J. MCGUIRE
CONTROLLER: ALLEN BRADLEY PLC 2/30 8.5 K

L1

MAIN PROCESSOR PANEL I/O RACK #1 OUTPUT SLOTS
OUTPUT CARD (CAT. NO. 1771-OA)
RACK NO. 1 ROUP NO. 3 MODULE: B

L1 COMMON FOR OUTPUTS
2310

LOADER HYDRAULICS FAILURE (OUTPUT FOR FT-2 SEQUENCE PANEL)
21.1
OUT
013/10
TERMINAL # 01310

INTERLOCKS TO FLAT TOP CONVEYOR FT-2 SEQUENCE PANEL

L2 COMMON FOR INTERLOCKS
2311

CARRIER NOT POSITIONED FOR LOADING OPERATION (OUTPUT FOR FT-2 PANEL)
22.1
OUT
013/11
TERMINAL # 01311

AIR STOP #7 FAILED TO CLOSE AFTER CARRIER TO BE LOADED (FOR FT-2)
23.2
OUT
013/14
TERMINAL # 01314

AIR STOP #1 FAILED TO CLOSE AFTER THE LOADED CARRIER (FOR FT-2)
23.3
OUT
013/15
TERMINAL # 01315

SPARE
—
OUT
013/16
TERMINAL # 01316
SPARE
2311

SPARE
—
OUT
013/17
TERMINAL # 01317
SPARE
2311

L2 COMMON FOR OUTPUTS
2311

Figure 10-5. Continued: b) Output Connection. (Courtesy of WRB Associates, Troy, MI).

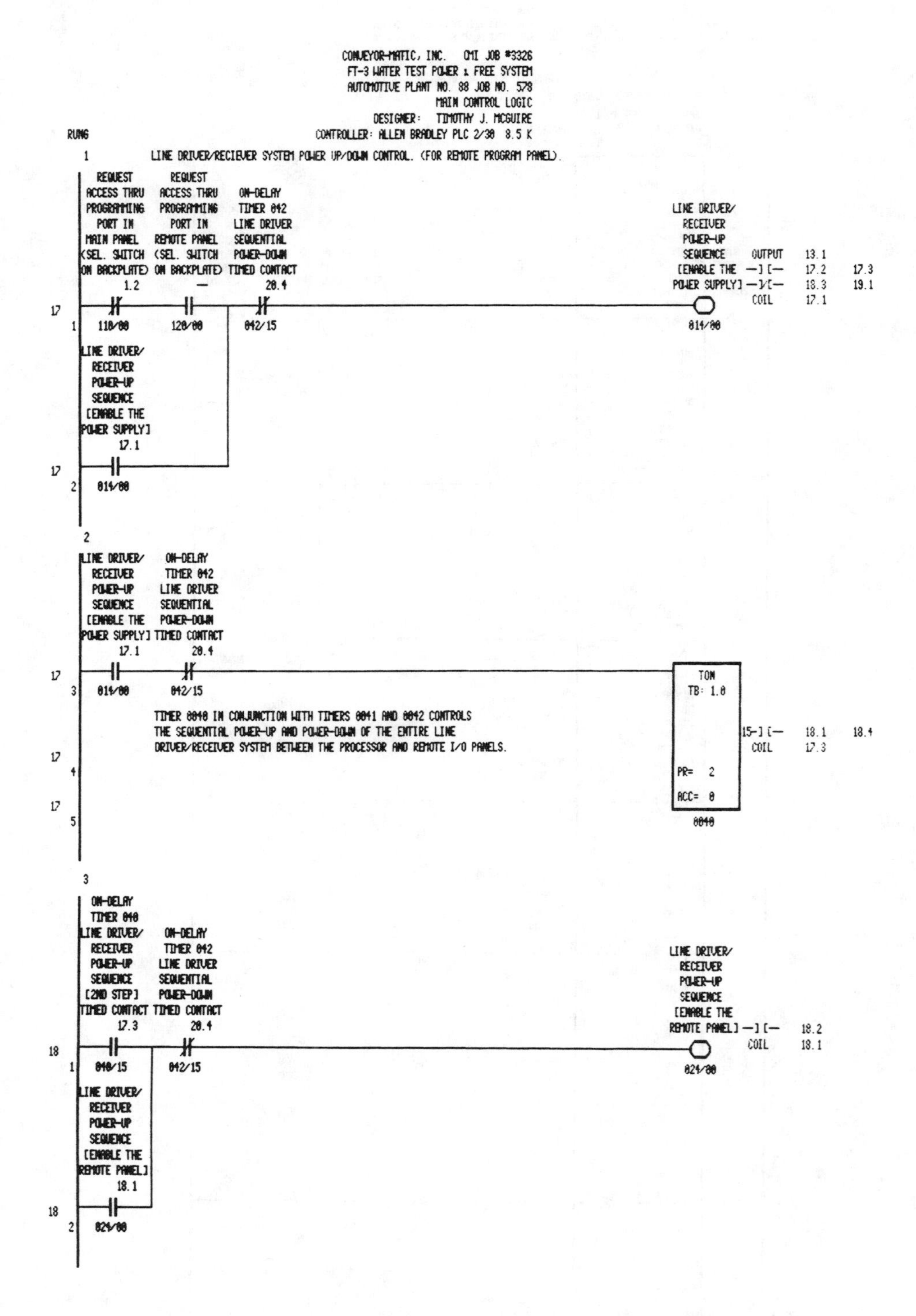

Figure 10-6. Sample printout of a control program using the Ladders documentation package. (Courtesy of WRB Associates, Troy, MI).

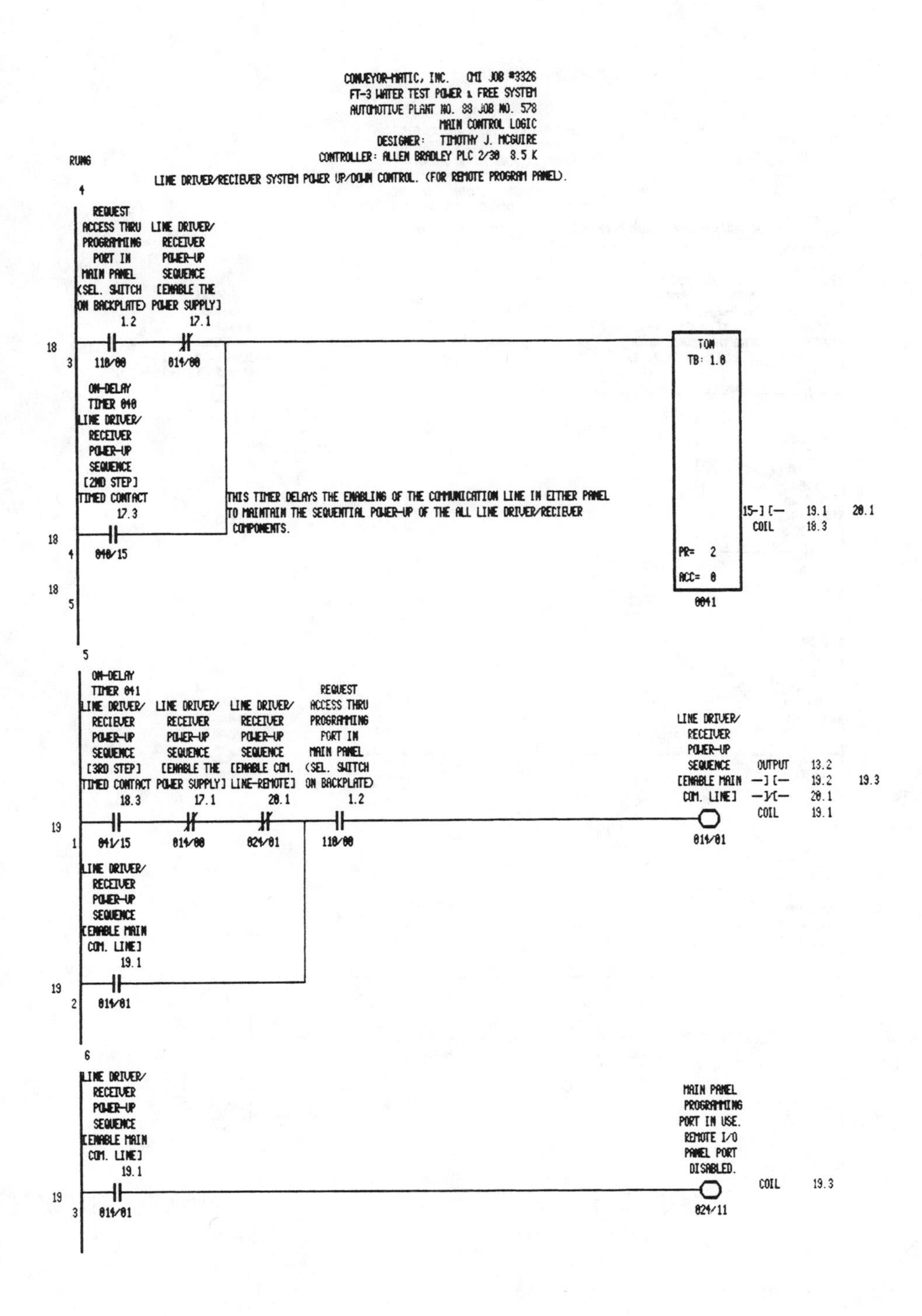

Figure 10-6. Continued.

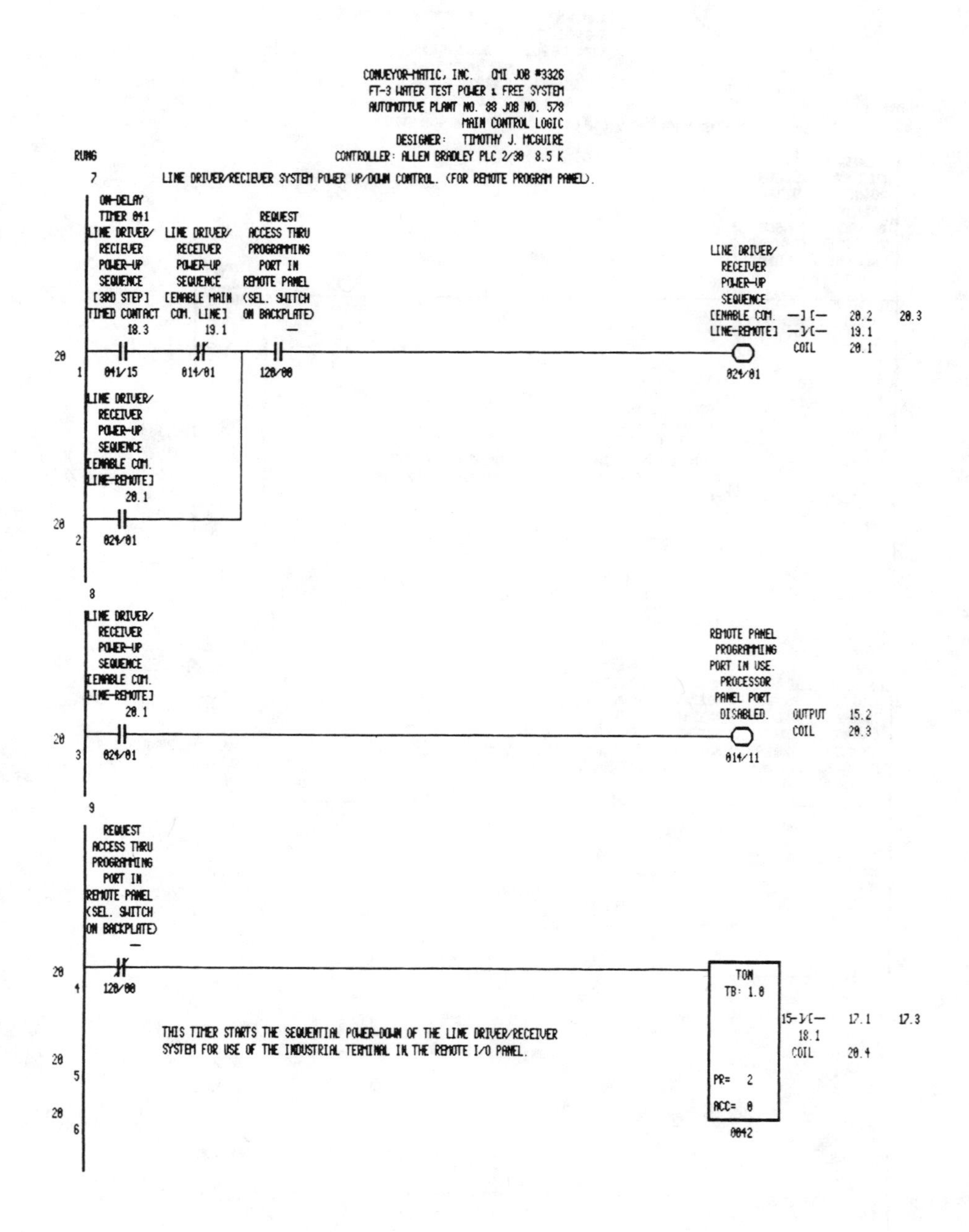

Figure 10-6. Continued.

Chapter

—11—

IMPLEMENTING AND PROGRAMMING THE PLC SYSTEM

"I must Create a System, or be enslaved by another Man's; I will not Reason and Compare; my business is to Create."

—William Blake

The implementation of a control program involves many steps as far as organization and order is concerned. Many of us generally jump into programming rungs, sequences of complex control, etc. without concern about all factors that must be taken into consideration. We believe that by reading this chapter, you will be aware of where a good starting point for programming a PLC lies.

The organization and control approach presented here should be used as a reference for your particular application. Nobody can really teach you how to program a specific application problem, but rather how to try to solve it using the techniques and power available in a programmable controller. Several examples of PLC programming are presented for various control scenarios, whether digital or analog, simple or complex. The more complex the example, the more organization is required. Several programming tips are provided for frequently encountered problems at the end of this chapter.

11-1 CONTROL DEFINITION

A user should begin the problem-solving process by defining the control task and making sure of what needs to be done. This definition information provides a basis for determining the programmable controller operations that must be performed. The definition of the control to be performed should be done by those who are familiar with the operation of the machine or process. This will help minimize possible errors due to misunderstanding of the process.

A control task definition usually occurs at many levels. Individuals within each department involved must be consulted to determine what inputs are required or have to be provided, so that everybody understands what is to be done in the project. For example, in a project involving the automation of a manufacturing plant in which materials are to be retrieved from the warehouse and sent to the automatic packaging area, personnel from both the warehouse and packaging areas must collaborate with the engineering group during the system definition. In case there are data reporting requirements, management should also be involved.

If a task is currently done manually or through relay logic, steps of this procedure should be reviewed to determine what improvements if any are possible. Although relay logic can be directly implemented in a PLC, it is advisable, when possible, to redesign the procedure to meet current needs of the application and to take advantage of the capabilities that a programmable controller offers.

The factors to be considered when defining the task are all closely related to the success or failure of the resultant program. This relationship will be revealed in the ability to provide correct control of the machine or process.

11-2 CONTROL STRATEGY

After the control task has been defined, the planning for its solution can begin. This procedure commonly involves determining the sequence of processing steps that must take place within a program to produce the output controls. This part of the program development is known as the development of an algorithm.

The term algorithm may be new or strange to some readers—it need not be. Each of us follows algorithms to accomplish certain tasks in our daily lives. The procedure that a person follows to get from home to school or work is an algorithm; the person exits the house, gets into the car, starts the engine, and so on; in the last of a finite number of steps, the destination is reached.

The strategy implementation for the control task using a PLC closely follows the development of the algorithm. The user must implement the control from a given set of basic instructions and produce the solution, or answer, in a finite number of such instructions. In most cases, it is possible to develop an algorithm to solve a problem. If doing so becomes difficult, it may be that further definition is needed. In this case, a return to the problem definition step may be required. For instance, we cannot specify how to get from where we are to Bullfrog County, Nevada unless we know both where we are and where Bullfrog County is. If a particular method of transportation is required, we need to be told that information as part of the problem definition. If there is a time constraint, we need to know that also. All of these requirements form part of the problem definition.

A fundamental rule in defining the program strategy is: *think first, program later.* Consider alternative approaches to solving the problem. Allow time to polish the approach (solution algorithm) before trying to program the control function. Adopting this philosophy will shorten the programming time, reduce debugging time, accelerate the start-up, and permit the focusing of attention on design when designing and on programming when programming.

During the formulation stage of the strategies, the user will be faced with a new application or a modernization of an existing process or machine. Regardless of which application must be done, the user will have to review the sequence of events that takes place, and through the addition or deletion of steps, optimize the control. Input and output considerations should be addressed, and a knowledge of what field devices the PLC will control is required.

11-3 IMPLEMENTATION GUIDELINES

A programmable controller is a powerful machine that can do only what it is told to do, and do it only as it is told to do so. It receives all of its directions from the control program, the set of instructions or solution algorithms created by the programmer.

A successful PLC control program depends greatly on how organized the user is. There are many ways to approach a problem, but if the application is approached in a systematic manner, the probability of making mistakes lessens.

The techniques used for the implementation of the control program are subject to the individuals involved in the programming. However, it is always recommended that certain guidelines be followed. Table 11-1 shows two approach guidelines that have proven to be very useful for implementing a programmable controller system. One of the approaches is directed towards new systems, while the other is for modernizing or upgrading the control system for existing systems that are functioning without a PLC, e.g. electro mechanical control or individual analog loop controllers.

New Applications	Modernizations
• Understand the desired functional description of the system	• Understand the actual process or machine function
• Review possible control method(s) and optimize the process operation	• Review machine logic of operation and optimize when possible
• Flowchart the process operation	• Assign real I/O and internal addresses to inputs and outputs
• Implement the flowchart by using logic diagrams or relay logic symbology	• Translate relay ladder diagram to PLC coding
• Assigning real I/O addresses and internal addresses to inputs and outputs	
• Translate the logic implementation to PLC coding	

Table 11-1. Approach guidelines.

As mentioned previously, understanding the process or machine operation is the first step in a systematic approach to solving the control problem. For new applications, the planning of strategy will follow the problem definition. Reviewing strategies for new applications, as well as revising the actual method of control for a modernization, will help detect or minimize possible errors that were introduced during the planning stages.

The differences in approach between a new or a modernization project become apparent during the programming stage. In a modernization project, the user ordinarily thoroughly understands the operation of the machine or process and what needs to be controlled. The sequence of events is usually defined by an existing relay ladder diagram, like the one shown in Figure 11-1, that can be translated into PLC ladder diagrams almost directly. However, a procedure should be followed to avoid possible errors and to maintain organization.

New applications usually begin with specifications given to the person or persons who will design and install the control system. These specifications are translated into a written description that explains the possible forms of control. The written explanation should be in simple terms so that confusion is avoided.

11-4 PROGRAMMING ORGANIZATION AND IMPLEMENTATION

This section deals with some important details that must be performed before any PLC program is started. As mentioned previously, organization is a key word when it comes to programming and implementing a control solution. The larger the project, the more organization is needed, especially when a group of people is involved.

A key to a control solution also depends on the ability to implement the solution. The user must know what is to be controlled by the PLC system, choose the correct equipment for the job (hardware and software) and understand the PLC. Once all these preliminary details have been covered, we may begin sketching the control program solution. All the work performed during this time is not in vain, but rather it

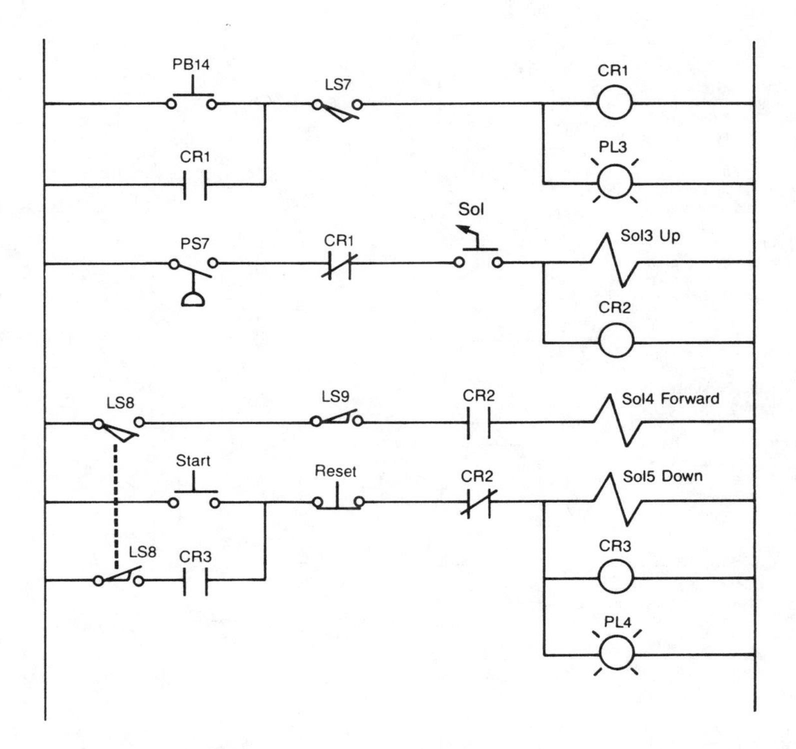

Figure 11-1. Typical electromechanical relay circuit diagram.

forms an important part of the system or project documentation. Most of us do not like to go back and document a system once it works, especially if we do not remember how we got it to work in the first place! Therefore, keeping this organization from the beginning will pay off later in the project.

Flowcharting and Generating Output Sequences

Flowcharting is a technique often used in planning a program after a written description is made. A flowchart is a pictorial representation that serves as a means of recording, analyzing, and communicating problem information. Broad concepts, as well as minor details and their relationship to each other, are readily apparent. Sequences and relationships that are hard to extract from general descriptions become obvious when displayed on a flowchart. Even the flowchart symbols themselves have specific meanings, which aid in the interpretation of the solution algorithm. Figure 11-2 illustrates the most common flowchart symbols and their meaning.

Flowcharts usually describe the process of operation in a sequential manner. A simple flowchart is illustrated in Figure 11-3. Each step in the chart performs an operation, whether an input/output, decision, or processing of data.

The main flowchart itself does not need to be complex and long; instead, it should point out functions to be performed (e.g. compute engineering units from analog input counts). Several smaller flowcharts may be used to implement the functions specified in the main flowchart.

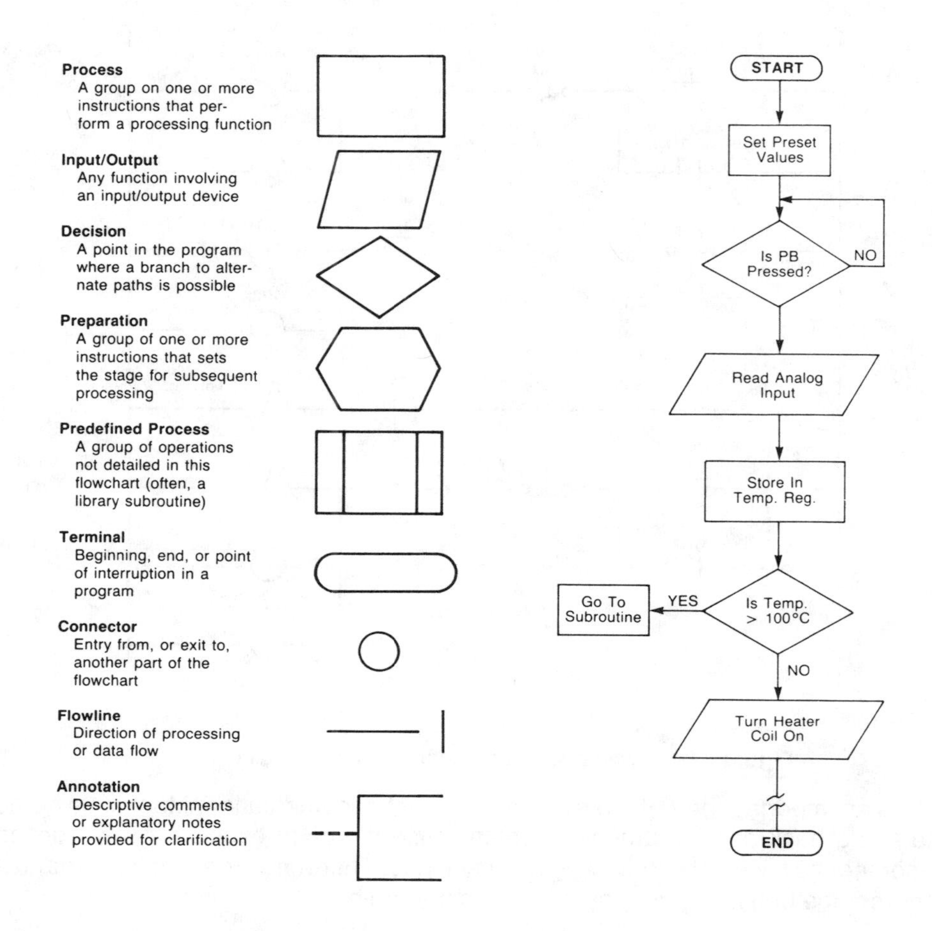

Figure 11-2. Common flow chart symbols. **Figure11-3.** Sample of a simple flow chart.

Once the flowchart is completed, the logic sequences can be obtained in one of two ways. First, *logic gates* specifying the input conditions, whether real or internal, can be used to describe a particular output sequence. Second, *PLC contact symbology* can be used directly to implement the logic necessary to represent an output rung. Figure 11-4 illustrates these two methods. Users should employ whichever method they feel most comfortable with, perhaps even an hybrid or mixed combination of both methods if desired (see Figure 11-5). Logic gate diagrams however, may be found more appropriate if the controller uses a Boolean instruction set.

The inputs and outputs marked with an X in Figure 11-4b may be used to indicate real I/O in the system. If no marking is present, the I/O point can be interpreted as an internal. The designations for the actual input signals can be the actual devices (e.g. LS1, PB10, AUTO, etc.) or symbolic letters or numbers that are associated with each of the field elements. A short description of the sequence can be helpful later during programming and is strongly recommended during this stage.

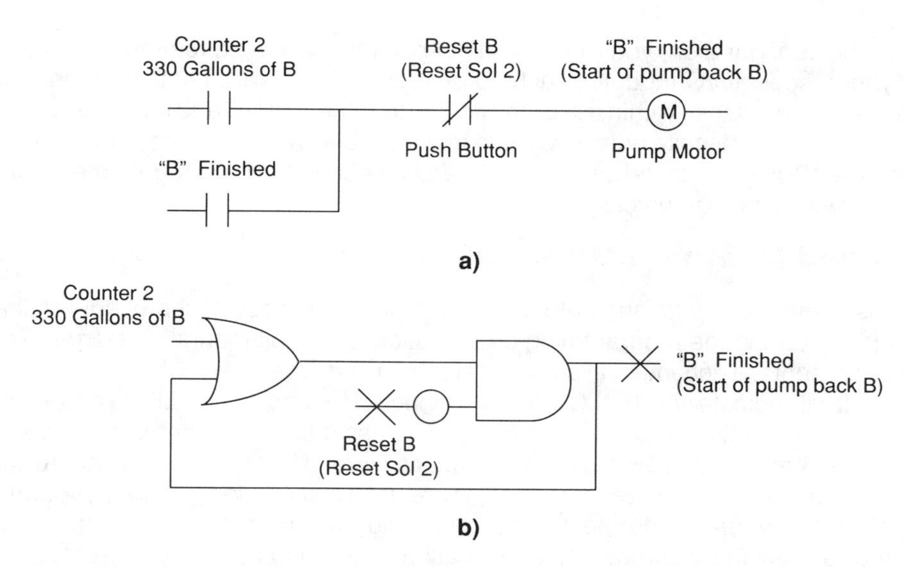

Figure 11-4. a) PLC contact symbology and **b)** logic gates representation.

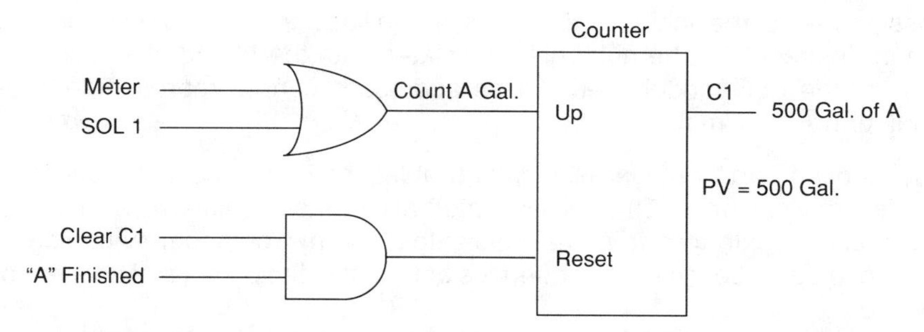

Figure 11-5. Combination method of logic gates and PLC symbology.

Configuring the PLC System

In a PLC application, like in many other computer system applications, several things may seem to need consideration at the same time. Configuration of the PLC system is one of them. The PLC configuration defines which I/O module will be used with which type of input and output signal and where the modules will be located in the local or remote rack enclosures.

Remember that the location of a module will define the address of those inputs and outputs that we are going to use in our control program. In Chapter 4 (pages 78-80), we presented a power supply loading example in which we discussed the current requirements for a set of I/O modules. We presented the option of choosing the power supply and mentioned that if there is possibility of more I/O interfaces to be added to the rack, then the modules' position would have to be changed, therefore changing all the addresses. If something like this sneaks out from under us "late" in the project, we would be upset, it would create a mess and, most importantly, it would take time to fix.

During the configuration, you must try to keep in mind possible future expansions, any type of special I/O modules such as fast response or wire fault input, and the placement of interfaces within a rack (all AC I/O together, all DC and low level analog together, etc.). All of these details will help towards achieving a better system design. In Chapter 10, we discussed system configuration documentation requirements and ways to represent it graphically.

Real and Internal I/O Assignment

The assignment of inputs and outputs is one of the most important procedures that takes place during the programming organization and implementation stages. The I/O assignment will document in an orderly fashion and organize what has been done thus far. It will indicate which PLC input is connected to which input device and which PLC output will drive which output device. The assignment of internals, including timers, counters, and MCR's, also takes place here. These assignments are the actual contact and coil representations that are used in the ladder diagram program. In applications where electromechanical relay diagrams are available (e.g. modernization of a machine or process), identification of real I/O can be done simply by circling the devices and then assigning the I/O addresses (see Example 11-1).

The assignment of real inputs and outputs, as well as internals, can be tabulated as shown in Figure 11-6. The numbers associated with the I/O address assignment depend on the PLC model used. These addresses can be represented in octal, decimal, or hexadecimal.

The description part of these assignment tables is used to describe the input or output field devices (real I/O) as well as internal outputs use (internals). The I/O assignment can be extracted from the logic gate diagrams, or ladder symbology, that were used to describe the logic sequences or from the circle marks of an electromechanical diagram.

The table of assignments is closely related to the input/output connection diagram shown in Figure 11-7. Although industry standards for input and output representations vary among users, inputs and outputs are typically represented by squares and diamonds, respectively. This I/O connection diagram forms part of the documentation package.

A conscious grouping of associated inputs and outputs is recommended during the I/O assignment. This grouping will allow monitoring and manipulation of a group of I/O (through I/O registers) simultaneously. For instance, if 16 motors are to be started sequentially, their starting sequence can be viewed by monitoring the I/O register associated with the 16 mapped I/O points. Due to the modularity of the I/O system, it is also recommended that all the inputs or all the outputs be assigned at the same time. This practice will prevent assignment of an input address to an output module, and vice versa.

Module Type	I/O Address Rack	Group	Terminal	Description
Input	0 0 0 0	0 0 0 0	0 1 2 3	LS1—Position LS2—Detect Sel Switch—Select 1 PB1—Start
Output	0 0 0 0	0 0 0 0	4 5 6 7	PL1 PL2 Motor M1 Sol 1
Output	0 0	1 1	0 1	Sol 2 PL3

Figure 11-6a. Sample I/O address assignment document for real inputs and outputs.

Device	Internal	Description
CR7	1010	CR7 Replacement
TDR10	T200	Timer on Delay 12 sec
CR10	1011	CR10 Replacement
CR14	1012	CR14 Replacement
—	1013	Set-up Interlock
•	•	•
•	•	•
•	•	•

Figure 11-6b. Sample address assignment document for internal outputs.

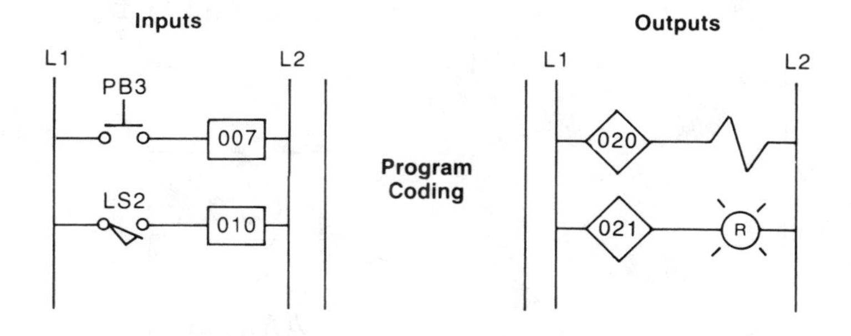

Figure 11-7. Typical I/O connection diagram.

Example 11-1

For the circuit shown in Figure 11-8, a) identify the real inputs and outputs by circling each; b) assign the I/O addresses; c) assign the internal addresses (if required); and d) draw the I/O connection diagram.

Note: Assume that the PLC has modularity of 8 points per module and that there are eight module slots per rack. The master rack is number 0. Inputs and outputs can take any address as long as the correct module is used. The PLC decodes whether an input or output module is connected in any slot. The number system is octal. Internals start at address 1000_8.

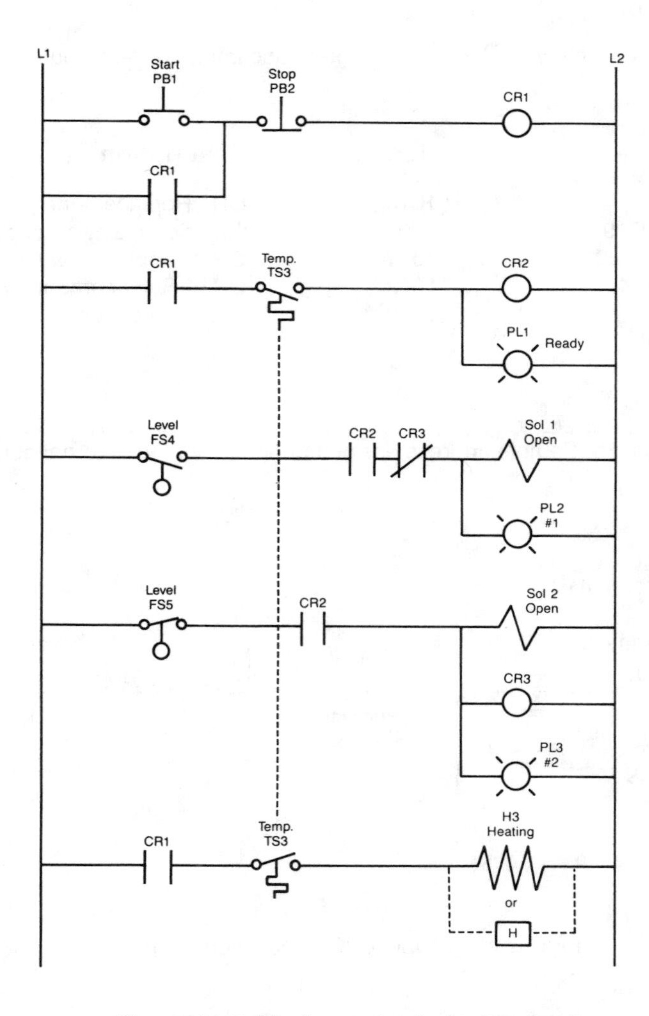

Figure 11-8. Electromechanical relay circuit.

Solution

a) The real input and output connections are circled and shown in Figure 11-9. Note that the temperature switch TS3 is circled twice even though it is only *one* device. In the address assignment, only one of them is referenced and only one of them is wired to an input module.

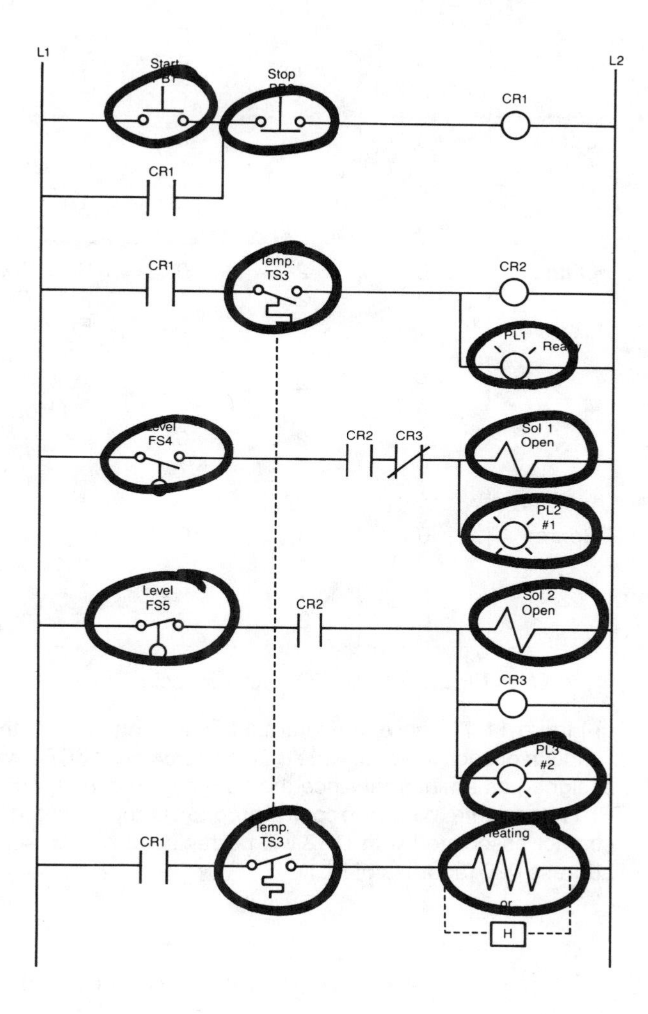

Figure 11-9. Identification (circled) of real I/O.

b) Figure 11-10 illustrates the assignment of inputs and outputs. It was started by assigning all inputs and all outputs. Spare I/O locations were left in case of future use.

Module Type	I/O Address Rack	Group	Terminal	Description
Input	0	0	0	Start PB 1
	0	0	1	Stop PS 1
	0	0	2	Temp TS 3
	0	0	3	Level FS 4
	0	0	4	Level FS 5
	0	0	5	—
	0	0	6	—
	0	0	7	—
Spare	0	1	0	Not Used
		⋮		
	0	1	7	
Output	0	2	0	PL 1 Ready
	0	2	1	Sol 1 Open
	0	2	2	PL 2 #1
	0	2	3	Sol 2 Open
	0	2	4	PL 3 #2
	0	2	5	H3 Heating
	0	2	6	—
	0	2	7	—
Spare	0	3	0	Not Used
		⋮		
	0	3	7	

Figure 11-10. I/O address assignment.

c) Figure 11-11 shows the internal I/O assignment and the description of each internal. Note that the control relay CR2 was not assigned as an internal since the output rung corresponding to PL1 is the same. When the control program is implemented, every contact associated with CR2 will be replaced by contacts with address 020 (address of PL1).

Device	Internal	Description
CR1	1000	Control Relay CR 1
CR2	—	Same as PL1 Ready
CR3	—	Same as Sol 2 Open

Figure 11-11. Internal output assignment.

d) The I/O connection diagram is illustrated in Figure 11-12. This diagram is based on the I/O assignment (section b). Note that only one of the temperature switches, the normally open TS3, was connected as an input. Programming of each switch in the logic program would be based on the connection of normally open (see Chapter 9, page 225 for more on input connections).

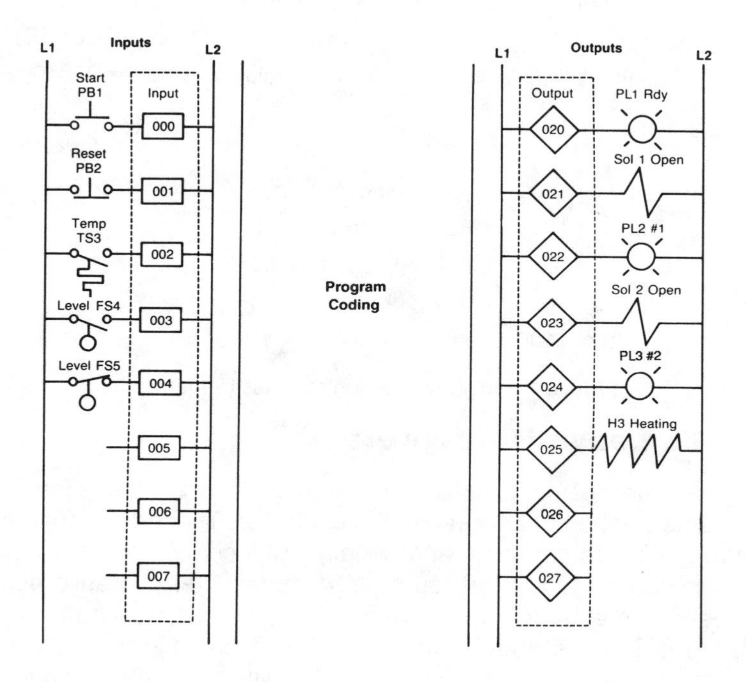

Figure 11-12. I/O connection diagram.

Register Address Assignment

The address assignment for the registers used in the control program is another important organizational details that must be enforced. It is not a tedious task by any means, but rather a disciplinary approach to assigning registers for unique uses in the program.

The easiest way to assign registers is to keep a listing of all the registers available in the PLC, and, as they are used, tabulate the registers' contents as well as their description and function. Figure 11-13 illustrates a sample listing of the first 15 registers of a PLC system, from address 2000_8 to address 2016_8.

Register	Contents	Description
2000	Analog input	Temperature input Temp 3 (inside)
2001	Analog input	Temperature input TEMP 4 (outside)
2002	spare	-
2003	spare	-
2004	TWS input	Set point (SP1) input from TWS panel 1
2005	TWS input	Set point volume (V1) from TWS panel 2
2006	constant 2350	Timer constant of 23.5 sec (.01 sec TB)
2007	Accumulated (C)	Accumulated value for counter R2010
2010	spare	-
2011	spare	-
2012	Constant 1000	Beginning of look-up table #1
2013	Constant 1010	Look-up value #2
2014	Constant 1023	Look-up value #3
2015	Constant 1089	Look-up value #4
2016	Constant 1100	Look-up value #5

Figure 11-13. Sample Register Assignment.

Portions To Leave Hardwired

During the assignment of inputs and outputs, it is necessary to make the decision as to what elements should not be wired to the controller. These elements will remain part of the magnetic control logic and usually will include elements that are not frequently switched off after start, such as compressors, hydraulic pumps, and others. Elements like emergency stops (e.g. rope switches, pushbuttons, etc.) and master start pushbuttons should also be left hardwired, principally for safety purposes. If for some reason the controller is faulty and there is an emergency situation, the system can be shut down without PLC intervention.

Figure 11-14 illustrates an example of components that are typically left hardwired. Note that in this diagram the PLC fault contact (or watchdog timer contact) is wired in series with other emergency conditions. These contacts are closed when the controller is operating correctly and would open when a fault occurs and are available for use by the system designer.

In the diagram shown in Figure 11-14 note that if any of the emergency situations occurs, including a PLC malfunction, the power (L1) to the rest of the circuits (i.e. I/O modules) will be inhibited. The turning OFF of the SCR will open the SCR contacts which allow power to flow; therefore, no power will be allowed to flow. Furthermore, a set of normally closed PLC fault contacts are also used in the hardwired section to alert personnel of a system failure due to a PLC malfunction. It is a good practice to implement this type of alarm signaling not only to the main PLC rack where the CPU is located, but also to each remote I/O rack location. This allows subsystem failures to be annunciated promptly so that solutions to the problem can be accomplished without endangering personnel.

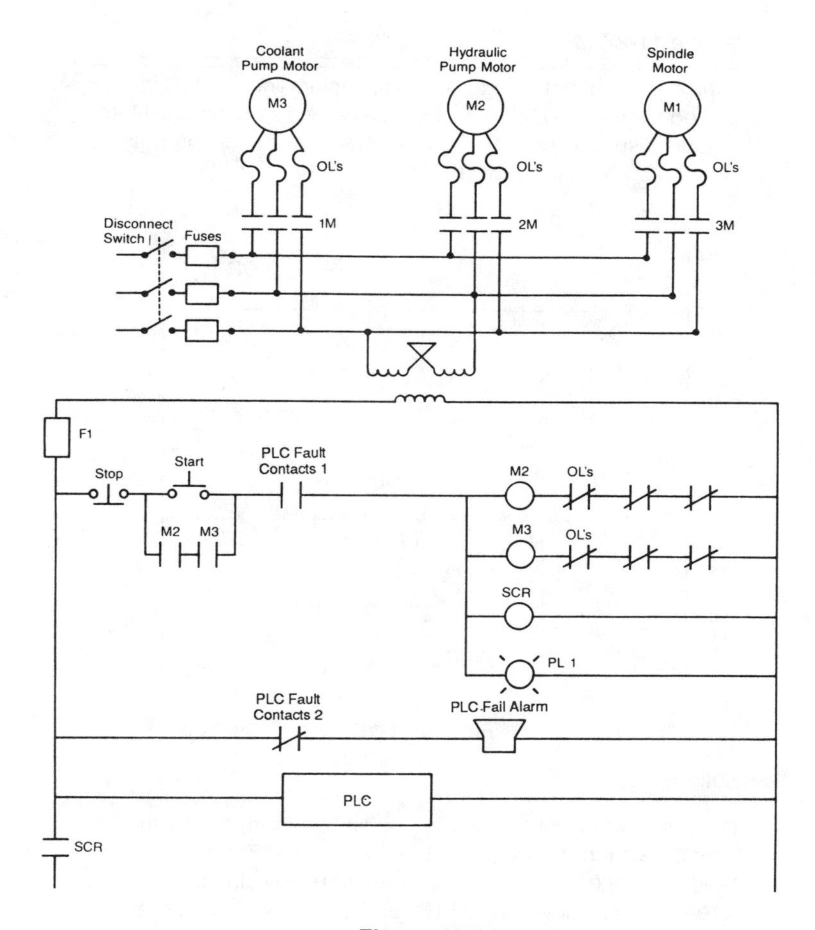

Figure 11-14

Special Cases of Input Device Programming

There are some circuits and input connections in programmable controller installations which require attention during the programming stages. One that we have covered in Chapter 9 was the programming of a normally closed input device. Remember that the programming of a device is closely related to how we want that device to behave in the control program.

An input device wired normally open can be programmed to act as a normally open or as a normally closed device. The same rule applies for a normally closed input. In general, if a device is wired normally closed and it is needed to act as a normally closed input, the reference address of the NC input is programmed normally open. As will be seen in the following example, a normally closed device in a hardwired circuit (e.g. CR3 normally closed contact in Figure 11-8, Example 11-1) is programmed as normally closed when replaced in the PLC control program since it is not referenced as an input; therefore it is not being evaluated as a real input device.

Example 11-2

For the circuit shown in Figure 11-15, implement the PLC ladder circuit equivalent and tabulate the I/O address assignment. Note: for inputs use addresses 10_8 through 47_8 and start outputs at address 50_8. Internals start at 100_8.

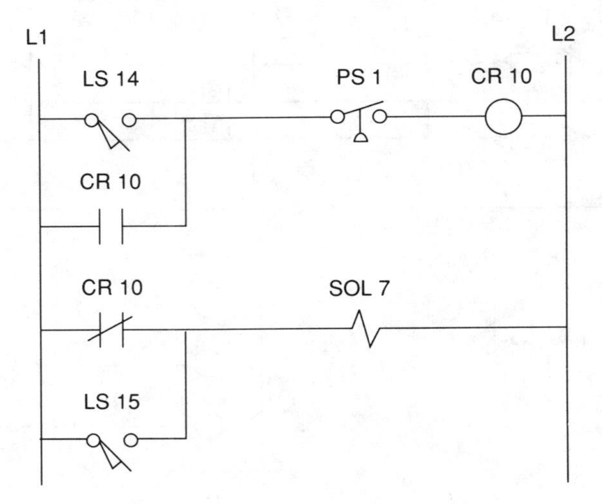

Figure 11-15

Solution

The ladder equivalent circuit is shown in Figure 11-16 for the address assignment described in Figure 11-17. The normally closed contact (CR10) is programmed normally closed since it is referenced by the internal coil 100 and it is required to operate as normally closed.

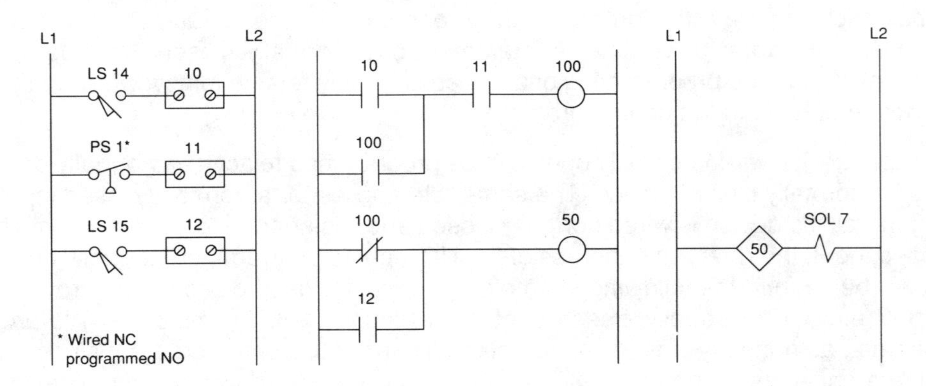

Figure 11-16

I/O Address	Device	Type
10	LS14	Input
11	PS1	Input
12	LS15	Input
50	SOL 7	Output
100	CR10	Internal

Figure 11-17. I/O and internal I/O address assignment.

Another circuit the programmer must be aware of is the MCR. In electromechanical circuit diagrams an MCR coil controls several rungs in the circuit by means of switching ON or OFF the power to those rungs; there is no definite end of the MCR except when the circuit is followed all the way through. When the hardwired ladder circuit is translated into PLC symbology, the END MCR instruction must be placed after the last rung the MCR is supposed to control. Figure 11-18 illustrates a partial ladder rung that has this type of MCR condition. The END MCR instruction should be placed in the PLC program after the rung containing the PL3 output.

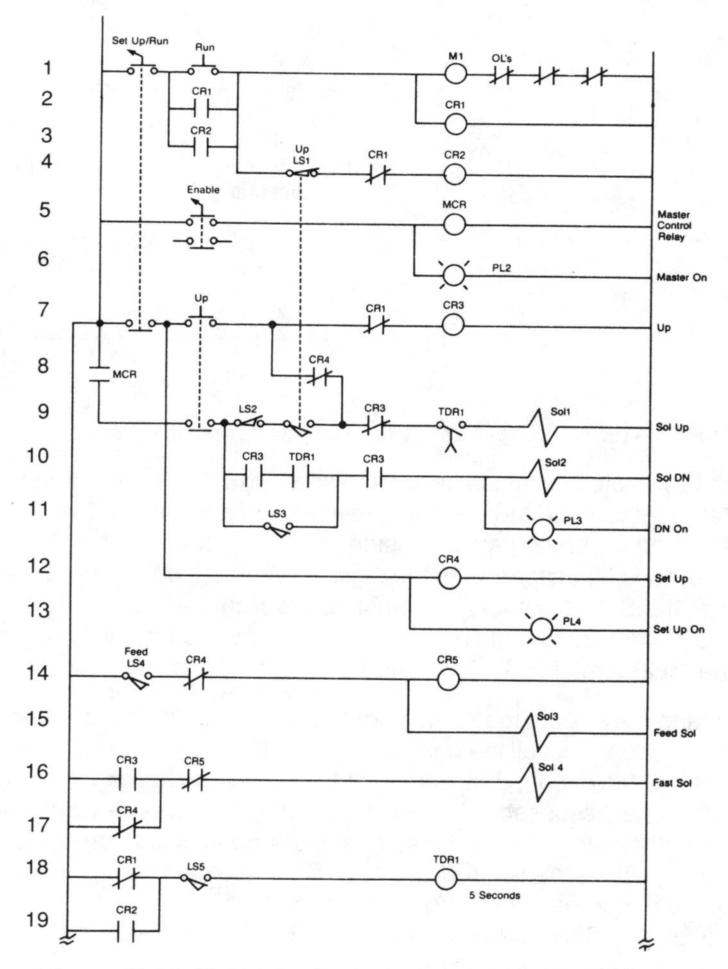

Figure 11-18. Electromechanical relay circuit with MCR fencing.

Another condition not so apparent in this diagram is the possibility of bidirectional power flow (8th line) by the normally closed CR4 contact, the instantaneous contact of TDR1 (10th line) and of course the programming of LS1 depending upon which contact, the NC or NO, is connected to the PLC.

To solve the bidirectional flow we would have to know, according to the machine function, whether or not CR4 influences the two output rungs it is connected to, that is, the CR3 control relay output and the solenoid SOL1 output (see Section 9-3).

In a PLC, instantaneous timer contacts are generally unavailable. To implement the instantaneous timer contacts (contacts that close or open once the timer is enabled) the programmer must use an internal output to trap the timer and use the internal's contact as instantaneous contacts as well as to drive the logic of the timer. Figure 11-19 illustrates this timer trap technique (without address assignment).

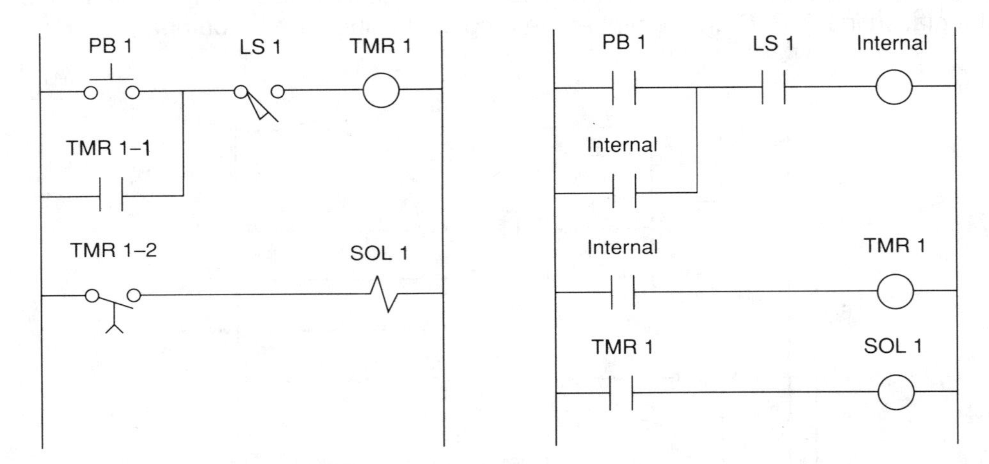

Figure 11-19. Trapping of TMR1 (left) using internal outputs (right).

Another critical characteristic of the circuit shown in Figure 11-18 is the forward path of the CR4 (NC contact) and the previous devices, Up and Set Up/Run, connected in series (line 7 and 8), driving a rung inside the MCR fenced zone. The solenoid SOL1 can be turned ON according to the logic, even in the case where the MCR is OFF. Therefore the SOL 1 output rung must be programmed (in the PLC program) outside the MCR zone and should also be able to be turned ON by the logic section included in the MCR circuit (LS1, LS2, etc.).

When a logic rung may seem too confusing, the best thing to do is to isolate it from the others. Then reconstruct all the possible logic paths from the right, starting at the output, to the left of the circuit, thus ending at the beginning of the rung. If a section of a rung, like the one discussed in the previous paragraph, has direct connection or interaction to another rung, the best way to avoid problems is to create an internal output at the point where the two rungs cross. Then use the internal output to drive the rest of the logic. For the circuit shown in Figure 11-18, this cross point would be at the normally closed CR4 and between the CR3 (NC) and the LS1 (NO) in line 9.

Program Coding/Translation

Program coding is the process of writing or rewriting the logic or relay diagram into PLC ladder program form. This ladder program is the actual logic that will implement the control of the machine or process and is stored in the application memory. The ease of program coding is directly related to how orderly the previous stages (assignment, etc.) have been done. Each element in the PLC ladder program has an address assigned to it according to the I/O assignment document. Figure 11-20 shows a sample program that is generated from the logic or relay diagram (internal coil 1000 replaces the control relay).

Note that the coding is a PLC representation of the logic, whether it comes from a new application or a modernization. This coding process is examined closer in the next two sections, where several programming examples are presented.

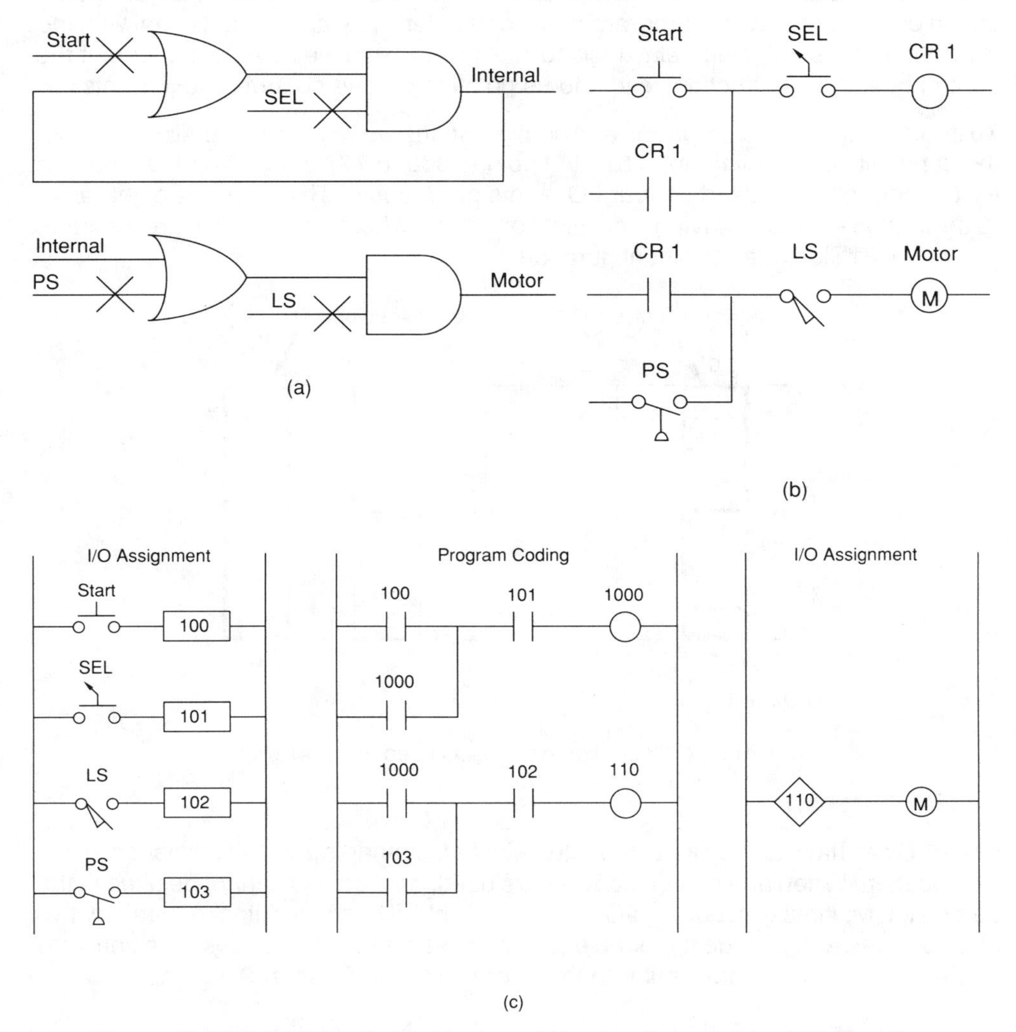

Figure 11-20. translation from **a)** logic and **b)** relay diagram into **c)** PLC coding.

11-5 DISCRETE I/O CONTROL PROGRAMMING

Introduction to Control Programming and PLC Description

In this section we will cover several programming examples that deal with replacement of relay systems or so called modernizations, and with examples that relate to new implementations of PLC control. The examples presented will deal primarily with ON/OFF types of control while the next section will cover more analog I/O interaction.

Modernization applications often involve the transformation of a machine or process control from conventional relay logic to a programmable controller. The relay panels that contain the control logic are usually characterized by maintenance problems involving contact chatter, contact welding, and other electromechanical problems. Changes targeted to improve performance in the machine can also call for optimization of the control with a programmable controller. New applications deal with the fact that there is no actual relay logic to replace; instead we have to implement the control based on instructions given to us pointing out the control requirements.

To implement the PLC programs and configurations we will use a mid-size PLC with the capability of handling up to 512 I/O points (000 to 777 octal). The I/O structure for the controller is based on four I/O points per module. There can be eight racks (0 through 7) each one having eight slots or groups where the modules are inserted. Figure 11-21 illustrates this configuration.

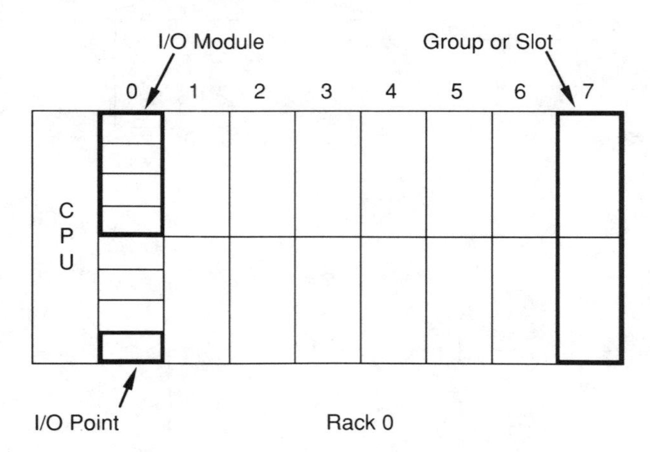

Figure 11-21. Illustration of the PLC used in the example.

The PLC can have analog input modules with four channels and can be placed in any slot location. When analog I/O modules are used, discrete I/O cannot be used in the same slot. Multiplexed register I/O is also available. These modules require any two slot positions and provide the enable (select) lines for the I/O devices. The software instructions available are similar to those presented in Chapter 9.

Addresses 000 through 777 octal represent either input or output device connections mapped to the I/O table. The first digit of the address represents the rack number, the second one the slot, and the third the terminal connection in the slot. The PLC detects whether an input or output is inserted in a slot.

Internal outputs may be used from point addresses 1000_8 to 2777_8. Register storage starts at register 3000_8 through register 4777_8. There are two types of timers and counters formats that can be chosen: ladder format or block format. Timers require the use of internal outputs to specify the ON delay output. The ladder format timers are identified by placing a T in front of the internal address. Block format timers only need to specify the internal address in the block's output coil.

Throughout the examples presented in this and the next section, we will try to use addresses 000_8 through 027_8 for discrete inputs and 030_8 through 047_8 for discrete outputs. Analog I/O will be placed in the last slot of the master rack (0) whenever possible. During the development of some of these examples you will find that the assignment of internals and registers is a task that is sometimes performed parallel to the programming stages. The assignment of internals and registers is an important detail that will be accomplished according to your own organizational ability.

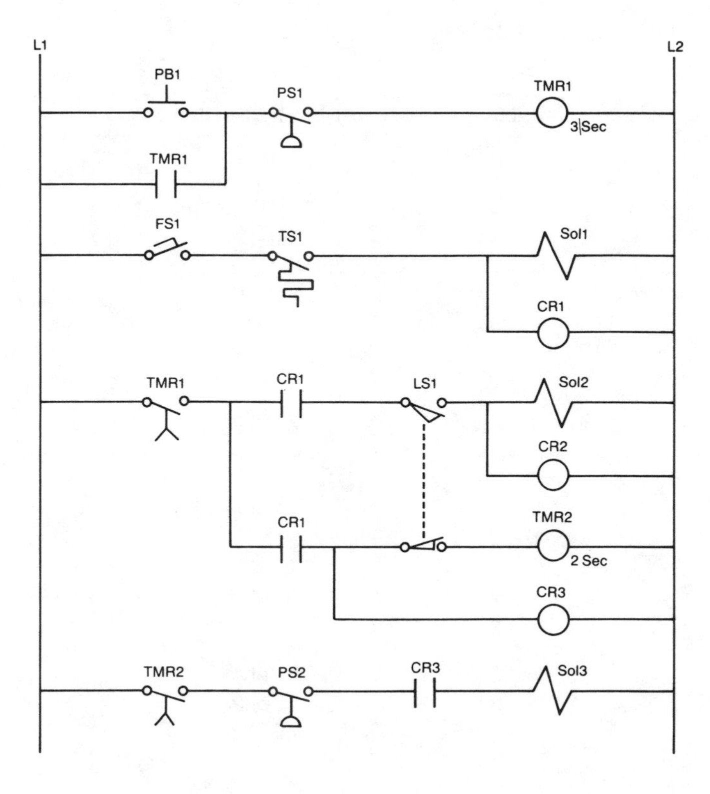

Figure 11-22. Simple relay diagram.

Simple Relay Replacement

This relay example involves the PLC implementation of the electromechanical circuit shown in Figure 11-22. As can be seen, the hardware timer TMR1 requires instantaneous contacts (first rung) which are used to latch the rung. If the instantaneous TMR1 contacts were implemented using the PLC time delay contact, then PB1 would have to be pushed for the required time preset of the timer to latch the rung. This instantaneous contact will be implemented by *trapping* the timer with an internal output.

Figures 11-23 and 11-24 show the I/O address assignments and internal output assignments for the real I/O in the circuit of Figure 11-22. The register assignment is tabulated in Figure 11-25. Note that the control relay CR1 and CR2 are not replaced with internals since the solenoids SOL1 and SOL2 output addresses (030 and 031) can be used. That is, everywhere we see CR1 or CR2 contacts we replace them with addresses 030 and 031 respectively. The limit switch LS1 is connected to the PLC input interface using the normally open LS1; the normally closed LS1 is implemented in the program by programming the LS1 NO reference with an examine OFF instruction. Figure 11-26 illustrates the PLC program coding solution.

	I/O Address			
Module Type	**Rack**	**Group**	**Terminal**	**Description**
Input	0	0	0	PB 1
	0	0	1	PS 1
	0	0	2	FS 1
	0	0	3	TS 1
Input	0	0	4	LS 1
	0	0	5	PS 2
	0	0	6	—
	0	0	7	—
	0	1	0	
		1	1	
		⋮		Not Used
	0	2	6	
	0	2	7	
Output	0	3	0	Sol 1
	0	3	1	Sol 2
	0	3	2	Sol 3
	0	3	3	—

Figure 11-23. I/O address assignment document for real inputs and outputs

Device	Internal	Description
TMR1	1000	Use to trap TMR1
CR1	—	Same as Sol 1 (030)
CR2	—	Same as Sol 2 (031)
TMR1	1001	Timer TMR1
TMR2	1002	Timer TMR2
CR3	1003	Replace CR3

Figure 11-24. Internal address assignment document.

Register	Description
4000	Preset timer count for 3 sec
4001	Accumulated count timer 1001
4002	Preset timer count for 2 sec
4003	Accumulated count timer 1002

Figure 11-25. Register assignment.

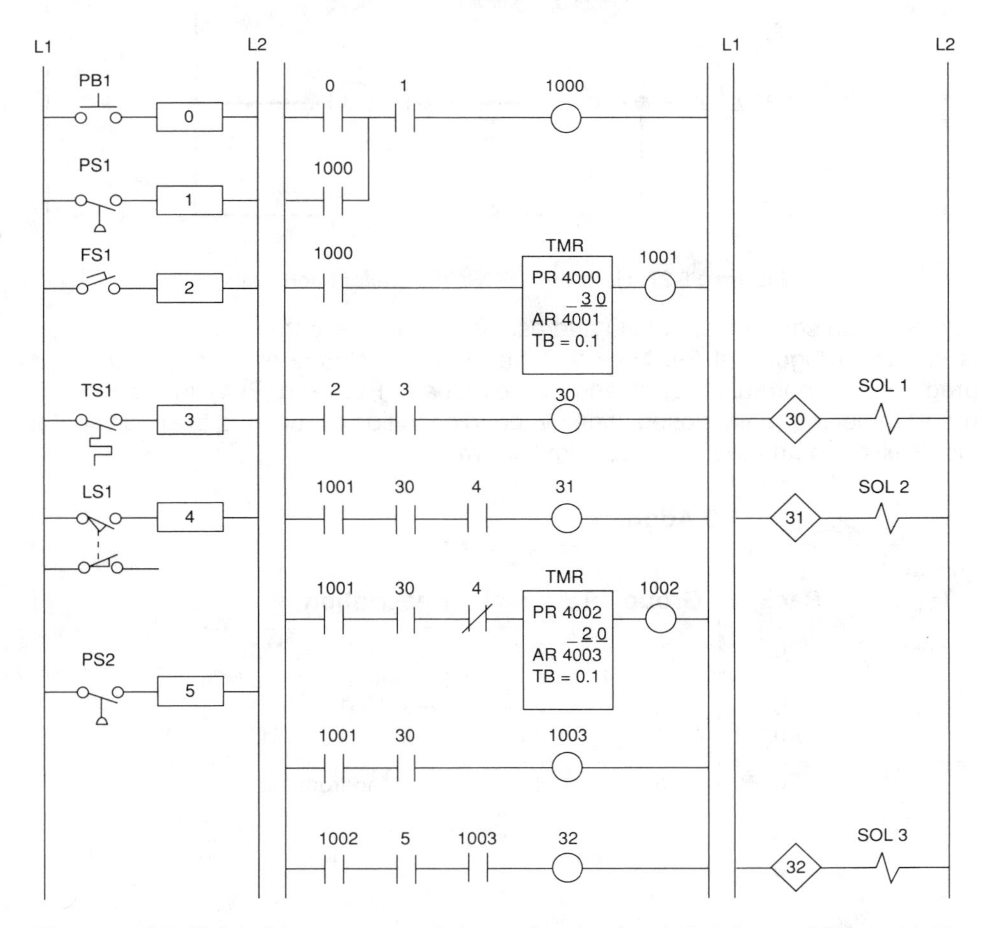

Figure 11-26. PLC ladder program solution to the simple relay diagram of Figure 11-22.

Forward/Reverse Motor Interlocking

We are required to implement the circuit shown in Figure 11-27. The forward and reverse pushbuttons are to be wired to the PLC using the normally open side of the pushbuttons. In addition to the electrical ladder to PLC ladder logic translation, we are required to add the overload contacts from the motors to the logic and to indicate with a light if an overload condition has occurred.

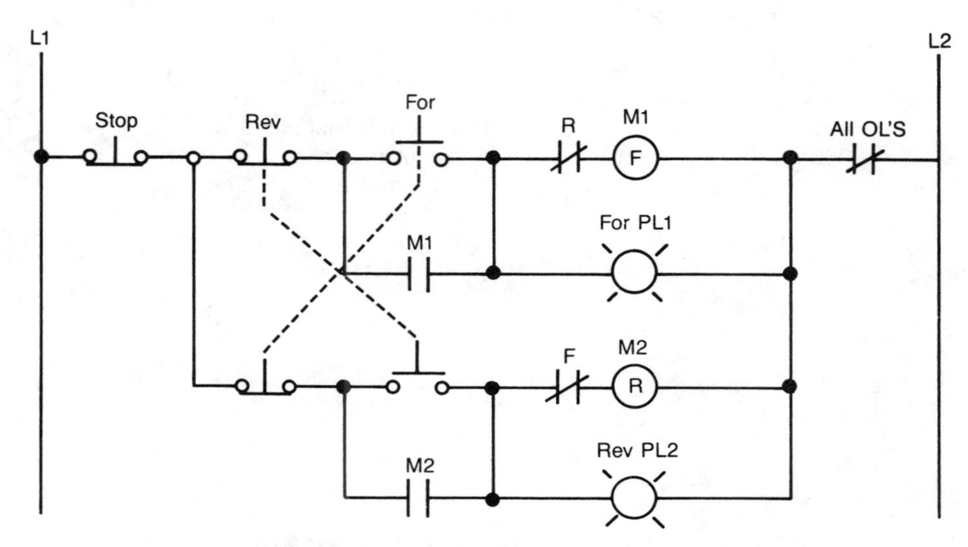

Figure 11-27. Hardwired FOR/REV motor interlocking.

Figure 11-28 shows the real I/O address assignment and the PLC implementation is shown in Figure 11-29. Note that the normally closed overload contacts are programmed normally open, and the overload (PL3 and PL4) indications are programmed normally closed. If an overload condition occurs, the overload contact input will open and the overload pilot light will be lit.

	I/O Address			
Module Type	**Rack**	**Group**	**Terminal**	**Description**
Input	0	0	0	Stop PB (wired NC)
	0	0	1	Reverse PB (wired NO)
	0	0	2	Forward PB (wired NO)
	0	0	3	Overload from M1
Input	0	0	4	Overload from M2
	0	0	5	
	0	0	6	
	0	0	7	

Figure 11-28. I/O address assignment document for real inputs and outputs

Figure 11-28 continued.

Module Type	I/O Address Rack	Group	Terminal	Description
	0	1	0	
	0	1	1	
		⋮		Not Used
	0	2	6	
	0	2	7	
Output	0	3	0	Motor M1 (FWD)
	0	3	1	Forward PL1
	0	3	2	Motor M2 (REV)
	0	3	3	ReversePL2
Output	0	3	4	M1 Overload Condition PL3
	0	3	5	M2 Overload Condition PL4
	0	3	6	
	0	3	7	

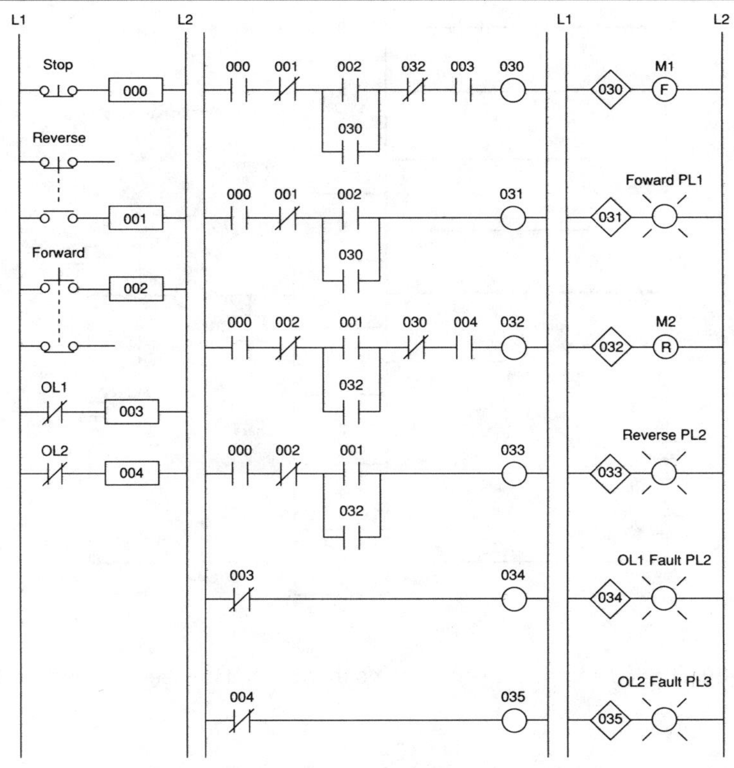

Figure 11-29. PLC implementation of the Forward/Reverse circuit.

AC Motor Drive Interface, Interlocking and Connections

A very widely used application of PLCs is the speed control of AC motors using variable speed (VS) drives. Figure 11-30 illustrates a diagram which shows the connections of an operators' station used to manually control the VS drive. The programmable controller implementation will allow the automatic control of the motor speed by changing the analog output voltage (0 to 10 VDC) of an analog interface.

The operator's station is composed of a speed potentiometer (speed regulator), forward/reverse direction selection, run/jog, and start and stop pushbuttons. These inputs will be implemented in the PLC program with the exception of the potentiometer, which will be replaced by the analog output. The inputs to the PLC will be added to the installation—the operator's manual station will not be used. The logic to start, stop and interlock the forward/reverse commands will be implemented in the PLC.

The I/O address assignment is shown in Figure 11-31. Figure 11-32 illustrates the connection diagram from the PLC to the VS drive's terminal block (TB1). The use of the contact output interface is required to switch the forward or reverse signal since

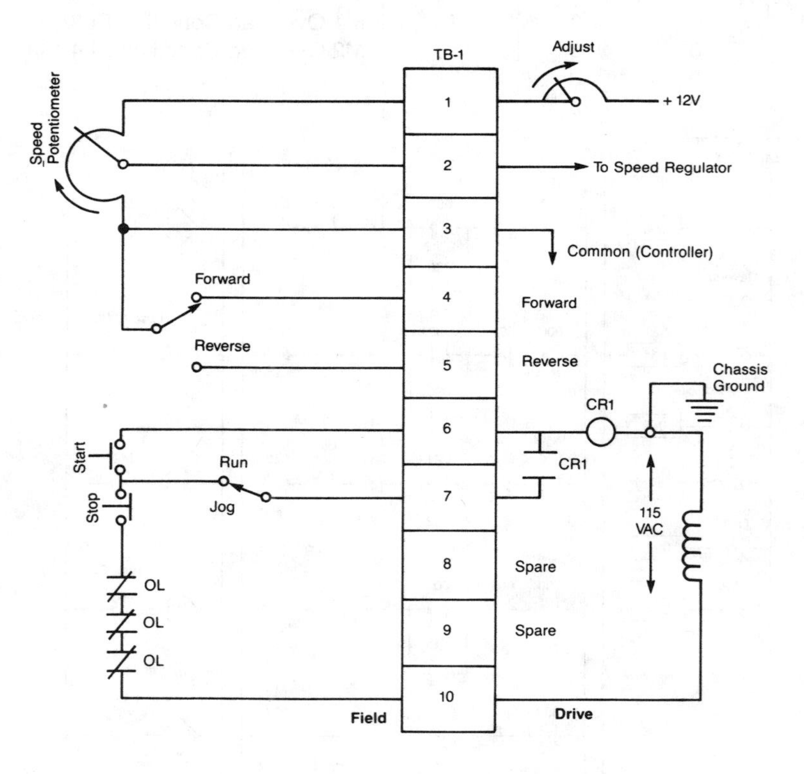

Figure 11-30. Block diagram of the operator's manual station used to control a VS drive.

Module Type	I/O Address Rack	Group	Terminal	Description
Input	0	0	0	Start
	0	0	1	Stop
	0	0	2	Forward/Reverse selector
	0	0	3	Run/Jog selector
		⋮		
Output 115 VAC	0	3	0	Drive enable (L1 from drive)
	0	3	1	
	0	3	2	
	0	3	3	
		⋮		
Output Contact	0	3	4	Forward
	0	3	5	Reverse
	0	3	6	
	0	3	7	
		⋮		
Analog Output	0	7	0	Analog Speed Reference 0-10 VDC
	0	7	1	
	0	7	2	
	0	7	3	

Figure 11-31. I/O address assignment document for real inputs and outputs for the AC motor drive interface.

the common is being switched. The power (L1) that will be used to turn the drive ON once the start PB is pushed is provided by the drive. The drive's 115 VAC signal is used so that the signal is in the same circuit and to avoid the possibility of having different commons (L2) in the drive (the start/stop common is not the same as the controller's common).

Figure 11-33 shows the PLC ladder program that will replace the manual operator station. The forward and reverse inputs are interlocked so that only one of them can be ON at a time (mutually exclusive). If the jog is selected, each time the start PB is pushed the motor will be energized to run at the speed set by the analog output. The analog output connection is illustrated in simple terms by allowing the output to be enabled when the drive is started. Register 4000 has the value in counts for the analog output to the drive. Internal 1000 is used in the block transfer to indicate the completion of the instruction.

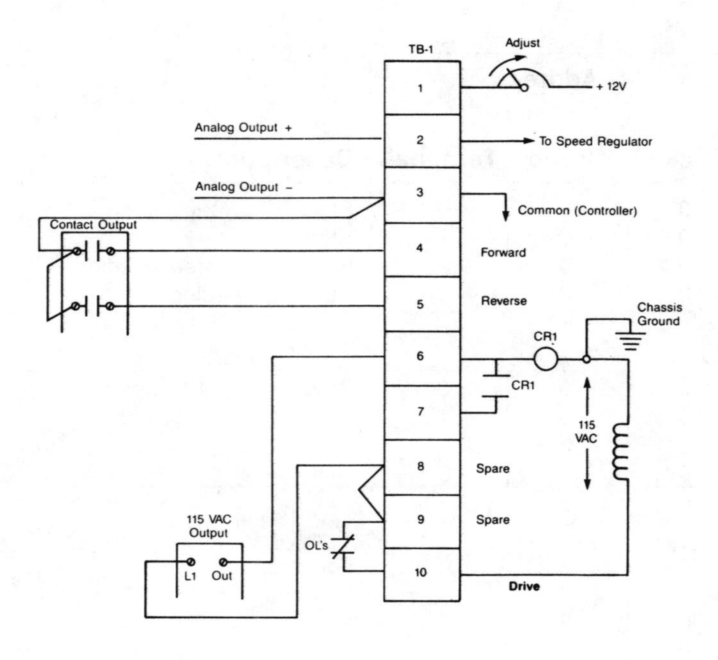

Figure 11-32. Connection diagram to the VS drive's terminal block TB-1.

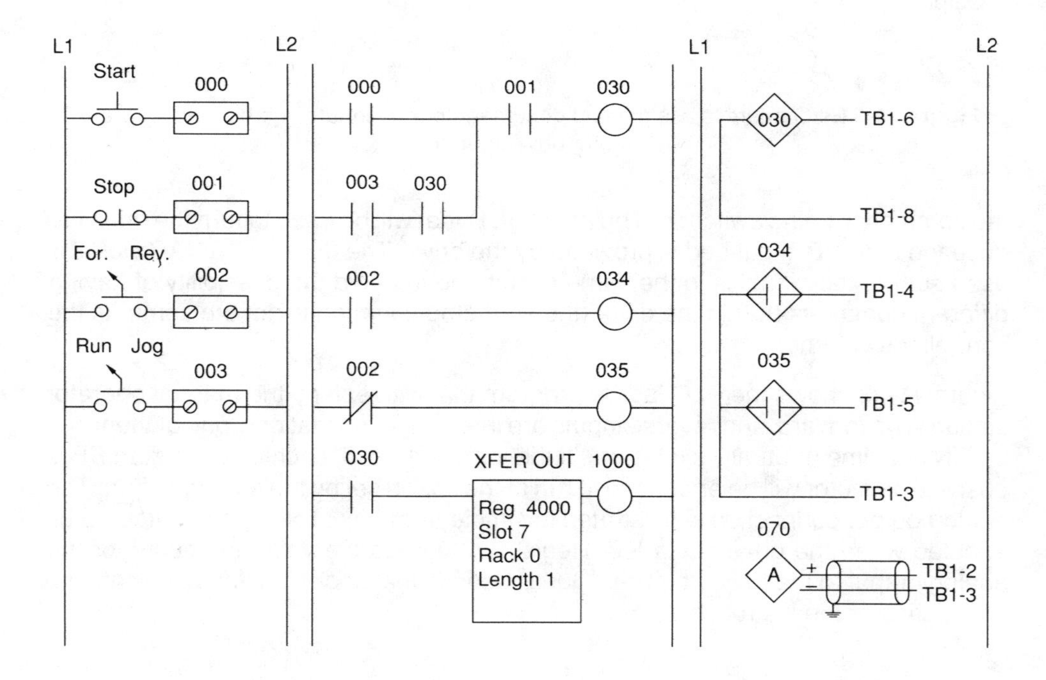

Figure 11-33. PLC ladder implementation for the VS drive.

There are occasions where the ability to run under automatic or manual control is required (AUTO/MAN). To implement the connections using the operator manual station in conjunction with the automatic PLC control, several additional hardwired connections must be made. The simplest and least expensive way is to use a selector switch (e.g. four pole, single throw single break selector switch). This switch will select either of the AUTO/MAN options. Figure 11-34 illustrates these connections. Note that the start, stop, run/jog, potentiometer and forward/reverse field devices shown are from the operator manual station. The devices connected to PLC interfaces under the same names (Figure 11-33) are used in the control program. If the AUTO/MAN switch is in automatic, the PLC will control the drive. If the switch is in manual the manual station will control the drive.

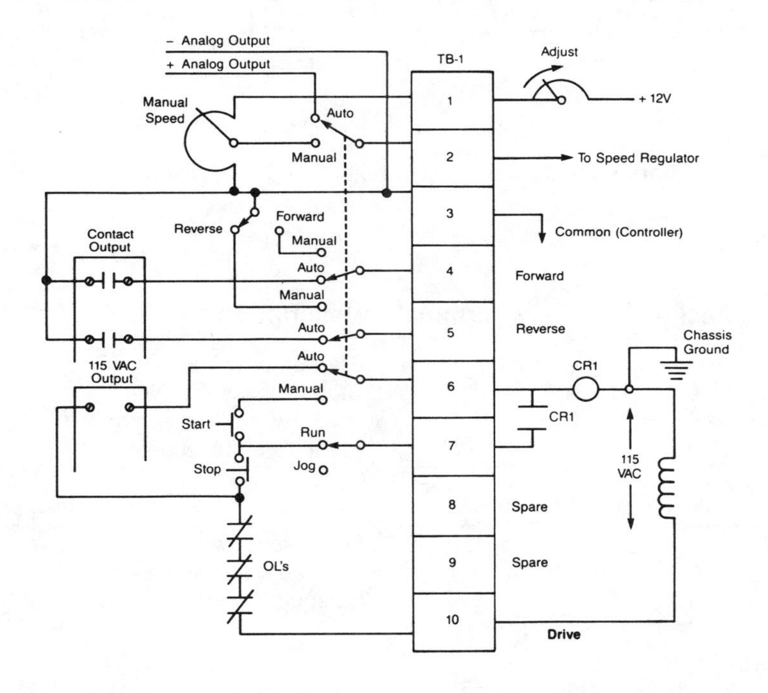

Figure 11-34. Connection diagram of AUTO/MAN operation of the VS drive.

Continuous Bottle Filling Control

In this example (shown in Figure 11-35), we are to implement a control program that will detect the position of a bottle via a limit switch, wait 0.5 seconds, and then fill the bottle until the photo sensor detects the filled position. After the bottle is filled, it will wait 0.7 seconds to continue to the next bottle. The start and stop circuits will also be included for the outfeed motor and for the start of the process. The I/O address assignment is shown in Figure 11-36 and the internal and register assignments are shown in Figure 11-37a) and b). The start and stop process signals are included in the assignment.

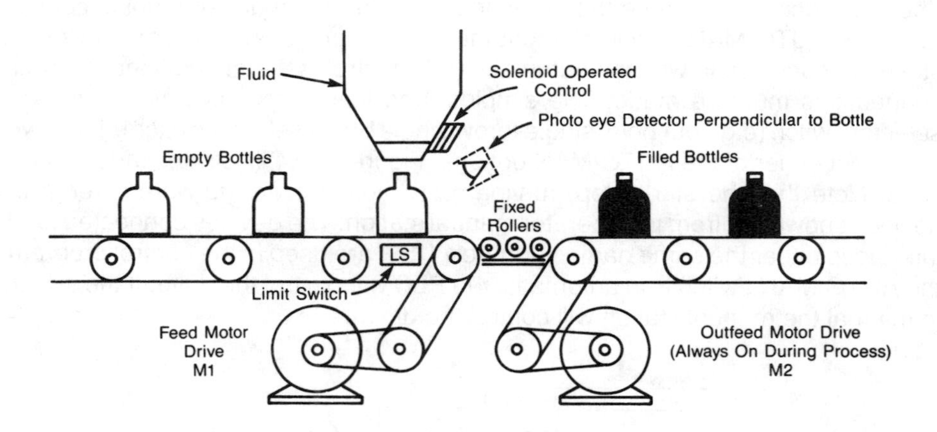

Figure 11-35. Illustration of bottle filling application.

Module Type	I/O Address Rack	Group	Terminal	Description
Input	0	0	0	Start process PB
	0	0	1	Stop Process PB (NC)
	0	0	2	Limit Switch (Position detect)
	0	0	3	Photo Eye (Level detect)
		⋮		
Output	0	3	0	Feed Motor M1
	0	3	1	Outfeed Motor M2 (System ON)
	0	3	2	Solenoid Control
	0	3	3	

Figure 11-36. Real I/O assignment.

Device	Internal	Description
Timer	1001	Timer for 0.5 sec delay after position detect
Timer	1002	Timer for 0.7 sec delay after level detect
—	1003	Bottle filled, timed out, feed motor M1

Figure 11-37a. Internal output assignment.

Register	Description
4000	Preset value 5, time base 0.1 sec (1001)
4001	Accumulated value for 1001
4002	Preset value 7, time base 0.1 sec (1002)
4003	Accumulated value for 1002

Figure 11-37b. Register assignment.

The PLC ladder implementation is shown in Figure 11-38. Once the start PB is pushed, the outfeed motor (output 031) will be ON until the stop PB is pushed. The feed motor M1 is energized once the system starts (M2 ON) and is stopped when the limit switch detects the correct position of the bottle.

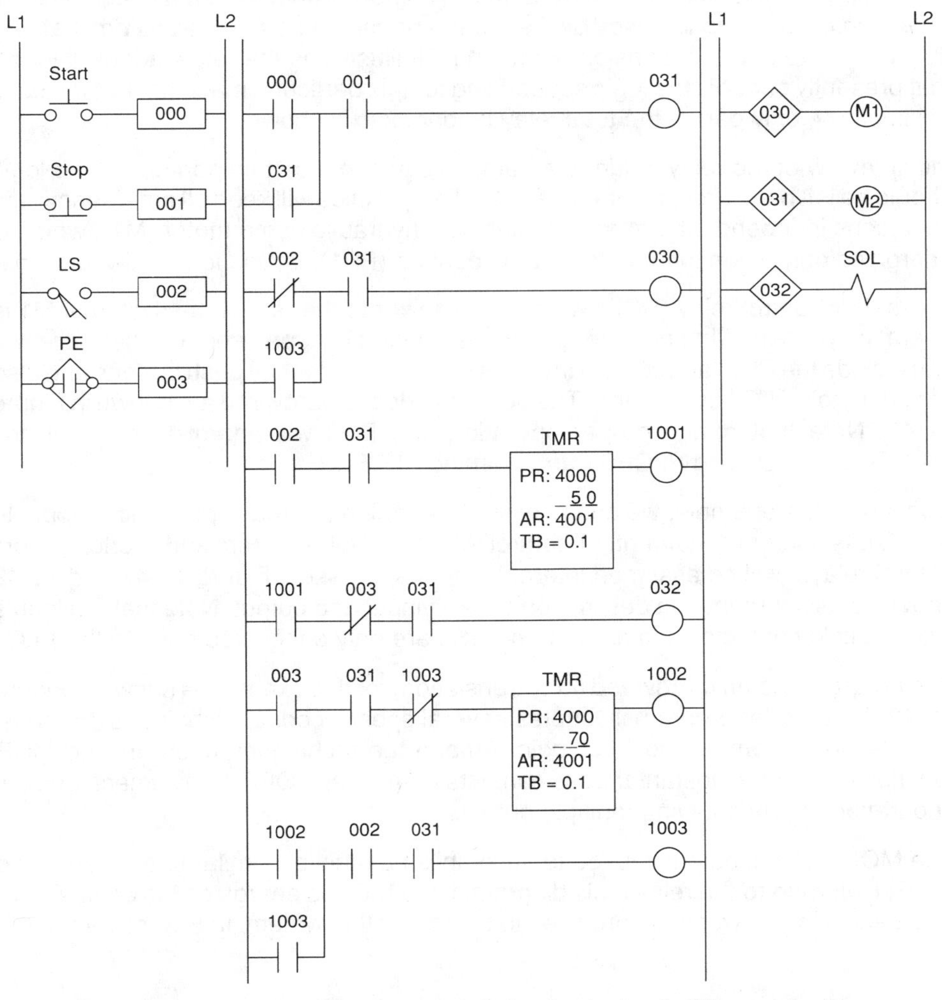

Figure 11-38. PLC ladder diagram solution for the bottle filling application.

Once the bottle is in position and 0.5 seconds have elapsed, the solenoid (032) is energized to open the valve and remains ON until the photoeye (PE) detects the proper level. The bottle remains in position for 0.7 seconds when the feed motor is started by energizing internal 1003. The feed motor remains ON until another bottle is detected by the limit switch.

Large Relay System Replacement Modernization

This example presents a modernization in which the machine control is to be changed from hardwired relay logic to PLC programmed logic. The field devices to be used are the same, with the exception of those that can be implemented in the controller (e.g. timers, control relays, interlocks, etc.). The objectives of modernizing the control of this machine are: a more *reliable control system, less energy consumption, less space used* by the control panel, and a *flexible system* that can accommodate future expansion. Figure 11-39 illustrates the relay ladder diagram that presently controls the logic sequencing for this particular machine. For the sake of simplicity, only part of the total relay ladder logic is shown.

Initial review of the relay ladder diagram indicates that certain portions of the logic should be left hardwired (lines 1, 2, and 3). This practice will keep all emergency stop conditions independent of the controller. The hydraulic pump motor (M1), which is energized only when the master start is pushed (PB1), is also left hardwired.

The safety control relay (SCR) will provide power to the rest of the system if M1 is operating properly, and no emergency push button is depressed. Furthermore, we can include the PLC fault contact in series with the emergency push buttons and also connect it to a PLC failure alarm. The portions to leave hardwired are shown in Figure 11-40. Note that during proper operation, the PLC will energize the fault coil (PLCFC), thus closing PLCFC1 and opening PLCFC2.

Continuing the example, we can now start assigning the real inputs and outputs to the I/O assignment document. All control relays, as well as timers and interlocks from control relays, will be assigned internal output addresses. Figure 11-41 and 11-42 show the assignment and description of each input and output. Note that the inputs with multiple contacts such as LS4 and SS3 are only connected once to the PLC.

The program coding (hardwired relay translation) for this example is shown in Figure 11-43. This ladder program illustrates several special coding techniques that must be used to implement the PLC logic. Among the techniques used are the MCR function in software, instantaneous contacts from timers, OFF-delay timers, and the separation of rungs having multiple outputs.

The MCR internal output is used to accomplish a function similar to the hardwired MCR. Referring to the relay logic diagram, if the MCR is energized, its contacts will close and allow power to be provided to the rest of the system. In PLC software, this

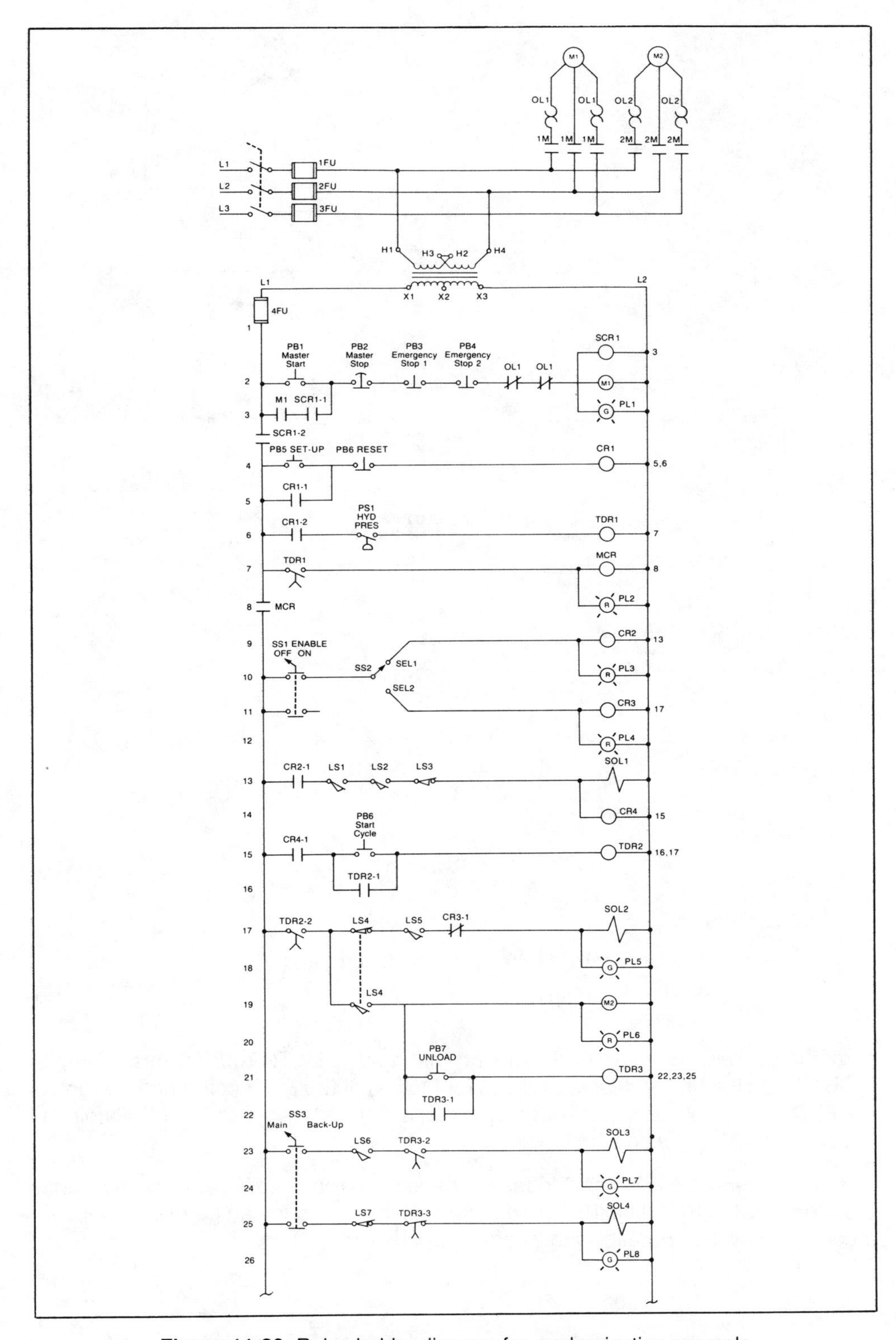

Figure 11-39. Relay ladder diagram for modernization example.

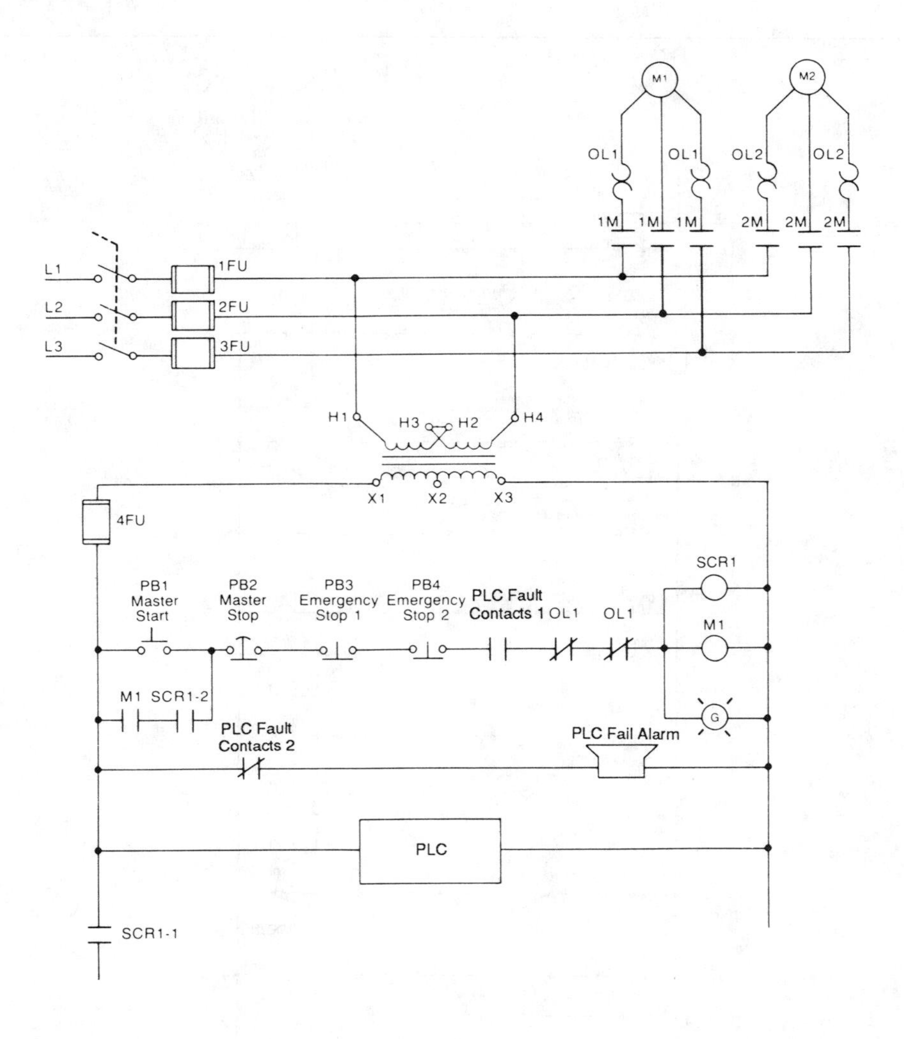

Figure 11-40. Portions left hardwired.

same function is accomplished by using the internal MCR1700 (for this example, MCR1700 is the first available address for MCRs). If the MCR coil is not energized, the PLC will not execute the ladder logic that is "fenced" between the MCR coil and the END MCR instruction.

The control relay CR2, in line 9, does not need to be replaced with an internal, since the "contacts" of PL3 (line 10) can be used. This technique can be used whenever a control relay is in parallel with a real output device.

Module Type	I/O Address Rack	Group	Terminal	Description
Input	0	0	0	PB 5 —Set up PB
	0	0	1	PB 6 —Reset (Wired NC)
	0	0	2	PS 1 —Hydraulic Pressure Switch
	0	0	3	SS 1 —Enable Selector Switch (NC contact left unconnected)
Input	0	0	4	Sel 1 —Select 1 position
	0	0	5	Sel 2 —Select 2 position
	0	0	6	LS 1 —Limit Switch Up-Position 1
	0	0	7	LS 2 —Limit Switch Up-Position 2
Input	0	1	0	LS 3 —Location Set
	0	1	1	PB 6 —Start Load Cycle
	0	1	2	LS 4 —Trap (Wired NC)
	0	1	3	LS 5 —Position Switch
Input	0	1	4	PB 7 —Unload PB
	0	1	5	SS 3 —Main/Back-up (Wired to PLC NO)
	0	1	6	LS 6 —Max. Length Detect
	0	1	7	LS 7 —Min. Length Back-up
	0	2	0	
	0	2	1	
	0	2	2	Spare
	0	2	3	
	0	2	4	
	0	2	5	
	0	2	6	Spare
	0	2	7	
Output	0	3	0	PL 2 —Set-up OK
	0	3	1	PL 3 —Select-1
	0	3	2	PL 4 —Select-2
	0	3	3	Sol 1 —Advance FWD
Output	0	3	4	Sol 2 —Engage
	0	3	5	PL 5 —Engage ON
	0	3	6	M 2 —Run Motor
	0	3	7	PL 6 —Motor Run On
Output	0	4	0	Sol 3 —Fast Stop
	0	4	1	PL 7 —Fast Stop On
	0	4	2	Sol 4 —Unload with Back-up
	0	4	3	PL 8 —Back-up On

Figure 11-41. Real I/O address assignment document for modernization example.

Device	Internal	Description
CR1	1000	CR-1
TDR1	T2000	Timer preset 10 sec Register 3000 ACC Register 3001
MCR	MCR1700	First MCR Address
CR2	—	Same as PL3 Address
CR3	—	Same as PL4 Address
CR4	—	Same as Sol 1
—	1001	To set up internal for instantaneous contact of TDR2
TDR2	T2001	Timer preset 5 sec Register 4002 ACC Register 4003
—	1002	To set up internal for instantaneous contact of TDR3
TDR3	T2002	Timer preset 12 sec Register 4004 ACC Register 4005

Figure 11-42. Internal output address assignment document for modernization example.

The separation of the coils in lines 17 and 18 of the hardwired logic is done, since the PLC used here does not allow rungs with multiple outputs. Using separate rungs for each output is also a good practice.

The normally-closed inputs that are connected to the input modules are programmed normally-open, as explained in the previous sections. The limit switch LS4 has two contacts (NO and NC in lines 17 and 19 of Figure 11-39). However, only one set of contacts needs to be connected to the controller. In this example, we have selected the normally-closed LS4. Although the normally-open contact is not connected to the controller, its hardwired function can still be achieved by programming LS4 as a normally-closed ladder contact (refer to programming an input device in Chapter 9).

Applications such as this one also require timers that have instantaneous contacts, which are not available in most PLCs. An instantaneous contact is one which opens or closes when the timer is enabled. To overcome not having an instantaneous contact, an internal coil is used. Line 15 in the hardwired logic shows that if PB6 is pressed and CR4 is closed, the timer TDR2 should start timing, and contact TDR2-1 would seal PB6. This arrangement cannot be implemented in the PLC without special considerations. If we use the software timer contacts, the timer will not seal until it has timed-out. If PB6 is released, the timer will be reset because PB6 is not sealed. A solution to the problem is accomplished by using internal coil 1001 to seal PB6 and start timing T2001 (TDR2). This technique is shown in the PLC program coding in lines 9, 10, and 11. The time delay contacts of T2001 are used for the ON delays.

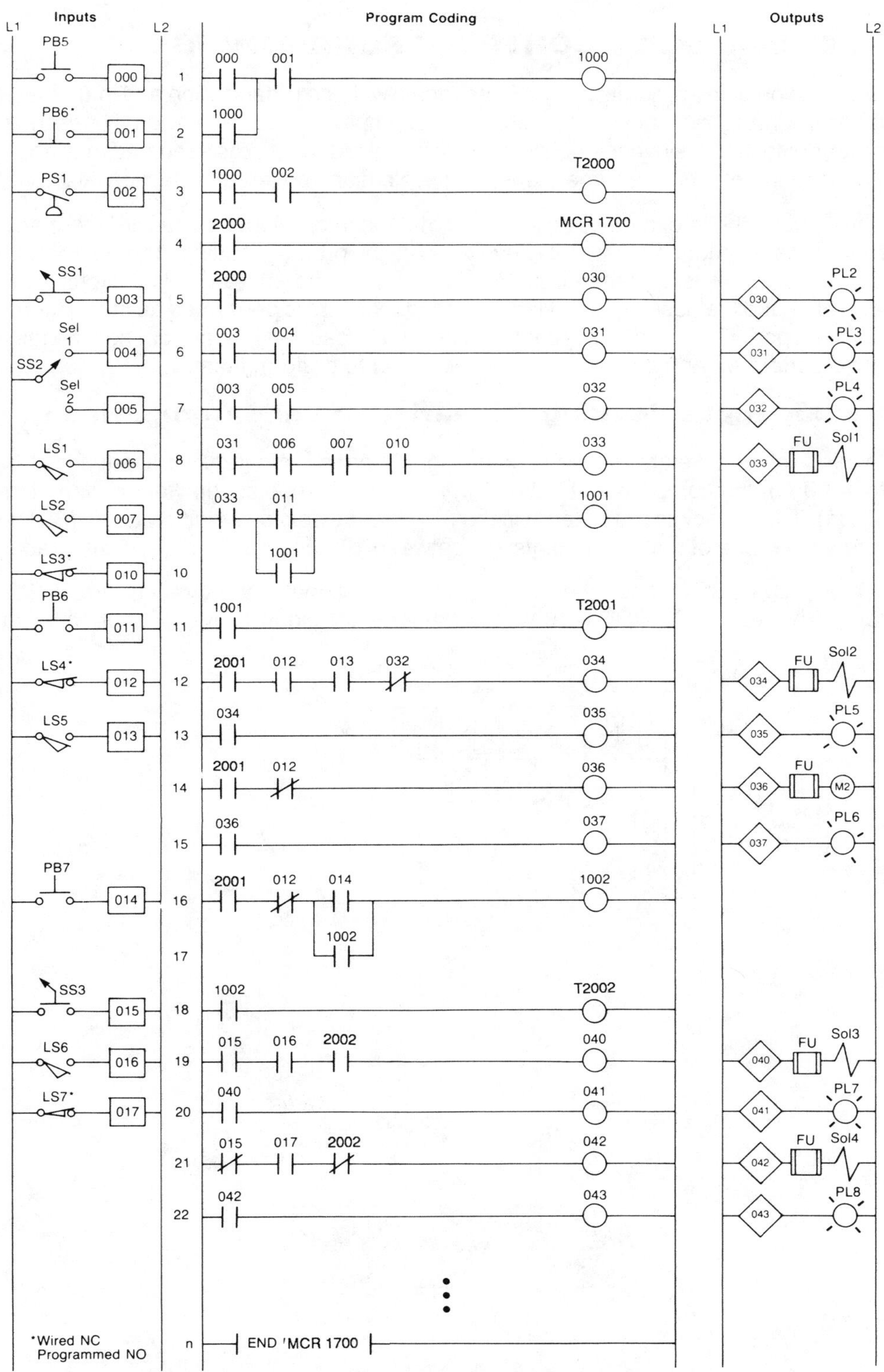

Figure 11-43. PLC program coding for modernization example.

11-6 ANALOG I/O CONTROL PROGRAMMING

This section of the programming chapter deals with the organization and implementation of analog readings and control. The examples presented generally describe new additions to a system—no existing electrical ladder logic is assumed available. The implementations assume the PLC specifications covered in Section 11-5.

Throughout these examples you will find that the internal address assignment and register assignment is most likely developed as the program coding from a flow chart or logic diagram is implemented. This is true since no prior electrical circuits are available and the assignment of internals and registers is performed as the program is developed. Flow charting becomes important because it globally defines the task to be performed and the order in which we wish to perform them.

Analog Input Reading, Linearization and Comparison

In this example, we are to take an input signal from a temperature transducer (0°C to 1000°C) and compare that value with two alarm set points (low alarm and high alarm). The set points are entered in the PLC via two sets of 4-digit switches (BCD). The valid range of these set points are between 100 and 850 degrees Centigrade.

The analog input module receives a signal, proportional to the temperature, which ranges between -10 VDC to +10 VDC. Whenever the signal is over or under the two

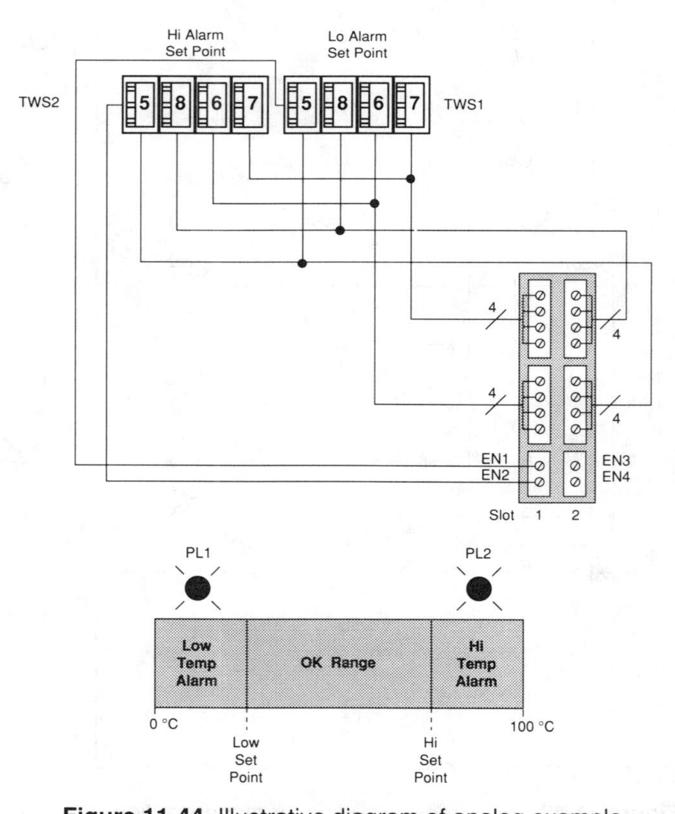

Figure 11-44. Illustrative diagram of analog example.

set points, an indicator light must be illuminated. Figure 11-44 illustrates a simple diagram of the elements used in this system example. The TWS are connected to a register input module with multiplexing (MUX) capability.

The analog input relationship between counts and degrees Centigrade is shown in Figure 11-45 (reference Chapter 7). A flow chart of the required steps for this example is illustrated in Figure 11-46. The subroutines used in the program are flow charted in Figure 11-47.

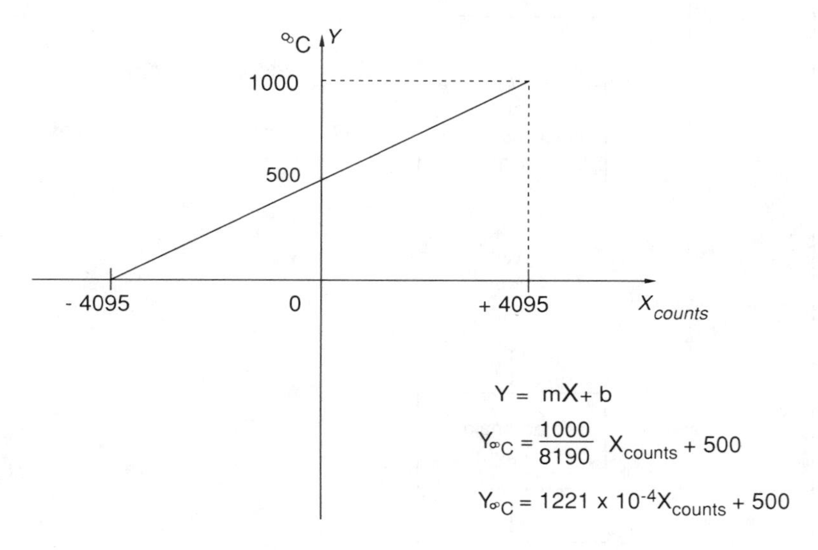

Figure 11-45.

The I/O and internal output address assignment is shown in Figure 11-48 and 11-49 respectively. The register assignment is listed in Figure 11-50. The PLC circuit implementation is shown in Figure 11-51; the subroutine circuits are shown in Figure 11-52 and 11-53. The block transfer input instruction, used in the reading of TWS, selects the slot location 1 which reads 8 bits (2 digits) and automatically goes to the next slot (slot 2) to get the other 8 bits.

Analog Position Reading from an LVDT

An LVDT (see Chapter 12) is being used to provide position feedback for a moving mechanism of a machine. The LVDT has a range of ±10 inches from its null position; therefore the effective total range is 20 inches from a zero reference. The LVDT provides ±10VDC and is connected to an analog input module which transforms the voltage into counts ranging from -4095 to +4095 for the -10 to +10 VDC swing. Figure 11-54 illustrates the system block diagram.

Once the machine is started by a start PB, the moving piece is to move to a "virtual" starting position (V.P.) defined by a set of 4-digit TWS. The TWS settings may range from 00.00 to 20.00; the decimal point is to be implemented in the controller. When the machine is started, the moving part must go to the virtual position. The machine cycle continues and once it is finished, the moving piece must return to the virtual position. The machine cycle may end at either side of the V.P. setting.

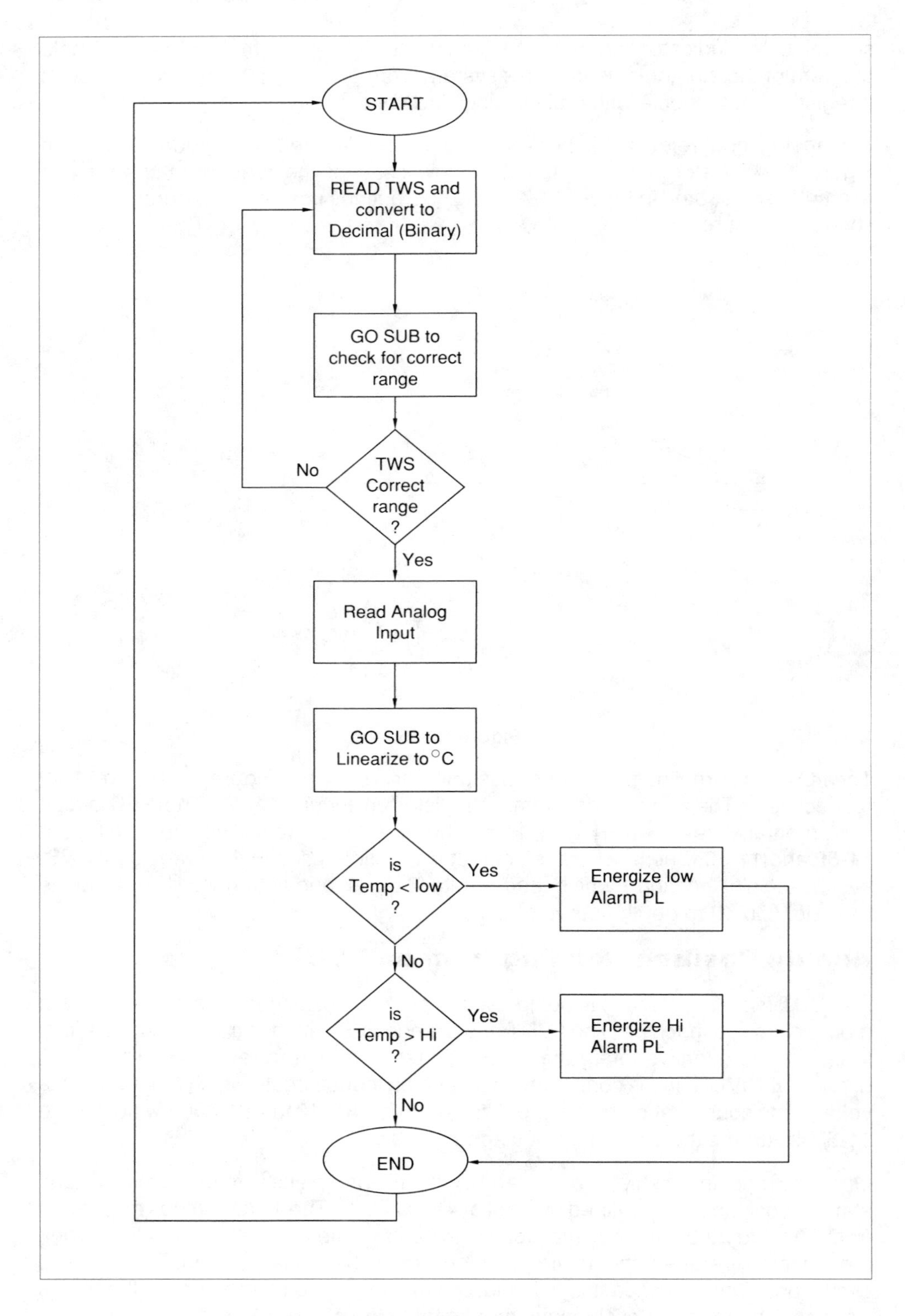

Figure 11-46. Main program flowchart.

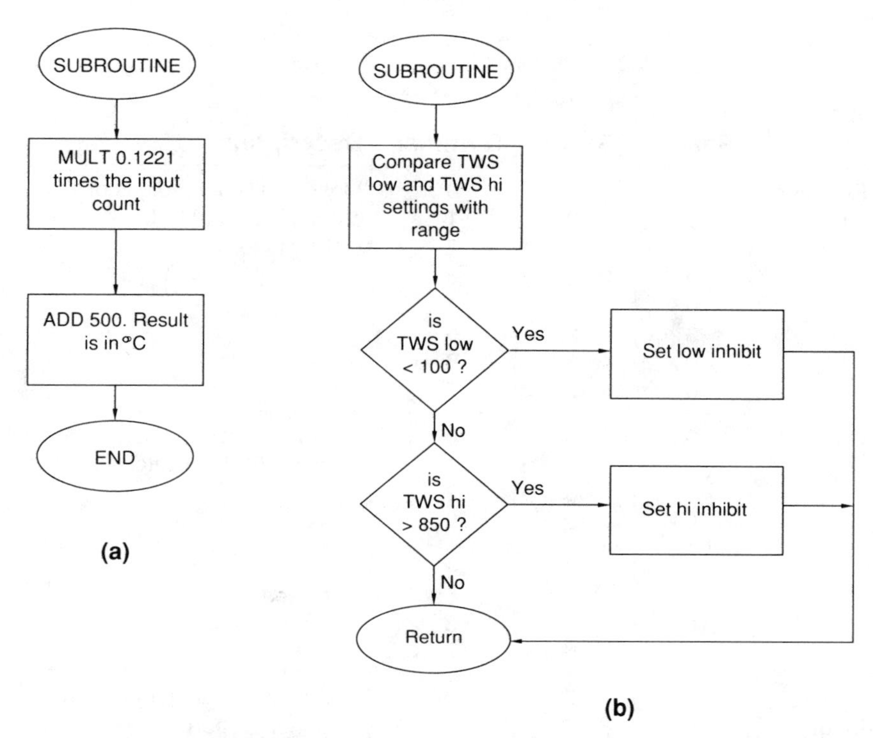

Figure 11-47. Subroutine flowcharts **a)** linearize to °C; **b)** check for correct range.

Module Type	I/O Address Rack	Group	Terminal	Description
Input	0	0	0	Start the analog reading and TWS input
	0	0	1	
	0	0	2	
	0	0	3	
Output	0	0	4	Not Used
	0	0	5	
	0	0	6	
	0	0	7	
Register Input (low byte)	0	1	0	Least significant 2 digits of TWS channels (1 and 2) MUX in these eight bits.
	0	1	1	
	0	1	2	
	0	1	3	
	0	1	4	
	0	1	5	
	0	1	6	
	0	1	7	

Figure 11-48. Real I/O address assignment.

Module Type	I/O Address Rack	Group	Terminal	Description
Register Input (hi byte)	0	2	0	Most significant 2 digits of TWS channels (1 and 2) MUX in these eight bits.
	0	2	1	
	0	2	2	
	0	2	3	
	0	2	4	
	0	2	5	
	0	2	6	
	0	2	7	
Output	0	3	0	Low alarm PL1 indicator
	0	3	1	Hi alarm PL2 indicator
	0	3	2	
	0	3	3	
	0	3	4	
	0		5	Not Used
	0		6	
	0	6	7	
Analog Input	0	7	0	Channel 1 Analog Input Temp
	0	7	1	Channel 2 Spare
	0	7	2	Channel 3 Spare
	0	7	3	Channel 4 Spare

Figure 11-48 continued.

Device	Internal	Description
—	1000	Xfer In TWS (MUX) enabled
—	1001	BCD to Binary conversion enabled
—	1002	Xfer In Analog input enabled
—	1003	Compare Temp with Lo Alarm
—	1004	Lo Alarm Condition
—	1005	Low Temp Alarm PL1
—	1006	Compare Temp with Hi Alarm
—	1007	Hi Alarm Condition
—	1010	Hi Temp Alarm PL2
	⋮	
—	1100	Go To Subroutine to check for valid ranges. Enabled.
—	1101	Low Temp input less than valid range
—	1102	Compare for hi range enabled.
—	1103	Hi Temp input more than valid range
—	1104	Less than low range latched
—	1105	More than hi range latched
—	1106	TWS inputs are OK range
	⋮	
—	1200	Go to Subroutine to linearize input counts to °C
—	1201	Addition of 500 enabled

Figure 11-49. Internal output assignment.

Register	Description
4000	Low limit alarm TWS (BCD)
4001	Hi limit alarm TWS (BCD)
⋮	
4010	Low limit alarm Decimal °C
4011	Hi limit alarm Decimal °C
⋮	
4100	Analog input counts (Temperature)
4101	Use in multiplication connection
4102	Result temperature in °C

Figure 11-50. Register usage assignment.

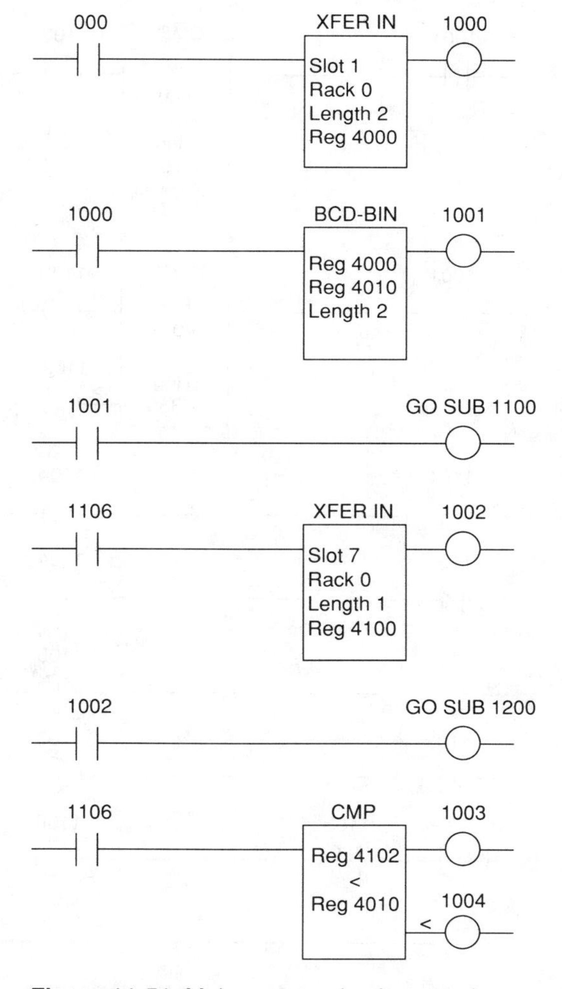

Figure 11-51. Main program implementation.

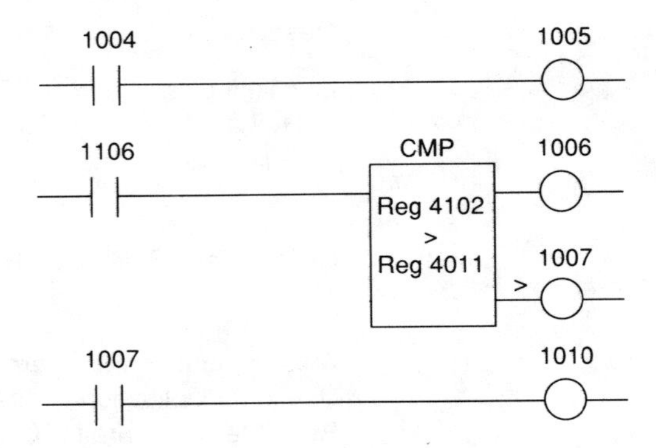

Figure 11-51 continued.

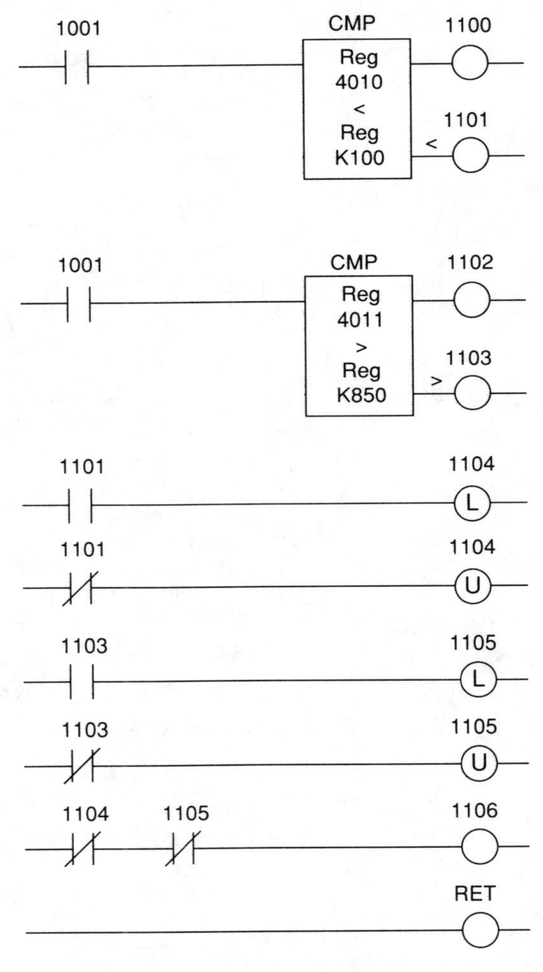

Figure 11-52. Subroutine 1100, check for valid TWS range and convert to decimal. Inhibit operation if not in valid range.

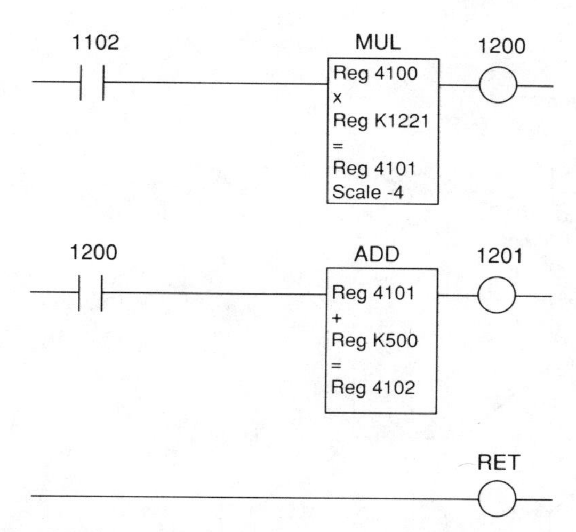

Figure 11-53. Subroutine 1200, convert analog counts to degrees according to linearization formula.

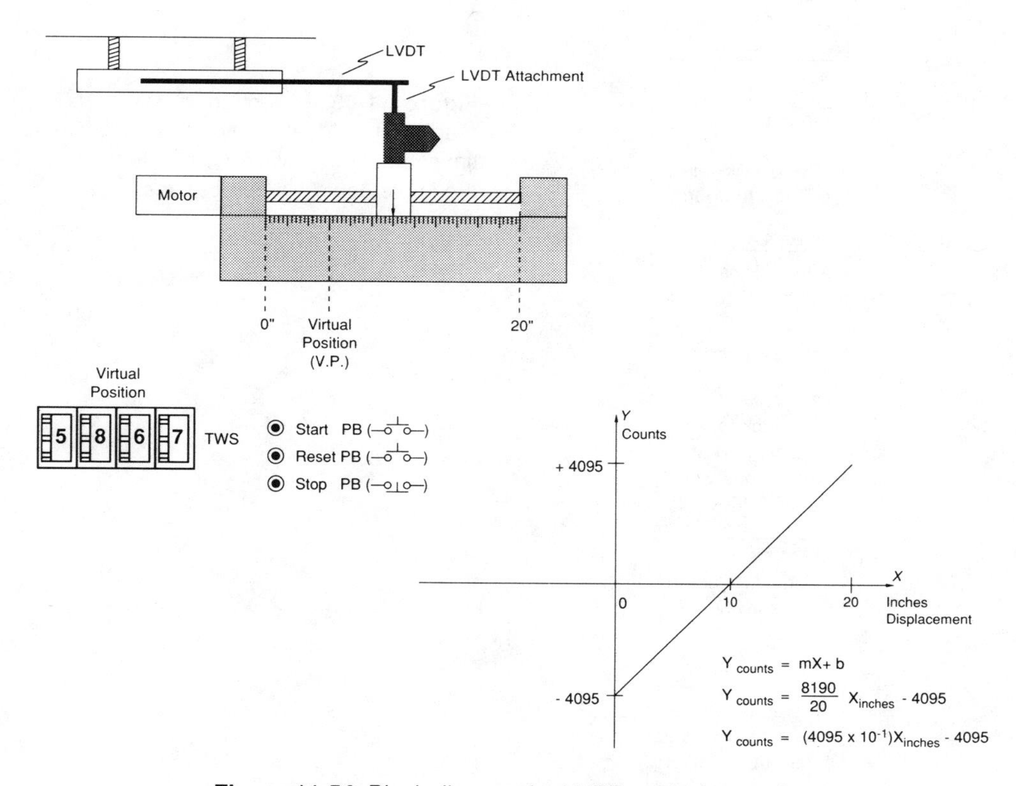

Figure 11-54. Block diagram for LVDT analog example.

Figure 11-55 illustrates the function flow chart while the I/O address assignment, internal assignment and register assignment are shown in Figure 11-56, 11-57, and 11-58 respectively. The PLC program solution for this example is shown in Figure 11-59.

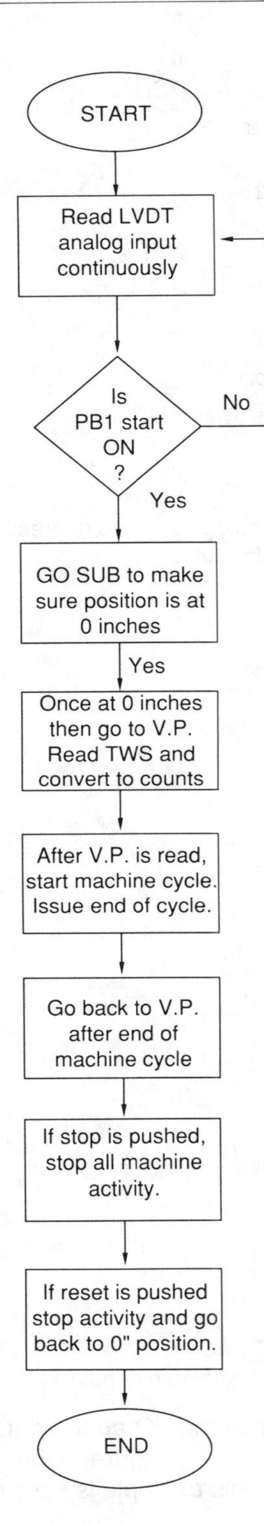

Figure 11-55. Flowchart of the LVDT reading and virtual position (V.P) calculations. Several flowchart block functions are implemented in subroutines.

Module Type	I/O Address Rack	Group	Terminal	Description
Input	0	0	0	Start PB1 to virtual position
	0	0	1	Stop PB2 stop machine (NC)
	0	0	2	Reset PB3 reset to 0" position
	0	0	3	
Register Input (High byte)	0	1	0	Most significant 2 digits of TWS Channel 1. Virtual Position (decimal points).
	0	1	1	
	0	1	2	
	0	1	3	
	0	1	4	
	0	1	5	
	0	1	6	
	0	1	7	
Register Input (Low byte)	0	2	0	Least significant 2 digits of TWS Channel 1. Virtual Position (decimal points).
	0	2	1	
	0	2	2	
	0	2	3	
	0	2	4	
	0	2	5	
	0	2	6	
	0	2	7	
Output	0	3	0	Forward command
	0	3	1	Reverse command
	0	3	2	
	0	3	3	
		⋮		
Analog Input	0	7	0	Channel 1 LVDT Analog input
	0	7	1	Channel 2 Spare
	0	7	2	Channel 3 Spare
	0	7	3	Channel 4 Spare

Figure 11-56. Real I/O assignment.

Device	Internal	Description
—	1000	Start machine command
—	1001	LVDT analog input enabled
—	1100	Latch for enable to go to subroutine
—	1150	Compare LVDT position with 0 inches
—	1151	Position reached
—	1152	Energize reverse motor command from this sub.
—	1153	One shot position 0" found
—	1200	Latch to enable to go subroutine
—	1250	Read TWS block enable
—	1251	Convert output from BCD-Binary (decimal)

Figure 11-57. Internal output assignment.

Device	Internal	Description
—	1252	Multiply (according to equation) enable
—	1253	Subtract enabled
—	1254	Compare enabled
—	1255	V.P. found (1254 ON)
—	1256	Energize forward motor from this sub.
—	1257	One shot position V.P. found
—	1300	Latch to enable to go to subroutine
—	1350	Compare LVDT with V.P. (≥)
—	1351	V.P. found from≥
—	1352	Compare LVDT with V.P. (≤)
—	1353	V.P. found from ≤
—	1354	Latch found V,P. from ≥
—	1355	One shot found V.P.
—	1356	Latch found V.P. from≤
—	1357	One shot found V.P.
—	1360	Reverse motor from this sub
—	1361	Forward motor from this sub
—	1362	One shot found V.P. from ≤ or from ≥
—	1400	Latch a reset PB condition to go to sub
—	1700	Latch to go to machine cycle
—	1750	Go sub machine cycle
—	1777	End of cycle signal

Figure 11-57 continued.

Register	Description
4000	TWS value in BCD. Virtual Position
4001	TWS value in binary after conversion
4002	Subtraction of -4095
4003	Virtual postion in counts (equation)
4100	LVDT analog value in counts

Figure 11-58. Register usage assignment.

The implementation of the flow chart is primarily done with the use of subroutines to facilitate interlocking and programming. A latch instruction is used to call a subroutine; it enables the program to go to the subroutine until the operation has been performed. Once the subroutine function is finished, the subroutine sends back, upon returning, an unlatch signal signifying the end of the subroutine. This unlatch signal is also used to trigger the execution of the next subroutine.

The subroutine codes are shown in Figures 11-60, 11-61, and 11-62. Note that in Figure 11-60 (check for 0" position) the compare instruction checks for the LVDT count to be less or equal than the constant -4090 instead of strictly equal to -4095. Once the LVDT passes -4090 counts it is assumed that the position is at 0 inches.

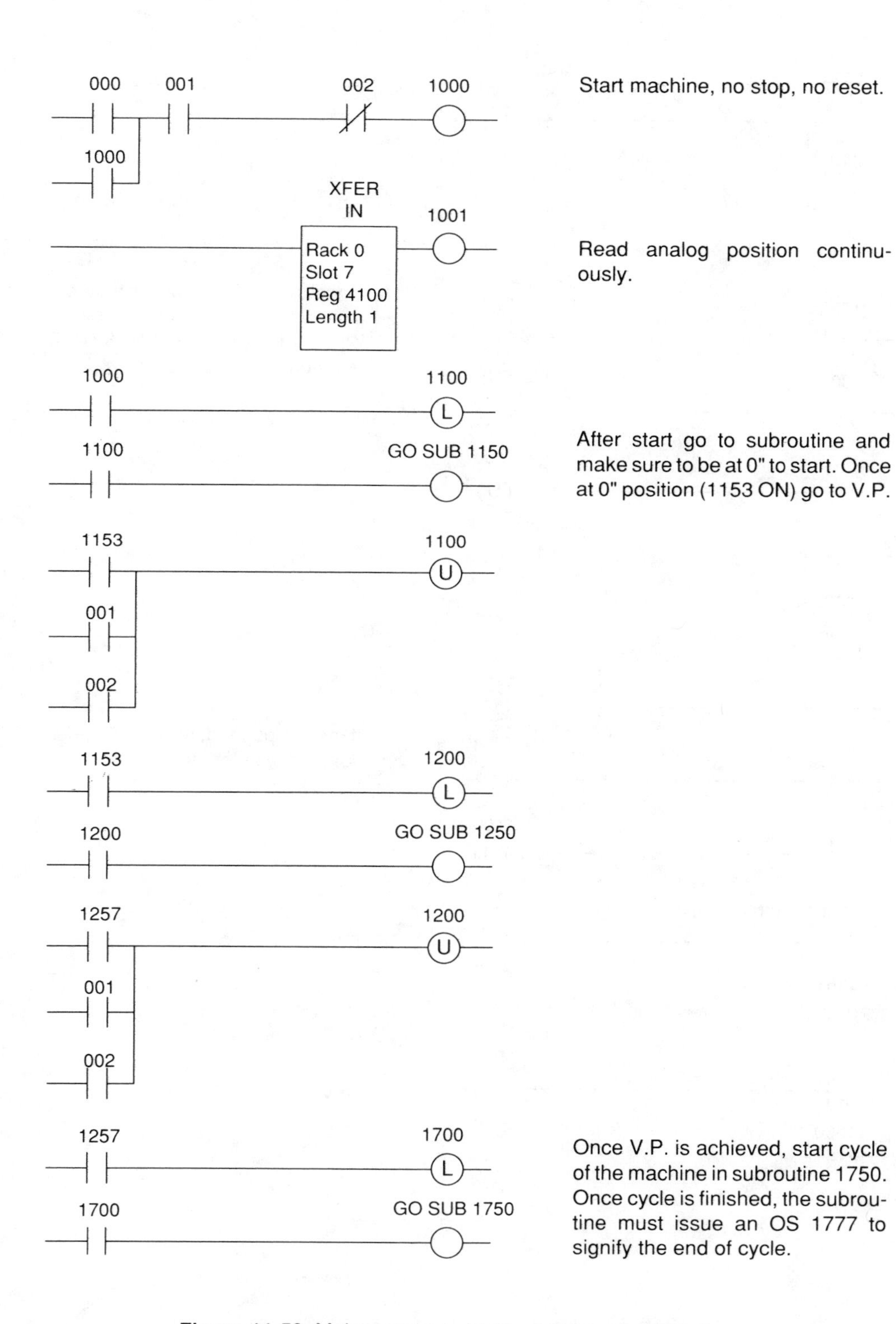

Figure 11-59. Main program which calls several subroutines.

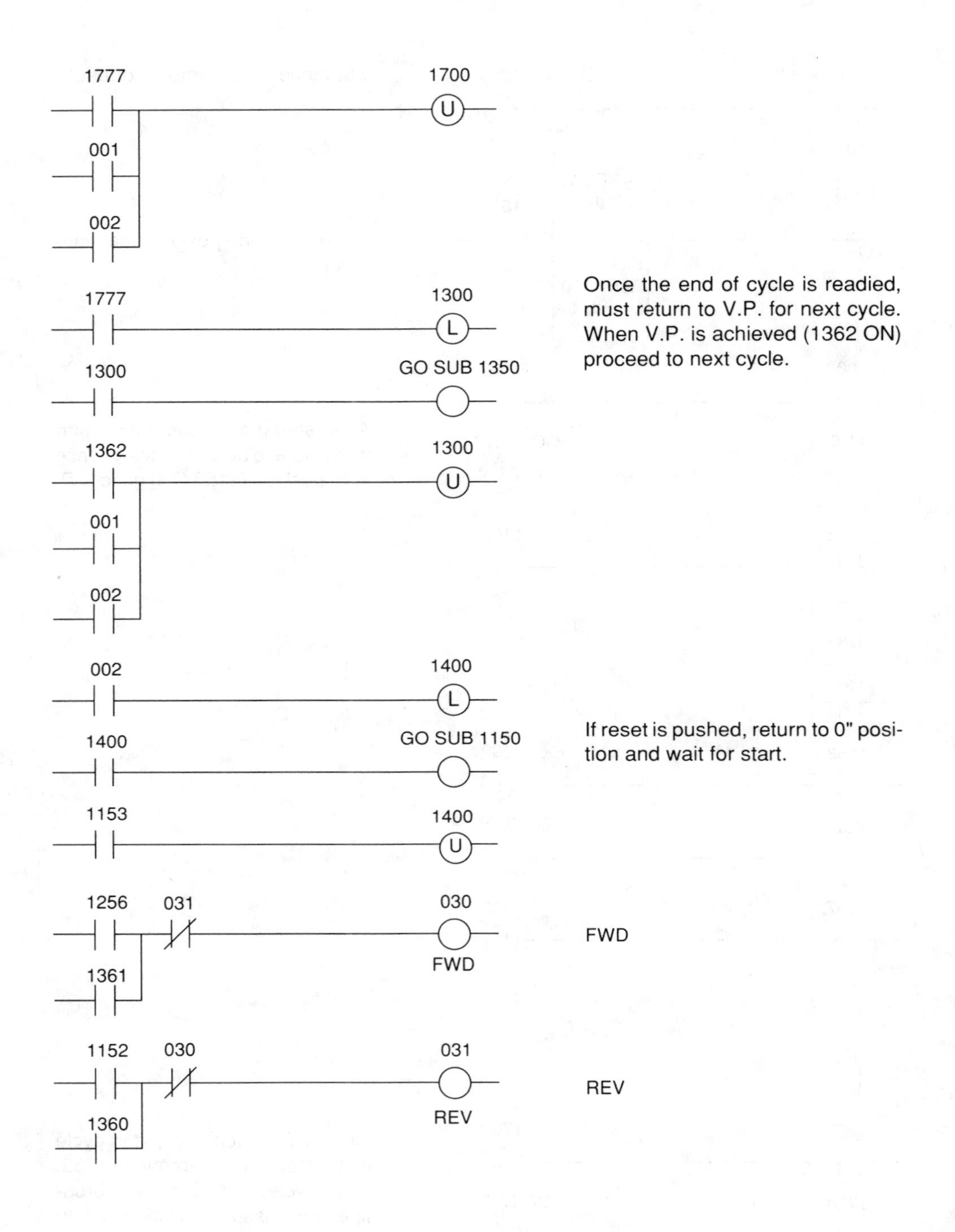

Figure 11-59 continued.

In Figure 11-61, the use of scale multiplication allows the computation of Virtual Position with two decimal points (10^{-2}) to be multiplied by the 4095 x 10^{-1} (409.5) constant; the final scale is 10^{-3}. This routine allows the motor to move the part to the virtual position as specified by the LVDT. Once V.P. has been reached, the system is ready (output 1257 one shot) to start the machine cycle. It is assumed here that the machine cycle subroutine will have to return, when finished, an *end of cycle* signal (output 1777).

When the end of cycle has been performed the PLC will tell the motor to move forward or reverse depending upon the moving part position at the end of cycle. The interlocking performed by output rungs 1354 and 1355 allow the motor movement (reverse) if the part was at more than the V.P. Rungs 1356 and 1357 perform a similar function if the part had been at a position less than the V.P.

The one shot circuits used are necessary to prevent the system from turning the motor forward and reverse until the part is at exactly the virtual position in counts. Analog count signals may jump one or two counts in either direction (up or down) which could cause an instability by having the forward and reverse signals fight with each other. The logic that is employed in this subroutine will turn the V.P. position around once the moving part is at or just over the virtual position. After the part is stopped at V.P., both forward and reverse motor commands from the subroutine are inhibited.

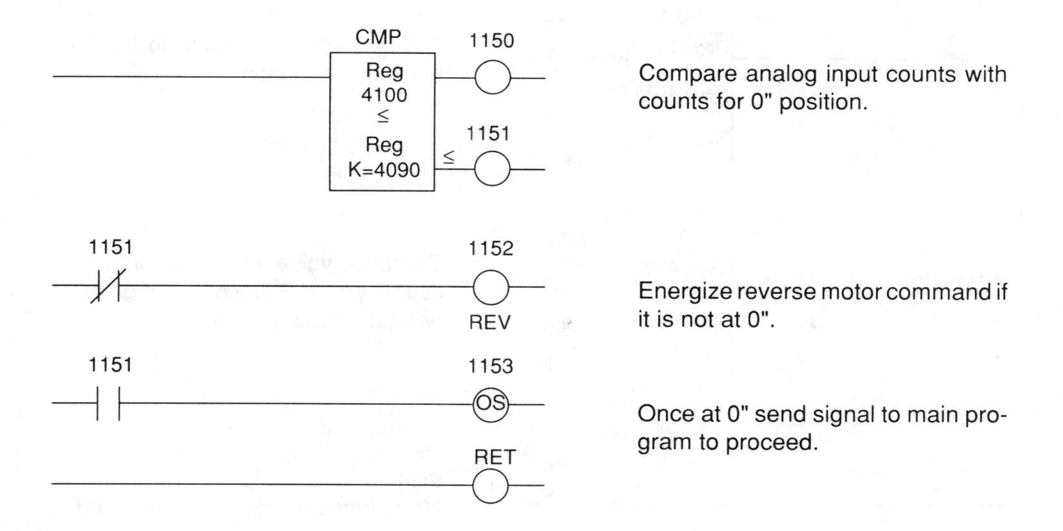

Figure 11-60. Subroutine 1150 brings moving part to 0" position.

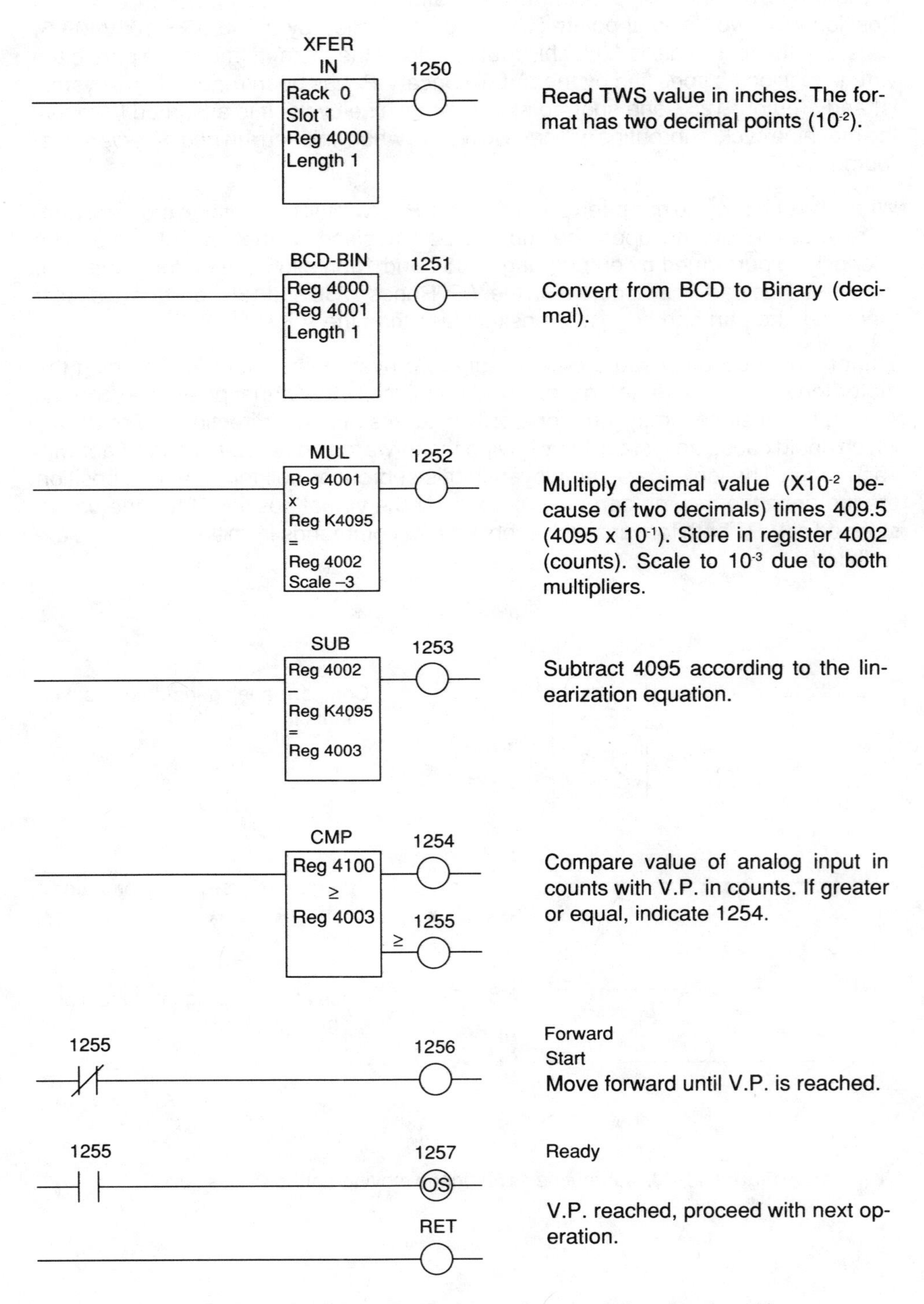
XFER
IN
1250
Rack 0
Slot 1
Reg 4000
Length 1
Read TWS value in inches. The format has two decimal points (10^{-2}).
BCD-BIN
1251
Reg 4000
Reg 4001
Length 1
Convert from BCD to Binary (decimal).
MUL
1252
Reg 4001
x
Reg K4095
=
Reg 4002
Scale –3
Multiply decimal value ($X10^{-2}$ because of two decimals) times 409.5 (4095×10^{-1}). Store in register 4002 (counts). Scale to 10^{-3} due to both multipliers.
SUB
1253
Reg 4002
–
Reg K4095
=
Reg 4003
Subtract 4095 according to the linearization equation.
CMP
1254
Reg 4100
≥
Reg 4003
1255
≥
Compare value of analog input in counts with V.P. in counts. If greater or equal, indicate 1254.
1255
1256
Forward
Start
Move forward until V.P. is reached.
1255
1257
OS
Ready
RET
V.P. reached, proceed with next operation.

Figure 11-61. Subroutine 1250 moves the part to the virtual position. It also reads the V.P. from TWS.

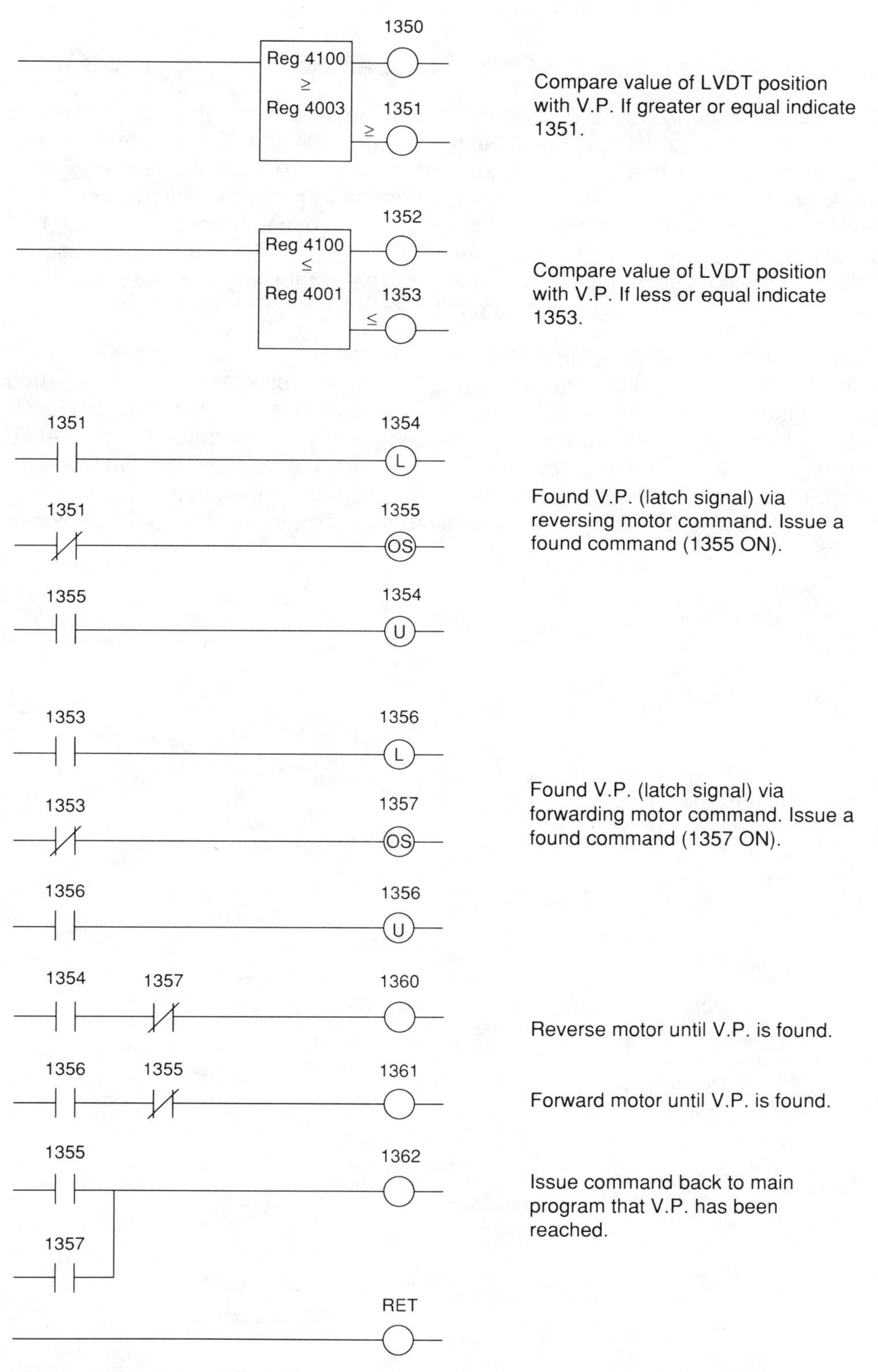

Figure 11-62. Subroutine 1350 returns the part to the virtual position after the end of cycle regardless of where the part is located.

Linear Interpolation of Non-Linear Analog Input Signal

There are PLC applications in which the analog input signal received does not have a linear relationship with the signal being measured. That is, the ratio of change in the measurement variable is not the same throughout the measurement range. For instance, a pressure transducer measuring hydraulic pressure may not provide a linear equation which represents psi changes versus voltage changes (and therefore input counts). Sometimes these non-linearities are created by the system which is being controlled. The non-linearity problem can be circumvented by the use of table look-up and linear interpolation methods based on pre-measured values.

This example involves a system that uses a pressure transducer which provides a 0 to +10 VDC output. The range of the pressure measurement is from 0 psi to 1000 psi. During measurement tests however, it was found that the relationship associated with the measurement was not linear. Figure 11-63 illustrates the difference between the supposed linearity curve and the actual measurements made. The analog input module transforms the signal into counts ranging from 0 to +4095 for 0 to 10 VDC. The test counts taken for different psi pressure values are shown in Figure 11-64.

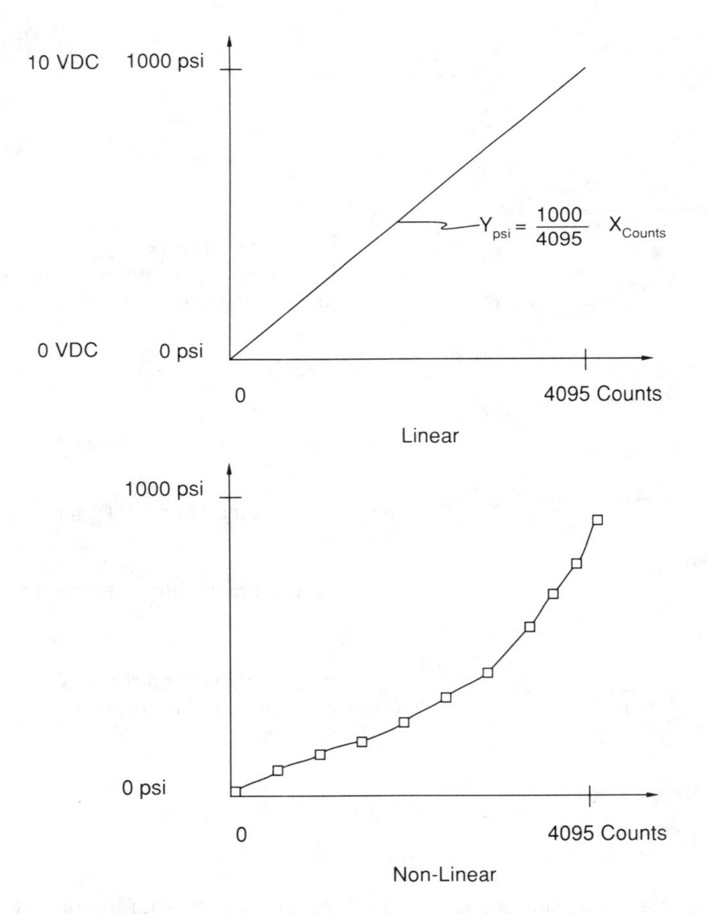

Figure 11-63. Linear behavior (top); actual non-linear samples (bottom).

psi Measurement	Analog Input Counts
0	0
50	600
100	1200
150	1500
200	1950
250	2280
300	2900
400	3300
600	3700
800	3820
1000	4090

Figure 11-64. Tabulation of psi samples and their corresponding counts.

Let's assume that our control algorithm calls for the input measurement to be converted to engineering units (in this case psi). Since a linearity calculation based on the linear equation is not possible, our solution to obtain psi values is to perform by estimating a pressure according to an input count reading. The linear interpolation is computed in the PLC by looking through a table (group of contiguous registers) for a psi value equivalent to our counts.

The two tables that will be used are the psi measurements and their corresponding counts. The count table starts at register location 3000 and the psi values at 3100. Figure 11-65 shows these two tables along with the corresponding pointer. The pointer (register 4000) is used to "point" at a register in the table according to the offset specified by the pointer (Table to Register Instruction). For instance, if the pointer value is 3 (Reg 4000 = 3), then it will be pointing to registers 3002 and 3102. Note that the contents of the pointer register are in decimal while the registers are in octal. The flow chart for the look-up and interpolation procedure is shown in Figure 11-66.

psi Table		Counts Table		Pointer
Register	psi	Register	Counts	REG 4000
3100	0	3000	0	1
3101	50	3001	600	2
3102	100	3002	1200	3
3103	150	3003	1500	4
3104	200	3004	1950	5
3105	250	3005	2280	6
3106	300	3006	2900	7
3107	400	3007	3300	8
3110	600	3010	3700	9
3111	800	3011	3820	10
3112	1000	3012	4090	11

Figure 11-65. Look-up table to be used with their corresponding psi and count values.

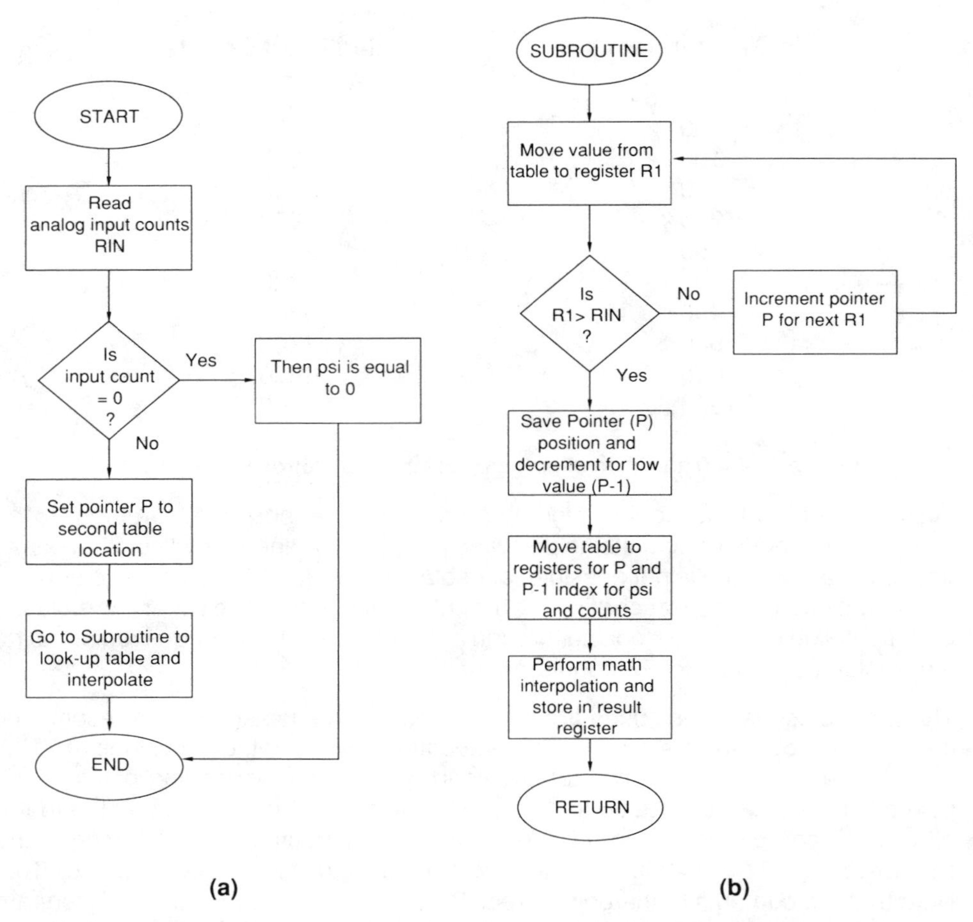

Figure 11-66. a) Main program flowchart and **b)** Interpolation subroutine flowchart.

The register assignment table for this example is shown in Figure 11-67. The internal output assignment is illustrated in Figure 11-68. The only real input that is included in this section of the example is the analog input module with address 070.

Figure 11-69 shows the method used to calculate the interpolation with the contents of each register used. The PLC program shown in Figure 11-70 implements the linear interpolation by means of finding the hi and lo count and psi values via transfers of table to register instructions. Once the values in a register are found to be greater than the actual analog counts, the pointer 4000 is stored in register 4050 (point P) and then decremented and stored back into 4000 (point P-1). Using the two pointers, the high and low values for the counts and psi are obtained via table to register instructions. If the actual value (analog counts) is more than the value pointed by the pointer, the pointer is incremented (add +1) and a new table value is tested.

In the software program presented, we take into consideration the fact that the actual count value may be zero counts. In this case, the equivalent psi is 0 and the program does not go into the subroutine. Otherwise, the program would enter into a loop error.

Register	Description
3000	Look-up Table Storage. Counts
3012	
•	
•	
•	
3100	Look-up Table Storage. psi
3112	
•	
•	
•	
4000	Pointer P-1
•	
•	
•	
4050	Pointer P
•	
•	
•	
4100	Analog Input Counts (Pressure)
•	
•	
•	
4150	psi result register
•	
•	
•	
4400	Low psi register $R_{psi\text{-}lo}$
•	
•	
•	
4450	Hi psi register $R_{psi\text{-}hi}$
•	
•	
•	
4500	Low count register $R_{CTS\text{-}LO}$
•	
•	
•	
4550	Hi count register $R_{CTS\text{-}HI}$
•	
•	
•	
4600	Temporary Register (Subtract)
4601	Temporary Register (Subtract)
4602	Temporary Register (Multiply)
4603	Temporary Register (Multiply)
4604	Temporary Register (Divide)

Figure 11-67. Register assignment.

Internal	Description
1000	Analog XFER enabled
1001	Compare for input = 0 counts
1002	Input counts are 0
1003	Move constant 0 to computed psi register
1004	Move constant 2 as pointer (enabled)

Figure 11-68. Internal output assignment.

Figure 11-68 continued.

Internal	Description
.	
.	
.	
1100	Go sub. Move table to register
1101	Compare enabled
1102	Hi counts ≥ input counts (Hi Cts)
1103	Increment (ADD) pointer enabled
.	
.	
.	
1200	Go to calculate math. Store Pointer
1201	Subtract 1 from pointer enabled
1202	Move table to register enabled (Low Cts)
1203	Move table to register enabled (Low psi)
1204	Move table to register enabled (Hi psi)
1205	Subtract enabled
1206	Subtract enabled
1207	Multiply enabled
1210	Subtract enabled
1211	Divide enabled
1212	Add enabled

Counts		Psi		Pointer Register
Register	Contents	Register	Contents	
4500	Low Counts	4400	Low psi	4000 @ P-1
4100	Actual Counts	4150	Computed psi	
4550	Hi Counts	4450	Hi psi	4050 @ P

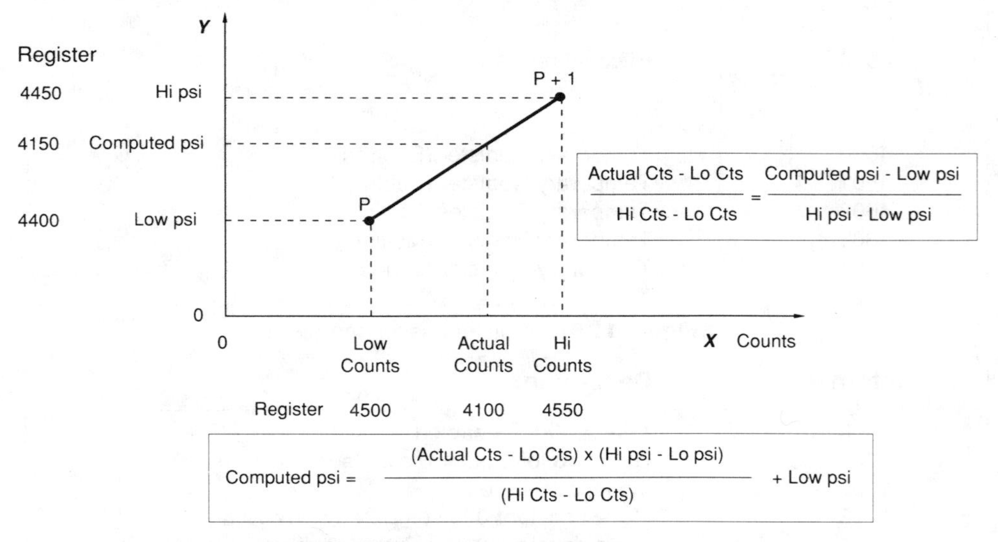

Figure 11-69. Interpolation method implemented.

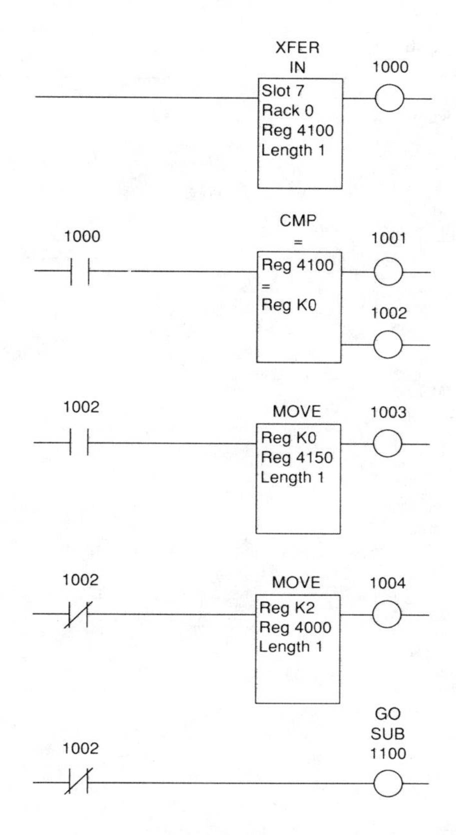

Figure 11-70a. Main program.

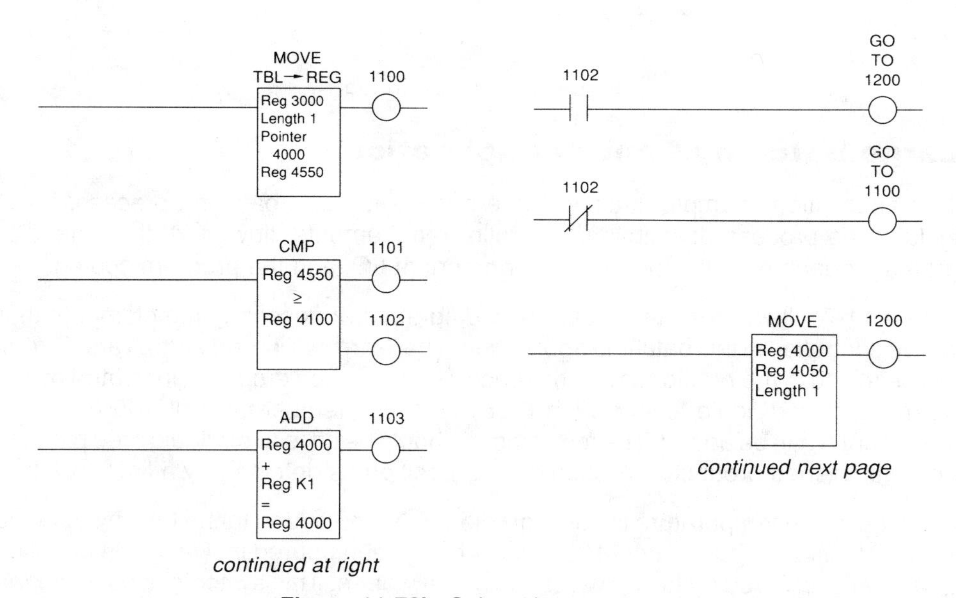

Figure 11-70b. Subroutine program.

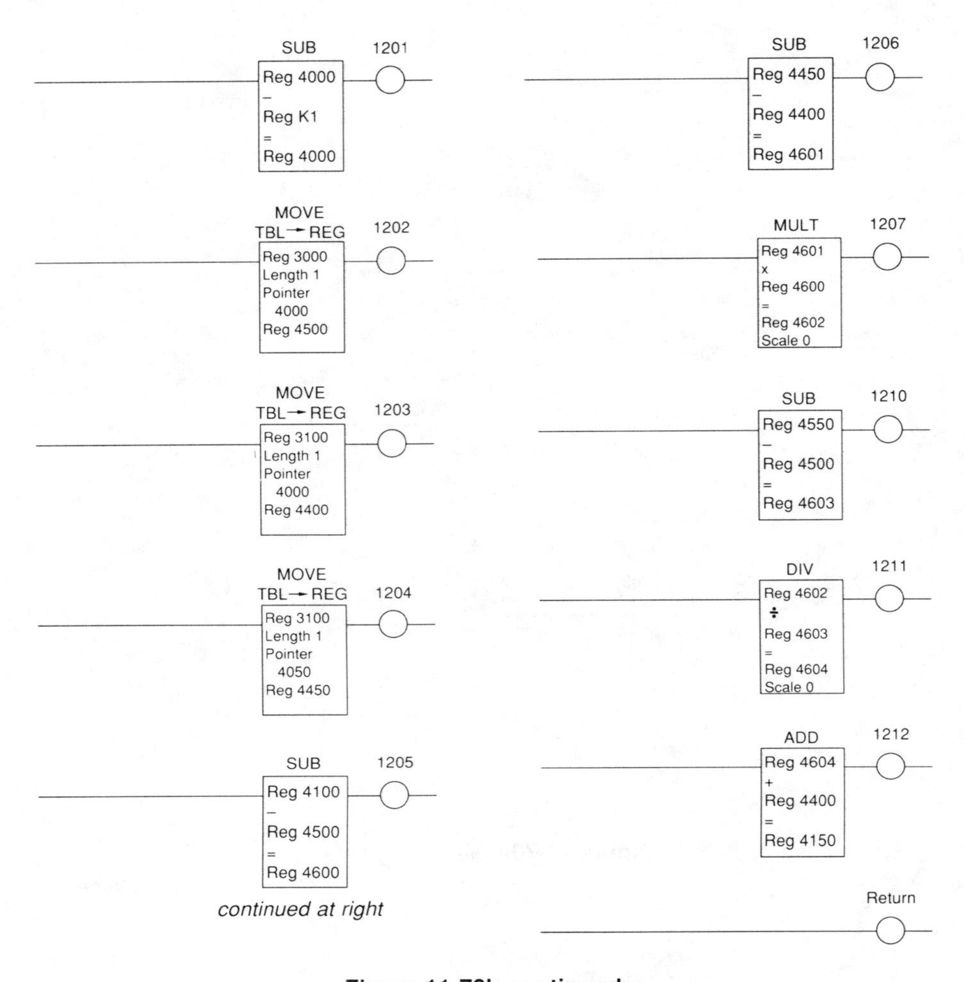

Figure 11-70b continued.

Large Batching Control Application

This application example deals with the automation of a batching process. It will include the process description, controller requirements, flowchart, the logic diagrams for each output sequence, assignment of I/O, and the program coding.

Figure 11-71 illustrates the process flow diagram that describes the elements that will be controlled in this batching application. There are two ingredients, A and B, that will be mixed in the reactor tank. The reactor tank must be empty as indicated by the normally closed liquid level switch (LLS) and at a temperature of 100°C before ingredient A can be added. The mixer motor should be off to avoid liquid precipitation, and the finished product tank should be in position as detected by a limit switch.

Once the initial temperature is set, ingredient A will be added to the tank by opening solenoid valve 1 (SOL1), until 100 gallons have been poured in. Detection of gallon quantity is achieved by LLS1, which is normally open. This switch closes when the proper level is reached. At this point, ingredient B is added by opening SOL2. The

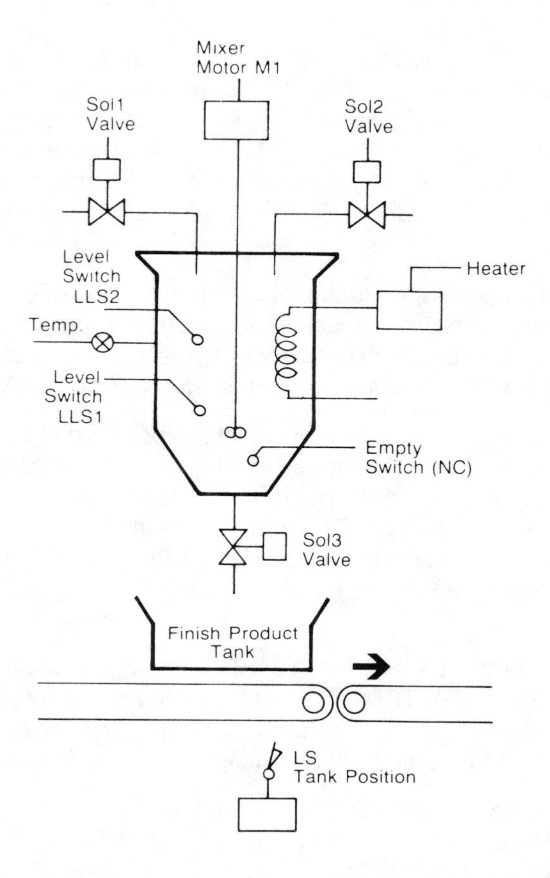

Figure 11-71. Process diagram for batching control.

quantity required is 400 gallons, and it is detected by LLS2 (500-100=400 gal.). The temperature should be kept at 100°C, +/-10%, during the add ingredients step. If the temperature drops, the heater should be turned ON automatically, while the process continues.

After both ingredients have been added to the reactor tank, the temperature should be raised to 300°C (+/-10%). Then the mixer should be turned ON for 20 minutes. The temperature should be controlled automatically at predefined set points during the process.

The drain valve should be activated by SOL3 after the mixing is completed. This operation should reset the process until another tank is placed in position, and the cycle is started again. Pilot light indicators should be incorporated in the system as a means of alerting the operator to the status of the batching process.

This application requires the capability of reading analog signals from the process. In this case, the voltage comes from a temperature transducer (0-10 volts), which has a range of 0°C to 500°C (50°C/volt). The temperature is controlled by the ON/OFF control of the heater coil. Standard 110 VAC input and output modules are also required.

A process flowchart is shown in Figure 11-72. It illustrates in global terms what has been described in the definition of the control task and serves as a preparation for the logic diagrams. The initial implementation of the logic required to control each of the process sequences can be accomplished with the use of logic diagrams. These diagrams represent the necessary conditions that will enable a rung to be energized. The real I/O is marked with an X. Figure 11-73 shows the logic diagrams for this example.

The first logic diagram represents the initial requirements to start the process. The start pushbutton, when pressed, enables the start mix output; if the tank is in position, SOL3 is closed and the stop push button is not pressed. The pilot lights PL1 and PL2 indicate that the tank is in position, and the system START MIX signal is enabled.

Logic diagram 2 starts the setting of the initial temperature (T1) at 100°C. The logic indicates that the mixer motor (M1) must be off, START MIX enabled, and the reactor tank empty. The READY TO MIX input is an interlock from logic diagram 6, and it is used in logic diagram 2 to disable T1 when T2 is being set. Note, the EMPTY signal is used in the OR function with the initial SET T1 to insure that even when the tank is still adding ingredients, the temperature control will continue to maintain the temperature at T1.

Logic diagram 3 controls the READY TO ADD signal, which allows ingredient A to be added. Here, the output TEMP OK1 (T1=100°C) of the block indicates that the temperature has been reached, and that START MIX signal is still enabled; thus, the process is ready to add the first ingredient.

In logic diagram 4, the READY TO ADD A enables SOL1 to open. This action occurs while LLS1 is still opened (less than 100 gallons), and the drain valve (SOL3) is not energized. When the liquid level is reached, LLS1 closes and according to the logic, it de-energizes SOL1. The second part in the logic diagram indicates that the add ingredient A step is finished.

In logic diagram 5, the SOL2 is opened to add ingredient B until LLS2 is closed (500 gallon level) indicating that 400 additional gallons have been added to the reactor tank. The remainder of the logic indicates that the add ingredient B step is finished.

Logic diagram 6 shows that once both ingredients are in the reactor tank (finish A, and finish B, both ON), the READY TO MIX control signal is enabled. This condition will start a new temperature control block to raise the temperature to 300°C and will disable the other temperature control (T1).

In logic diagram 7, after the temperature is at 300°C, and the READY TO MIX (SET T2) is on, the mixer will turn on. At the same time, the timer is enabled, and after 20 minutes (1200 sec.), it times-out and resets the mixer motor logic. The timer output represents the FINISH MIXING signal, used to energize SOL3, which opens the drain valve to discharge the mixed ingredients (logic diagram 8). The valve remains open until the empty switch is back to its normal state (closed).

The logic diagram 9 is used to turn the heater on if the temperature is low. The heater can be turned on from either of the two temperature control block outputs. The sequences 10, 11, and 12 are provided to indicate to the operator the status of the temperature inside the tank.

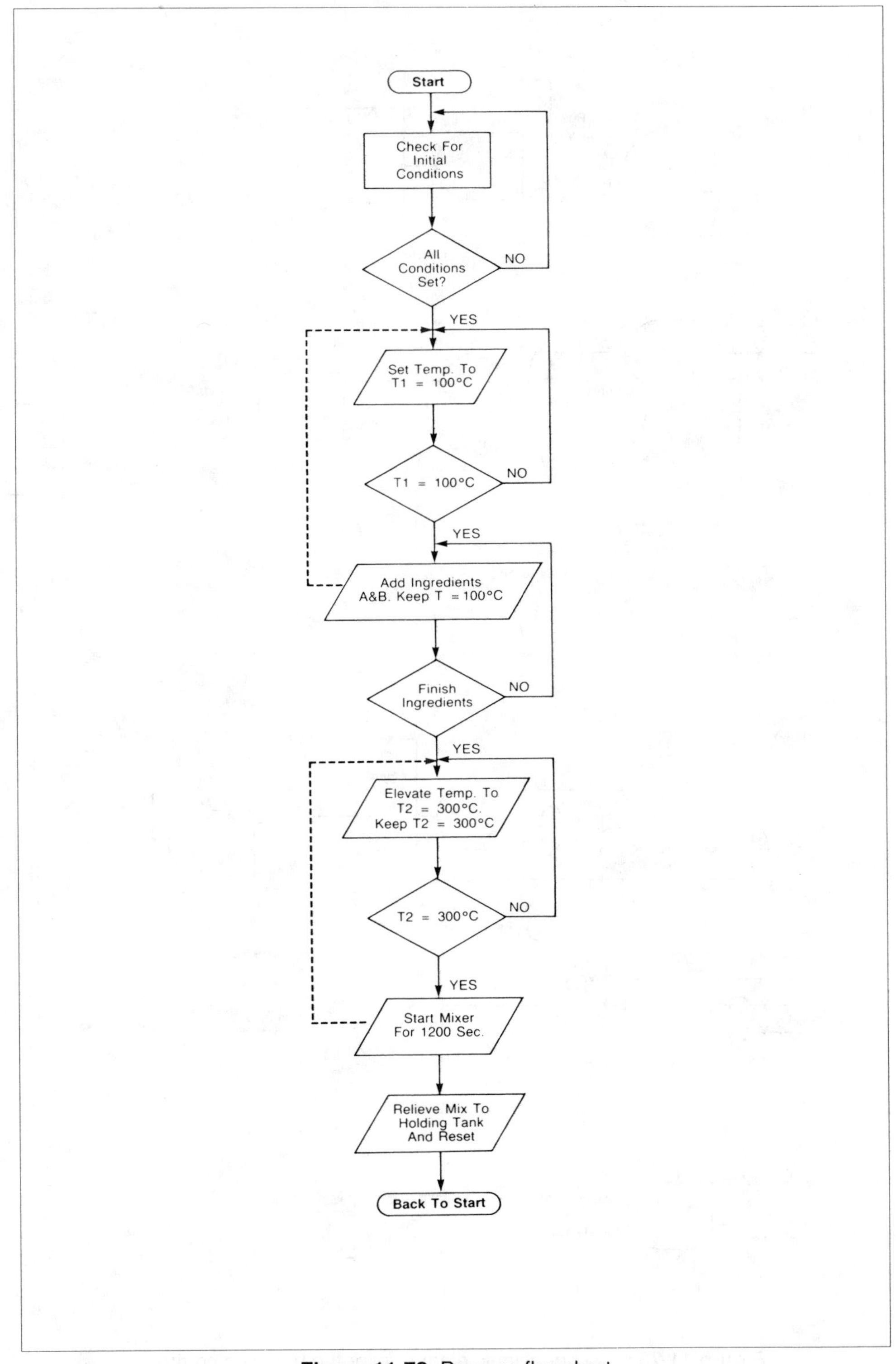

Figure 11-72. Process flowchart.

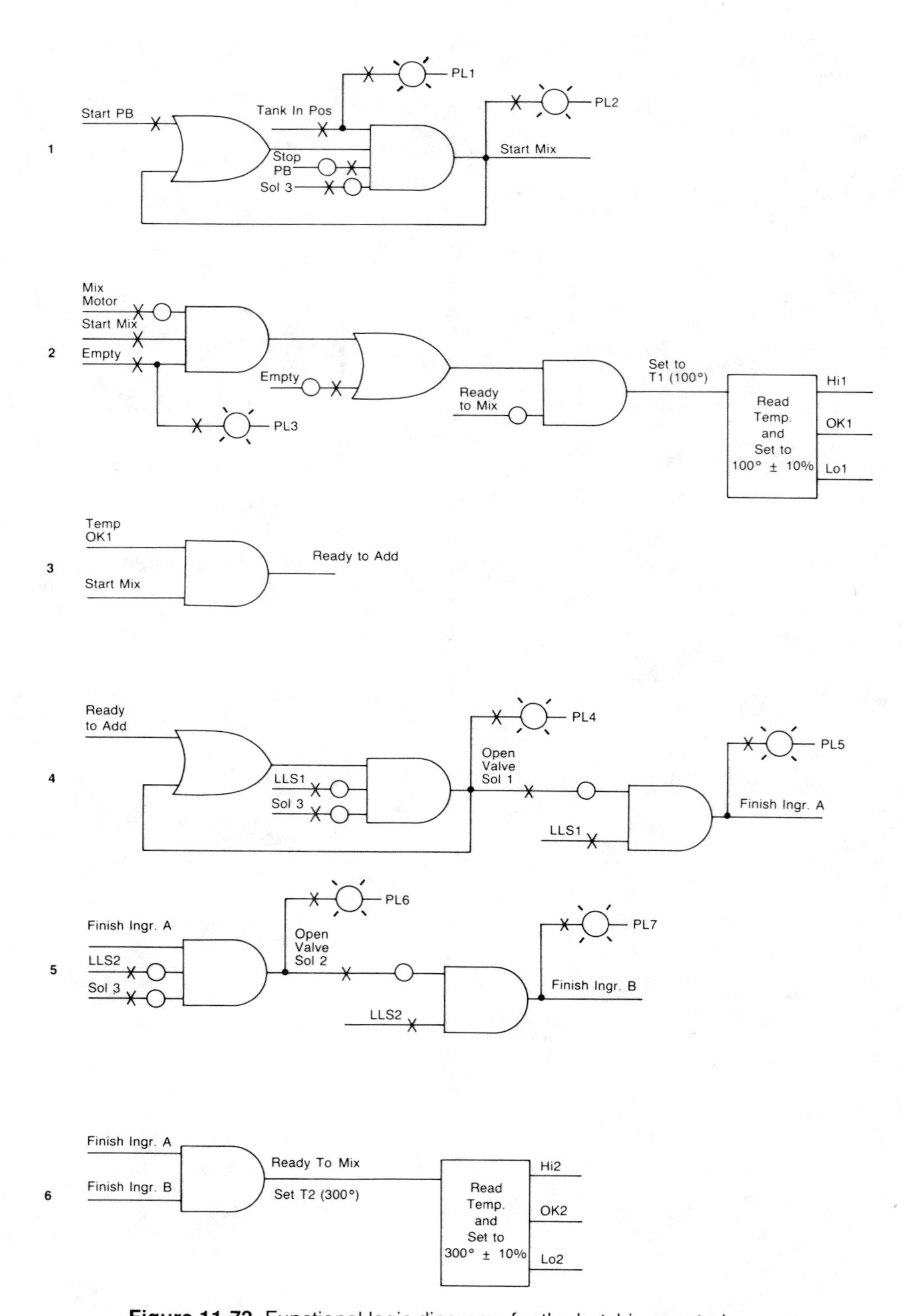

Figure 11-73. Functional logic diagrams for the batching control.

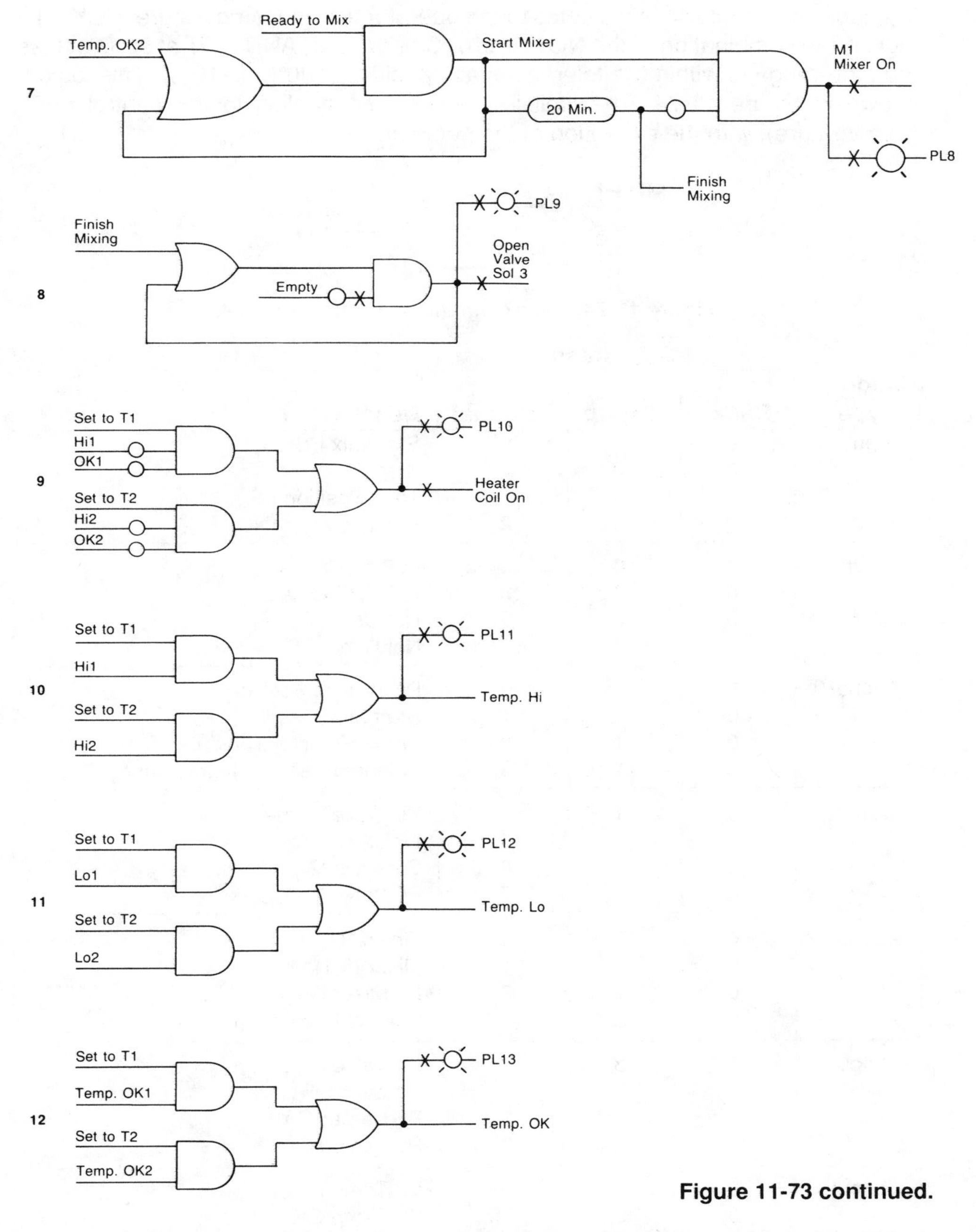

Figure 11-73 continued.

The logic for reading the temperature will be done using "compare" functional block instructions. Once the command, or logic, indicates control of temperature, the compare block will be enabled, and three comparisons will be done to detect more than 110°C, 100°C, and less than 90°C.

It will be necessary to compare using a limit (LIM) compare function, since it is required that the ingredients be added at 100°C. The output of this block will be OK1.

The logic for the pilot light indicates to the operator that the temperature is OK. This logic is the combination of the NOT at 110 °C or greater, AND NOT at 90°C or less; thus the range is within the tolerances as specified (100°C +/-10%). This logic is shown in Figure 11-74. The limit instruction also applies to the control of T2 (temperature), with the exception of the setpoint.

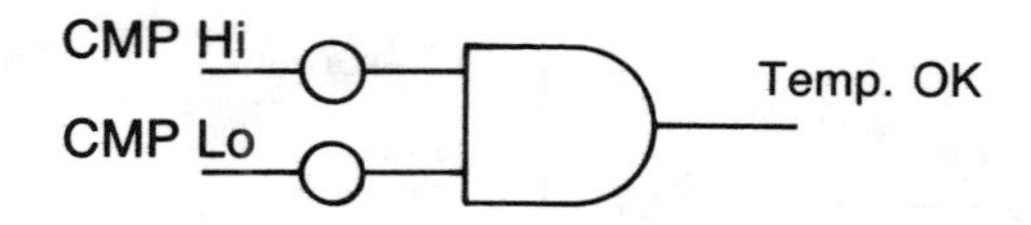

Figure 11-74. Logic diagram for TEMP OK signal.

Module Type	I/O Address Rack	Group	Terminal	Description
Input	0	0	0	Start mix PB
	0	0	1	Stop PB
	0	0	2	Tank Position LS
	0	0	3	Empty Switch (NC)
Input	0	0	4	LLS 1 (100 gal)
	0	0	5	LLS 2 (500 gal)
	0	0	6	Not Used
	0	0	7	Not Used
Output	0	1	0	PL Tank in Position
	0	1	1	Start Mix
	0	1	2	Empty Reactor Tank
	0	1	3	Solenoid Valve 1 (Ingredient A)
Output	0	1	4	PL Valve 1 Open
	0	1	5	Finish A
	0	1	6	Sol Valve 2
	0	1	7	PL Valve 2
Output	0	2	0	Finish B
	0	2	1	Mixer (M1)
	0	2	2	PL Mixer On
	0	2	3	Solenoid Valve 3 (Drain)
Output	0	3	0	PL Valve 3 Open
	0	3	1	Heater Coil
	0	3	2	PL Heater On
	0	3	3	PL Temp H
Output	0	3	0	PL Temp Lo
	0	3	1	PL Temp OK
	0	3	2	Not Used
	0	3	3	Not Used
Analog Input	0	3	4	Input 34 Connected to Transmitter
	0	3	5	Corresponds to I/O Register 3034
	0	3	6	Input 36 left as is
	0	3	7	Register 3036

Figure 11-75. Real I/O address assignment.

The assignment of real I/O can begin by addressing the real inputs and outputs. These inputs and outputs are tabulated in the assignment table. The assignment of I/O for this application example is illustrated in Figure 11-75. Note that the modularity for the digital I/O is four points per module. The analog module contains two input channels that occupy one half of a group (four locations).

The assignment of internals is shown in Figure 11-76. Here, several internal coil addresses are tabulated to represent control relay conditions related to the logic diagram. The coils associated with the compare functional blocks are internals used to describe the temperature conditions, such as HI and LO. The analog value of the temperature is stored in I/O register 3034 and will be compared to storage registers that hold the equivalent values of the temperature ranges.

Device	Internal	Description
Logic	1000	Set to T1
CMP Block	1001	CMP Hi 1
LIM	1002	CMP for range OK1
CMP Block	1003	CMP Lo 1
Logic	1004	TEMP OK 1
Logic	1005	READY TO ADD
Logic	1006	READY TO MIX/SET to T2
CMP Block	1007	CMP Hi 2
LIM	1010	CMP for range OK 2
CMP Block	1011	CMP Lo 2
Logic	1012	TEMP OK 2
Logic	1013	START MIXER
Timer	2000	Timer preset 20 min (1200 sec) Register 4000 ACC Register 4001

Figure 11-76. Internal output assignment.

Translation of the logic diagrams into PLC diagrams is the next step after the tabulation of the input and output assignments. The coding of the program is done by following the logic diagram sequences previously specified and also by using the I/O and internal tables as references for the addresses. The program coding for this example is shown in Figure 11-77.

Note that the ladder logic shown in the program coding is the implementation of each logic diagram. The internals are assigned as specified in the internal assignment table. Several storage registers are added in the compare blocks to hold the preset values. These values correspond to the equivalent temperature set points used, including the tolerances (i.e. 110°C, 100°C, 300°C, 270°C, etc.).

The voltage from the temperature transmitter is 0 to 10V, representing 0 to 500°C. Each volt represents a change of 50°C. The controller used in this example receives the voltage and converts it to a count ranging from 0000 to 9999. This is proportional to the voltage and, therefore, to the temperature.

The first set point (register 4000) contains the value 2200 which indicates a count proportional to 110°C (100°C +/-10%); register 4003 contains 1800, equivalent to 90°C (100°C -10%). Registers 4001 and 4002 contain values of 2040 (102°C) and 1960 (98°C) respectively. These values are used to detect a small range in which the temperature is very close to 100°C to start the ingredient addition. We do not compare the value to 100°C because this value may never be exactly 100°C at a particular time while reading the analog value. These two registers are used in a LIM compare block that detects when the temperature is between 98°C and 102°C. The preset values of the other compare blocks are specified in the same manner.

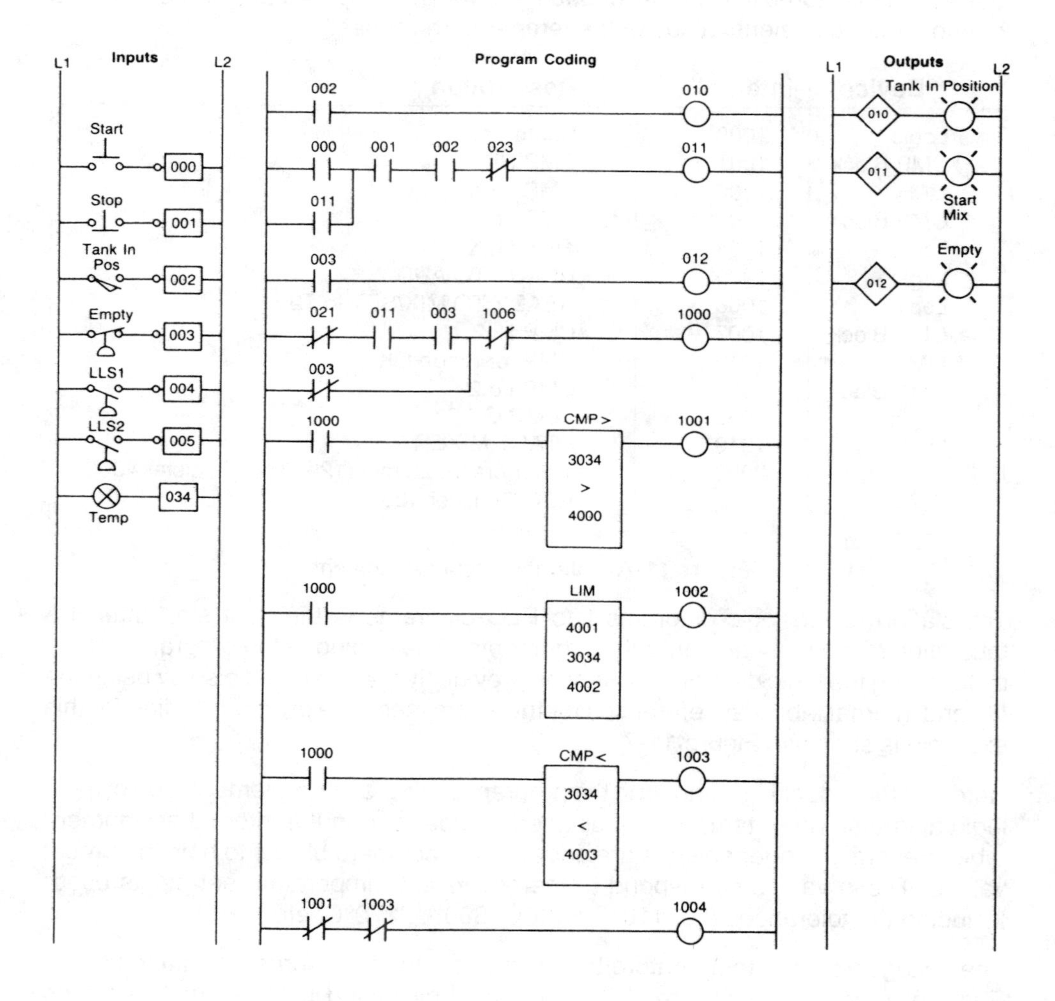

Figure 11-77. Program coding for batching control.

Figure 11-77 continued.

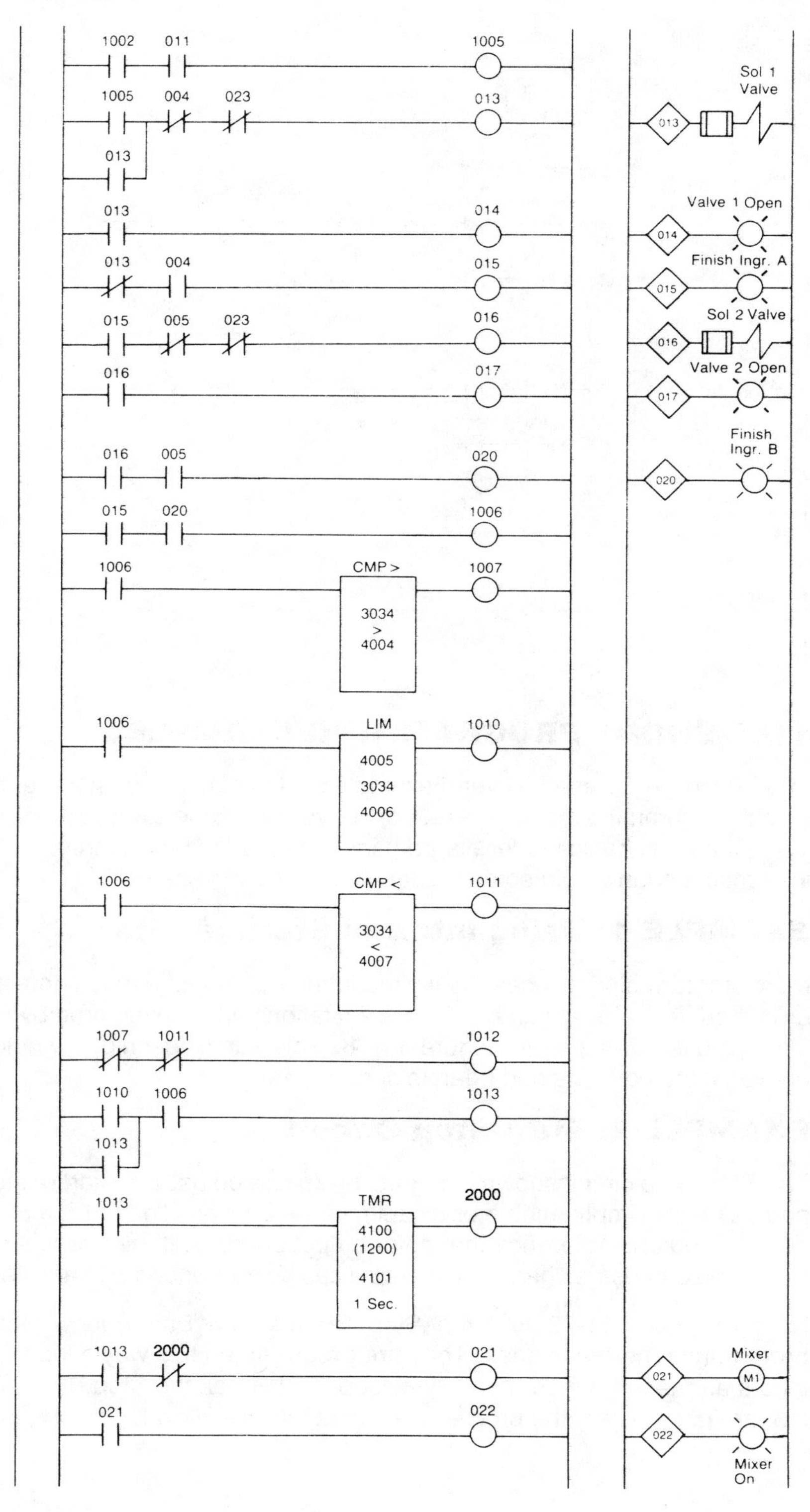

Figure 11-77 continued.

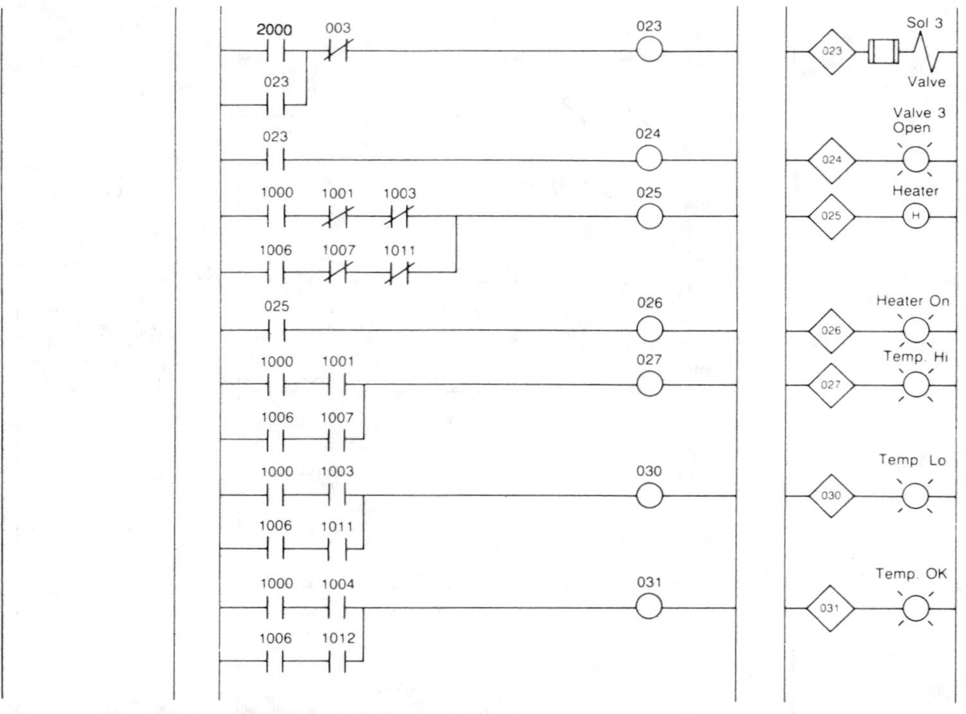

11-7 SHORT PROGRAMMING EXAMPLES

This section will present several examples of logic networks that are often required when programming the controller. For convenience, the examples are implemented using the most basic ladder diagram instructions, and therefore may require more instructions than would some higher level instruction sets.

EXAMPLE 1: Using Internal Storage Bits

Most programming devices are limited in the number of series contacts, or parallel branches, that a rung can have. This limitation can be overcome by using internal storage bits, as shown in Figure 11-78. This same technique would have been applied if the contacts had been in series.

EXAMPLE 2: Start/Stop Circuit

The Start/Stop circuit shown in Figure 11-79 can be used to start or stop a motor or process or to simply enable or disable some function. To start a motor, the ladder output need only reference the motor output address. If the intent is only to detect that some process is enabled, the output can be referenced with an internal address.

Note in Figure 11-79 that the Stop PB and the Emergency Stop inputs are programmed normally-open. They are programmed this way, since these types of inputs are usually wired normally-closed. As long as the Stop PB, and the E. Stop PB, are not pushed, the programmed contacts will allow logic continuity. Since the

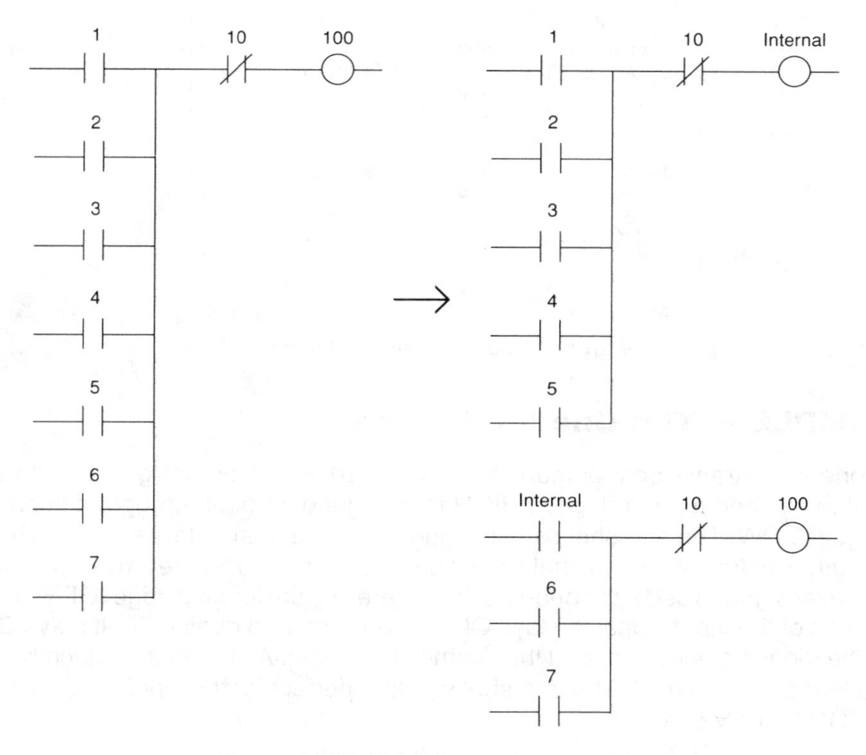

Figure 11-78. Using internal storage bits.

Figure 11-79. Start/Stop circuit.

Start pushbutton (normally-open) is a momentary device (allows continuity only when closed), a contact from the motor output is used to seal-in the circuit. Often, the seal-in contact is an input from the motor starter contacts.

EXAMPLE 3: Exclusive-OR Circuit

The Exclusive-OR circuit in Figure 11-80 is used when it is necessary to prevent an output from energizing if two conditions, which can activate the output independently, occur simultaneously.

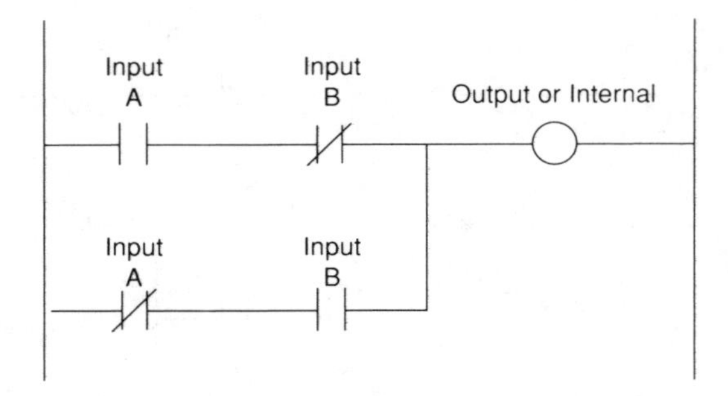

Figure 11-80. Exclusive OR circuit.

EXAMPLE 4: The One-Shot Signal

The one-shot (transitional output) in Figure 11-81 is a program generated pulse output that, when triggered, goes HIGH for the duration of one program scan and then goes LOW. The one-shot can be triggered from a momentary signal such as a pushbutton, or from an output that comes on and stays on for some time (e.g. motor). Whichever signal is used, the one-shot is triggered by the leading edge (OFF-to-ON) transition of the input signal. It stays ON for one scan and goes OFF. It stays OFF, until the trigger is activated, and then comes ON again. A one-shot is typically used as a clear or reset signal. The one-shot signal is perfect for this application, since it stays ON for only one scan.

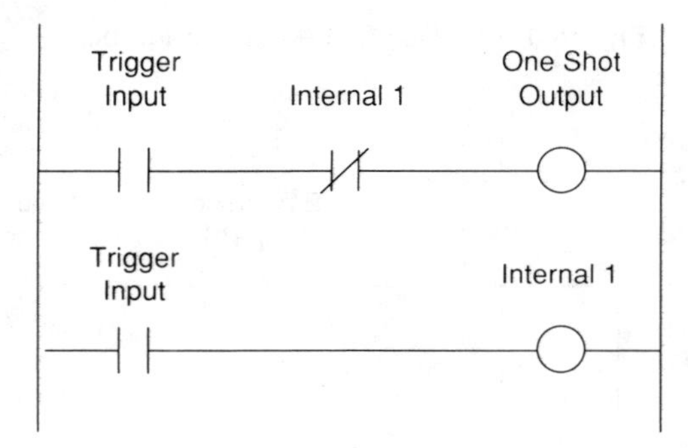

Figure 11-81. One-Shot pulse circuit.

Example 5: Trailing Edge One-Shot

The trailing edge one-shot (shown in Figure 11-82) is implemented in cases where it is required to generate a pulse with a one scan duration. The trigger for this pulse is the trailing edge of the trigger pulse.

Example 6: Initialization Using An MCR

The logic circuit shown in Figure 11-83 can be used when it is necessary to set up several parameters during an initialization period. Typically, this initialization is done only once during the program, either when the system is first powered up, or when power is reapplied after a power loss. The parameters that are usually initialized are timer and counter preset values, high and low limit setpoint values, or any other preset or starting values.

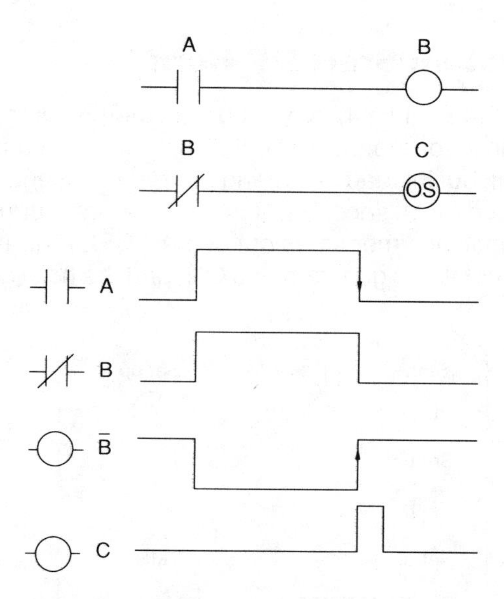

Figure 11-82. Trailing Edge One-Shot.

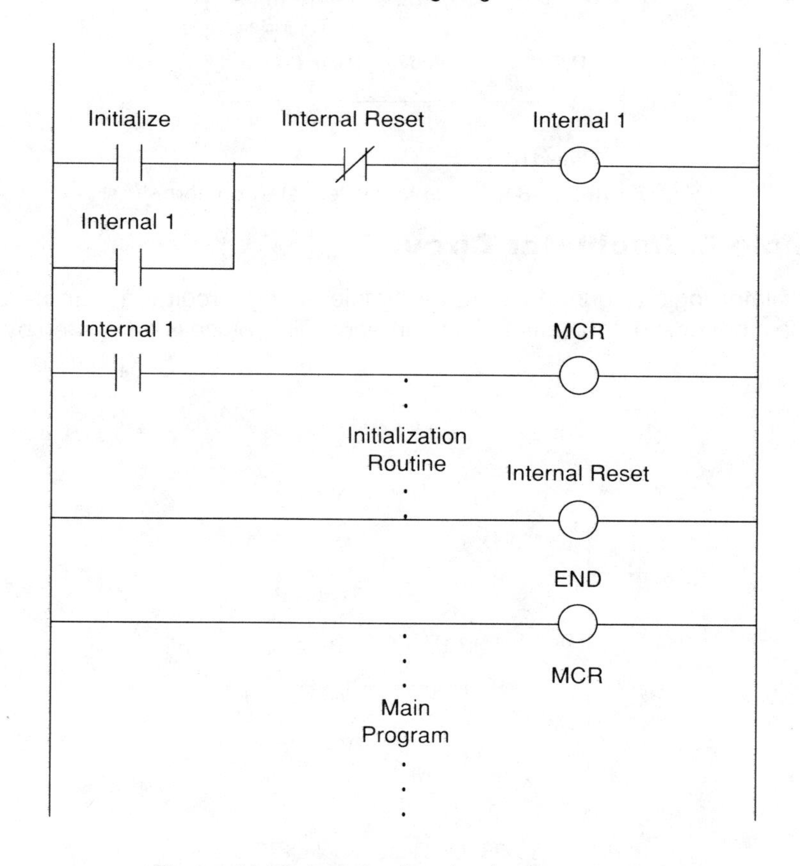

Figure 11-83. Initialization circuit using an MCR.

Example 7: System Start-UP Horn

The start horn logic (Figure 11-84) is often used when moving equipment is about to be started (e.g. conveyor motors). The SET-UP signal in this example is similar to the start/stop circuit, but instead of starting the system, it enables the timer, which allows the horn to sound for 10 seconds. Note that the horn sounds when the START input is closed, and until the timer times out or the RESET input opens. The system can be started if the SET-UP signal remains ON, and the horn delay timer times out.

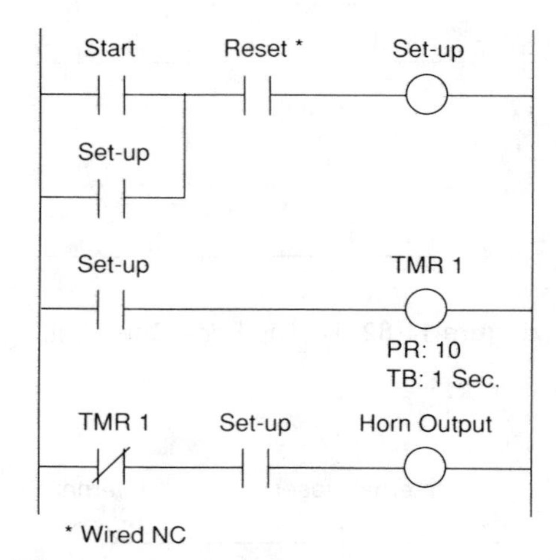

Figure 11-84. Circuit for system start-up horn.

Example 8: Oscillator Circuit

The oscillator logic (Figure 11-85) is a simple timing circuit that can be used to generate a periodic output pulse of any duration. This pulse is generated by TMR1.

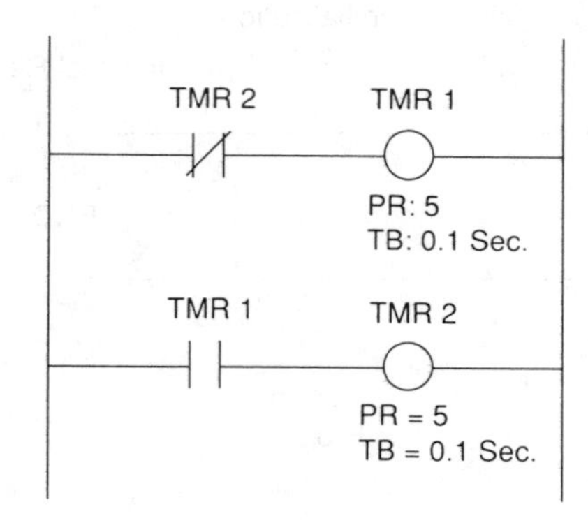

Figure 11-85. Oscillator (ON/OFF) circuit.

Example 9: Annunciator Flasher Circuit

The flasher circuit, shown in Figure 11-86, is used to toggle an output ON and OFF continually. The "oscillator circuit" (from example 7) output (TMR1) is programmed in series with the alarm condition. As long as the alarm condition is TRUE, the annunciator output will flash. The output in this case would be a pilot light; however, this same logic could be used for a horn that is pulsed during the alarm condition. Note that any number of alarm conditions could be programmed using the same flasher circuit.

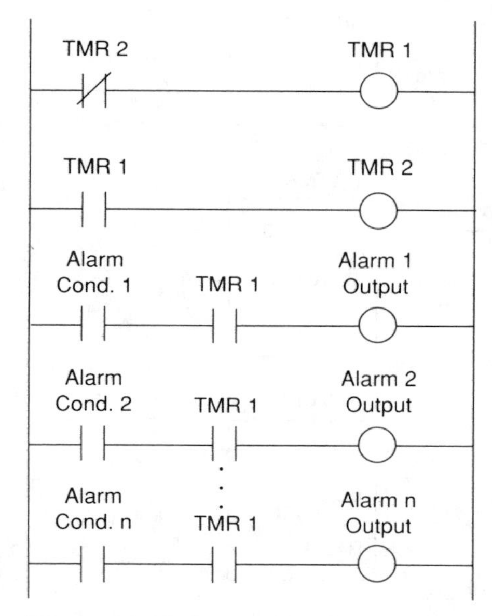

Figure 11-86. Annunciator flasher circuit.

Example 10: Self-Resetting Timer

The self-resetting timer (Figure 11-87) will provide a one scan pulse each time the timer is energized. The repetition of this pulse is determined by the specified preset value of the timer.

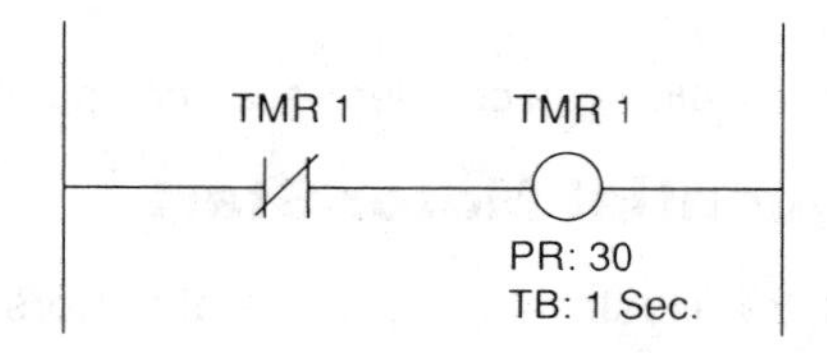

Figure 11-87. Self-resetting timer circuit.

Example 11: Scan Counter—Scan Time Computation

The scan time can be computed using the circuit shown in Figure 11-88. This short program counts the number of two scans that occur in a time interval defined by the timer (10 seconds). Once the time sample elapses, the number of scan counts is multiplied by two; The time of 10 seconds is divided by the number of total scans. The result register is scaled so that the number in the result represents milliseconds.

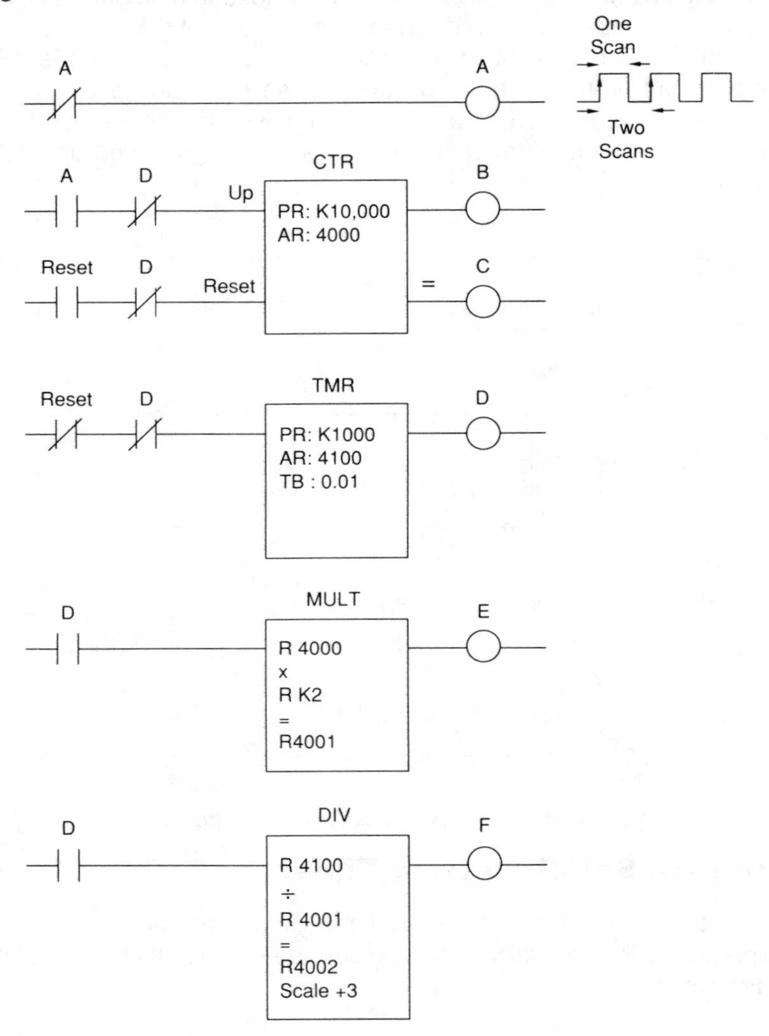

Figure 11-88. Scan counter—scan computation.

Example 12: Sequential Motor Start

This example (Figure 11-89) illustrates how several motors or other devices could be started sequentially, as opposed to all at once. For simplicity in this example, we use an ON-delay timer to delay the start of each motor. However, this approach would be impractical for starting a large number of motors. If a large number of motors are to be started, other techniques that do not require as many timers as motors are used (e.g. shift registers, self-resetting timers, oscillator circuits, etc.).

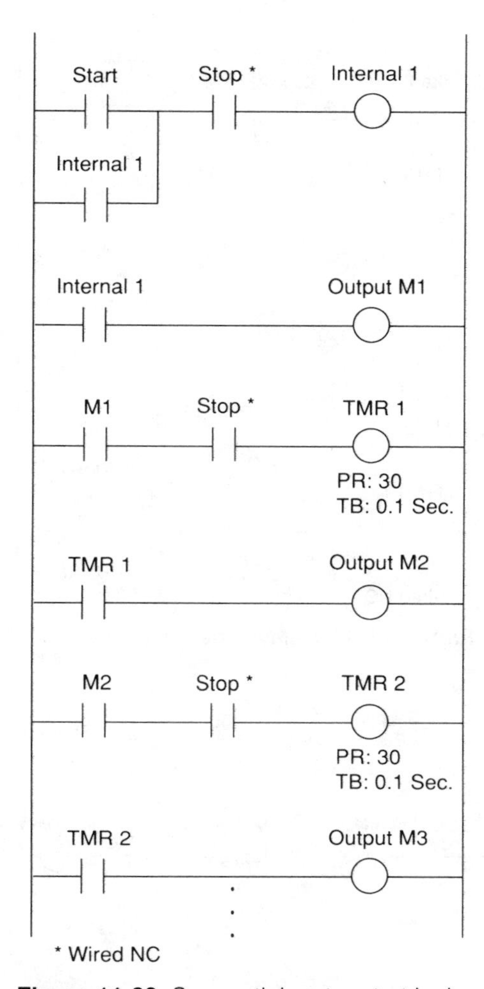

Figure 11-89. Sequential motor-start logic.

Example 13: Delayed De-Energize Device

This example (Figure 11-90) illustrates the use of the OFF-delay timer to de-energize a motor or any device after a delayed period. Note that the output of the timer is originally HIGH, thus maintaining the TMR1 contact closed. When the STOP MOTOR1 pushbutton (wired NC) is depressed while the motor is running, the internal output is energized, which enables the OFF-delay timer. When the timer times out, the contacts open, and the motor is de-energized.

Example 14: A 24-Hour Clock

The 24 hour clock has many applications, but is generally used to display the time of day or to determine the time that a report is generated. The logic used to implement the clock is shown in Figure 11-91. It consists of three counters: one counts 60seconds, another counts 60 minutes, and the third counts 24 hours. The time can be displayed by outputting the accumulator register values of each counter to seven-segment BCD displays.

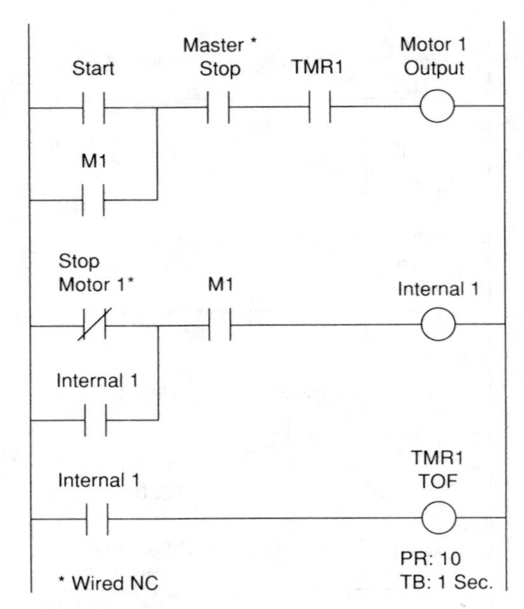

Figure 11-90. Delayed de-energize circuit.

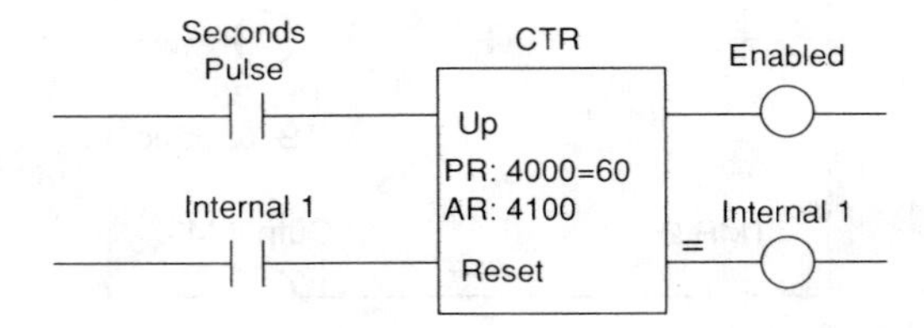

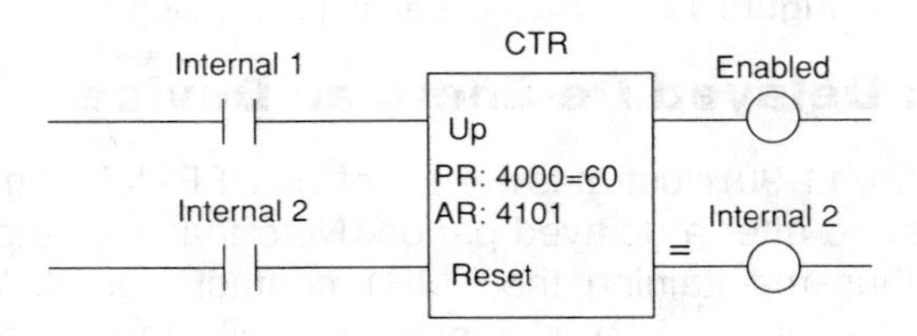

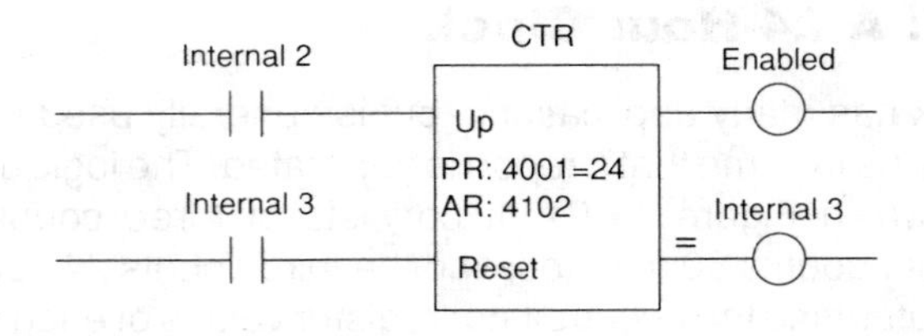

Figure 11-91. A 24-hour clock circuit.

Example 15: Counting Beyond The Maximum Count

Depending on the application, it may be necessary to count events that will exceed the maximum allowable number that can be held in a register. The maximum count in most controllers is either 9999 (BCD) or 32767 (binary). To count beyond 9999 simply involves cascading two counters in which the output of the first counter is used as the input of the UP-Count of the second counter. If 32767 is the maximum count, the same approach would result in an erroneous count. The first register would contain a value of 1 (after 32767 is reached), and the second register would contain 00000. This result would indicate a count of 100000 instead of the actual count of 32768.

A solution to this situation is to set the preset value of the first counter to 9999 and use a second counter to register each time 10000 counts occur. The following sequences in Figure 11-92 illustrate this technique.

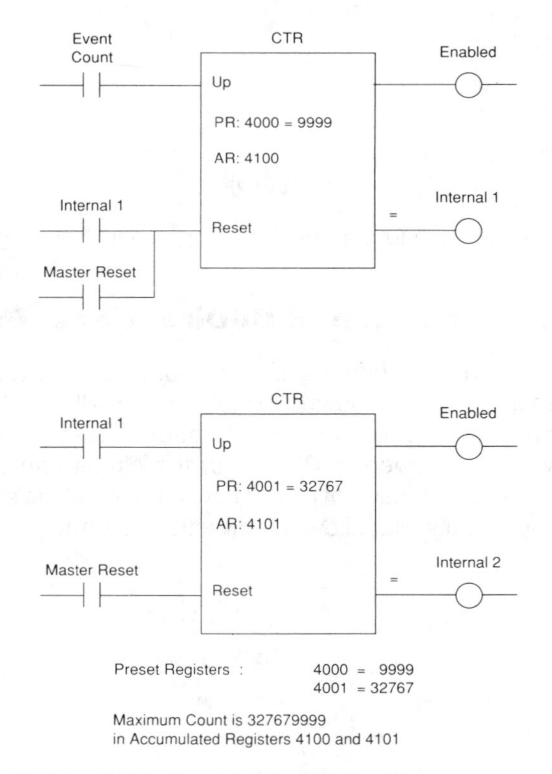

Figure 11-92. Counting beyond maximum count.

Example 16: Push-to-Start/Push-to-Stop Circuit

Often, it is desirable to have a single pushbutton perform the start (enable) and stop (disable) functions. In this example (Figure 11-93) when the pushbutton (PB1) is depressed for the first time, internal output 2 goes HIGH(ON) and remains HIGH. If the pushbutton is depressed again, internal output 2 goes LOW (OFF). The second logic rung detects the first time PB is pressed, while the first rung detects the second time that the button is depressed. A simplified timing diagram shows the operation of this circuit.

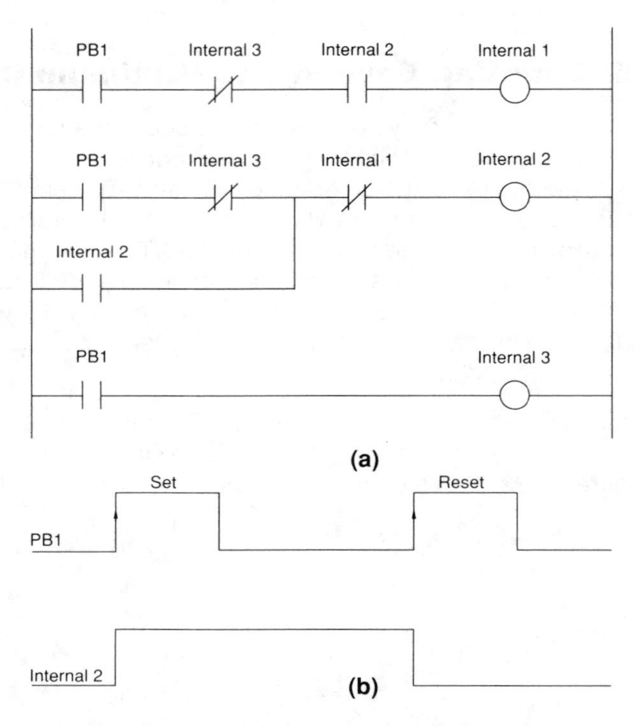

Figure 11-93. a) Push-to-Start/Push-to-Stop circuit; **b)** timing diagram.

Example 17: Elimination of Bi-Directional Power Flow

On occasion, when converting relay-logic to program logic, we will find relay circuits that have been designed to allow power flow bi-directionally as shown in Figure 11-94a. Power can flow down through CR1 or up through CR1 to make a complete path. Bi-directional flow is not allowed in PLCs, but the circuit can be restructured to establish a circuit for each direction of power flow. The result, as shown in Figure 11-94b, is two separate circuits that allow uni-directional, left-to-right power flow.

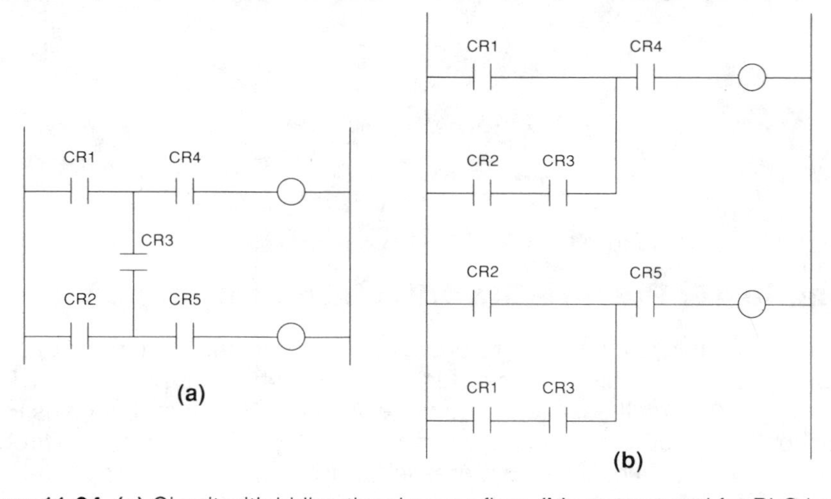

Figure 11-94. (a) Circuit with bidirectional power flow **(b)** restructured for PLC logic.

Chapter

—12—

DATA MEASUREMENTS AND TRANSDUCERS

"We work not only to produce but to give value to time."

—Delacroix

12-1 BASIC MEASUREMENT CONCEPTS

As in any engineering discipline, the user must know how to use and apply different field devices that are connected to measurement equipment or measurement instruments. By using these instruments the user will undoubtedly encounter measurement errors; therefore, it is very important to know how to deal with and interpret any expected and unexpected data. Let's take a look at how data can be interpreted.

Data Interpretation

Data interpretation and representation is a very important subject when dealing with on-line process control operations. This data provides the controller system with important information about the inner workings of the process. Therefore, every user must understand clearly what data is being collected, and how to interpret it in a correct manner. This will give us knowledge of how the control is to be applied so that the process will behave in a predictable manner.

Process data gathering can be clearly understood by taking a look at how instrumentation and data collection devices interpret data readings. There are four main ways to interpret data sampling readings, each one depicting a different *meaning.* These four methods include:

- Mean
- Median
- Mode
- Standard Deviation

Mean. Mean can be defined as the average value of readings that have been taken. This value is useful in cases where prior determination of future or expected readings are required. For the sake of simplicity, let's take an instrument which provides a signal ranging from 2 mV to 20 mV at time intervals. The readings the instrument takes continuously may form a set that is used to provide the final value when the reading must be reported. This final value could be the mean of the set of readings taken when the instrument was not reporting data. Suppose the instrument last reported 13 mV, and another value will be reported in 10 seconds. Meanwhile, data is being collected, say, every 2 seconds, having values of 14 mV, 14.5 mV, 15 mV, 14.7 mV and 14.8 mV. At the new reporting time, the instrument will provide the mean of the entire reading set which is 14.6 mV. The mean of the reading X, or $\overline{X}$, for the instrument can be defined as :

$$\overline{X} = \frac{X_1 + X_2 + X_3 + \dots + X_n}{n} \qquad \text{or} \qquad \overline{X} = \frac{\sum_{n=1}^{i} X_n}{n}$$

where ***n*** is the total number of readings (i.e. 5), or:

$$\overline{X} = \frac{14 \text{ mV} + 14.5 \text{ mV} + 15 \text{ mV} + 14.7 \text{ mV} + 14.8 \text{ mV}}{5}$$

$$\overline{X} = 14.6 \text{ mV}$$

Median. The median is the middle value of all the readings, in ascending order. In the previous example, we had readings of 14 mV, 14.5 mV, 14.7 mV, 14.8 mV, and 15 mV in ascending order. Therefore the median is 14.7 mV.

$$M = X\left(\frac{m+1}{2}\right) \quad \text{for an odd number of samples, or}$$

$$M = \frac{X\left(\frac{m}{2}\right) + X\left(\frac{m}{2}+1\right)}{2} \quad \text{for an even number of samples}$$

where m is the total number of readings. For example, the value X_3 corresponds to the third (m = 3) value in ascending orderwhich is 14.7 mV. Note that the second median equation is made up of the mean of the two center values.

The median calculation provides statistical information about measurements taken which can be more error tolerant than the mean calculations. For example, if in the above example we had had a reading of 20 mV due to an error created by noise in the system, instead of the 14 mV reading, then the mean would have been pushed to 15.8 mV, whereas the median would have been 14.7 mV. The median value is not affected very much by extreme deviation caused by errors the measurement.

Mode. The mode is defined as the sample value that is most consistent in the data sample; that is, the value which is repeated the most. There can be more than one mode in a sample. Going back to our instrumentation measurement example and observe that if we had taken readings of 14 mV, 14.5 mV, 14 mV, 14.5 mV and 14.5mV, our mode value would have been 14.5 mV. If six readings had been taken and the sixth one was 14 mV, then two mode values would have existed, the 14.5 mV and 14 mV.

The mode value is likely to occur in discrete processes where events are not broken down into infinitesimal readings, as in the measurement of temperature. It will be highly unlikely to obtain a significant mode value in PLC counts read from an analog input module because changes in counts are very easily introduced in the readings. However, mean and median values could be very valuable in determining errors in measurements.

Standard Deviation. Often, it is not sufficient to know the mean value of a set of readings taken in a process but also how these readings are spread out from the mean. For instance, in our instrument reading, if we had sample readings of 9 mV, 9.5 mV, 15 mV, 19.7 mV and 19.8 mV, the mean would also have been 14.6 mV, yet the readings would have been more spread out. Standard deviation (σ) measures this spread and can be expressed as:

$$\sigma = \sqrt{\frac{\sum_{n=1 \text{ to } i} (\overline{X} - X_n)^2}{n-1}}$$

where x is the calculated mean, and n is the number corresponding to each reading starting at 1 and ending at i (ith reading is the last). This formula essentially computes the existing deviation of each sample from the mean. The larger the standard deviation value, the more spread out the values (samples) from the mean. For our instrument readings the standard deviation will have a value of $\sigma = 5.25$.

The standard deviation can add valuable information about the data sample, thus helping make a more quantitative evaluation of the sample measurements. When the data collected (samples) is evenly distributed around its mean in a *bell form*, it is said to have a *normal distribution* or *Gaussian distribution* (see Figure 12-1). In this situation, several conclusions can be obtained (see Figure 12-2):

- 68% of all readings lie within $\pm 1\sigma$
- 95% of all readings lie within $\pm 2\sigma$
- 99% of all readings lie within $\pm 3\sigma$

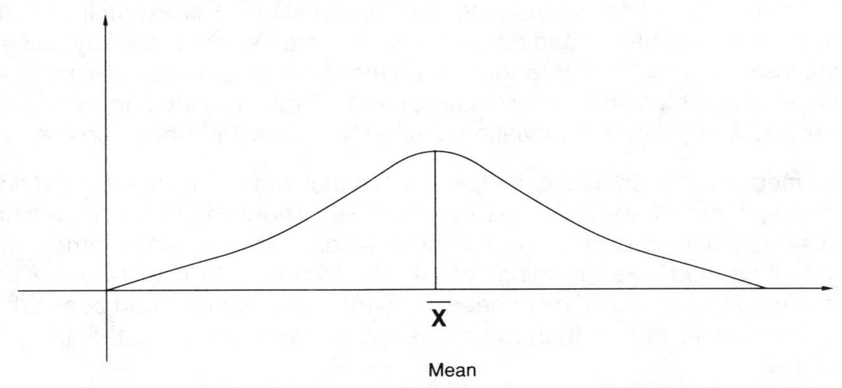

Figure 12-1. Normal distribution curve.

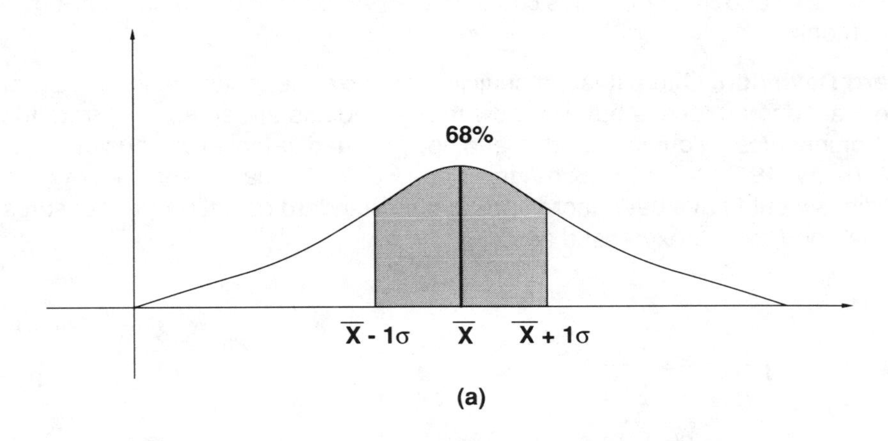

Figure 12-2. Percentage distribution as a function of σ spreads.

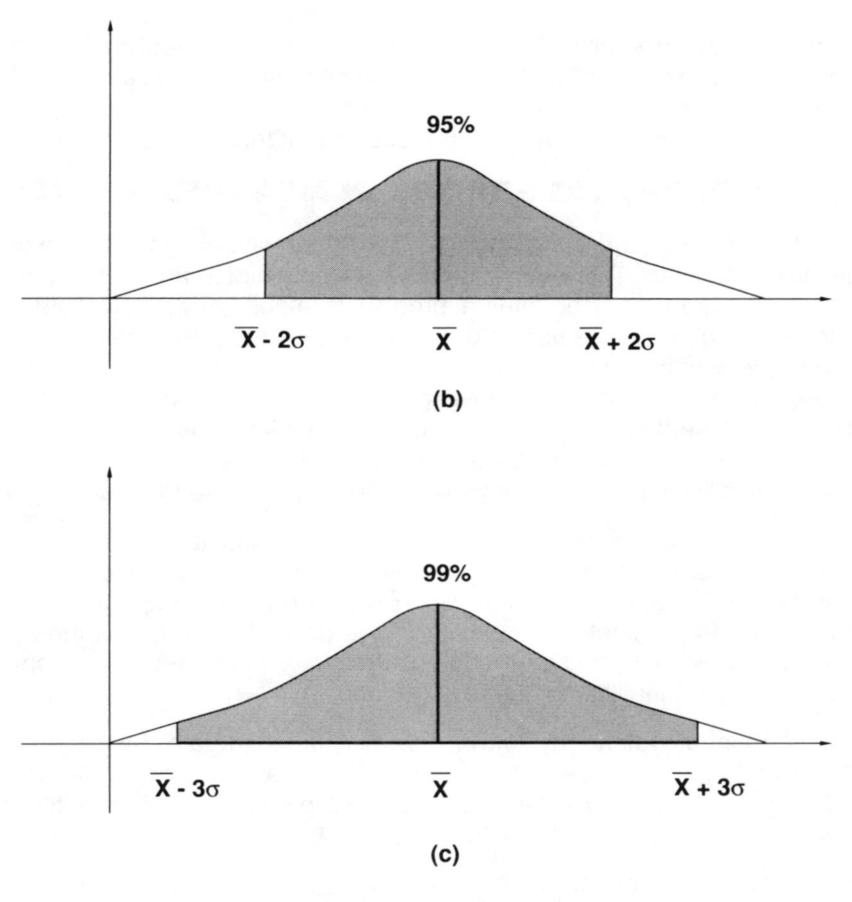

Figure 12-2. Continued.

The standard deviation in a Gaussian distribution measurement gives us the ability to make a more quantitative determination of how the data is spread. For example, if a set of reactor vessels in a continuous process has a temperature loop that maintains a 358°C temperature with a standard deviation of 40°C and another control loop keeps a 358°C temperature with a standard deviation of 20°C, we know that the latter provides us with a more peaked graph about the mean. In fact 68% of the temperature readings lie between 338°C and 378°C while in the other loop the readings lie between 318°C and 398°C.

Measurement Errors

Any time we are controlling a system which produces a finished good, the possibility of the introduction of error is present. This error can be introduced because of equipment malfunction or misreadings of process variables. A measurement error can be defined as the variation or deviation of a reading from the true or expected value. Errors can be classified in three categories:

- Gross Errors
- System Errors
- Random Errors

Gross errors are those resulting from human miscalculation, system errors are those resulting from the instrument itself or the environment, and random errors are those resulting from unexpected actions in the process line. Table 12-1 shows some of the examples and methods of estimation and reduction of these three error types.

12-2 INTERPRETING ERRORS IN MEASUREMENTS

Interpretation of errors and use of this information can be beneficial for the control of a machine or process. This interpretation can be obtained as an anticipation of the outcome (error prediction) or after a product is made (error detection). Error prediction is, of course, more useful, but in reality it is harder to implement. On the other hand, detection of an error after a product is made is easier to implement because a final product can be checked against a reference model that matches all specifications. Nevertheless, the latter is better to use than none at all. For instance, it is better to stop production of a machined piece because it does not meet the customer's specifications rather than ship to the customer a bad product.

Interpretation of error can be implemented using statistical analysis. This type of statistical data analysis is in fact part of the foundation of an artificial intelligence based system. The system continuously monitors the process, before and after, and adjusts production parameters accordingly. The data measurements the system performs are stored in what is referred to as the *global data base* (see Chapter 13 for its use in artificial intelligence based systems).

In automated control systems, system errors are usually generated by the controlling system and by the process itself. Random errors are created by unknown events that occur during the process. System errors can be predicted and corrected while random errors can be detected and acted upon.

System errors may be generated from several events forming a combined error composed of a mix of several propagation errors throughout the process. Other system errors are generated from guarantee errors which are provided to us, perhaps by our supplies or raw materials, before the process. Random errors must apply the science of statistical analysis in the detection and reaction ofthe induced error.

Interpreting Combined Errors

Combined errors can be defined as a final error caused by the interaction of two or more independent variables, each one causing a different error. This interaction is propagated through the system and therefore may also be called *propagation error.* By calculating statistical data of sample before propagation, and knowing that a final product sample must be within a specified average and standard deviation, we can predict the outcome of the final product and possibly make corrections throughout the process.

The probabilities of an outcome formed by several variables (e.g. materials going into a batching process) are directly related to the average value of each variable. For instance, if we have two ingredients, A and B, and the average weight is $\overline{A}$ and $\overline{B}$, then the final nominal mix value of both materials would be $\overline{A} + \overline{B}$. This outcome

	Gross Errors	System Errors	Random Errors
Example	Reading the incorrect scale. False instrument readings. Setting the zero adjustment improperly. Applying the wrong formula. Inexact recording of data. Incorrect instrument settings.	Instrument not calibrated properly. Not accounting for nonlinearities. Worn-out parts. Loss of power due to improper application of technique.	Changes in the environment such as pressure and temperature. Variation of materials used in a process line. Vibration in a conveyor line. Any disturbancy.
Prediction	No prediction can be drawn.	Compare readings to those of standard calculations. Determination of an error so that a cumulative error can be measured and expected.	Application of statistical analysis by data collection.
Prevention or Reduction	Paying more attention to correct displays. Having different persons take the same readings. Take multiple readings from an instrument. Be aware of instrument capabilities.	Monitoring the consistency of the technique. Emphasize regular maintenance of machinery and/or instruments. Making sure that instruments are calibrated on a regular basis.	Physical special provisions should be made to instruments or process line so that it can withstand disturbances. When certain types of error are anticipated, the use of statistical analysis can predict the best possible reading evaluation.

Table 12-1. Error types, examples and prevention.

is the addition of both $\overline{A}$ and $\overline{B}$ because the operation to be performed is a blending or mixing which implies that the quantities are added. The final outcome is related to the *equation* that governs that process being performed. In actual circumstances, the hardest thing to obtain is the real equation of the process, and usually it is only approximated.

The standard deviation tells how spread out from the mean the samples are located. The standard deviation of the outcome product is very beneficial because it can tell us in advance how the final product is spread out about its mean in relation to each of its component variables. In our blending example, the average weight outcome (W) is represented by:

$$W = \overline{A} + \overline{B}$$

where A and B are the two means of the ingredient products. If the distribution follows the normal (bell) curve, we can say that:

- 68% of all samples are $W \pm 1\sigma_w$ or $A + B \pm 1\sigma_w$
- 95% of all samples are $W \pm 2\sigma_w$ or $A + B \pm 2\sigma_w$
- 99% or all samples are $W \pm 3\sigma_w$ or $A + B \pm 3\sigma_w$

where σ_w is the standard deviation of the final product.

Now we have to find a relationship between the ingredients A and B, and the required σ_w. To obtain a form that can be used with two or more input variables, let's say that the equation that governs the making of the final product and/or process is defined by a function K.

After numerous sample observations (n^{th}) the final product formula will be of the form:

$$K_n = K\,(A_n, B_n)$$

Now we can conclude that the best possible value for the outcome of the function will be of the form:

$$K = K\,(\overline{A}, \overline{B})$$

where the final outcome is a function of the two averages A and B. Let's define any deviation of a sample observation from the mean as

$$\Delta K_n = K_n - K\,(\overline{A}, \overline{B}) \text{ or}$$

$$\Delta K_n = K(A_n, B_n) - K(\overline{A}, \overline{B})$$

if the deviation from the mean is 0 ($\Delta K_n = 0$), implying that the value of the nth observation were that of the mean, then we would have

$$K(A_n, B_n) = K(\overline{A}, \overline{B})$$

Based on differential calculus theory, we can transform the ΔK_n term into partial derivations as:

$$\Delta K_n = \frac{\partial K}{\partial A}\,\Delta A_n + \frac{\partial K}{\partial B}\,\Delta B_n$$

By taking the average value of the sum of the squares and performing the square root of the right hand term we have:

$$\sigma K = \sqrt{\left(\frac{\partial K}{\partial A}\right)^2 \sigma_A^{\,2} + \left(\frac{\partial K}{\partial B}\right)^2 \sigma_B^{\,2}}$$

where σK is the standard deviation of the final outcome, and σA and σB are the standard deviations of the independent variables A and B. The other terms are the partial derivative of the function. This term indicates that a prediction of the standard deviation of a function (product) can be approximated beforehand by knowing the standard deviations of the independent variables and the function of the process itself. The following example illustrates the uses of this function.

Example 12-1

Pellets, shaped as spheres, are being produced in a manufacturing plant. These pellets are heated for a period of time to accommodate specific changes in the sphere size. After numerous observations, quality control has determined that the radius has a mean R = 1.0 inches and a standard deviation of $\sigma_R = 0.0008$ inches. The pellet material weight has a value of W = 0.15 lbs/in^3 and a standard deviation of $\sigma_w = 0.00082$ lbs/in^3.

a) Find the probable sphere weight of the final product and its standard deviation. **b)** Make suggestions on how this information could be used.

Solution

a) The total weight (W_t) of the sphere can be calculated as volume x weight (V x W)

$$V = \frac{4}{3}\pi R^3$$

$$W_t = \frac{4}{3}\pi R^3 \cdot W$$

Therefore, the final total weight at normal process conditions would be:

$$W_t = \frac{4}{3} \times 3.1416X\ (1.0)^3 \times 0.15$$

$$W_t = 0.628 \text{ lbs}$$

The standard deviation is calculated using the formula:

$$\sigma_W = \sqrt{\left(\frac{\partial W_t}{\partial W}\right)^2 \left(\sigma_W\right)^2 + \left(\frac{\partial W_t}{\partial R}\right)^2 \left(\sigma_R\right)^2}$$

$$\sigma_W = \sqrt{(\tfrac{4}{3}\pi R^3)^2(\sigma_W)^2 + (4W\pi R^2)^2(\sigma_R)^2}$$

$$\sigma_W = \sqrt{(17.545)(6 \times 10^{-7}) + (3.553)(7 \times 10^{-7})}$$

$$\sigma_W = \quad 0.003607 \text{ lbs.}$$

b) The information calculated shows that, based on the samples performed for average radius (after) and average weight (before) of the produced part, the standard deviation of the finished product can be estimated at 0.206 lbs. If this value is within the required range specified by quality control, the product would be acceptable. On the other hand, if the value for average final weight is detected to be greater or smaller producing an unacceptable standard deviation, the process could be altered (e.g. more or less heat) to control the radius (by expansion) and therefore the shape of the sphere so that the weight is adjusted to within the desired standard deviation. The process adjustment would of course require that a definition of how much required heat is needed to alter the shape and size of the pellets. To implement this type of system, it would be necessary to have the ability of measuring samples during the manufacturing process via transducers and other equipment.

Interpreting Guarantee Errors

Guarantee errors are values that come from known specifications which define that a product or material will be between a specified arithmetic deviation from the mean. For example, if a supplier specifies that a metal part used in a manufacturing assembly line has a length of 26 centimeters and a guarantee deviation (error) of less than 0.1%, then its supplied parts are within 26 cm $\pm$ 0.026 cm. Moreover, if the manufacturer specifies a $\pm 3\sigma$ standard deviation from samples performed, it would mean that 99% of the parts will be within $\pm$0.026 cm arithmetic deviation from the mean of 26 cm.

To anticipate the possible value (outcome) using guarantee limits, we apply the arithmetic worst case to the nominal value. The following example illustrates how two variables are manipulated according to its guarantee values to obtain the outcome's worst case condition error tolerance.

Example 12-2

An electric heater is based on a resistance value of 150 ohms and a current control system. The resistor has a guarantee deviation of $\pm$0.15% of total resistance; while the current, being controlled from a PLC's analog output, has a $\pm$0.1% guarantee limit at 4.5 amps. Find the nominal power at the heater and the error (deviation from the true mean) as guaranteed by the limits.

Solution

The equation which describes the power dissipation is:

$$P = I^2R$$

where I is the current flowing through resistance R. The guarantee limits of the component are:

$$I = 4.5 \text{ amps} \pm 0.1\% = 4.5 \pm 0.0045 \text{ amps}$$

$$R = 150 \text{ ohms} \pm 0.15\% = 150 \pm 0.225 \text{ ohms}$$

The variation of power caused by each of the guarantee error limits are ΔP_I and ΔP_R (power variations due to current and resistance respectively).

$$\Delta P_I = \frac{\partial P}{\partial I}\Delta I = 2IR\,\Delta I$$

$$= 2(4.5)(150)(\pm 0.0045)$$

$$= \pm 6.075 \text{ watts}$$

$$\Delta P_R = \frac{\partial P}{\partial R}\Delta R = I^2 \Delta R$$

$$= (4.5)^2(\pm 0.225)$$

$$= \pm 4.556 \text{ watts}$$

The nominal power calculation is:

$$P = I^2R = (4.5)2(150) = 3037.5 \text{ watts}$$

Adding the power variations to the nominal value (arithmetic worst case) we get the expected worst case value:

$$P = P_{nominal} \pm \Delta P_I \pm \Delta P_R$$

$$= 3037.5 \pm 6.075 \pm 4.556 \text{ watts}$$

$$= 3037.5 \pm 10.631 \text{ watts}$$

$$= 3037.5 \text{ watts} \pm 0.35\%$$

The outcome power changes based on guarantee variable errors is 0.35% of the total power.

12-3 IMPLEMENTATION OF TRANSDUCER MEASUREMENTS

This section deals primarily with two measuring techniques or methods which are used in the implementation of transducer circuits. A knowledge of the inner workings of transducer measurement circuits will give you a better perspective of not only how they are used, but also where functional error may be encountered when measurement problems arise. These two techniques involve the use of *bridge circuits* and *linear variable differential transformer* mechanisms (LVDT).

Either of these two techniques can be used to implement what we define as a transducer. For instance, to detect pressure and changes in pressure, we may use a strain gauge which is based on the bridge circuit technique or we could use a Bourdon tube which is based on the LVDT mechanism technique.

Bridge Circuits Techniques

Bridge circuits are used primarily when the changing parameter in the measurement is a resistive element. The resistance changes in the bridge circuit, depending on how the bridge is configured, will provide voltage or current changes in its output proportional to the changes in the resistive measurement element. Figure 12-3 illustrates a bridge circuit.

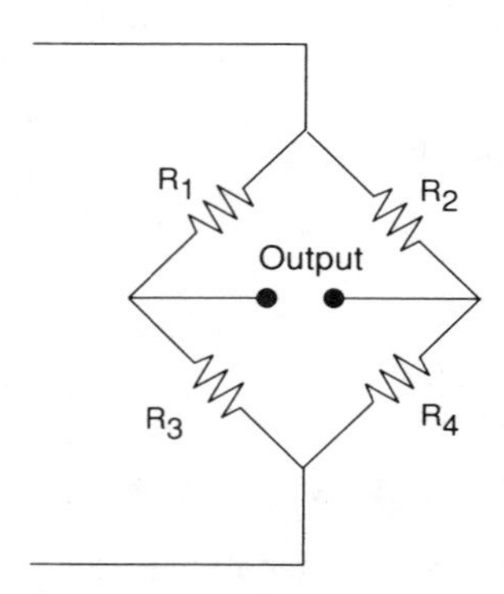

Figure 12-3. Simple bridge configuration.

Voltage Sensitive Bridge. The voltage sensitive bridge, as its name implies, senses a voltage differential at the output of the bridge which in turn will be proportional to the resistance change in the bridge. This resistance change generally creates what is called a *bridge imbalance*. Figure 12-4 illustrates the voltage sensitive bridge, where D is the detector device and R_D is its resistance. The value of R_D for a voltage sensitive bridge is very high. Such resistance could be the input impedance of an amplifier module of a PLC. The following example illustrates the bridge relationship between the resistors in the bridge. Note that the bridge imbalance is created by the changes in resistance of R_4 (measuring element); other resistors have fixed known values.

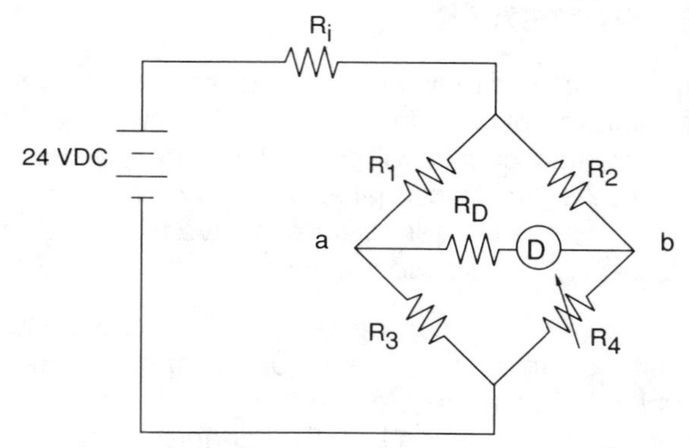

Figure 12-4. Voltage sensitive bridge configuration.

Example 12-3

For the voltage sensitive circuit shown in Figure 12-4, a) find the equation that describes the voltage differential measurement between point a and point b, and b) the bridge resistance ratio when the voltage differential is zero (balance state).

Solution

a) Assuming that $R_D = \infty$ (very large) and the exitation voltage impedance $R_i = 0$, the voltage at point Va and Vb are:

$$Va = \left(\frac{R_3}{R_1 + R_3}\right) V$$

$$V_b = \left(\frac{R_4}{R_2 + R_4}\right) V$$

The voltage difference between point a and point b is:

$$V = V_a - V_b$$

$$V = V\left(\frac{R_3}{R_1 + R_3} - \frac{R_4}{R_2 + R_4}\right)$$

$$V = V\left(\frac{R_2R_3 - R_1R_4}{(R_1 + R_3)(R_2 + R_4)}\right)$$

b) When the differential voltage $\Delta V = 0$, we get:

$$0 = R_2R_3 - R_1R_4$$

$$R_1R_4 = R_2R_3$$

therefore, the bridge resistance ratio is:

$$\frac{R_1}{R_2} = \frac{R_3}{R_4}$$

where R_4 is the measuring resistance element.

Current Sensitive Bridge. The current sensitive bridge creates a current flow change through the output of the bridge, i.e. between point a and point b (refer to Figure 12-4). The current flow is the result of bridge imbalances created by resistance changes in the measuring element. Other resistors in the bridge are of known and fixed values.

When current changes are being measured, the detecting device (D) has a very low resistance R_D. Therefore allowing current to flow from point a to b through the detector. Typical devices that have very low impedance include galvanometers and low input impedance current amplifier interfaces (PLC modules).

The equation that describes the current that flows through the detector due to bridge imbalances is defined by:

$$I_D = \frac{V R_4}{[R_i (1 + \frac{R_2}{R_1}) + R_2 + R_{4B}]\ [R_D (1 + \frac{R_3}{R_1}) + R_3 + R_{4B}]}$$

The term R_{4B} is the resistance value when the bridge is balanced. The term ΔR_4 is the change (absolute value) between R_{4B} and the new resistance value due to the measuring element change. The following example illustrates how this equation is used to obtain a current proportional to the change in resistance.

Example 12-4

A thermistor with a nominal resistance of 10KΩ is being employed in a bridge configuration to measure small changes in temperature (see Figure 12-5). An amplifier input module is being used to measure the small changes in current and has an input impedance of 300 Ω. What is the current to be measured if the changes in temperature equate to 10% change in resistance?

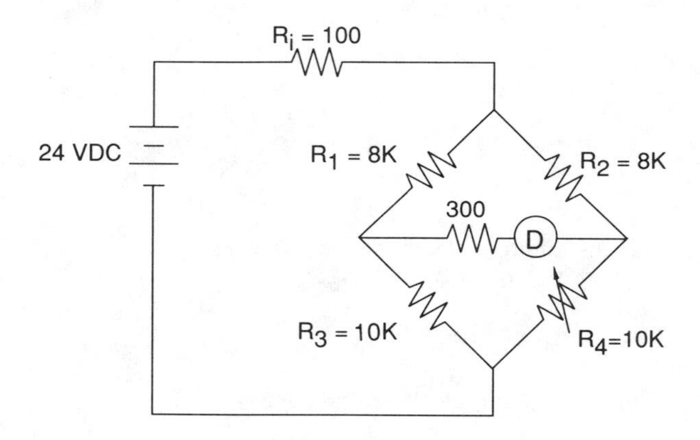

Figure 12-5. Thermistor used in a bridge circuit.

Solution

The value of the thermistor resistance due to the temperature change is 10% which will translate into an R4 value of 10KΩ + 1KΩ or 11KΩ. The difference in thermistor resistance is calculated as:

$$\Delta R_4 = |R_4 - R_{4B}|$$

$$= |11K\Omega - 10K\Omega|$$

$$= 1K\Omega$$

The current measurement is defined by:

$$I_D = \frac{V R_4}{[R_i (1 + \frac{R_2}{R_1}) + R_2 + R_{4B}]\ [R_D (1 + \frac{R_3}{R_1}) + R_3 + R_{4B}]}$$

$$I_D = \frac{24 \times 1K}{[100\ (1 + \frac{8K}{8K}) + 8K + 10K]\ [300\ (1 + \frac{10K}{8K}) + 10K + 10K]}$$

$$I_D = \frac{24K}{(18.2K)\ (20.675K)} = 0.06378 \text{ mA}$$

LVDT Techniques

The Linear Variable Differential Transformer (LVDT) is an electromechanical mechanism which provides a voltage reference which is proportional to a movement or displacement of a core inside of a coil. Figure 12-6 illustrates a cutaway of an LVDT.

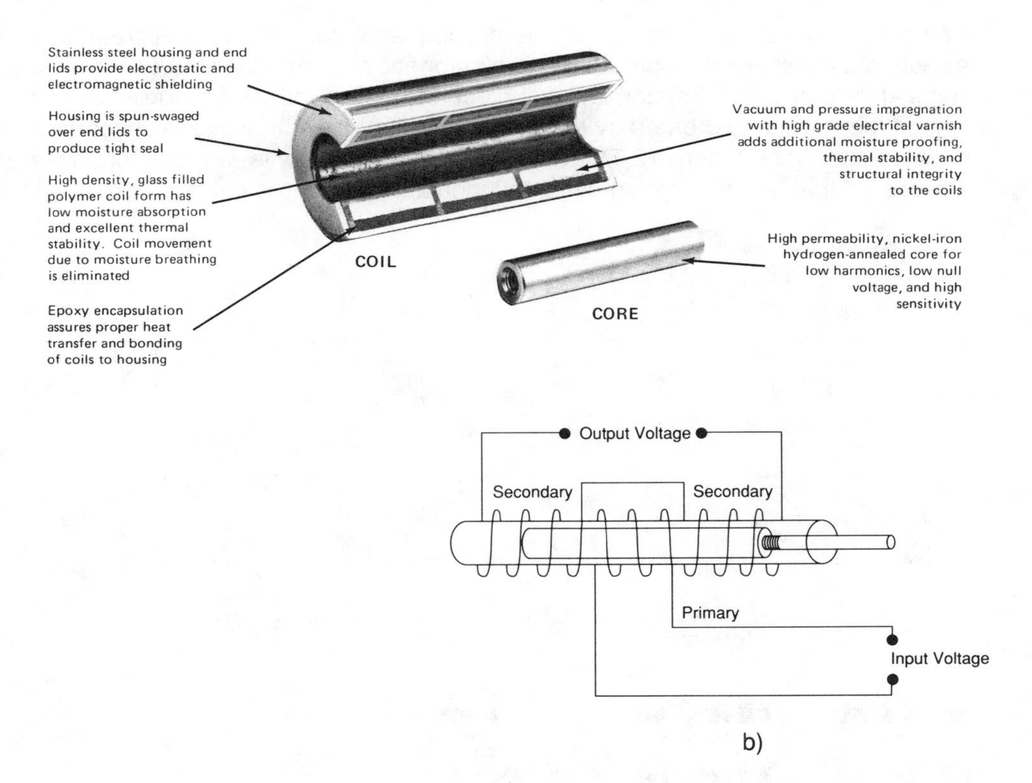

Figure 12-6. a) Cutaway of LVDT; b) diagram of LVDT. (Courtesy of Schaevitz Engineering. Pennsauken, NJ).

Transducer	Uses or Measurements
Load Cell	Torque, force, weight, tensile testing
Bourdon Tube	Pressure, fluid pressure, volume porosity
Bellows, Diaphragm	Pressure, fluid pressure, low range pressure measurements
LVDT Lineal	Lineal displacement, level measurement
Manometers	Measuring height of a column of a liquid of known density
LVDT Gauge	Heads Machine tool, mechanical dial position indicators, gauging and displacement (very small) measurements
Accelerometer	Acceleration in one or more exes, servo positioning applications, scismic systems
Inclinometers	Level sensing, incline
Variable Reluctance Proximity	Proximity detection of ferromagnetic objects

Table 12-2. Transducers that may use LVDT mechanisms.

The LVDT principal mechanism is used in the implementation of many transducers. These transducers transform the measurement variable or measurement (e.g. pressure) into a linear actuation which is attached to the LVDT's core. Therefore, the measurement will be proportional to the output voltage of the LVDT. Table 12-2 shows different transducers which could use the LVDT mechanism. Let's now take a short look at the operation of the LVDT mechanism.

When the primary coil has an AC voltage applied, an induced voltage is created in the secondary coils. As the core (made of a magnetic material) of the LVDT moves, voltage at the output of the secondary coil will change. This voltage change is caused by the induced voltage created by the core movement and the way the secondary coils are wound (see Figure 12-7). Each coil in the secondary is wound in opposite directions so that, as the cores moves, the induced voltage will change polarity.

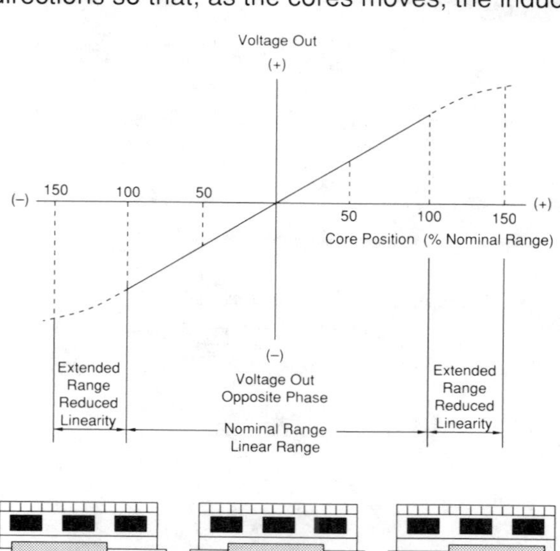

Figure 12-7. LVDT core movement and output voltage. (Courtesy of Schaevitz Engineering, Pennsauken, NJ).

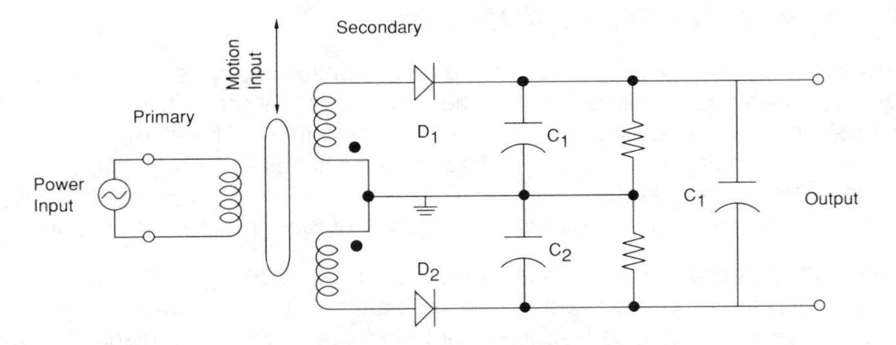

Figure 12-8. Demodulator circuit used to rectify the LVDT output signal.

Modern LVDTs provide demodulation or rectification circuits to convert the secondary output into a DC voltage signal. This voltage is, of course, proportional to the core movement within its linear range. The resultant voltage when the core is at its starting position is +V and when it is at the end position is -V. When the core is at the middle, a null or zero volt output is created. Figure 12-8 illustrates a simple demodulator or rectifier circuit for an LVDT.

Example 12-5

Represent graphically the position of an LVDT core which has a range of 20 inches total displacement and an output of ±10VDC.

Solution

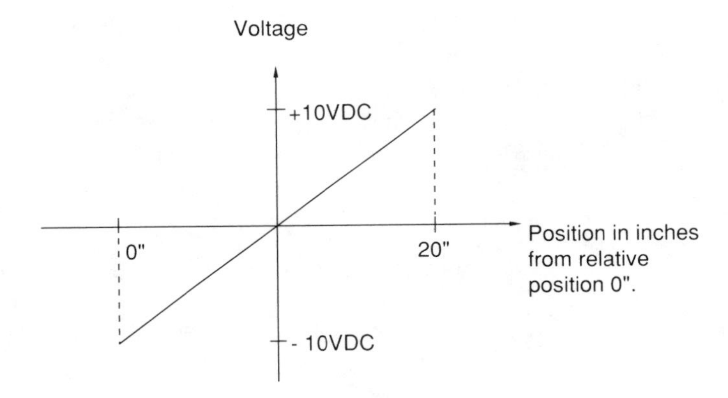

Core Position	Linear Position	Voltage Output
Core at -100%	0"	- 10 VDC
Core at 0% (null)	10"	0 VDC
Core at +100%	20"	+10 VDC

12-4 THERMAL TRANSDUCERS

Thermal transducers are used to sense and monitor changes in temperatures. These temperature changes may be caused by the process itself (exothermic process) or by induced heat/cool process control inputs. There are two primary subdivisions of thermal transducers. The first one is based on internal *resistance* changes due to temperature variations; the second one is based on *voltage* differential—by the transducer itself—as a result of the temperature variations.

These transducers provide at their output, after conditioning, voltage or current signals proportional to the temperature measurement range. Depending on the transducer and the PLC used, special input modules or analog input interfaces are used to input the temperature data to the controller. An understanding of thermal transducer operation will give you an inside look on how and where to use them in a process control application. There are many types of thermal transducers in the marketplace; however, we are going to cover the ones most commonly used which are *Resistance Temperature Detectors (RTDs), Thermistors* and *Thermocouples.* The RTD and thermistor are grouped under the internal resistance subdivision while the thermocouple belongs to the voltage differential group.

Resistance Temperature Detectors (RTD)

RTDs are temperature transducers made of wire conductive elements. The most common elements used are platinum, nickel, copper, and nickel-iron. These conductive wires are coiled around an insulator which serves as support, and covered with a protective sheath material (protecting tube). Figure 12-9 illustrates a simple construction diagram of an RTD. In an RTD, the resistance of the conductive wires *increases* linearly with increases in the temperature being measured; for this reason, RTDs are said to have a *positive temperature coefficient.*

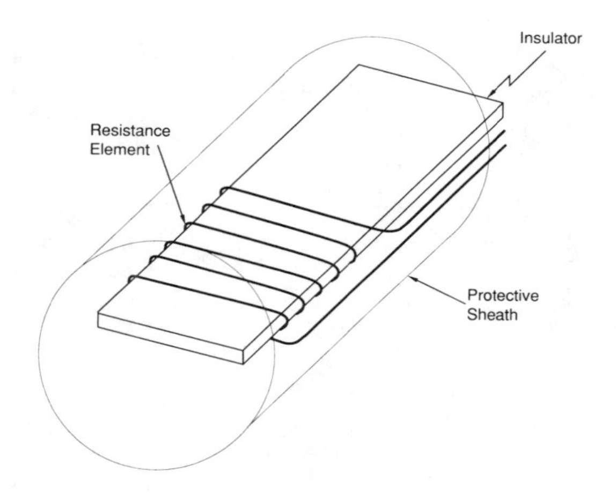

Figure 12-9. RTD construction diagram.

RTDs are generally used in a bridge circuit configuration (signal conditioning). Figure 12-10 illustrates an RTD in a bridge circuit. As mentioned in the previous section, the bridge circuit will give an output proportional to the changes in resistance. Since the RTD is the variable resistor in the bridge (reacting to temperature changes) the bridge output will be proportional to the temperature being measured.

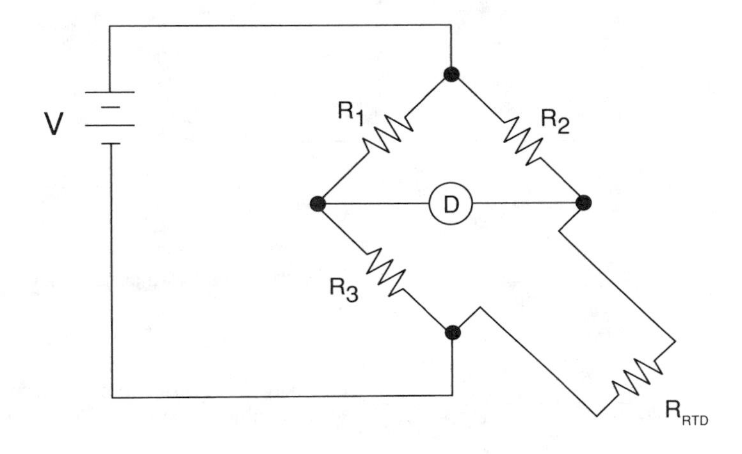

Figure 12-10. RTD in a bridge circuit configuration.

As can be seen in Figure 12-10, the RTD element may be located away from the bridge circuit—its signal conditioner. The user must be aware of the lead wire resistance created by the wire connecting the RTD with the bridge circuit. The lead wire resistance causes the total resistance in the RTD arm of the bridge to increase since the lead wire resistance is added to the RTD resistance. If proper compensation is not provided to the RTD circuit, a measurement error will be present.

Figure 12-11 illustrates a typical wire compensation method used to balance lead wire resistance. Referencing Figure 12-11, the lead resistance R_{L1} and R_{L2} are identical because they are made of the same material. These two resistances R_{L1} and R_{L2} are added to R_2 and to R_{RTD} (the RTD) respectively so that the wire resistance is added to two adjacent sides of the bridge, therefore compensating for the resistance of the lead wire in the RTD measurement. The equation of the bridge before and after compensation is described on the following paragraph. Note that R_{L3} has no influence on the bridge circuit since it is connected to the detector (e.g. input module, amplifier, etc.).

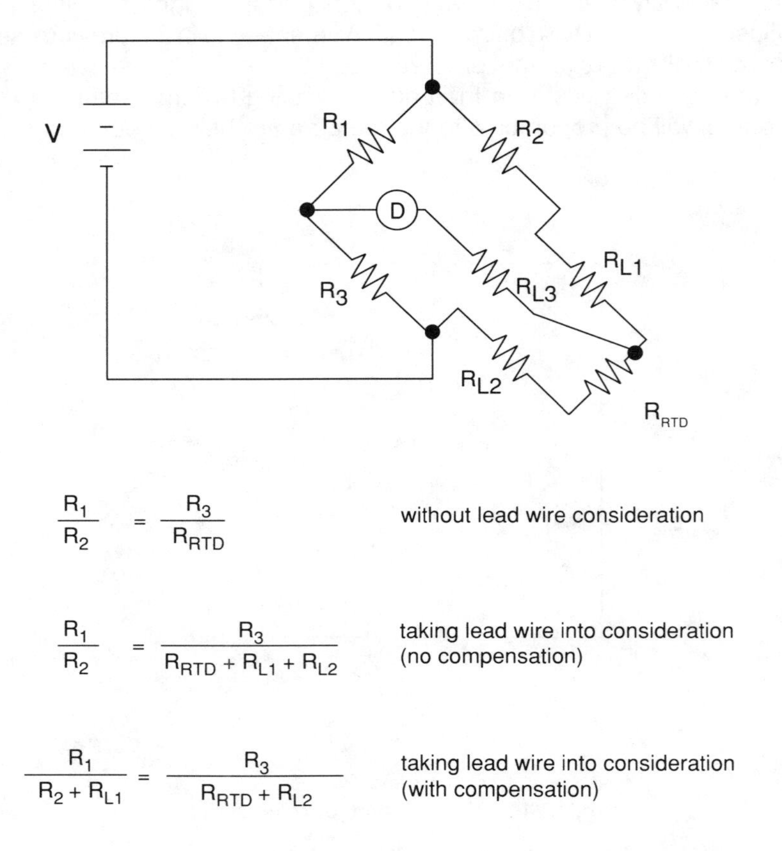

Figure 12-11. RTD bridge configuration with lead wire compensation.

The changes in RTD resistance are of course proportional to changes in temperature. These resistance changes are based in the following equation:

$$R_T = R_{To} [1 + \alpha_1(T - T_0) + \alpha_2(T - T_0)^2]$$

where R_T is the change in resistance at temperature T, R_{To} is the RTD resistance at a reference temperature point T_0 (e.g. Copper 10Ω at 25°C) and α_1 and α_2 are constants per degree °C that vary with the RTD material.

When RTDs are connected to a PLC's RTD input module, the interface computes the temperature (T) directly for the changes in resistance R_T. All the table values based on the equation for the RTD's type inputs are stored in the module. Lead wire compensation connections are also provided in the RTD input module.

If an RTD is used with a standard analog input module, the user must design the bridge circuit as well as an amplifier to get the signal to match that of the input module range (e.g. 0 to 10 VDC). The computation of temperature would have to be performed in the PLC by obtaining the linear temperature versus voltage curve utilizing the resistance versus temperature curve and computing the temperature using the Temperature vs. VDC equation or look-up table linear interpolation for the input count value analog input voltage. This technique could be used with any transducer which uses the bridge technique of signal detection (e.g. thermistors, strain gauges, etc.). If the transducer is linear in its temperature detection versus resistance, an equation can be used to compute the temperature in the PLC. If the transducer is not linear in its temperature detection range, linear interpolation must be performed by using a look-up table. Chapter 7 deals with the handling of linear equations in analog readings while examples of linear interpolations of analog readings are presented in Chapter 11. A simple interpolation example is presented in the thermocouple section.

Thermistors

Like RTDs, thermistors are temperature transducers which exhibit changes in internal resistance that are proportional to changes in temperature. Thermistors are made of semiconductor material such as oxides of cobalt, nickel, manganese, iron and titanium. Figure 12-12 illustrates several thermistors. These semiconductor materials experience an opposite temperature versus resistance behavior as those of the RTD conducting materials. As the temperature increases, the resistance of the thermistor decreases; therefore the thermistor is said to have a negative temperature coefficient. Although most thermistors have negative coefficients, it is possible to find thermistors with a positive temperature coefficient.

Thermistors can be classified in two major groups: bead thermistors and metallized surface thermistors. Under each of these two classifications fall several types of thermistors as shown in Table 12-3. Each of these two groups of thermistors offers advantages and disadvantages as shown in Table 12-4.

Bead Type Thermistors	Metallized Surface Contact Thermistor
Bare beads	Discs
Glass-coated beads	Chips
Glass probes	Flakes
Glass rods (axial lead probes)	Rods
	Washers
Bead-in-glass Tube or enclosure	Wafers

Table 12-3. Classification of thermistors. (Courtesy of Thermometrics, Inc., Edison, NJ).

Type	Advantages	Disadvantages
Bead Type Thermistors	• Good to excellent stability, Leadwires, are strain relieved in glass hermetic seal. • High operating and storage temperatures. • Smaller sizes available. • Fast response times.	• Normally broad resistance tolerances, high cost for close tolerances • Medium to low dissipation constants. • Matched pairs or resistive padding are required for interchangeability.
Metallized Surface Contact Type Thermistors	• Normally tighter tolerances, lower cost for close tolerances. • Low cost, single units for interchangeability. • Medium dissipation constants.	• Moderate to good stability. Difficult to obtain high stability without hermetic seal. • Limited operating and storage temperatures. • Medium sizes available. • Medium response times.

Table 12-4. Advantages and disadvantages of thermistor types. (Courtesy of Thermometrics Inc., Edison, NJ).

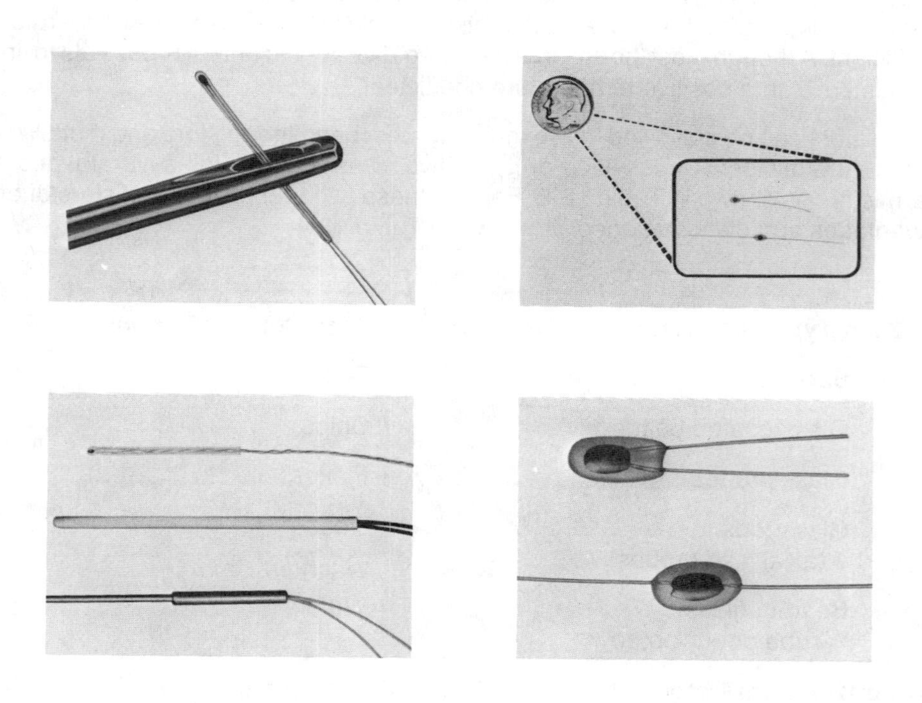

Figure 12-12. Different types of thermistors. (Courtesy of Thermometrics Inc., Edison, NJ).

Thermistors experience a much greater change in resistance than RTDs. Figure 12-13 illustrates a graph of temperature versus resistance ratio (R_T/R @ 25°C, R_T is resistance at temperature T). As can be seen, there are large changes of resistance with relatively small increases in temperature. An advantage to these abrupt changes in resistance is that for certain temperature ranges, the thermistor can provide better resolution. Therefore, thermistors provide more accurate readings when the span of measurement is narrow or small. For instance, a thermistor with 1MΩ at 25°C will have a resistance of approximately 300KΩ at 50°C, giving a 700KΩ for a 25°C span (28KΩ change per °C). This type of resolution may be required in energy management applications where narrow temperature spans are provided and accurate control measurements are required.

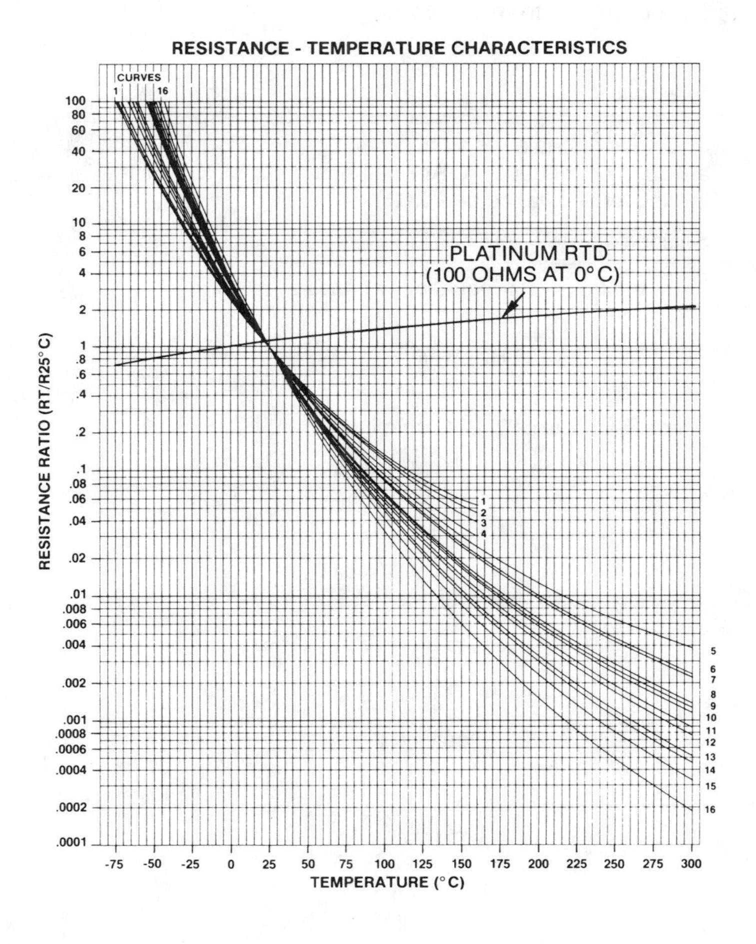

Figure 12-13. Temperature vs. resistance curve (R_T/R @ 25°C). (Courtesy of Thermometrics Inc., Edison, NJ).

Thermistors' resistance as a function of temperature can be defined by:

$$R_T = R_O\, e^{\beta\left(\frac{1}{T} - \frac{1}{T_O}\right)}$$

where R_T is the thermistor resistance at absolute temperature T (°C +273), T_0 is the temperature reference (absolute °K). And β is constant (in degrees absolute) that depends on the type of thermistor used (between 3400 to 4000). As can be seen from the equation and from the resistance ratio graph (Figure 12-13) the resistance changes are proportional to the natural logarithm of its temperature, thus indicating rapid changes in resistance for the changes in temperature. When using thermistors, the linear area to be used as a temperature measurement range should not exceed around 100° to 150°C so that linearity can be maintained. Table 12-5 illustrates a table comparison between thermistors and RTDs.

Example 12-6

A thermistor has a resistance value of 100KΩ at 3 °C and has a β constant of 3900 °K. At what temperature (°C) will the thermistor resistance be 20KΩ?

Solution

From the thermistor equation we have that:

$$R_T = R_O\, e^{\beta\left(\frac{1}{T} - \frac{1}{T_O}\right)}$$

where T_0 is 3°C or 276°K (absolute).

$$R_T = 100K\, e^{3900\left(\frac{1}{T} - \frac{1}{276}\right)}$$

$$20K = 100K\, e^{3900\left(\frac{1}{T} - \frac{1}{276}\right)}$$

$$\frac{1}{5} = e^{3900\left(\frac{1}{T} - \frac{1}{276}\right)}$$

$$\ln 0.2 = 3900\left(\frac{1}{T} - \frac{1}{276}\right)$$

$$\frac{1}{T} = \frac{-1.61}{3900} + \frac{1}{276}$$

$$T = 311.5°K$$

Therefore the temperature in degrees C will be:

$$311.5 - 273 = 38.5°C.$$

Type	Advantages	Disadvantages
Thermistors	• Fast Response • Small size • High resistances eliminate most lead resistance problems • Rugged, not affected by shock or vibration • Lower cost.	• Nonlinear • Narrow span for any singla input • Interchangability is limited unless matched pairs are used
RTD	• Linear over wide operating range • Wide temperature operating range • High tempeature operating range • Interchangeable over wide range • Better stability at high temperature	• Low sensitivity • Higher cost • No point sensing • Affected by shock and vibration • Requires 3 or 4 wire wire operation • Can be affected by contact resistance

Table 12-5. Comparison table between thermistors and RTDs. (Courtesy of Thermometrics Inc., Edison, NJ).

Thermocouples

Thermocouples are temperature measuring devices which are made of two different metals (bimetallic). The thermocouple construction, shown in Figure 12-14, is such that when the two metals are joined together at different temperatures, a voltage differential is created. This voltage generation is known as the Seebeck effect. Temperature T_1, hot junction, is the temperature being measured while T_2, or cold junction, is the reference temperature. As temperature T_1 increases, the voltage differential (emf) between material A and B will also increase in proportion to the temperature. Table 12-6 shows a thermocouple comparison, recommended environmental applications, and types of materials used.

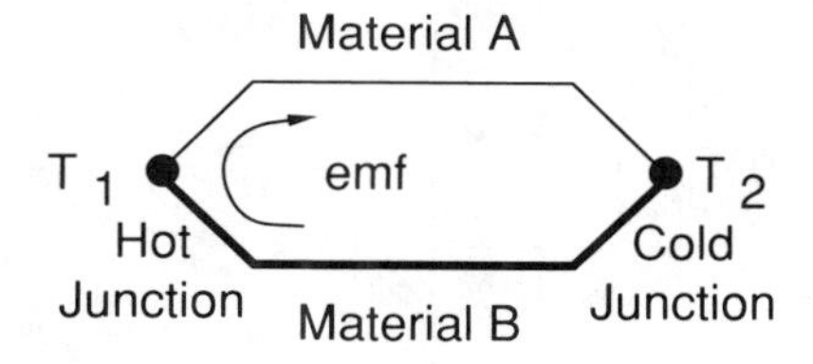

Figure 12-14. Thermocouple diagram.

ISA Type Designation	Positive wire (Numbers = Percentages)	Negative wire (Numbers = Percentages)	Millivolts per °F*	Recommended Temp °F* Min.	Recommended Temp °F* Max.	Scale Linearity	Atmosphere Environment Recommended	Favorable Points	Less Favorable Points
B	Pt70-RH30	Pt94-Rh6	.0003-.006	32	3380	Same as for type R couple	Inert or slow oxidizing	-	-
E	Chromel	Constantan	.015-.042	-320	1830	Good	Oxidizing	Highest emf/°F	Larger drift than other base metal couples
J	Iron	Constantan	.014-.035	-320	1400	Good; nearly linear from 300-800	Reducing	Most economical	-
K	Chromel	Alumel	.009-.024	-310	2500	Good; most linear of all T/C	Oxidizing	Most linear	More expensive than T or J
R	Pt87-Rh13	Platinum	.003-.008	0	3100	Good at high temps. Poor below 1000°F	Oxidizing	Small size, fast response	More expensive than type K
S	Pt90-Rh10	Platinum	.003-.007	0	3200	Same as R	Oxidizing	Same as R	More expensive than type K
T	Copper	Constantan	.008-.035	-310	750	Good but crowded at low end	Oxidizing or Reducing	Good resis. to corrosion from moisture	Limited Temperature
Y	Iron	Constantan	.022-.033	-200	1800	About same as type J	Reducing	-	Not Industrial standard
-	Tungsten	W74-Re26	.001-.012	0	4200	Same as R	Inert or vacuum	High temperature	Brittle, hard to handle, expensive
-	W94-Re6	W74-Re26	.001-.010	0	4200	Same as R	Inert or vacuum	Same as above	Slightly less brittle than above
-	Copper	Gold-Cobalt	.0005-.025	-450	0	Reasonable above 60°K	-	Good output at very low temp.	Expensive lab. type T/C
-	Ir40-Rh60	Iridium	.001-.004	0	3800	Same as R	Inert	-	Brittle, expensive

$^{*}{}^{\circ}C = \frac{{}^{\circ}F - 32}{1.8}$

Table 12-6. Thermocouple Comparisons. (From Instrument Engineers' Handbook: Process Measurement by B.G. Liptak and K. Venczel. Copyright 1982. Reprinted with permission of the publisher, Chilton Book Company, Radnor, PA).

The reference junction (cold junction) is specified at 0°C. Therefore, all standard thermocouple tables depicting the thermocouple voltage output are based on the 0°C cold junction reference. Figure 12-15 illustrates graphically how to keep the reference junction at 0°C. Although in theory the reference junction should be at 0°C (standard), its use in industrial applications it is certainly not practical. Therefore, to implement thermocouple readings based on the standard tables (at 0°C reference), cold junction compensation must be performed. To see how compensation is performed, let's look at an example.

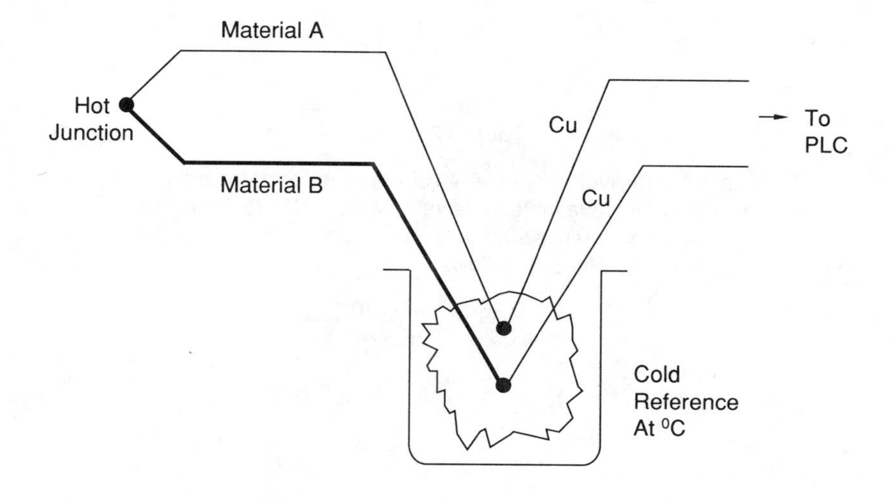

Figure 12-15. Thermocouple kept at 0 °C (ice bath).

Example 12-7

A Chromel-Constantan (type E) thermocouple has a reading of 16.42 mV at its output. The reference junction is at 46°F (Figure 12-16). Find the temperature T at the hot junction (temperature being measured). Utilize compensation to obtain the correct temperature reading.

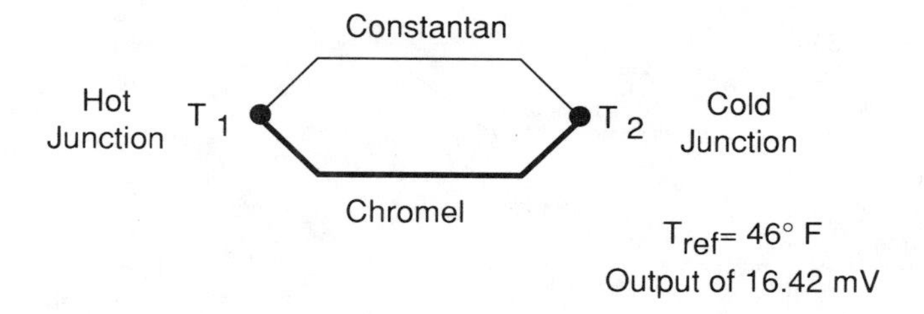

Figure 12-16. Constantan and Chromel thermocouple diagram.

Solution

Since the reference temperature is not at 0°C, compensation for a temperature reading must be performed. To obtain the compensation we must first go to the type E thermocouple table (referenced at 0°C, 32°F) and get the millivolt reading for 46°F. The values we get are shown in Table 12-7

Temperature	Millivolt Output
40°F	0.26 mV
46°F	?
50°F	0.59 mV

Table 12-7

To obtain the millivolt output at 46°F we must perform a linear interpolation of the available table values. Referring to Figure 12-17, we obtain the interpolated value by finding the proportional ratio of the listed values as follows:

$$\frac{X - 0.26}{0.59 - 0.26} = \frac{46 - 40}{50 - 40}$$

$$\frac{X - 0.26}{0.33} = \frac{6}{10}$$

$$X = (0.33)\,(0.6) + 0.26$$

$$X = 0.458 \text{ mV}$$

The value 0.458 mV is an offset value that must be added to the reading of the thermocouple output to compensate for cold junction readings. The total reading should be:

$$\text{Output} = 16.42 \text{ mV} + 0.458 \text{ mV}$$

$$= 16.878 \text{ mV}$$

(cold junction compensated)

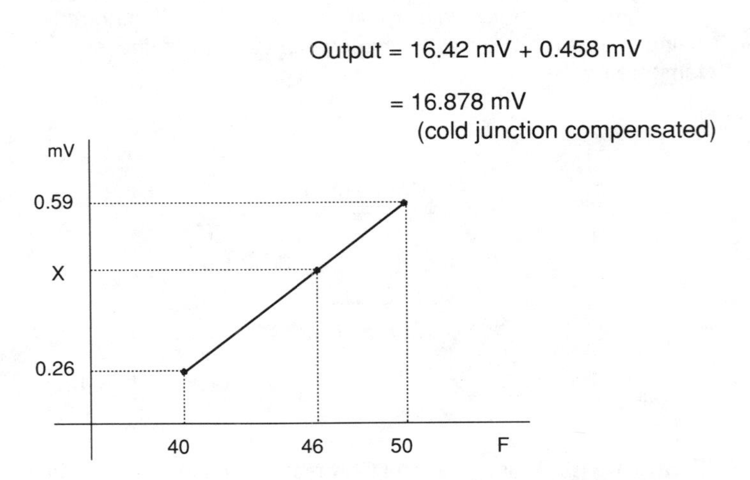

Figure 12-17. Interpolation for compensation example.

To obtain the temperature for a 16.878 mV we go back to the type E thermocouple table and find the closest value in the 16.878 mV range. These values are shown in Table 12-8.

Temperature	Millivolt Output
470°F	16.68 mV
?	16.878 mV
480°F	17.10 mV

Table 12-8

Again, since there is no exact value for the millivolt reading, linear interpolation of the two known table values must be performed. The temperature T is calculated as:

$$\frac{470 - 480}{16.68 - 17.10} = \frac{470 - T}{16.68 - 16.878}$$

$$\frac{-10}{-0.42} = \frac{470 - T}{-0.198}$$

$$T = (0.198)(23.809) + 470$$

$$T = 474.71 \text{ °F}$$

(hot junction temperature with cold junction compensation)

Thermocouple connections are implemented using lead wires as shown in Figure 12-18. Optimum lead wires are made of the same thermocouple material to maintain the thermocouple characteristics and avoid lead wire resistance. In practice however, the lead wires used are copper wires since the thermocouple material wires are too expensive and distances are generally long. Copper wires provide low resistance, thus minimizing lead wire resistance.

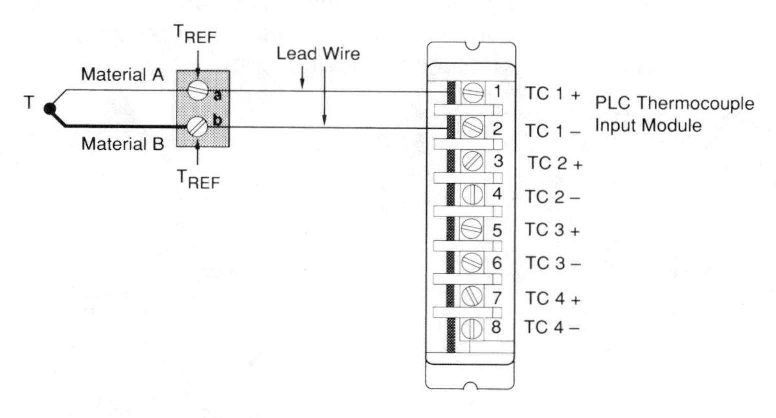

Figure 12-18. Thermocouple connection to a PLC module.

The temperature at the reference points (a and b in Figure 12-18) must be maintained at the same value. Therefore, special shielded cable must be used so that the temperature in the cable materials is kept the same all the way to the PLC input. PLCs that have thermocouple input interfaces provide cold junction compensation at the module and computes the necessary temperature adjustments. The reference temperature in the module is usually read via a thermistor since the span of the reference temperature is rather narrow. The PLC essentially performs the computations described in Example 12-7. All the thermocouple tables are stored in the module's memory.

Increases in thermocouple resolution can be accomplished by connecting several thermocouples in series thus forming what is known as *thermopiles*. Thermopiles generate larger output voltages, therefore reducing sensitivity requirements for the measuring device. Figure 12-19 illustrates a thermopiling arrangement. The voltage measurement of the thermopile will be the total voltage sum at the three hot junctions and the average will be that value divided by three. Note that each of the thermocouples must be isolated from one another to avoid thermal reaction among them as well as to avoid a thermocouple emf short. The reference cold junctions must also be maintained at the same temperature T_{Ref}.

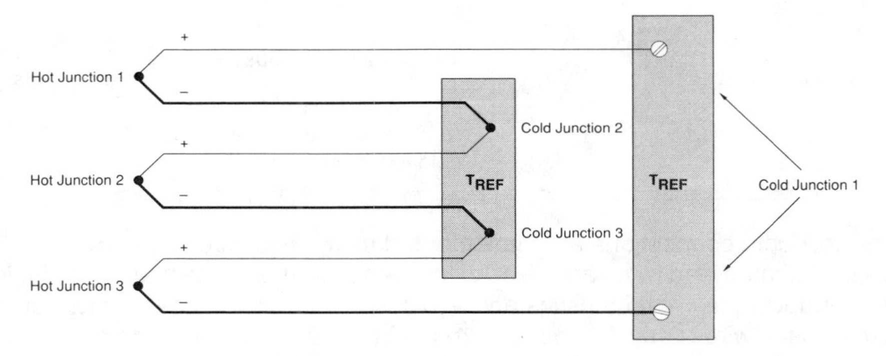

Figure 12-19. Example of thermopile arrangement.

Other thermocouple group configurations allow the measurement of temperature difference as well as direct average readings from several thermocouples. An illustration of these two configurations is shown in Figure 12-20.

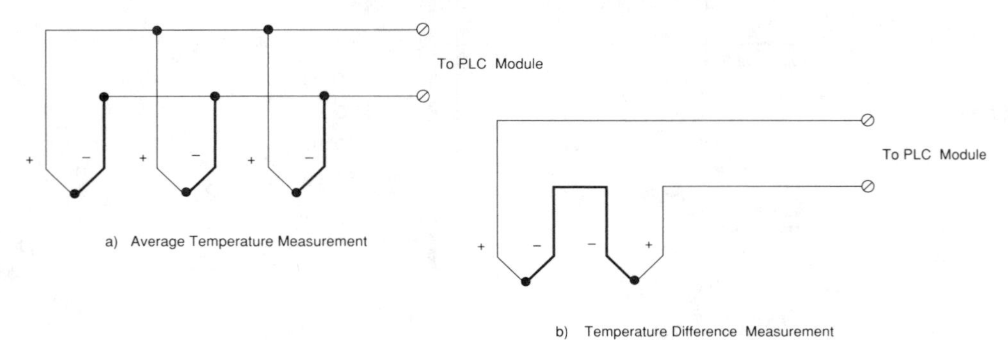

Figure 12-20. Thermocouple configurations a)average temperature and b) temperature differential.

Example 12-7

Three thermocouples type B are connected in a thermopile configuration and will be input to a thermocouple input module in a PLC system. Illustrate the thermopiling configuration and connection to the interface and the type of cable used. The distance from the 2000°F furnace where the thermocouples are connected to the module is 300 feet. Cold junction compensation is provided at the interface by the module's measuring thermistor.

Solution

Figure 12-21 illustrates a simplified diagram of the wiring configuration. The cable used is shielded cable type B thermocouple material and it is brought all the way to the module since the cold junction compensation is performed at the module. Please note that all reference temperatures should be the same.

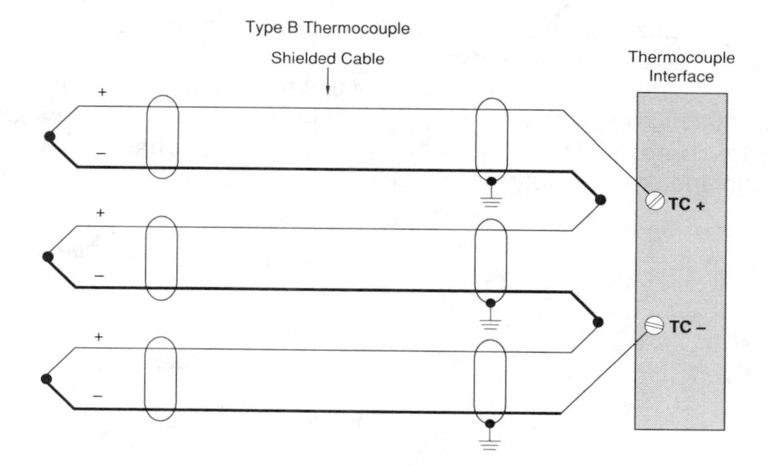

Figure 12-21. Thermopiling connected to a PLC module.

12-5 DISPLACEMENT TRANSDUCERS

This section on displacement transducers will deal primarily with the use of the LVDT and the potentiometer. The LVDT's operation as a mechanism was discussed in Section 12-3 in detail. Other displacement transducers that are used to provide feedback information, such as encoders and leadscrews were discussed in Chapter 8 and are dealt with as position and motion related feedback devices.

LVDT

As mentioned previously, the LVDT is based on the principle of a movable core inside a wound coil which generates voltage changes depending upon the position of the core. Therefore, attachment of a rod or similar element to the core provides the means for measuring linear displacement of the rod along a path.

The LVDT is designed to measure displacement of any type; whether it is caused by induced pressure, force or just linear displacement. There are LVDTs capable of measuring displacements from ±0.05 inches (±1.27 mm) to ±10 inches (±254 mm). The voltage output from the LVDT is generally provided at ±10 VDC for any range.

Due to the symmetry of the LVDT, the null repeatability is extremely stable, therefore making the LVDT a great null position indicator in closed loop control systems. The LVDT provides excellent resolution since the most minute movement of the core will produce an output. This resolution in movement is primarily due to very low friction inherent in the LVDT design.

Potentiometer

The potentiometer is perhaps the simplest displacement transducer available. Its principle is based on resistance changes due to movement of the potentiometer *wiper* (see Figure 12-22). The potentiometer is powered by a voltage source and the wiper provides an output voltage proportional to the movement of the attached element. This voltage output is related to the voltage drop between the top section of the potentiometer and the resistance accumulated by the wiper position. Potentiometer transducers, however, tend to have problems with friction in the wiper arm, limited resolution in the wirewound unit, mechanical problems due to wear, and are sensitive to vibration. On the other hand, potentiometers have a wide range of applications and are relatively inexpensive.

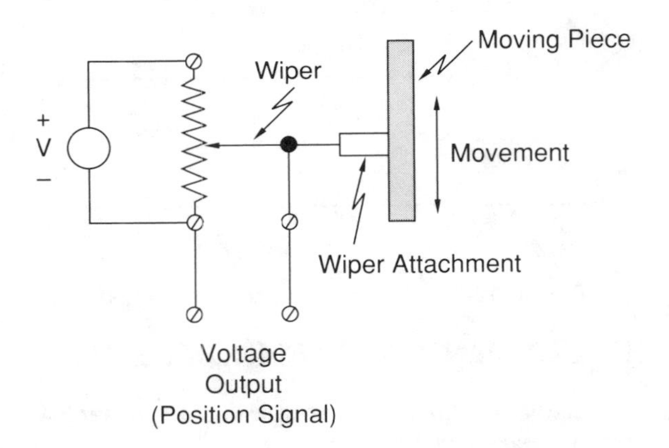

Figure 12-22. Illustration of a potentiometer application.

12-6 PRESSURE TRANSDUCERS

Pressure transducers transform the force per unit area exerted on its surroundings into a proportional electrical signal after signal conditioning has been applied. Pressure measurements are also utilized in other transducer measurements such as flow and strain.

Two of the most common pressure transducers are the strain gauge and the load cell, which is based on the strain gauge principle. The signal conditioning generally utilized with these two pressure transducers is the bridge circuit technique.

Strain Gauges

The strain gauge is a mechanical transducer which is used to measure body deformation, or strain, due to the force applied to the area of a rigid body. Strain gauges can be used in different applications such as flow measurement where pressure differential measurements are used. The strain gauge is also used in direct strain measurements where stress is being applied to a rigid body.

Strain gauges are based on the principle of resistance changes of its wires due to an applied force. They are made of metal wire such as copper, iron, platinum, etc., or semiconductor material such as silicon and germanium. The semiconductor type gauge exhibits a more sensitive response since they provide a greater change in resistance due to the deformation caused by the applied force.

There are two main categories of strain gauges: *bonded* and *unbonded*. These two types are illustrated in Figure 12-23. The bonded strain gauge is attached directly to the rigid body on a thin layer of synthetic thermosetting resin (epoxy) where the stress

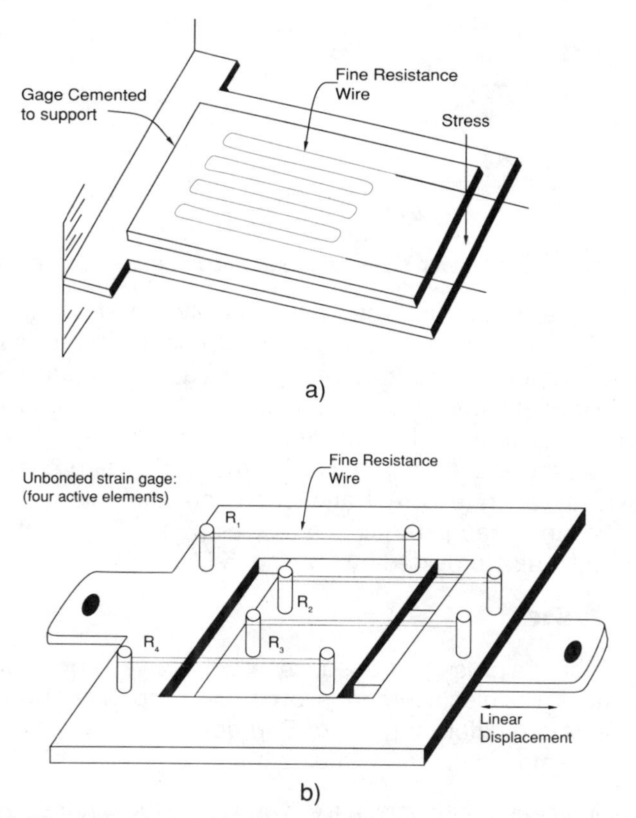

Figure 12-23. a) Bonded strain gauge and b) Unbonded strain gauge.

is being applied. Unbonded strain gauges operate under the same principle; however, there is a moving part which moves with the force applied. This movement in turn changes the resistance of its wires, therefore creating a voltage differential (due to force) in a bridge circuit.

Strain gauge circuits are affected by changes in temperature and therefore require a temperature compensation. This compensation is incorporated by adding a *dummy* gauge to the bridge conditioning circuit (see Figure 12-24). As the temperature increases, the resistance in the measuring gauge changes; however, the dummy (made of the same material) is connected to the bridge, its resistance also changes but does not affect the output of the bridge. Note that the changes in resistance in the gauge due to force are only attributable to the active strain gauge since stress is only measured in one direction.

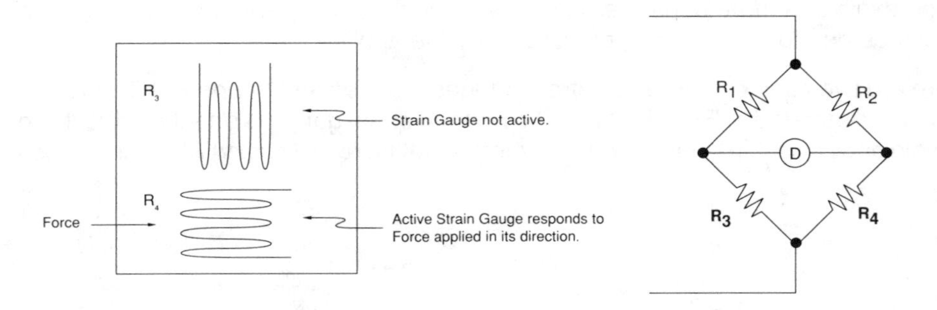

Figure 12-24. Strain gauge with dummy gauge to compensate for temperature changes.

The signals provided by the strain gauge are usually very small, and therefore require amplification. Some programmable controller manufacturers, such as Cincinnati Milacron, provide interfacing to strain gauges via a strain gauge module (Figure 12-25). The strain gauge module can receive signals from two strain gauges and have two amplified signal outputs. The amplified signals can be input to an analog input module as a pressure control variable for monitoring and control. The amplifier module provides separate SPAN and ZERO adjustments for field calibration, if necessary. The strain gauge requires a precision ±10 VDC level excitation from an excitation module, also provided by Cincinnati Milacron.

Bourdon Tube

The Bourdon tube is a pressure transducer which converts pressure measurements into displacement. The displacement is, of course, proportional to the pressure being measured. There are different types of Bourdon tubes: the spiral, helical, twisted, and, perhaps the most common C-tube type.

The pressure inlet of the tube is at a fixed end while the other end of the tube is free to move. Figure 12-26 illustrates the commonly used Bourdon C-tube. The mechanism used to convert the pressure into an electrical signal is the LVDT since the

Figure 12-25. Strain gauge input module. (Courtesy of Cincinnati Milacron).

intermediate conversion is displacement. Therefore, the Bourdon tube is said to convert pressure to an electrical signal proportional to the pressure via the LVDT (linear displacement to electrical signal).

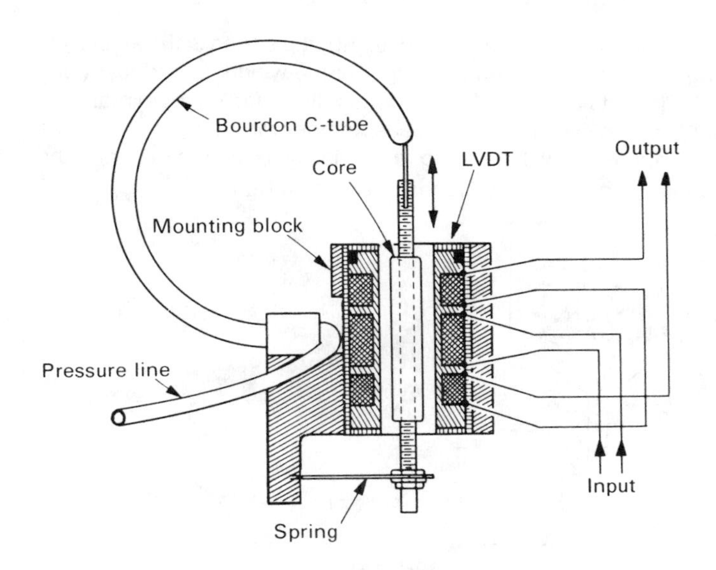

Figure 12-26. Illustration of the Bourdon C tube.
(Courtesy of Schaevitz Engineering, Pennsauken, NJ).

Load Cells

Load cells are force or weight transducers based on a direct application of bonded strain gauges. These devices measure deformations produced by a weight application. Applications of load cells are numerous, especially when measurements such as hopper support weighing in dry or liquid mixing systems are required (see next sections example).

12-7 FLOW TRANSDUCERS

Flow transducers, as the name implies, are used to measure the flow of materials in a process. This flow of materials can be in the form of a solid, gas, or liquid. All flow control applications utilize the term Q or *rate of flow* to observe flow measurements in a system. In this section we will look at the rate of flow term for each of the materials and the transducer used to measure them.

Solid Flow Transducer and Measurements

The measurement of solid flow is generally depicted by the use of strain gauge based load cell transducers to measure the weight of a product. Solid flow frequently integrates the use of a conveyor or belt product transporter in combination with the load cell. The units generally used for this type of flow measurement are lbs/min (Kg/min). Figure 12-27 illustrates an example of how a load cell can be used to measure flow of solids. The equation that describes the flow (Q) is:

$$Q = \frac{\text{Mass x Speed}}{\text{Length of Weight Measurement}} = \frac{W\,V}{L}$$

where the mass is represented by W (weight), the speed is the velocity of the moving transporter (V) and the length of the weight transducer (load cell) is L. If the units used for W are pounds (lbs.), for V feet/min and for length feet (English system), the Q measurement will be in lbs/min. In the metric system mass W will be in Kg, the velocity in meters/min, and the length in meters, thus giving a Q measurement in Kg/min.

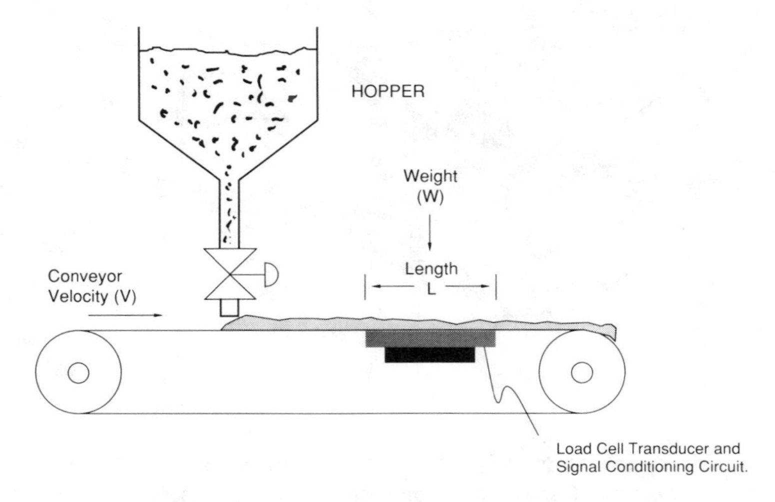

Figure 12-27. Example of load cell used to measure flow measurement.

Example 12-8

A conveyor is transporting material which is weighed on a platform 2 meters in length. The load cell is connected through a bridge circuit and an amplifier to an analog input module. The mass to be weighed is 50 Kgs of material. The required flow is 1200 Kgs/min. Find the speed which the conveyor must run to obtain the required flow rate and suggest how you could control the conveyor so that the flow rate is kept constant.

Solution

The velocity of the conveyor can be obtained from the flow rate equation as:

$$Q = \frac{W\,V}{L}$$

$$V = \frac{Q\,L}{W}$$

$$V = 48 \text{ m/min}$$

The speed of the conveyor could be controlled by having the PLC compute the flow rate and make changes to a motor or we could change the drive to the speed of the motor. We could also control an analog valve so that the hopper output is varied according to the required flow rate and speed.

Fluid Flow Transducer and Measurement

There are two primary ways to measure fluid flow. These two ways are by means of measuring pressure differentials or by the detection of the fluid motion. There are different *methods* of obtaining a pressure differential in the process line guiding the flow. The two most common ones are the *Venturi tubes* and the *orifice plates*. One of the most commonly found fluid flow transducers which employ fluid motion detection is the *turbine flow meter*. The latter fluid flow transducer transforms flow directly into electrical signals.

Pressure Based Flow Meter. The Venturi tube and the orifice plate methods are based on the Bernoulli effect which relates velocity of the flow with a pressure differential at two different points. The measurements of these pressures can be accomplished by using pressure transducers which will transform pressure into an electrical signal. The two most commonly used transducers are the strain gauge (covered in Section 12-5) and the Bourdon C tube. These two transducers use the bridge circuit and LVDT techniques respectively. If low pressures are being measured, bellows, diaphragms and capsules can be utilized to achieve better pressure reading resolution. These pressure transducers may use the LVDT technique. Figure 12-28 illustrates these low pressure transducers.

Figure 12-28. Low pressure transducers.
(Courtesy of Schaevitz Engineering, Pennsauken, NJ).

Figure 12-29 illustrates a diagram of the Venturi tube; the orifice plate flow transducer is shown in Figure 12-30. The pressure differential ΔP is equal to the difference in pressures P_1 and P_2 $(P_1 - P_2)$. This ΔP is also related to the velocity of the fluid through the Bernoulli effect. The velocity at point P_2 as a function of ΔP is expressed as:

$$V = k\sqrt{\Delta P}$$

$$\text{where } \Delta P = P_1 - P_2$$

where ΔP is the pressure differential measurement $(P_1 - P_2)$and K is a constant that includes the density of the fluid, the ratio of the cross-sectional areas at point P_1 and P_2 of the pipe and obstruction, temperature and other factors. To obtain the flow rate measurement, we get:

$$Q = \text{Velocity x Area} = VA$$

$$Q = V_1A_1 = V_2A_2$$

$$Q = VA = Ak\sqrt{\Delta P} = K\sqrt{\Delta P}$$

Where K is a new constant composed of k times the area A_2. The value of Q, flow rate, will give us volume per unit time of the flow (ft/min x ft^2 = ft^3/min).

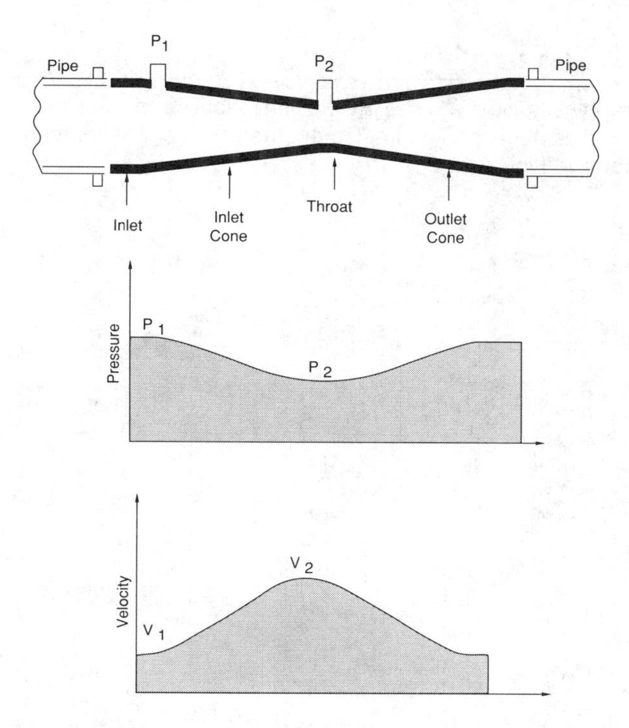

Figure 12-29. Venturi tube diagram and the pressure and velocities at P_1 and P_2.

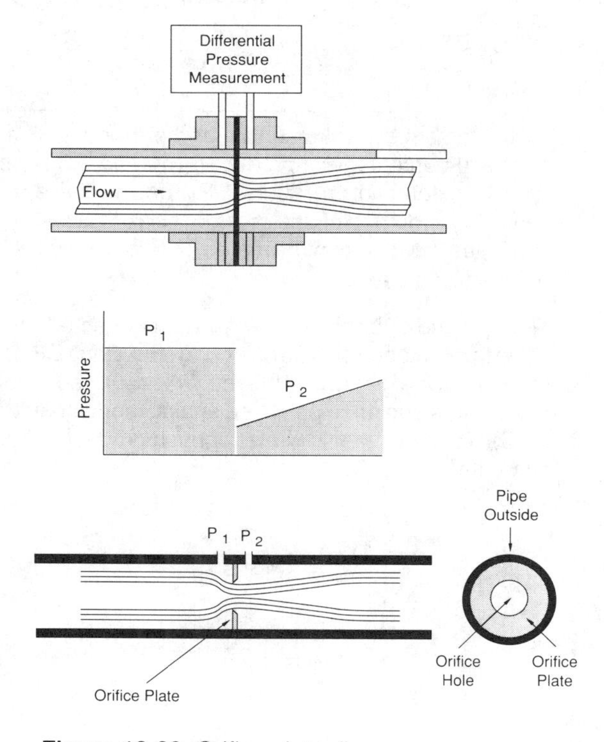

Figure 12-30. Orifice plate diagram.

Example 12-9

Graphically illustrate the PLC connections and functions that would be necessary to implement the ratio control computation shown in Figure 12-31.

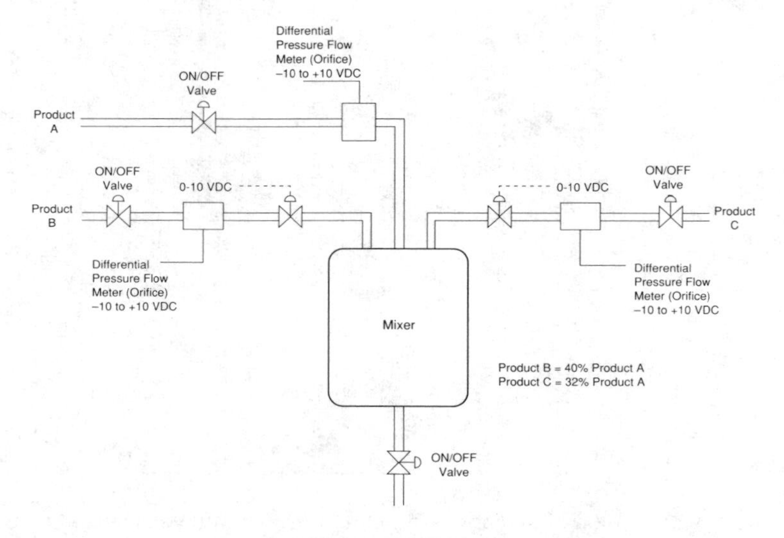

Figure 12-31

Solution

To implement the ratio control of the products B and C at the specified percentage of product A (*wild flow*) it is necessary to read the differential pressures (DP) from the orifice flow meter and control the output of the analog servo valves. Figure 12-32 illustrates the ratio control implementation using flow ratio as the process variable.

The discrete outputs (120 VAC) are connected to the ON/OFF valve to allow each of the products to flow. Each DP P & ID symbol represents the differential pressure measurement from the orifice flow meters and are input to the analog input modules (-10 to +10 VDC). To obtain the flow rate for each product we would have to compute:

$$Q_A = K_A\sqrt{\Delta P_A}$$

$$Q_B = K_B\sqrt{\Delta P_B}$$

$$Q_C = K_C\sqrt{\Delta P_C}$$

where K_A, K_B and K_C are the given constants. The square root value of the analog input should be taken after the input counts corresponding to the ΔPs have been computed to engineering units (through linearization, etc.). As the A product flows, the B and C flow are computed so that the ratio to A is maintained (B = 0.40A and C = 0.32A). The output control valves for product B and C will have to be controlled to maintain the proper ratio.

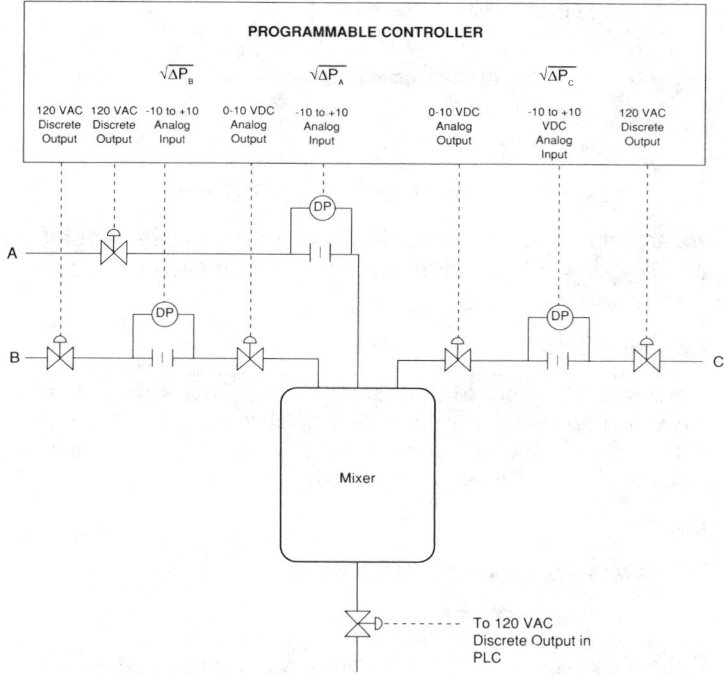

Figure 12-32. Solution diagram to Example 12-9.

Motion Detection Flow Meters. The turbine flow meter is perhaps one of the most common types of motion detection flow meters. These devices are available for use in applications where measuring liquid and gas flows as well as in situations where very low flow rates are encountered.

The turbine meter is composed of a multi-bladed rotor which is suspended in the liquid flow. As the fluid flow passes through the blades, a rotary motion is created. This rotary motion creates a magnetic flux that is picked up by a coil, where a small voltage is created (as low as 10 to 20 mV) and then amplified. The turbine meter is designed such that movement of its blades give pulses out proportional to the volume passing through the turbine. The output pulses generally provide gallons per minute (GPM) information. Some turbine meters may provide an analog output proportional to the flow rate being measured for that particular turbine. Figure 12-33 illustrates a simple diagram of the turbine flow meter.

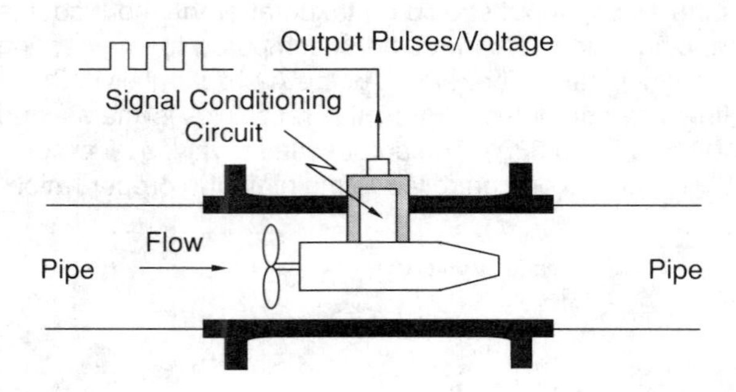

Figure 12-33. Illustration of a turbine flow meter.

Turbine meters are widely used in petrochemical and pipelines transfer of petroleum flows. In liquid Oxygen and Nitrogen gas metering applications, special types of turbine flow meters are also used.

Example 12-10

A programmable controller system receives an analog signal from a turbine flow meter. The flow rate is given at 60 GPM. The area of the pipe is 2 square inches; find the velocity of the flow to be displayed in feet per second on a four digit LED display.

Solution

The flow rate of the fluid is described by :

$$Q = VA$$

where Q is the flow rate, V is the velocity of the flow and A is the cross-sectional area of the pipe. The velocity of the flow is:

$$V = \frac{Q}{A}$$

Note that the units given must be converted to obtain the velocity in ft/sec. To convert gallons to cubic feet we first convert gallons to cubic meters and then to cubic feet.

$$1 \text{ gal} = 3.785 \times 10^{-3} \text{ m}^3$$

$$1 \text{ m}^3 = 35.31 \text{ ft}^3$$

Therefore

$$1 \text{ gal} = 3.785 \times 10^{-3} \times 35.31 \text{ ft}^3$$

and 60 GPM is equal to 1 gallon per second. The cross-sectional area of 2 square inches is equal to 2/144 square feet. So for the velocity (in ft/sec) we obtain:

$$V = \frac{1 \times (3.785 \times 10^{-3}) \times 35.31}{2/144} = 9.622 \text{ ft/sec}$$

To obtain velocity of a fluid in a pipe (in ft/sec) when the flow rate is given in GPM and the area is given in square inches. The following equations can be used:

$$V_{(ft/sec)} = \frac{Q_{GPM} \times 0.3208}{A_{(sq\ in.)}}$$

12-8 SUMMARY

In this chapter we have introduced basic measurement concepts which will help in the understanding of how data and errors are interpreted and analyzed. This information, which is based on statistical analysis, is very helpful when integrating an intelligent or knowledge-based PLC system.

Different techniques used to transform the physical measurement of the transducer sensor are also explained. The two most common ones are the bridge circuit and the LVDT mechanism. Transducers in general can be understood as being composed of several intermediate measurement and connection elements. For instance, a flow transducer may provide an electrical signal proportional to the flow by using a pressure based transducer such as a Bourdon tube which in turn uses an LVDT. Figure 12-34 illustrates a block diagram specifying a primary and secondary use of transducers that make up the final transducer as we know them.

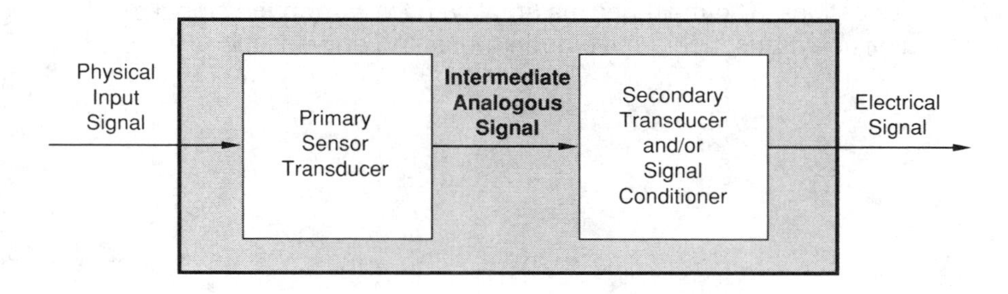

Figure 12-34. Illustration of a primary and secondary use of transducers.

The sections covering thermal transducers provide invaluable information that will be needed in any type of temperature measurement in a process. Depending on the temperature range being measured and the linearity required, the user can select the appropriate thermal transducer. Figure 12-35 shows a simple diagram which describes a range and relative output comparison of the RTD, thermistor and thermocouple transducers.

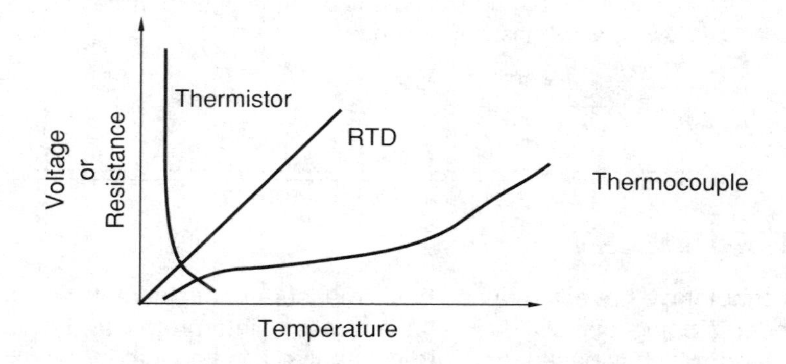

Figure 12-35. Thermistor, RTD, and thermocouple linearity curve comparison.

Displacement and pressure tranducers are presented and their operation explained. Flow measurements are covered for solid flow as well as liquid flow measurements. The principal method used to measure flow is the use of pressure differential methods and by detection of motion as in the case of turbine flow meters. Flow control applications most likely will require that the PLC system have the capability of square root computation. This square root instruction is needed since flow rate computations require the square root of the differential pressure.

In this chapter, we have focused on the most commonly found and used transducers in PLC applications. The mechanisms employed to perform the signal conditioning to most available transducers have been presented and described in detail.

Chapter

—13—

ARTIFICIAL INTELLIGENCE AND PLC SYSTEMS

"When we return from error, it is through knowing that we return."

—Saint Augustine

13-1 INTRODUCTION

Artificial intelligence (AI) is an area of computer science that has been in use for some time. In fact, the conceptual design of AI was first devised in the early 1960s. AI definitions vary among people in the computer society and, to some degree, is made difficult to perceive and understand. AI can be defined as a subfield of computer science which encompasses the creation or development of (computer) programs to solve tasks that require extensive knowledge.

The software programs that form an intelligent system (AI) are developed with the utilization of the knowledge of experts in the field where it is being applied. For instance, a food processing AI system that involves the making and packaging of a food product will have knowledge introduced from chemists, food technologists, packaging experts, maintenance departments, and others closely associated with the operation.

In this chapter we present different techniques of AI that can be implemented in a process control application utilizing a PLC based system. These methods and techniques will define the route to take when implementing AI into your application. The result will be a system that can successfully diagnose, control and predict outcomes based on the knowledge introduced and the program sophistication.

13-2 TYPES OF AI SYSTEMS

An exact definition of the types of artificial intelligence systems is very difficult to point out simply because many people assert different types of classifications and their scope of definitions is wide. For the purpose of this text, we are concerned with three types of AI systems:

- Diagnostic System
- Knowledge System
- Expert System

Each of these types have similar characteristics. One type of system evolves into the other primarily due to the degree of sophistication added. This sophistication is due to the size of data bases, and the extent of how the process data is compiled and interpreted.

Diagnostic Systems

Diagnostic systems can be considered as the lowest level of artificial intelligence implementation. They are based primarily on the detection of faults within an application without regard to how a solution to the problem can be accomplished. For example, a diagnostic system could detect a pump fault through the detection of loss of tank pressure or flow meter readings in a batching system.

The diagnostic system approach reaches a conclusion based on known facts (knowledge) introduced to its detection system (inferring techniques). This type of AI is used in applications which provide a small knowledge and data base structure. Decisions made are generally GO or NO-GO, and sometimes may provide the fault's probable causes.

Knowledge Systems

A knowledge-based AI system is, in reality, a progression of the diagnostic type. It not only detects faults and process behavior based on introduced knowledge, but also makes decisions on the process and/or a probable cause of a fault.

In the batching system mentioned before, a knowledge system would go beyond the diagnosing of a fault. It could provide suggestions on probable fault devices as well as deciding whether to continue the process (if the fault is non-critical) or to shut down (if the fault is critical). These decisions are based on the knowledge introduced and the set of rules that define each fault condition.

On the other hand, the fault detection could have been a false fault detection (false alarm). A knowledge system would check whether the elements providing the fault decision, i.e. flow meter, pressure transducer, are operating correctly. Comparisons are made between the observations (process feedback) and the procedures or measures taken to make these comparisons. For instance, if a fault does occur and it is a valid non-critical fault, the control system may issue a *continue process, stop after finished* and an *alert personnel* command.

Expert System

The expert system can be considered the top of the line AI application. It performs all the capabilities a knowledge system performs and more. An expert system provides the additional capability of analyzing process data using statistical analysis. The use of statistical data analysis provides the system with the ability to predict outcomes based on current process assessments. The outcome prediction may be a decision to continue a process in spite of a fault detection.

In our descriptive example used in the other two AI systems, an expert system may decide to continue the batching operation until the non-critical fault generates another fault. This decision may have been arrived at due to the fact that the average pressure sensed in the mixing reactor tank was within tolerance limits, i.e. readings observed about the mean, and continue in spite of the fact that the flow meter reported a loss of flow. The system continues production and alerts personnel on the basis that pump and flow meter feedback has been, perhaps, lost.

The knowledge that has to be introduced has more data, therefore generating more data verification (feedback information). The decisions to be made are harder to implement in the form of a software program since the decision trees involve more options and attributes.

The implementation of an expert AI system implies not only extra programming effort but also more hardware capability. The total system will need more transducers to check other transducers and field devices. The PLC will require perhaps the use of two or more processors to implement the control and intelligent program. The speed of the system must be fast to be able to operate in real time. Memory requirements will be larger since knowledge data must be incorporated and stored in the system.

13-3 ORGANIZATIONAL STRUCTURE OF AN AI SYSTEM

A typical artificial intelligence system is composed of three primary elements:

- Global Data Base
- Knowledge Data Base
- Inference Engine

Figure 13-1 illustrates a block diagram of an AI system architecture. As can be seen in the figure, the AI system must receive its knowledge from "someone" that thoroughly understands the process or machine being controlled. This individual, called the *expert*, must communicate all information about maintenance, fault causes known by experience, etc. to the person responsible for the system implementation. The person performing the implementation of the system is generally referred to as the *knowledge engineer*. The process of gathering data from the expert and transmitting it to the knowledge engineer is known as *knowledge acquisition*.

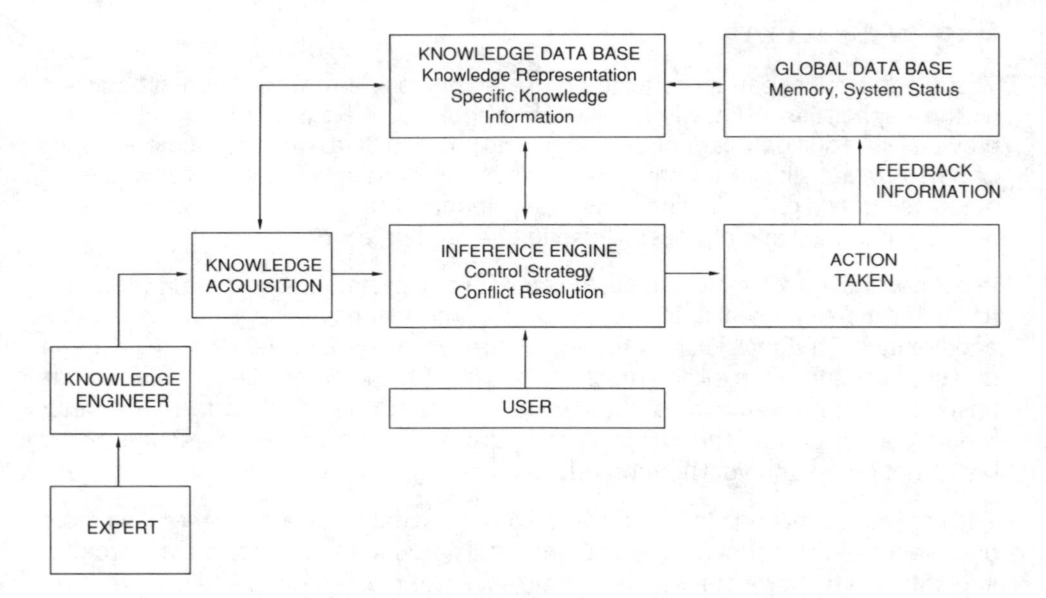

Figure 13-1. Artificial Intelligence system architecture.

Global Data Base

The global data base contains all available information pertaining to the system to be controlled. This information is concerned with the input and output data flow from the process. The global data base resembles a storage area where information is updated about the process. The data in this area can be accessed at any time to implement statistical analysis on past process control data, which in turn can be used to implement AI decisions.

The global data base is resident in the memory of the control system implementing AI. If a PLC is using a diagnostic AI function, the global data base will most likely be located in the storage area of the PLC's data table. If a PLC in conjunction with a computer or computer module is used to implement a knowledge-based AI system, the global data base will probably be located at the computer or computer module's memory or hard disk storage subsystem.

Knowledge Data Base

The knowledge data base is a section of the AI system where the information extracted from the expert is stored. This information may not deal only with the process control itself, but also with the fault cases and their probable causes and possible solutions.

As will be seen later, all the rules that govern the decisions to be made are stored in this section. The knowledge data base is stored in the part of the memory of the system that implements the artificial intelligence techniques. The more involved the AI system, the larger the knowledge base. The knowledge data base of a diagnostic system is less complex than that of a knowledge system; likewise, the knowledge system of a knowledge data base is less involved than that of the expert system's.

Inference Engine

The inference engine section of an AI system is where all decisions are made. This section uses the knowledge stored in the knowledge data base to arrive at the decision. The inference engine is also in constant interaction with the global data base to examine and test real time data as well as past history about the process.

The residence of the inference engine is at the main CPU; that is, in the CPU that is performing the AI computations. The inference engine can really be thought of as the main program that executes all applicable rules and decisions about the process. In a PLC based system, the inference engine may or may not be stored in the main CPU, depending upon the complexity of the system (i.e. diagnostic, knowledge or expert).

13-4 KNOWLEDGE REPRESENTATION—RULE BASED

Knowledge representation can be defined as how the complete artificial intelligence system strategy is organized. That is, how the knowledge engineer represents the expert's input. This representation is kept in the knowledge data base. As will be seen in the rule based section, the expert's knowledge is transformed into IF and THEN/ ELSE situations where action and decisions are made.

A control system that implements any type of artificial intelligence, whether diagnostic, knowledge or expert, executes the control strategy via a software control program in the inference engine.

Whenever a decision needs to be made because of a fault or other causes pointed out by the expert, the inference engine will go to the knowledge representation, and a decision on probable causes is obtained. How this decision is arrived at will most likely be a group of software subroutines.

Once an AI decision has been reached in the knowledge data base, the inference engine will determine actions to be taken. It is during this time, depending on the control strategy formulation (main program), that the inference engine may go to the global data base and inspect data for verification or more information.

Rule Based

The rule based knowledge representation deals with how the expert's knowledge is used to infer a decision. The rules used are IF something happens (antecedent) THEN take this action (consequent). For instance, if the expert is asked: *what happens when the volume in the tank drops?* and the response is: *the tank system is not working properly*, the knowledge engineer may implement it as a rule IF volume is less than a setpoint THEN annunciate a system malfunction due to loss of volume.

Rules can be as long and complex as needed by the process. Rules generated usually define the involvement of the AI system. For instance, a simple rule based system (few rules, not very complex) may be encountered when formulating a simple diagnostic implementation:

IF temp is less than setpoint THEN open steam valve

A more complex diagnostic would involve rules that depend on parent rules:

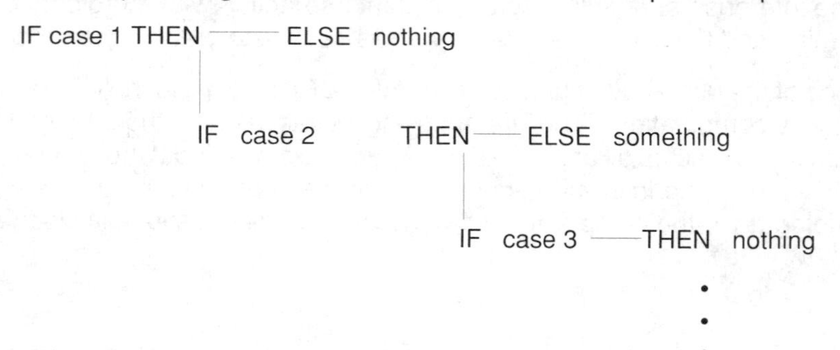

where each of the case conditions represents a particular measurement, comparison or situation. A slightly different degree of complexity in a rule based knowledge representation comes when the rule may have several probable causes—for example:

IF volume drops THEN { valve failure or / pump failure or / feedback failure }

the consequents would have to be further investigated to arrive at a total formal answer. Figure 13-2 illustrates a representation or formation of rules. The consequents arrived at in the knowledge representation can be used by the inference engine to obtain a better definition of, let's say, the problem cause as may be the case in a knowledge or expert AI system.

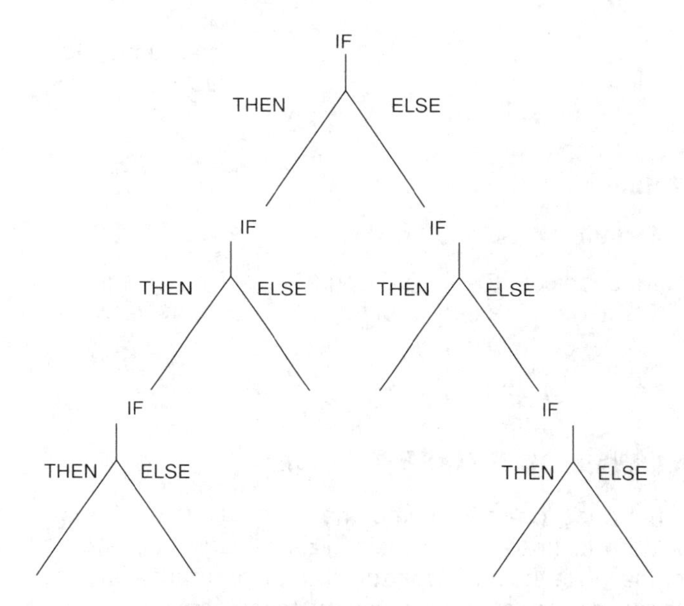

Figure 13-2. Rule formation.

Example 13-1

A PLC controlled carton box conveyor transports two sizes of boxes which, according to their sizes, are diverted to different palletizing operations. The activation of the diverter is via a solenoid actuator. Write the rules that can be used in a knowledge data base to detect a possible cause of a solenoid's malfunction.

Solution

A solenoid failure can have one or two results according to the conditions presented. These results are reflected in the coil (coil burnt out) and in the solenoid's mechanical linkage (mechanical damage). The following conditions and causes describe the two possible rules that lead to the fault.

Rule #1		
Result	**Condition**	**Cause**
Burnt Out Coil	Excessive Temp. is developed due to continuous high current input	—Low line voltage failure to pull plunger —High ambient temp —Mechanically blocked plunger —Operations too rapid

Rule #2		
Result	**Condition**	**Cause**
Mechanical Damage	Excessive Force exerted on the plunger	—Overvoltage —Reduced load

13-5 KNOWLEDGE INFERENCE

Knowledge inference deals with the methodology used for the gathering and analyzing of data to draw a conclusion. Knowledge inference can occur in the inference engine while the main control strategy program is being executed as well as in the knowledge data base while rules solutions are being compared or computed.

The way by which solutions are arrived at are dependent on the approach taken in the software program. Operator interaction on control problems found can add versatility to the finding of the proper solution. For example, if a system detects a failure due to a misreading in an inspection system, the operator may be alerted to the problem and advised of probable causes. Furthermore, it may wait for an operator's input (e.g. check for laser intensity in the receiver side. Is the laser beam reflecting at the correct angle?); the input from the operator can be used by the control system (in the inference engine) to come up with more intelligent solutions to the problem.

In small systems, knowledge inference is performed on a local basis. That is, the resident software comparing the inference engine is at the control system. Remember that the degree of AI involvement in the system will determine how much hardware will be required (e.g. computer modules, larger powerful PLCs, smaller PLCs with personal computers, etc.).

In large distributed intelligent systems, the knowledge inference will most likely be located at a main host in a hierarchical system. When all global data bases are in constant network communication, and knowledge inference information can be passed from one controller to another, the intelligent system is said to be using a *blackboard architectural structure.* In all types of intelligent systems, knowledge inference is implemented using certain methods of rule elaboration. Some of these methods include *forward chaining* and *backward chaining.* Statistical information can also be analyzed as part of the knowledge inferencing to get the resolutions or outcomes from the control system.

Blackboard Architecture

Large complex distributed control systems generally involve the interaction of several subsystems which maintain continuous communication over perhaps a local area network. When artificial intelligence is added to these large systems, the knowledge inferencing as well as global and knowledge data bases are distributed in the architecture of the control system. Blackboard architecture is the name given to this large system which utilizes several subsystems containing local global and knowledge data bases. Figure 13-3 illustrates a blackboard configuration of an intelligent control system.

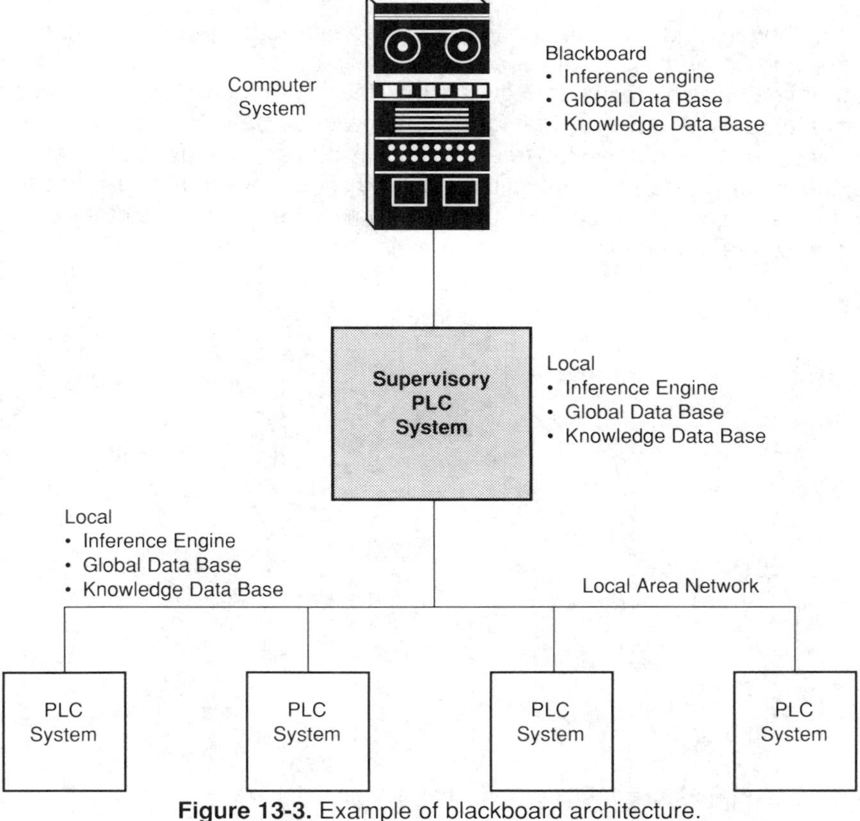

Figure 13-3. Example of blackboard architecture.

Each of the controllers in the network may or may not have a local inference engine, global data base, and knowledge data base, depending upon the degree of inferencing being accomplished on the local basis. In fact, a PLC controller in the subsystem level may even contain a computer module which helps it inference engine computations. The hierarchy of the control system may allow the supervisory PLC controller to poll each of the subsystem numbers and transfer all or part of their local global data base.

The host computer element in this control structure can be considered to hold the blackboard, a region where information from all subsystems via the supervisory PLC is stored. The inference engine of the host will most likely implement complex solutions according to its knowledge inference about the total control system.

Forward Chaining

Forward chaining can be defined as the method used to arrive at possible outcomes from given data inputs. For instance, in our Example 13-1, forward chaining can be structured as if the solenoid fails in the diverter, then it could create the following consequences: jamming the conveyor or misplacing boxes in the two palletizers. Typical forward chaining inferences generally receive process information via the global data base. Monitoring specific inputs in a control system can be thought of as the input part of a forward chaining approach.

Within the forward chaining method there are two different types of fact searching: *depth first* and *breadth first*. Both searches deal with how the outcome is obtained. The depth first search, shown in Figure 13-4, deals with the order of execution or evaluation of rules (A,B,D etc.) that form the knowledge base. The outcome based on the rules is searched down the tree on a priority basis. In the example mentioned above, the control system will detect the solenoid failure, evaluate a new rule to see if jamming has occurred. If the conveyor is jammed then it will find other possible consequences (e.g. material inside box may break, could cause a spill).

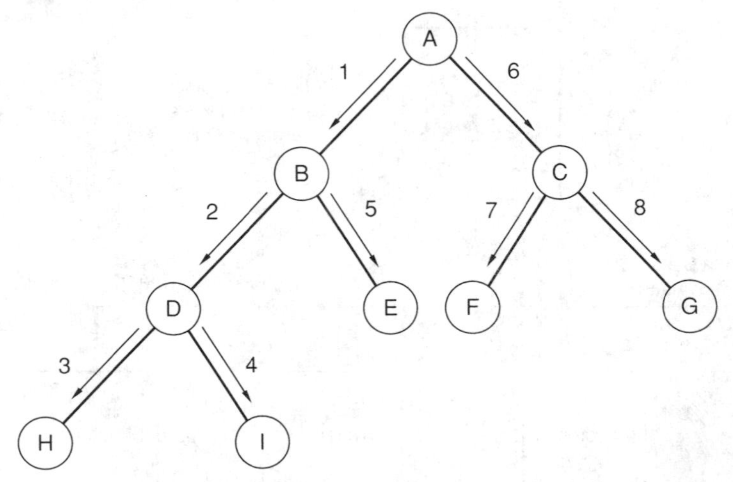

Figure 13-4. Forward chaining depth first search.

The breadth first method evaluates each rule in the same level of the tree (see Figure 13-5); that is, from A to B and C, etc. In our conveyor example, the evaluation of rules, after the solenoid failure, will go to the possible outcome of the jamming, and then to the possible palletizer misplacement, and so on.

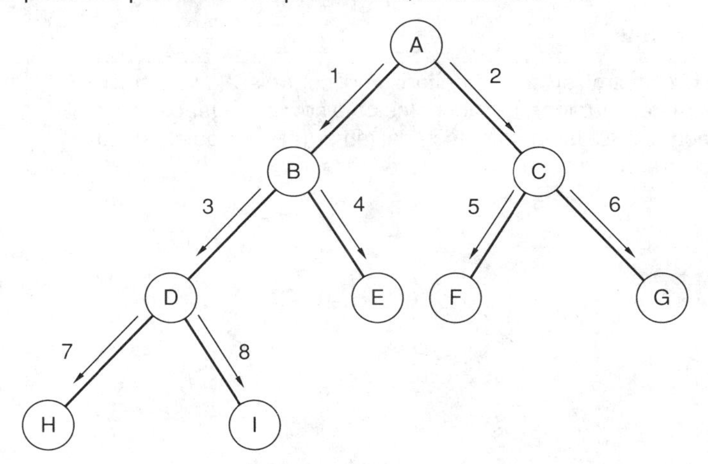

Figure 13-5. Forward chaining breadth first search.

Backward Chaining

Backward chaining deals with finding the causes of an outcome. For instance, in our Example 13-2, the example's solution are backward chaining evaluations since the reason for the solenoid's failure is searched and possible causes are reached. Basically, the backward chaining analyzes the consequence to obtain the antecedents.

Similar to forward chaining, the backward chaining can use the depth search or breadth search methods. In our conveyor example, after the solenoid failure occurs, a depth search will first check the conditions of one rule then proceed to check each possible cause from that condition. If the breadth search is employed, then the system will look to each of the condition rules and continue down the tree to obtain the causes for each of the conditions.

Statistical and Probability Analysis Inferencing

The use of statistical analysis and probabilities play an important role in an artificial intelligence system. This important aspect of AI is of particular interest in expert systems where prediction of outcomes are required. The data which is analyzed is usually stored in the system's global data base.

In Chapter 12, we covered data representation and interpretation of events taking place using statistical tools, in particular the mean, mode, median and standard deviation. These statistical computations can help determine a future outcome based on what has been happening in the current process. Decisions made, based on statistics, could be related to consequences derived from rules described in the knowledge representation. For instance, if an error fault was detected it does not necessarily mean that the fault actually occurred even though the feedback data transducer devices may be operating correctly. Using statistical analysis, the inference engine may decide not to apply the supposed control to the fault or advise personnel, but instead continue to monitor the situation more closely.

Example 13-2

Part of a control system monitors and controls a hydrostatic cooker under a temperature loop. Indicate how intelligence could be added to the system to detect real temperature problems and how to screen out false temperature faults.

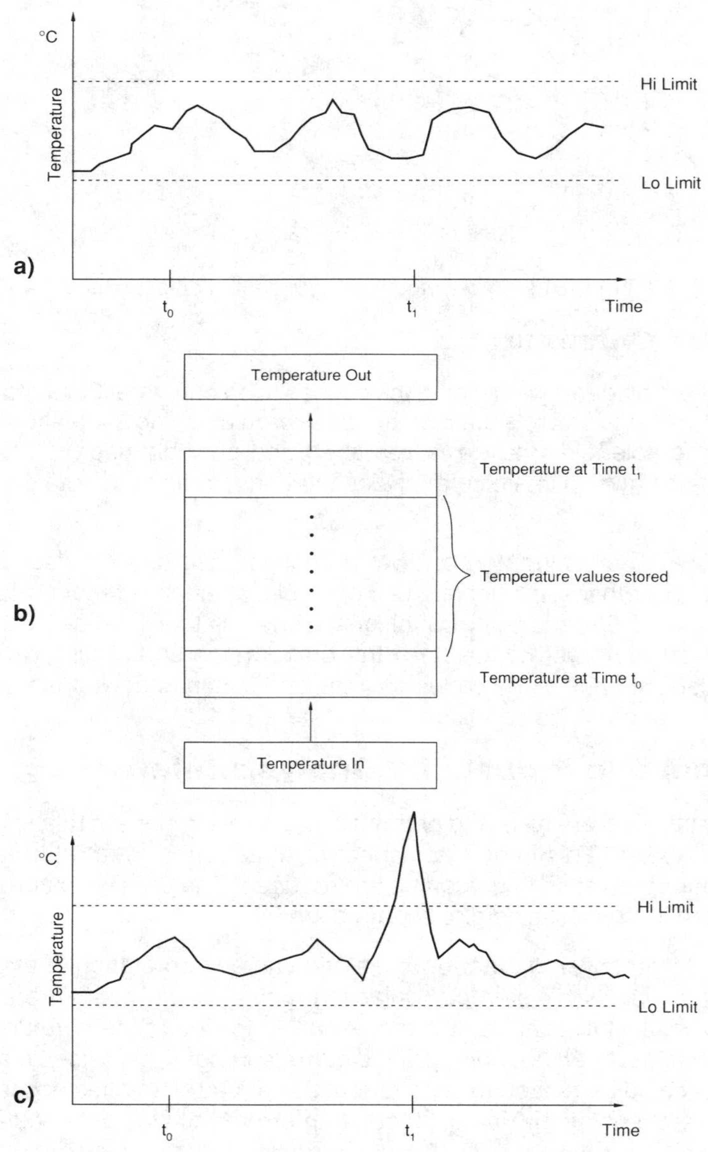

Figure 13-6. a) Mean of readings; **b)** FIFO storage method; **c)** Readings with a sample out of standard deviation range.

Solution

Let's assume that a profile of temperature readings can be described as shown in Figure 13-6a. We can store a window of the temperature profile from time T_0 to time T_1 continuously and accumulate and report an average temperature during the period. This temperature accumulation can be performed using FIFO instructions to a storage area of a fixed number of registers (see Figure 13-6b). Through the program, we can also compute the mean, median and standard deviations of current temperature readings.

If a high limit alarm is detected (see Figure 13-6c), a normal system would try to control the cooker by adjusting its temperature loop. However, the temperature fault may not have been caused by any other factor than a noise spike near the temperature transducer. An intelligent system could detect this sudden increase by recognizing that it is well beyond its mean and median readings over the past T_0 to T_1 period, therefore exhibiting a large standard deviation.

By implementing a rule that considers the statistics about the process, we can ignore the false alarm and not store the temperature reading value for our average calculations report. Furthermore, the global data base of the system can receive information concerning time of day, when the spike occurrence took place and level of spike reading for future use. The system must keep close track of the temperature increase in case it is a true alarm. Temperatures' rate of change computations would also help in the determination of a fault detection.

Probabilities can be used as an aid to determine or to approximate the possible cause of a fault in a diagnostic ruling. One of the most commonly used methods is the Baye's theorem of conditional probability. The use of conditional probability in an AI system is known as *conditional probability inferencing*. Remember that to employ probabilities in any system, history about the process must be maintained. The expert would generally provide this type of data about the process.

The probability of X event occurs based that Y already occurred (P (X/Y) based on the Baye's theorem is defined as follows:

$$P(X/Y) = \frac{P(Y/X)\,P(X)}{P(Y/X)\,P(X) + P(Y/\overline{X})\,P(\overline{X})}$$

where P(Y/X) is the probability of Y occuring when X has occurred and P(X) is the prior probability that X has occurred; $P(Y/\overline{X})$ indicates the conditional probability that Y occurs if X does not occur.

Example 13-3

Part of a conveyor system controls a solenoid operated diverter which sends two different boxes to two different repackaging areas. The system detects which box goes where by a decoding of several photoelectric eyes. Find the most probable cause of a conveyor fault when the boxes are going to the correct place. The material handling expert indicates that due to the size and type of boxes and environment, the following probabilities hold:

For the solenoid cause:

- Prior probability of solenoid fault is 20% (80% that it is not)
- There is a 35% probability that the boxes will go to the right place when the solenoid is faulty
- There is a 60% probability that the boxes will go to the right place when the solenoid is good.

For the photoeye cause:

- Prior probability of photoeye fault is 35% (65% that it is not)
- There is a 25% probability that the boxes go to the right place when the eye is faulty.
- There is a 45% probability that the boxes go to the right place when the eye is good.

Solution

We can include the expert's data into the knowledge representation, by calculating in terms of percentage,which element has a higher percentage probability of having occurred. This computation is as follows using the Baye's formula:

For solenoid:

P (S/B) = P (Solenoid faulty/boxes are going to right place)

$$P(S/B) = \frac{P(B/S) \times P(S)}{P(B/S) \times P(S) + P(B/\overline{S}) \times P(\overline{S})}$$

$$P(S/B) = \frac{0.35 \times 0.20}{0.35 \times 0.20 + 0.60 \times 0.80}$$

$$P(S/B) = 12.7\%$$

For photoeye:

$$P(E/B) = P\text{ (Solenoid faulty/boxes are going to right place)}$$

$$P(E/B) = \frac{P(B/E) \times P(E)}{P(B/E) \times P(E) + P(B/\overline{E}) \times P(\overline{E})}$$

$$P(E/B) = \frac{0.25 \times 0.35}{0.25 \times 0.35 + 0.45 \times 0.65}$$

$$P(E/B) = 23.02\%$$

The computations performed seem to indicate that a photoeye fault is most likely to have occurred in the conveyor system. The operator could be alerted and the system temporarily halted. Statistics of fault occurrence could be updated in the global data base so that they can be used in the future.

Conflict Resolution

A conflict is said to occur when a set of rules are triggered at the same time. The system is required to start execution of a first rule depending on priorities. For example, there are three facts that have taken place indicating a high temperature, low pressure and flow obstruction. These three facts by their own or in combination can trigger any rule consequents, for instance:

Rule 1: if high temperature, then start cooling procedure

Rule 2: if low pressure and flow obstruction, then open relief valve in main supply pipe.

Rule 3: if high temperature and low pressure, then open relief valve in main supply pipe and alert personnel in area.

As it can be seen, a decision to execute one of the three rules must first be made. Therefore, we have to select the rule which exhibits the greater priority, in this case that of rule number three. Other rule execution cases will depend on what the expert has asserted to be the sequence to be followed.

13-6 APPLICATION OF AI TECHNIQUES IN A BATCHING SYSTEM—FAULT DIAGNOSTIC

The example presented in this section is intended to illustrate how the methodology described in the previous section can be employed. For the sake of simplicity, we are not going to elaborate on the PLC program coding for the application, but we will describe the rules that will be used to define the knowledge representation.

The extent of artificial intelligence implementation is at the diagnostic level. This type of implementation can be accomplished in a PLC based control system. The method of rule definition is backward chaining; once a fault is detected, the cause of the fault is searched. In the batching system, the AI fault detection is implemented by only one of the two ingredients. The rules for the second ingredient would be similar to the first one.

Definition of the Process

There are two ingredients, A and B, to be mixed in a tank (Figure 13-7). Figures 13-8 through 13-12 show the flowcharts of the process as well as the steam valve versus temperature relationships. The process is as follows:

- A flow meter counts the number of pulses so that we can keep track of the ingredients (how many gallons).
- A pump motor provides the necessary pressure to send the ingredients through the line.
- Before any of the ingredients is poured into the tank the temperature inside the tank should be 100°C. This is accomplished by opening the steam valve 40% (solenoid actuated).
- A load cell based pressure transducer is used to read the volume inside the tank. This will be used to detect whether the ingredient is pouring into the tank in the event of a faulty signal.
- After the two ingredients are in the tank the temperature should be 800°C before the mixing. This is accomplished by opening the steam valve 100% until temperature reaches 800°C then the steam valve will remain 60% open to maintain 800°C.
- There are two thermoswitches that will detect the two desired temperatures and will serve as feedback in case of a fault.
- A steam valve is used to heat up the tank. This temperature is controlled via a temperature transducer that will give the desired temperature.
- A float switch is used to detect an empty tank.
- If the mix is not correct, it can be disposed of through an auxiliary valve.
- A discharge valve is provided to drain the desired solution (mix) into the next step of the process (mixing finished). When cooling down, the steam valve will go back to 40% to achieve the desired temperature of 100°C for the next batch.
- A motor is provided to agitate the two ingredients.

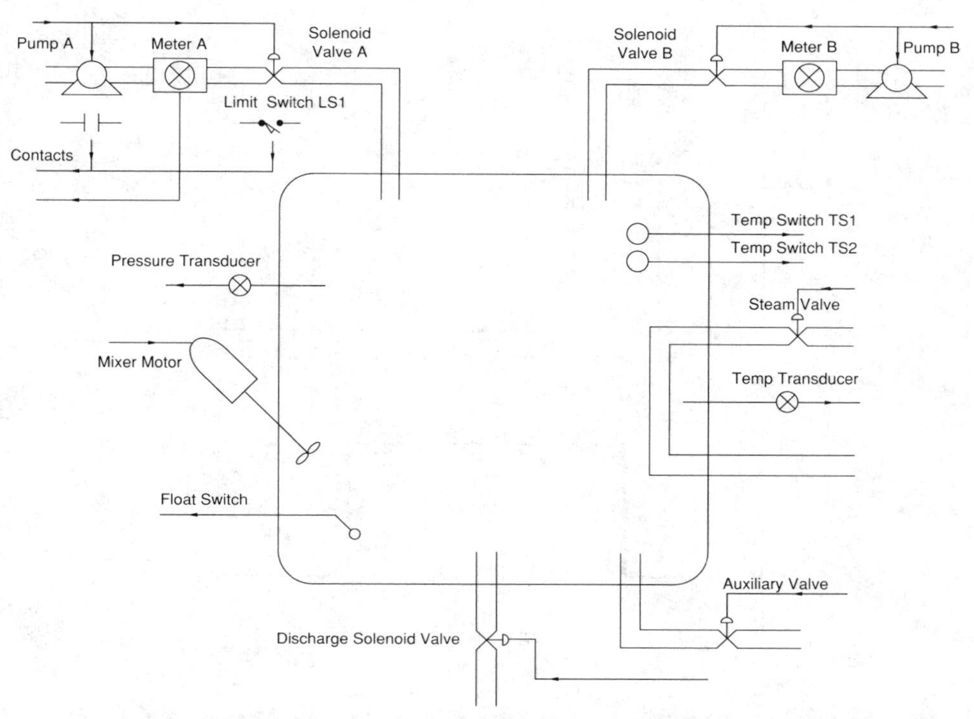

Figure 13-7. Batching system block diagram with feedback devices.

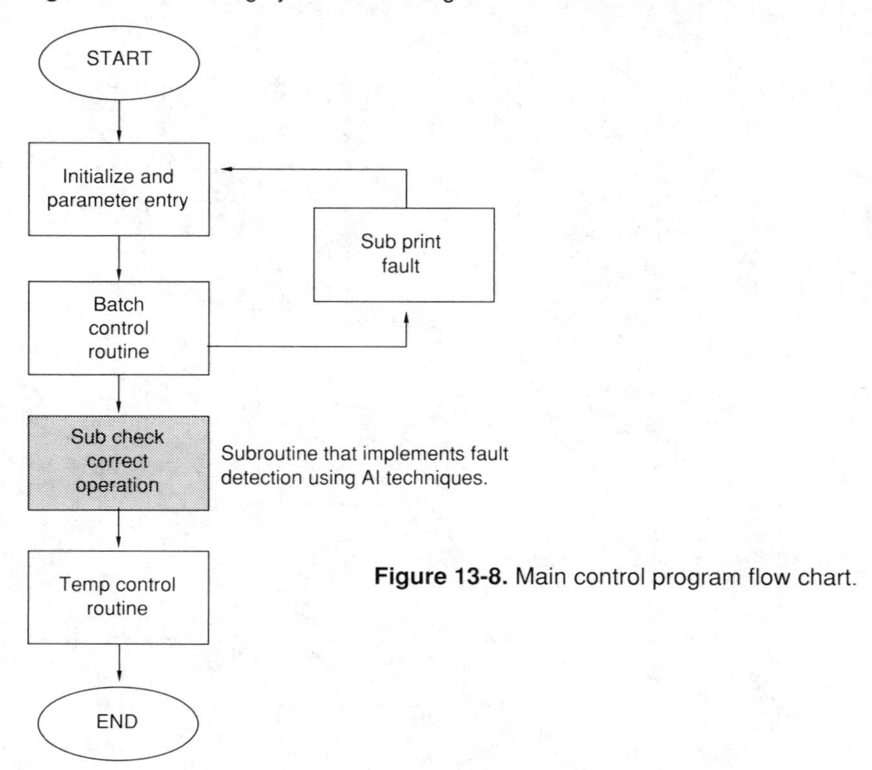

Figure 13-8. Main control program flow chart.

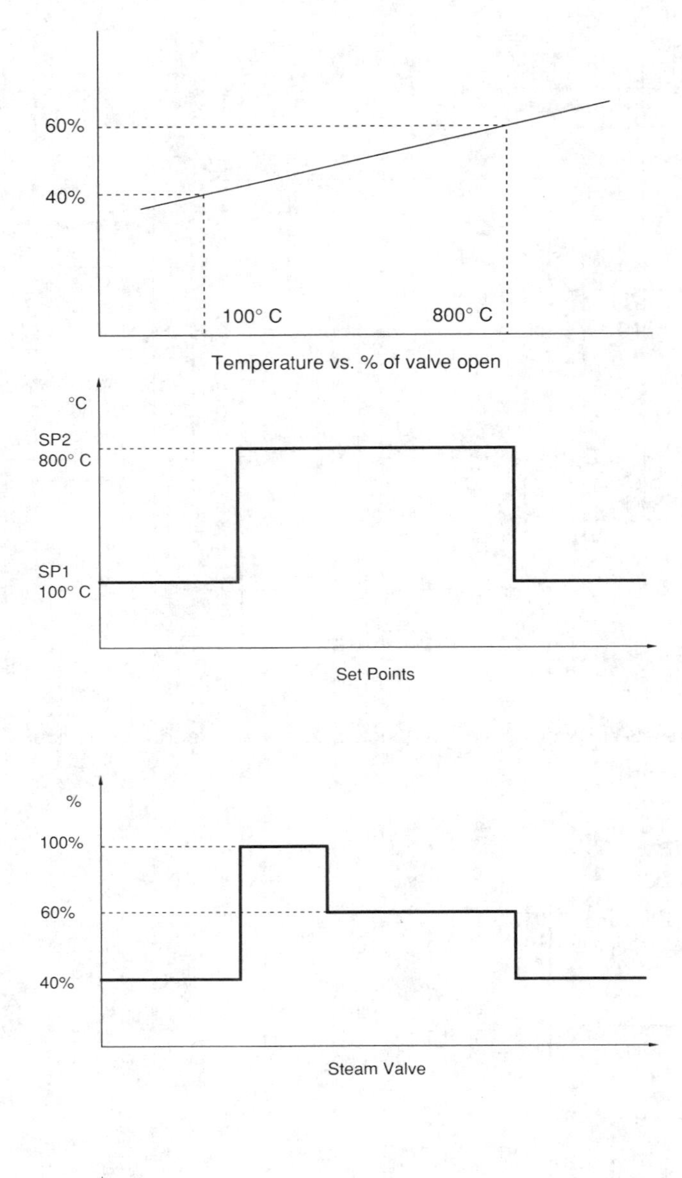

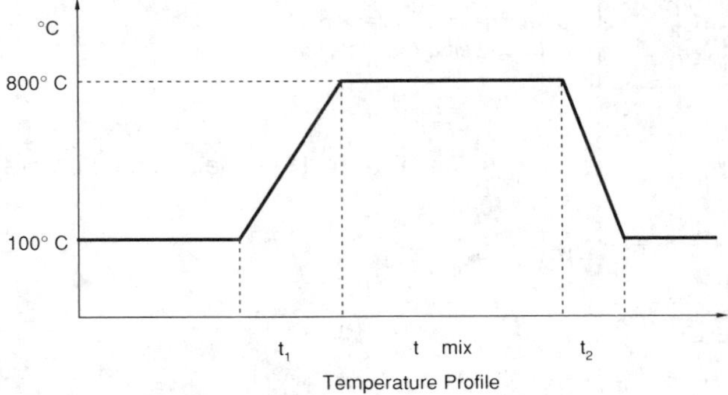

Figure 13-9. Temperature and steam valve relationship.

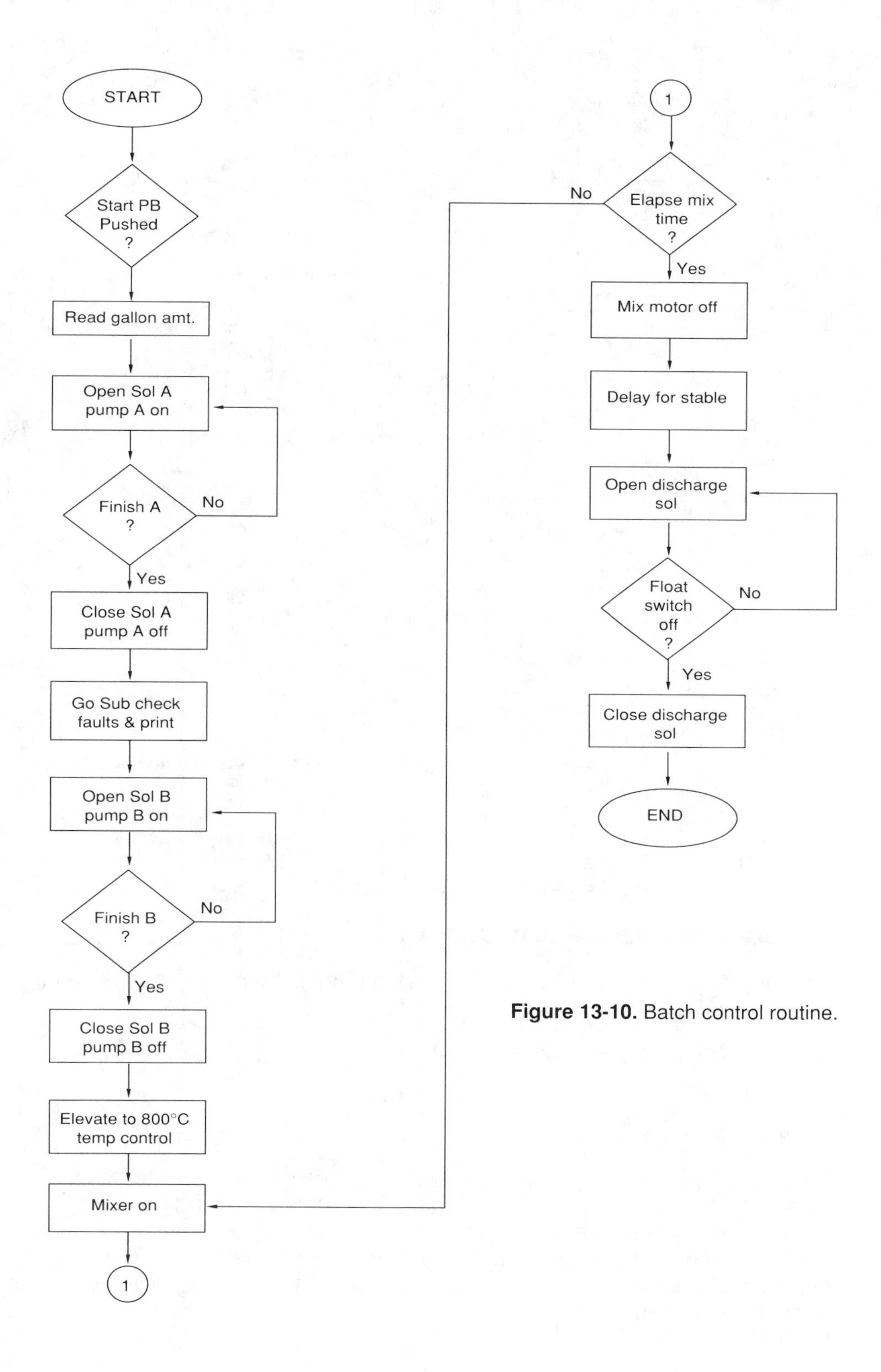

Figure 13-10. Batch control routine.

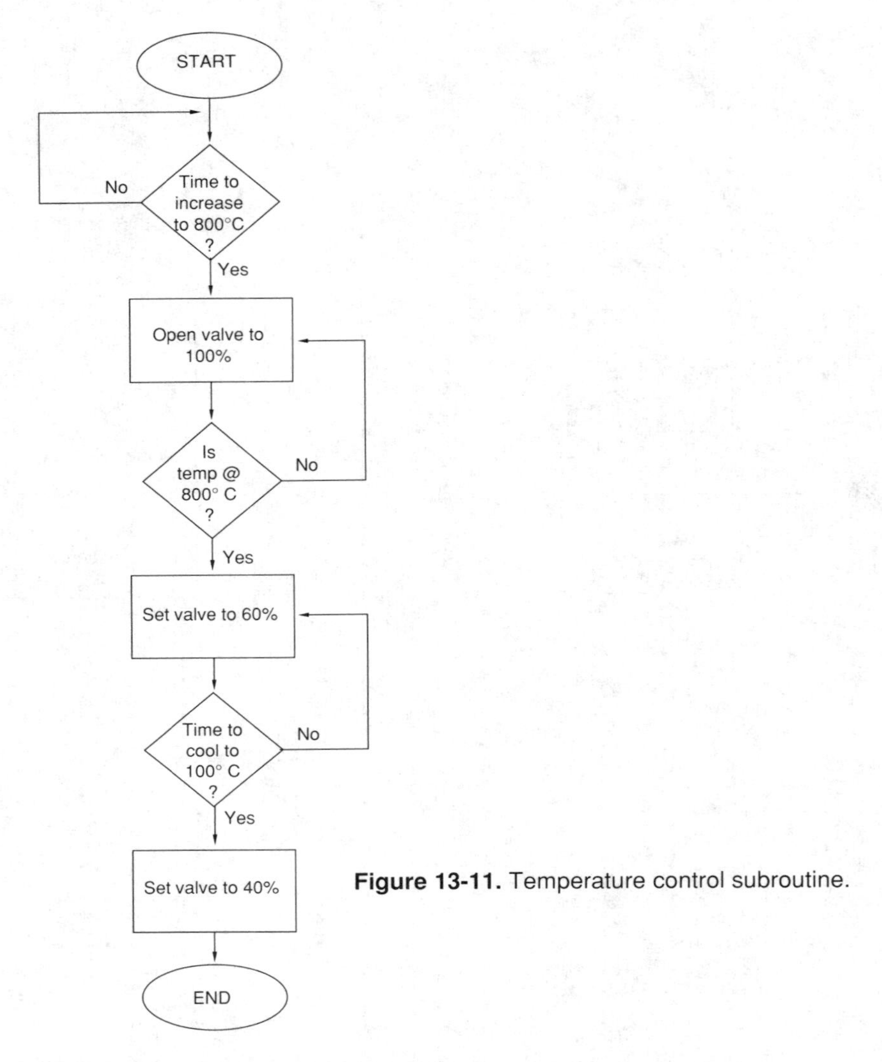

Figure 13-11. Temperature control subroutine.

Process Control Fault Detection

The fault detection that will be implemented in this system is subdivided in three major areas:

1. When the ingredients are being poured
2. During the elevation of temperature
3. During the cooling of the reactor vessel

For each of these areas a fault versus possible cause can be established. The detection of a fault is accomplished with feedback information from each of the controlling and measuring devices. This feedback information requires the incorporation of additional field devices to the system so that the faults can be verified via feedback. Table 13-2 shows the control devices and the feedback devices used to perform the system checks.

Control Devices	Feedback	Purpose
Valve	Limit Switch and Pressure Transducer	Check solenoid actuation in valve
Pump	Contacts and Pressure Transducer	Check pump operation
Flow meter	Pressure Transducer	Check ingredient flow
Steam Valve	Temperature Switch	Check steam valve

Table 13-2. Control and feedback devices used in batching system.

Rule Definitions

Based on the process control description and the possible failures we would like to detect, the following events and actions can take place.

0- If there is no fault, operation OK- proceed with entire process.

1- If there is a fault in the valve, then check pressure; if pressure is OK, then there is a fault in the limit switch. Continue process and stop. Fix fault.

If pressure is not OK then there is a fault in the valve. Abort process and alert operator. Open auxiliary valve. Fix fault.

2- If there is a fault in the pump, read pressure, if pressure is OK then there is a fault in the pump's contact; continue process and stop. Fix fault.

If pressure is not OK then there is a fault in the pump, abort process and alert operator. Open auxiliary valve. Fix fault.

3- If there is a fault in the valve and the pump, read pressure, if pressure is OK then fault is in limit switch and/or pump's contact. Continue process under pressure reading (back-up) and stop. Fix fault.

If pressure is not OK then there is a fault in the valve and/or pump. Abort process and alert operator. Open auxiliary valve. Fix fault.

4- If there is a fault in the meter, read pressure; if pressure is OK then there is a fault in the flow meter. Continue process and stop. Fix fault.

If pressure is not OK then there is a fault in the pump and/or valve. Abort process and alert operator. Open auxiliary valve. Fix fault.

5- If there is a fault in the meter and valve, read pressure; if pressure is OK then there is a fault in the flow meter and/or the limit switch. Continue process and stop. Fix fault.

If pressure is not OK then there is a fault in the valve and/or pump. Abort process and alert operator. Open auxiliary valve. Fix fault.

6- If there is a fault in the meter and the pump, read pressure; if pressure is OK then there is a fault in the meter and/or pump's contacts. Continue process and stop. Fix fault.

Fault	Device Malfunction	Possible Fault Cause		Feedback Decision Element
		Non-Critical i = 1 Stop at end	Critical i = 2 Abort batch	
$F_{1,i}$	Valve	Limit Switch	Valve	Pressure Transducer
$F_{2,i}$	Pump	Contacts	Pump	Pressure Transducer
$F_{3,i}$	Pump and Valve	Limit Switch and/or Contacts	Valve and/or Pump	Pressure Transducer
$F_{4,i}$	Flow meter	Flow meter	Valve and/or Pump	Pressure Transducer
$F_{5,i}$	Flow meter and valve	Flow meter and/or Limit Switch	Valve and/or Pump	Pressure Transducer
$F_{6,i}$	Flow meter and Pump	Flow meter and/or Contacts	Valve and/or Pump	Pressure Transducer
$F_{7,i}$	Flow meter, pump and valve	Flow meter and/or Contacts and/or Limit Switch	Valve and/or Pump	Pressure Transducer
$F_{8,i}$	Steam Valve (heating)	Temperature Transducer	Steam Valve	Temperature Switch TS2
$F_{9,i}$	Steam Valve (cooling)	Temperature Transducer or Steam Valve	—	Temperature Switch TS1

Table 13-1. Control device malfunctions and possible faults.

If pressure is not OK then there is a fault in the pump and/or valve. Abort process and alert operator. Open auxiliary valve. Fix fault.

7- If there is a fault in the pump and valve and flow meter, read pressure; if pressure is OK then there is a fault in the flow meter and/or pump's contacts and/or limit switch. Continue process and stop. Fix fault.

   If pressure is not OK then there is a fault in the valve and/or pump. Abort process and alert operator. Open auxiliary valve. Fix fault.

8- If there is a fault during the elevation of temperature to 800°C, and TS2 is ON in a predetermined time, then there is a fault in the temperature transducer. Continue process and stop.

   If TS2 does not respond in the predetermined period then there is a fault in the steam valve. Abort process and alert operator. Open auxiliary valve. Fix fault.

9- If there is a fault during the cooling of the tank back to 100°C, and TS1 is ON in a predetermined time, then there is a fault in the temperature transducer. Fix fault.

   If TS1 does not respond in the time period then there is a fault in the steam valve. Both faults are non-critical since the batch is finished and does not require an abort batch command. Fix fault.

We can define a set of faults F to represent the possible malfunctions that can be represented as:

$$F_{n,i} \text{ for } n = 0 \text{ to } 9,\ i = 1 \text{ to } 2$$

where $F_{0,i}$ is a correct operation (no-fault), and $F_{9,i}$ is a non-critical fault operation. For i = 1, we consider the fault to be non-critical and for i = 2 critical.

We can further describe the two types of faults that belong to the set $F_{n,i}$ non-critical (i = 1) or critical (i = 2).

$$F_{n,i} \in F_{n,1} \text{ or } F_{n,2} \text{ for } n = 1 \text{ to } 9$$

The actions taken for non-critical faults are: alarm the operation, continue process and stop at end of batch, indicate to operator possible faulty devices. For critical faults, the actions taken are: abort batch process, alarm to operator of critical fault, open auxiliary valve, indicate to operator possible faulty devices.

Application Summary

Applying AI techniques to a control system would most likely involve additional hardware and software to the system. The extent of complexity may vary upon how much fault detection is desired. In this example, rules for only one ingredient were presented, and although the second ingredient may be similar, the programming for it would still have to be performed and it may be lengthy.

Additional intelligence could be added to a similar system by storing data from the process. For instance, how many times the pump has been turned ON, and the contact status read back to the system, how many times the valve has been turned ON and OFF and the limit switch responded, etc. This information in conjunction with last time and type of failure, when and how it was fixed, and when the last maintenance was performed, would allow the system to identify cases in which two possible causes are generated for a single fault.

The added information has to be kept in the global data base and decisions can be made based on the probabilities assigned or calculated throughout several past process performances. Undoubtedly, the more intelligence added to a system, the more productive the system will be. This intelligence will mean less time to fix any possible downtime problems and will create a safer process environment.

Chapter

—14—

LOCAL AREA NETWORKS

"What a deal of talking there would be in the world if we desired at all costs to change the names of things into definitions."

—Lichtenberg

As control systems become more complex, they require more effective communication schemes between the various system components. Some machine and process control systems require programmable controllers to be interlocked so that data can be passed among them to accomplish the control task efficiently. Other systems require a communication scheme that will allow centralized functions, such as data acquisition, system monitoring, maintenance diagnostic and management production reporting, to be designed into a plant-wide system to provide maximum efficiency and productivity.

This chapter will present *local area networks* and the role they play in achieving factory integration. It also includes a comparison of several communication networks that are currently in use.

14-1 PRINCIPLES OF LOCAL AREA NETWORKS

Definition

The most general definition of a local area network (LAN) is "a high-speed medium-distance communication network". The maximum distance between two nodes on the network is usually given as at least one mile. Most definitions of a local network usually require that it supports at least 100 stations. The transmission speed ranges from 56 kilobaud to 10 megabaud. An industrial network is one which meets the following criteria:

- Capable of supporting real-time control
- High data integrity (error detection)
- High noise immunity
- High reliability in harsh environment
- Suitable for large installations

The other two most common types of local networks are business-system networks, such as Ethernet, and parallel-bus networks, such as Cluster/One. Business networks do not require as much noise immunity as industrial networks because they are intended to be used in an office environment. The access time requirements are also less stringent. A user of a business work station can easily wait a few seconds for information, but a machine being controlled by a PLC may need information within milliseconds. Parallel-bus networks are intended for microcomputers and minicomputers in office environments over short distances.

Advantages

Before local area networks came into use, two methods were used to communicate between PLCs. The first method was to connect an output card of one PLC to an input card of a second PLC via a pair of wires. This method, which provided only one bit of information per pair of wires, could be expensive to install and very cumbersome to use. For a large installation, wiring costs can be greatly reduced through the use of a local area network. In the second method, PLCs communicated through their programming ports via a central computer. Until recently, the computer usually had to be customer-supplied and programmed. This method limited the data throughput to the baud rate of the PLCs programming port (as do some local area networks) and has the disadvantage that if the central computer fails, the network becomes unusable (See Star Topology). Through the use of local area networks, large amounts of usable data can be exchanged among PLCs and any other hosts in a very efficient manner, through a dedicated communication link.

LAN Applications of the PLC

The most common applications of networks are centralized data acquisition and distributed control. Data collection and processing performed in individual controllers can burden the processor scan time, consume large amounts of memory, and tends to complicate the control logic program. These disadvantages can be eliminated by providing a data highway configuration in which all data is passed to a host computer that will perform any data processing. In distributed control applications, control functions once performed by a single controller are distributed among several controllers. Besides eliminating the disadvantage of dependence upon a single controller, performance and reliability are usually improved.

In order to use the distributed processing approach, the local area network and the PLCs attached to it must provide the following functions:

- Communication between programmable controllers
- Upload to a programmer or host computer from any PLC
- Download from a programmer or host computer to any PLC
- Read/Write I/O values/registers of any PLC
- Monitoring of PLC status and control of PLC operation

14-2 TOPOLOGIES—NETWORK GEOMETRY

The topology of a local area network defines the geometry of the network or how individual nodes are connected to the network. The major factors affected by topology are throughput, implementation cost, and reliability. The basic topologies used today are: *Star, Common Bus,* and *Ring.* It should be noted, however, that a large network, such as the one shown in Figure 14-1, may consist of a number of interconnected topologies.

Star

As previously mentioned, the first PLC networks consisted of a multi-port host computer with each of its ports connected to the programming port of a PLC. This arrangement is known as a "star" topology and is shown in Figure 14-2. The *network controller* shown can be either a computer, a PLC, or other intelligent host. Most

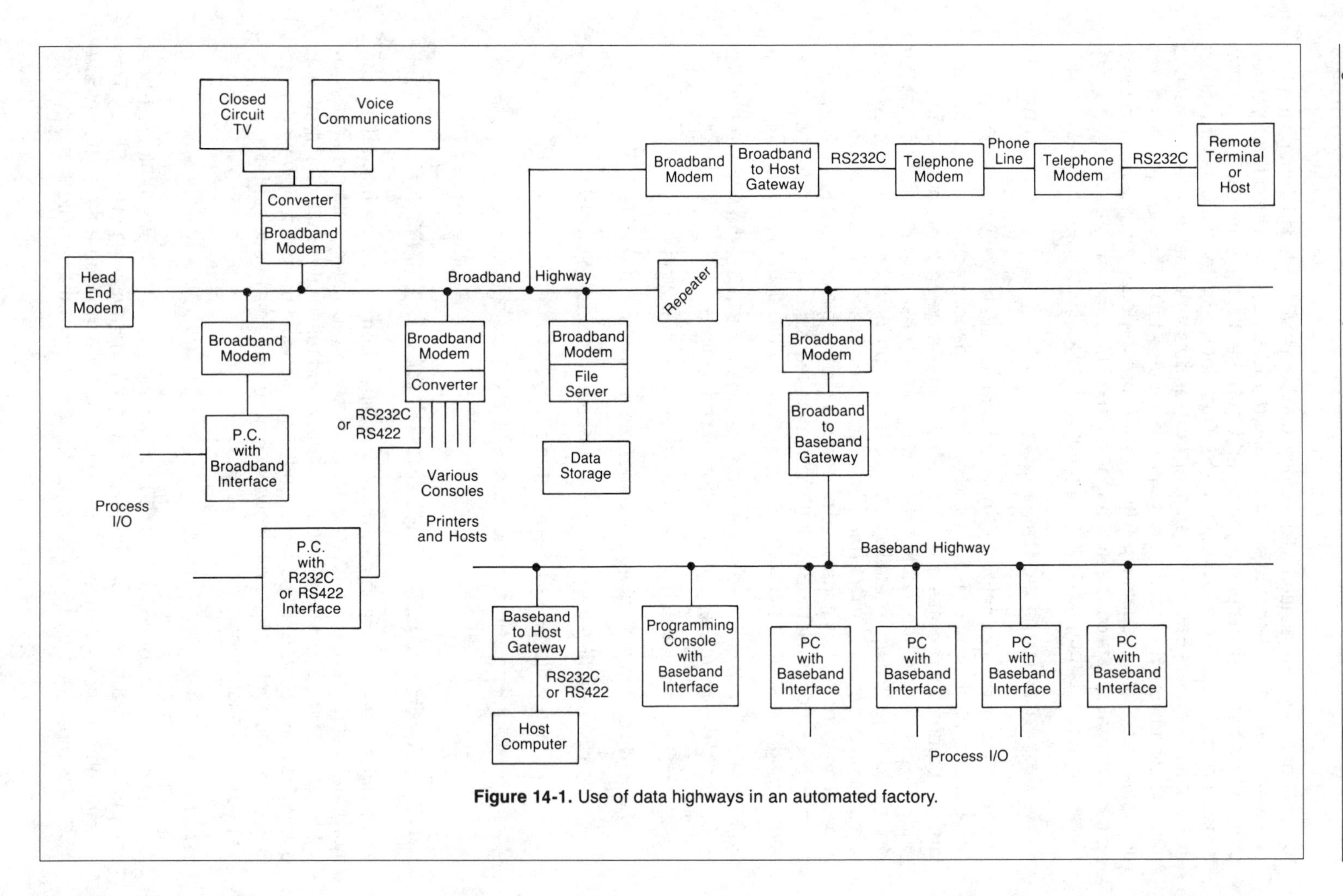

Figure 14-1. Use of data highways in an automated factory.

commercial computer installations are star networks in which many terminals are tied to a central computer. The main advantage of this topology is that it can be implemented with a simple point-to-point protocol. Each node can transmit whenever necessary. If error checking is not required, or a simple parity bit per character will suffice, then a "dumb" terminal can be a node. The star topology, however, has the following disadvantages:

- It does not lend itself to distributed processing due to dependence on a central node.
- The wiring costs are high for large installations.
- Messages between two nodes must pass through the central node, resulting in low throughput.
- There is no broadcast mode, which can lower throughput even more.
- Failure of the central node will bring down the network.

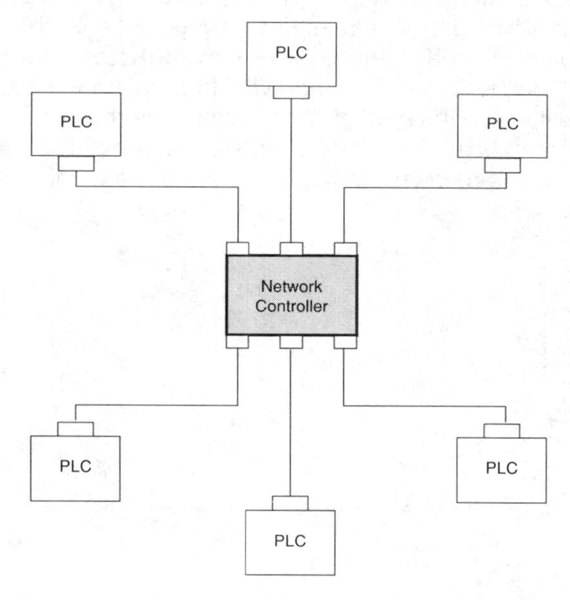

Figure 14-2. Star topology.

Common Bus

The common bus topology is characterized by a main trunkline in which individual nodes are connected to a PLC in a *multi-drop* fashion as shown in Figure 14-3. Contrasted to the star topology, communication can take place between any two nodes without passing information through a network controller. An inherent problem, however, is determining which node may transmit at any given time so that data collision is avoided. Several methods *(access methods)* of communication have been developed to solve this common bus problem.

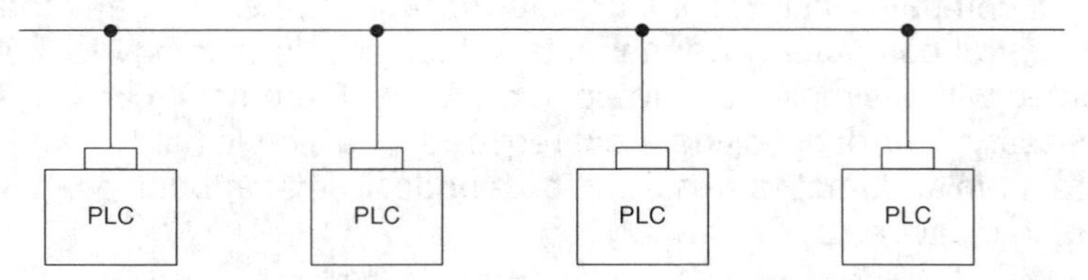

Figure 14-3. Bus topology with masterless access.

Common bus topologies are very applicable to distributed control applications since each station has equal independent control capability and can exchange information at any given time. In case a station must be added or removed from the network, very little reconfiguration is needed. The bus topology generally uses a coaxial cable, with proper terminators, as the communication media. The main disadvantage is the dependence of a shared bus to service all the nodes. A break in the trunkline could affect many nodes.

Another implementation of the bus topology consists of several slave controllers and a network controller which acts as a master (see Figure 14-4). In this configuration, the master sends data which is received by the slaves; if data is needed from a slave, the master will poll (address) the slave and wait for a response. No communication takes place unless it is first initiated by the master. The master/slave bus topology is usually implemented using two pairs of wires. On one pair of wires, the master sends, and all the slaves receive. On the other pair of wires, all the slaves transmit and the master receives.

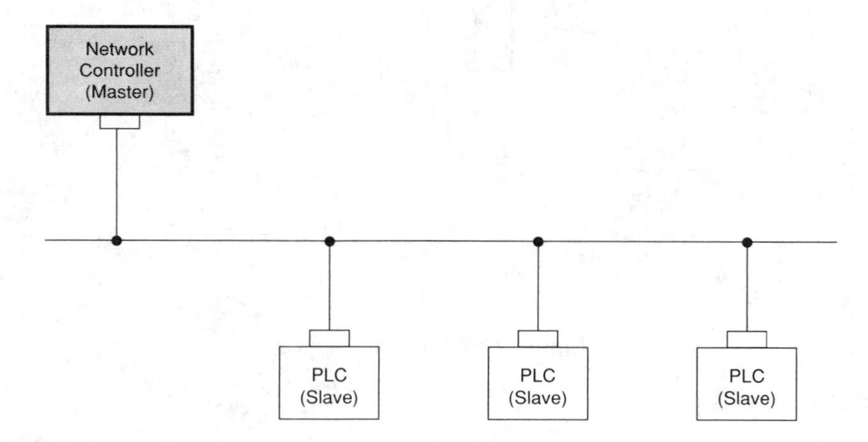

Figure 14-4. Bus topology with master/slave access.

Ring

The "ring" topology, shown in Figure 14-5, has not found its way into the industrial environment because failure of any node (not just the master) will bring down the network unless the failed node is bypassed. It is mentioned here, since it does not require multi-dropping and is, therefore, a good candidate for a fiber optic network.

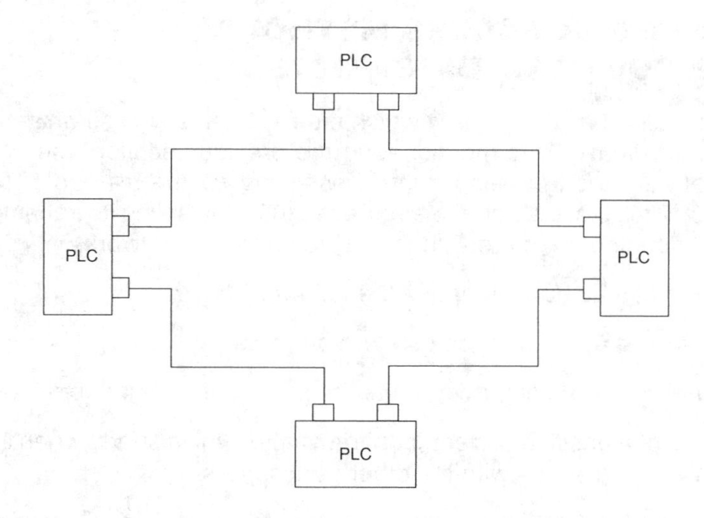

Figure 14-5. Ring topology.

Star-Shaped Ring

Some LAN manufacturers, like Proteon, have overcome the problem of node failure through the use of a *wire center*. The wire center, shown in Figure 14-6, allows failed nodes to be automatically bypassed. As can be seen, however, this requires twice as much wire as the star topology. Therefore, it would have to offer some other significant advantage (such as use of fiber optics) before being applied to large installations.

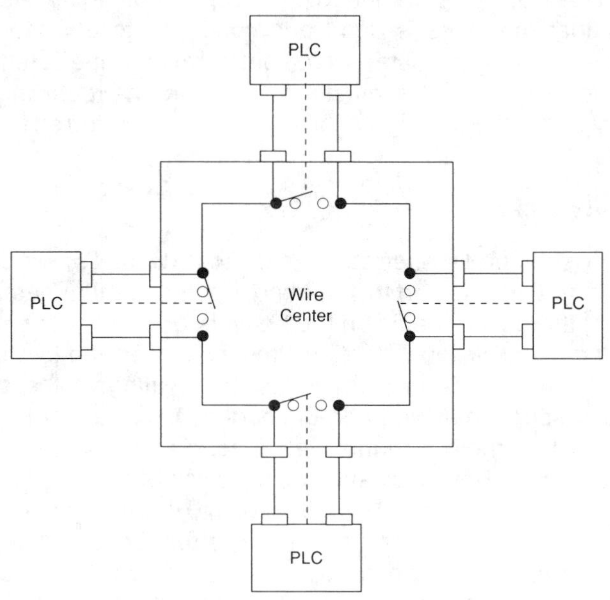

Figure 14-6. Star-shaped ring topology.

14-3 NETWORK ACCESS METHODS —NODE COMMUNICATION

The access method is a manner in which the PLC gains access of the network for transmitting information. As mentioned in the previous section, the bus topology requires that each node takes a turn transmitting on the medium. This process requires, first of all, that the node has some means of disabling its transmitter in such a way that it does not interfere with the operation of the network.

This operation can be done in one of the following ways:

- Use of a modem which can turn off its carrier
- Use of a transmitter which can be put into a high-independence state.
- Use of a passive current-loop transmitter which shorts when inactive and is wired in series with the other transmitters.

The most commonly used access methods are: *polling, collision detection,* and *token passing.*

Polling

The access method most often used in master-slave protocols is known as "polling". Polling is a technique by which each station (slave) is polled, or interrogated, in sequence by the master, to see if it has data to transmit. The master sends a message to a specific slave and waits a specific amount of time for the slave to respond. The slave should respond by sending data or a short message signifying that it has no data to send. If the slave does not respond within the allotted time, the master assumes that the slave is dead and continues to poll the other slaves. Interslave communication in a master/slave configuration is inefficient since polling requires data to be first sent to the master and then to the receiving slave. Since polling is used in master/slave configurations, it is generally referred to as the *master/ slave access method.*

Collision Detection

Collision detection is generally referred to as CSMA/CD for *Carrier Sense Multiple Access with Collision Detection.* In this method, each node that has a message to transmit waits until there is no traffic on the network and then transmits. While the node is transmitting, the collision-detection circuitry is checking for the presence of another transmitter. If a collision (two nodes transmitting at the same time) is detected, the transmitter is disabled and the node waits a random amount of time before trying again. This method works well as long as there is not an excessive amount of traffic on the network. However, each collision and retry takes time which cannot be used for transmission of data. Therefore, the throughput drops off and access time increases as traffic increases. For this reason, collision detection has not been popular in control networks. It is, however, quite popular in business applications. Collision detection can be used for data gathering and program maintenance in large systems and in real-time distributed control applications with a relatively small number of nodes.

Token Passing

Token passing is an access technique that is used to eliminate contention among PLC stations that are trying to gain access to the network. Token passing can be thought of as a form of distributed polling. A token is a message granting a polled station the exclusive but temporary right right to control the network (i.e. transmit information). This right, however, must be relinquished to a next designated node upon termination of transmission.

In a common bus network configuration using the token pass technique, each station is identified by an address. During operation, the token is passed from one station to the next in a sequential manner. The node which is transmitting the token also knows the address of the next station to receive the token. In token passing, one or more information packets containing source, destination, and control data circulate in the network. This information is received by each node and taken if needed. If the node has information to send it sends it in a new packet.

In a typical scenario, station 10 passes the token to station 15 (next address), which in turn will pass the token to station 18 (next address of 15). If a next station does not transmit to its successor within a fixed amount of time (token pass timeout), then the token-passing station assumes that the receiving station has failed. In this case the originating station starts polling addresses until it finds a station that accepts the token. For instance, if 15 fails, station 10 will poll 16 and 17 without response since they are not present in the network, and then polls 18, which responds to the token. The new receiving station will become the new successor and will remove the failed station from the network (18 will be 10's next address). The failed station will be patched-out.

The time required to pass the token entirely around the network is dependent on the number of nodes, and can be approximated by multiplying the token holding time by the number of nodes in the network. the token pass access method is preferred in applications requiring distributed control containing many nodes or having stringent response time requirements.

14-4 COMMUNICATION MEDIUM

If installed properly, most local area networks can be connected with any of the media discussed in this section. The installation includes the appropriate physical connectors and the correct electrical terminations. Media types commonly used for PLC networks include twisted-pair conductors, coaxial cables, and optical fibers. The type of media used and the number of nodes installed will affect the performance of the network (i.e. speed and distance). Figure 14-7 shows a comparison of different communication methods used with these media.

Twisted Pair Medium

Twisted pair has been used extensively in industry for point-to-point applications, over distances of up to 4000 feet, and transmission rates as high as 250 kilobaud.

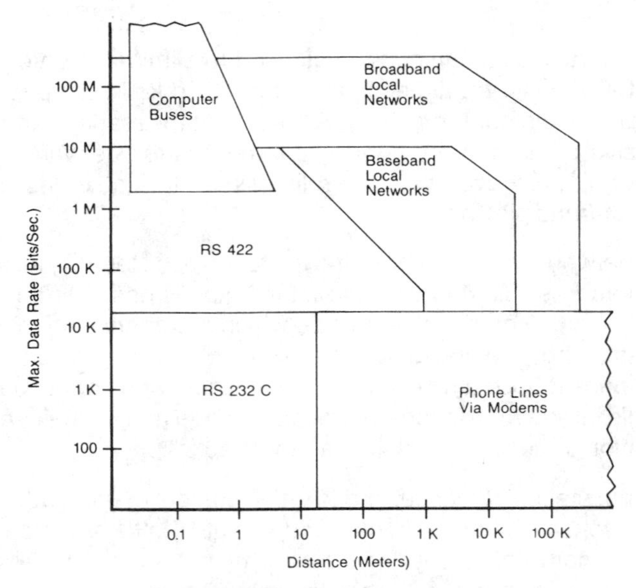

Figure 14-7. Communication comparisons.

Twisted pair is relatively inexpensive and has fair noise immunity, which is improvedif shielded. Performance drops off rapidly as nodes are added to a twisted pair bus.

Baseband Coax Medium

The performance limitations of twisted pair are mainly due to its nonuniformity. The characteristic impedance varies throughout the cable, making reflections difficult to reduce because there is no "right value" of termination resistor to use. Coaxial cable, however, is extremely uniform, thus eliminating reflections as a problem. The limiting factor then becomes capacitive and resistive loss. Baseband coax has been used in local area networks at speeds up to 2 megabaud and distances up to 18,000 feet. It is usually 3/8 inch in diameter.

Broadband Coax Medium

Broadband coax is 1/2 to 1 inch in diameter and has been used for years to carry cable television signals. It can support a transmission rate of 150 megabaud. Although this type of coax can be used to increase distance on a baseband network, it is intended to be used on a broadband network. These networks use frequency division multiplexing to provide many simultaneous channels. Each channel has a different RF carrier frequency. Broadband local area networks use just one of these channels and one of the access methods previously mentioned. The transmission rate on the channel is typically 1, 5 or 10 megabaud. Broadband local area networks are capable of covering as much as 30 miles through the use of bidirectional repeaters and can support thousands of nodes. One advantage of using broadband is that the network can be implemented on an existing broadband network. The other channels of such a network are typically used for video, computer access, and various monitor and control functions.

Each broadband channel consists of two channels—a high-frequency forward channel and a low-frequency return channel. If only two nodes need to communicate, one can transmit on the forward channel and the other can transmit on the return channel. In a multidrop network, a head-end modem is needed to retransmit the return channel on its corresponding forward channel. The repeaters amplify the forward channel signals in one direction and the return channel signals in the other direction. An example of a broadband network with a baseband subnetwork was shown in Figure 14-1.

Fiber Optics Medium

The main shortcoming of fiber optics today is that a low-loss tap has yet to be developed. This deficiency eliminates fiber optics for use in a large bus topology, but not from the star or ring topologies. In addition, fiber optics is three to four times more expensive than baseband coax, and optical couplers are several times more expensive than strictly electrical interfaces. Fiber optics does, however, have some impressive advantages. First, it is totally immune to all kinds of electrical interference. Second, it is small and light-weight. Transmission rates of up to 800 megabaud have been achieved at distances of 30,000 feet. In light of these facts, it would appear that fiber optics is bound to increase in industrial usage as the technology develops.

14-5 UNDERSTANDING NETWORK SPECIFICATIONS

The purpose of this section is to assist the reader in determining if a particular network being considered can support a given application. Table 14-1 has been provided for the purpose of comparing the networks offered by various vendors. It lists all of the readily available specifications for each network, but some specifications, such as response time, can be difficult to determine. Each aspect of the network which must be considered is examined in the following topics.

Maximum Number of Devices

For the system being considered, the system designer must determine how many nodes are required on the network and what type of device is to be used at each node. The device may be a PLC, a vendor-supplied programmer, a host computer, or an intelligent terminal. As discussed, it must be determined if the network will support each type of device and exactly how that device will be interfaced to the network (hardware and software). If the device is a PLC, the model of the PLC must also be chosen. It should be noted that some models may not have the capability for being interfaced to the network. The network must be capable of supporting the number of nodes required for the current application, plus some reasonable number of nodes for future expansion.

Maximum Length

The length of a data highway is generally specified in two parts: the maximum length of the main cable and the maximum length of each drop cable which is used between the device and the main cable. The values given in Table 14-1 are for main cables.

The drop lengths are usually in the range of 30 to 100 feet. It is recommended, however, that drop lengths be kept as short as possible, since any drop will introduce some reflection onto the network. The ideal case (electrically) is to run the main cable right to the device and back out again, even though this procedure increases wiring costs somewhat. Another important piece of information that must be obtained from the vendor is the type of cable which has to be used to achieve the specified length. If the system needs the maximum length, the proper type of cable must be used. If the system requires a much shorter cable, money can usually be saved by using a less expensive cable.

Response Time

Response time, as used in this book, means the time between an input transition at one node and the corresponding output transition at another node. As shown in the following equation, response time is the sum of the time required to perform the operations necessary to detect the input transition, transmit the information to the output node, and operate the output.

$$RT = IT + 2 \times ST1 + PT1 + AT + TT + PT2 + 2 \times ST2 + OT$$

where:

IT = input delay time: the electrical delay involved in detecting the input transition

ST1 = scan time for sending node

ST2 = scan time for receiving node

PT1 = processing time: for the sending node, between solving the program logic and becoming ready to transmit the data

PT2 = processing time: for the receiving node, between receiving the data and having data ready to be operated on by the program logic

AT = access time: time involved in becoming ready to transmit and transmitting

TT = transmission time: time to transmit the data (this is the only time which is directly proportional to baud rate)

OT = output delay time: the electrical delay involved in creating the output transition

The scan time includes I/O update time and any other overhead time, as well as program logic execution time, and can be easily defined as the time between I/O updates. The scan time is doubled in the above equation to include the case where the input signal changes just after the I/O update. In this case, the logic is first executed with the old information, the I/O update is done, then the logic is executed with the new information. This causes a two-scan delay.

I/O delay times and scan times are readily available values. Transmission time can be determined once the data rate and frame length are known. The data rate can be equal to the baud rate ,but it is usually less. Synchronous systems, which use Manchester encoding, have a data rate which is one half of the baud rate. Asynchronous systems have a data rate which is 80% of the baud rate due to the start and stop bits which accompany each 8 data bits. The access time and the two processing times are dependent upon the particular installation and generally have to be obtained from the manufacturer. If equipment is available, it is much easier and more accurate to determine the overall response time through actual measurements. Section 14-7 presents a simple procedure for performing this measurement.

It should be noted that the parameter which needs to be determined here is not the average response time, but the maximum response time. Therefore, steps must be taken to create a worst-case environment during response time measurements. Creating this scenario involves tasks such as downloading programs and monitoring points while taking the measurements, because this sort of activity increases PLC scan times and network access times.

Manufacturer	Network Name	Max. Nodes	Max. Length(ft.) Max. Baud Rate	Baud Rate (max)	Access Method	Comments
Allen-Bradley	Data Highway	64	10,000	56K	Token	
General Electric	GEnet	999	15,000	5M	Collision Detection	IEEE 802.3 Compatible.
Gould-Modicon	Modbus	247	15,000	19.2K	Master-Slave	Uses 2 Twisted Pairs, Shielded.
Gould-Modicon	Modway	250	15,000	1.544M	Token	
Industrial Solid State Controls	Copnet	254	32,000	115.2K	Master -Slave	Includes Interfaces to Allen-Bradley, Gould-Modicon and Texas Instruments PLCs. HDLC Protocol.
Measurex	Data-Freeway	63	10,000	1M	Collision Detection	
Reliance Electric	R-Net	255	12,000	800K	Token	ASCII/X3.28/HDLC Gateway. Uses HDLC Framing.
Square D	SY/Net	200	2000	500K	Timed Token	Worst case access = 600msec with 50 PLCs.
Texas Instruments	TIWAY I	254	10,000	115.2K	Master -Slave	HDLC Protocol
Texas Instruments	TIWAY II	2^{32}	32,000	5M	Token	Broadband IEEE 802.4 compatible
Westinghouse	WDPF	254	18,000	2M	Token	100msec Fixed Access Time.10,000 Points/Sec Throughput.
Westinghouse	Westnet	50	10,000	1M	Master -Slave	Gateway Interface to Westnet
Comments			Broadband networks cover unlimited distances using CATV repeaters.	Lower baud rates allow greater lengths		

Table 14-1. Sample network specifications.

Throughput

Some manufacturers specify the LAN throughput. This value usually represents the number of I/O points, which can be updated per second through the network. The throughput value is not enough to derive actual values for access time and data rate, even though it gives the system designer some idea of these values. In addition, throughput is sure to vary with system loading as a result of each node's processing time. Therefore, to have an accurate measurement of throughput, the conditions under which the measurement was taken must be known.

Devices Supported

When considering each device in the system, the designer must ask not only *"will the local area network support this device"*, but also *"what is involved in connecting the device to the network."* For user-supplied devices, the designer must also determine what support software will be required.

Programmable Controllers. All of the available networks support at least some of the manufacturer's PLCs. The PLC is generally connected to the network through a separately-purchased interface unit. The interface unit connects to the PLC either through a high-speed parallel bus or through the PLC's serial programming port. In the latter case, two additional terms must be added to the response time equation: the transmission time on the programming port and the programming port processing time.

Programming Devices. Although most manufacturers offer some type of programming device which can be connected to the network, some manufacturers do not. In this case, all programming must be done through the programming port of the individual PLC. A programming unit connected to the network provides centralized programming of any PLC on the network and also various monitoring and control functions if available.

Hosts. Host support usually means that a user-supplied host computer can be used to perform programming functions provided that it is programmed to conform to the manufacturer's protocol. It is usually connected to the network through a device called a "gateway". The gateway contains a network port and another port (usually RS-232), which is connected to the host. This gateway greatly simplifies the software which the user must write for the host because the host-to-gateway link requires only a simple point-to-point protocol, rather than the masterless multidrop protocol of the network. The gateway also provides the appropriate electrical interfaces. Since most computers can provide an RS-232 port, no additional hardware is usually required.

Intelligent Terminals. The type of intelligent terminal referred to here is actually a small host computer complete with operating system and mass storage. It is interfaced to the network in exactly the same way as a large host computer. Anyone considering using one of these terminals on a network should investigate the software requirements closely to determine if the terminal's operating system will support the requirements. Some operating systems, for instance, provide for the transmission of only ASCII data, not binary data.

Gateways. In addition to the host gateway previously mentioned, some manufacturers may provide gateways to other multidrop networks. There are also other types of host gateways; for instance, the host interface might be a high-speed RS-422 synchronous type. In this case, a protocol designed for synchronous use, such as HDLC, would probably be used.

Application Interface

The question to answer here is *"how does the application program being executed by each PLC allow it to share information with other PLCs?"* Most manufacturers provide at least one of the following methods.

- Reading of registers in other PLCs.
- Writing to registers in other PLCs.
- Reading and writing of network points or registers.

For example, the state of an input on one PLC can be detected by another PLC on the network through the use of a network coil and network contact. This configuration is described in Figure 14-8.

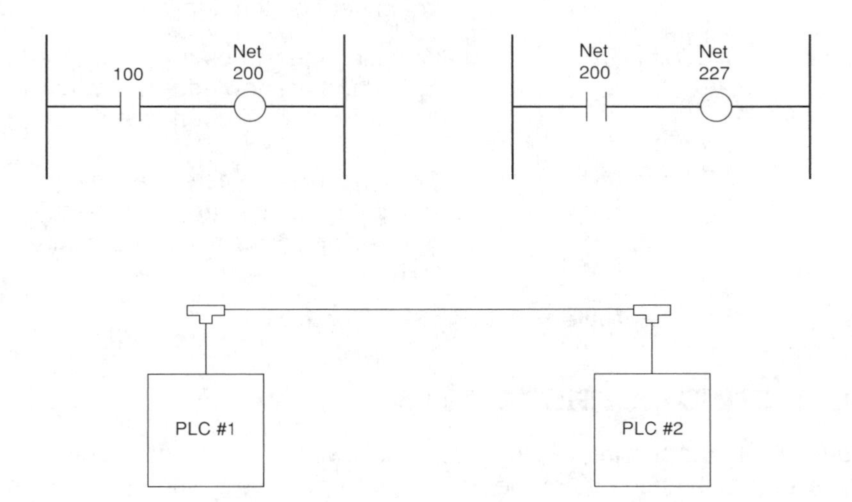

Figure 14-8. Input detection through a PLC network.

Whenever the network coil (Net 200) in PLC#1 is energized, the network contact Net 200 (—| |—)on PLC#2 will close. This contact can be used like any other contact in PLC#2's ladder program. It is up to the user to insure that each network coil is used by only one PLC on the network. Reading and writing of registers is usually done via functional blocks. Care must be taken to assure that the capabilities provided are sufficient to support the communications needs of the application. Chapter 9 shows some of the typical network instructions found in PLCs.

Layer	Level Name	Function
Layer 7	Application	The level seen by users; user interface.
Layer 6	Presentation	Controls functions requested by the user; data is restructure from other standard formats; code and data conversion.
Layer 5	Session	System to system connection; log-in and log-off is controlled here; establishing connection and disconnection.
Layer 4	Transport	Provides reliable data transfer between end devices; network connections for a given transmission are established by protocol.
Layer 3	Network	Outgoing messages are divided into packets; incoming packets are as-sembled into messages for higher levels. Establishes connections between equipment on the network.
Layer 2	Data Link	Outgoing messages are assembled into frame and acknowledgements; error detection or error connection is created.
Layer 1	Physical	Parameters such as signal voltage swing, bit duration and how electrical connections are established in this layer.

Table 14-2. ISO OSI reference model.

14-6 NETWORK PROTOCOLS

A *protocol* is a set of rules which must be followed if two or more devices are to communicate with each other. The protocol includes everything from the meaning of the data to the voltage levels on the wires. The protocols define how the following problems are to be handled:

- Communication line errors.
- Flow control to keep buffers from overflowing.
- Access by multiple devices.
- Failure detection.
- Data translation.
- Interpretation of messages.

ISO

In 1979, the International Standards Organization published its Open Systems Interconnection Reference Model. This model divides the various functions that protocols must perform into seven hierarchical layers. Each layer needs only to interface with its adjacent layers and is unaware of the existence of the other layers. The ISO's OSI reference model layers are tabulated in Table 14-2.

Strictly speaking, only layers 1 and 7 are absolutely required. The other layers need only to be added as more services are required (such as error-free delivery, routing, session control, data conversion, etc.). Most of today's networks contain layers 1, 2, and 7. As networks and protocols continue to develop, the other layers are quickly being implemented as required to allow connection to other networks.

IEEE 802

The IEEE Standards Project 802 was established by the IEEE Computer Society in 1980 for the purpose of developing a local network standard. Such a standard would allow equipment of different manufacturers to communicate via the local network. After studying all the users' requirements and all the manufacturers' desires, the committee decided to produce a standard that defines several types of local networks. Figure 14-9 shows the scope of the IEEE 802 standard. Industrial users and PLC manufacturers have been showing a great deal of interest in both the broadband and baseband versions of the token bus option, especially for the Manufacturing Automation Protocol.

Proprietary

All the LANs presented in Table 14-1 use protocols which were devised by the manufacturers. None of these protocols has been proposed as a standard, but some of the specifications are available to those who are interested in interfacing to those manufacturer's local area networks.

ANSI

The American National Standards Institute has produced standard protocols for layers 2 and 3 that have been widely accepted. The most well-known and widely used of these is X.25, which specifies layer 2 and 3. It has been implemented on many host computers to interface to publicly switched networks. The X3.28 standard specifies layer 2 for point-to-point or multidrop service with a variety of options. This standard is used by many PLC manufacturers on their programming ports.

EIA

The Electronic Industries Association produced the RS-232C standard to define the physical interface (layer 1) between a computer or terminal (DTE) and a modem (DCE). This standard was rapidly adopted by modem manufacturers and has provided users with a wide selection of compatible modems. Although the standard did not address interconnection of two DTEs (such as a computer and a terminal), it was adopted for this use by crossing the transmit and receive wires in the cable.

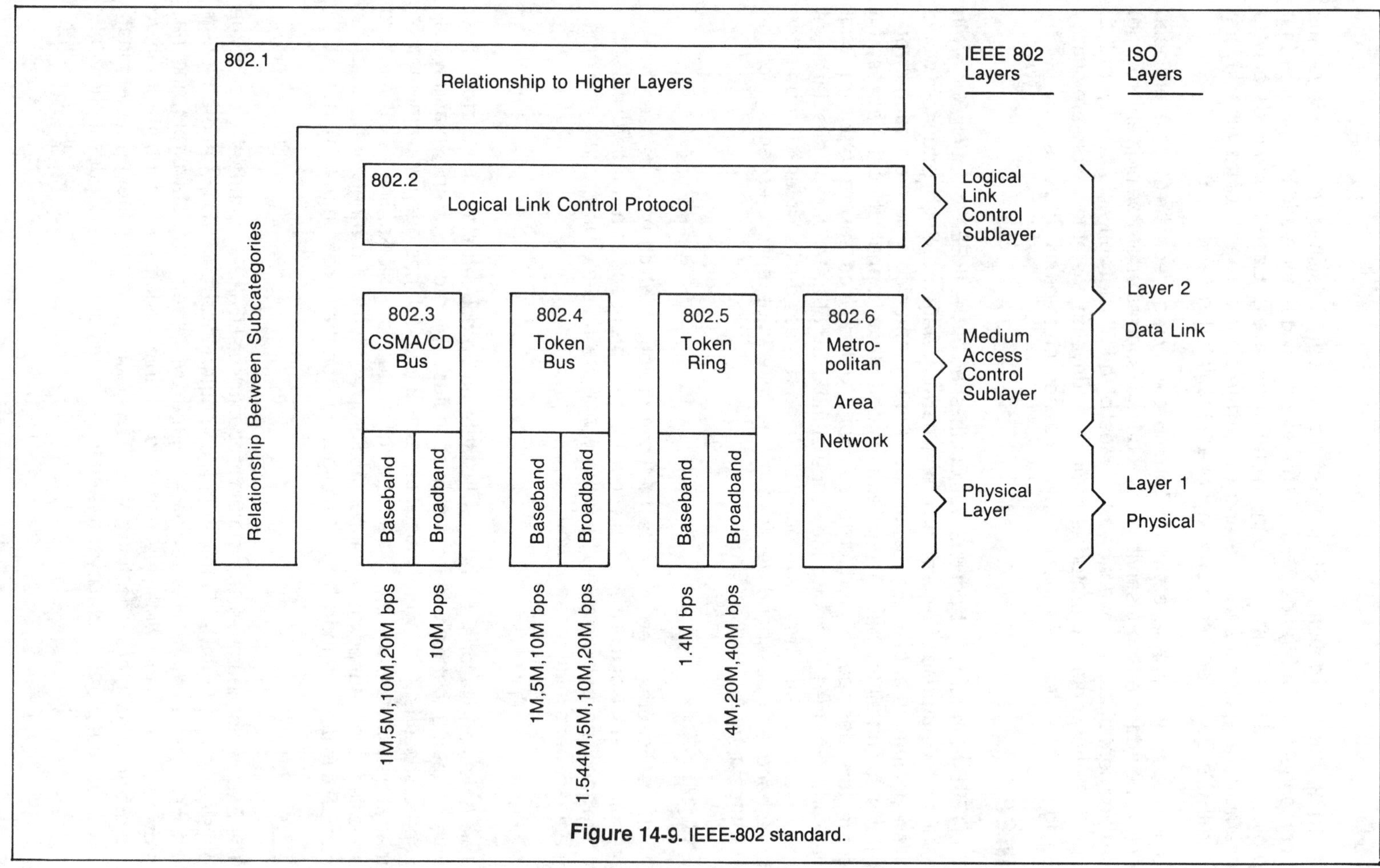

Figure 14-9. IEEE-802 standard.

The maximum transmission rate is 19.2 kilobaud. Although the specified maximum distance is only 50 feet, it has been used for distances up to several hundred feet. The RS-422 standard defines a physical layer which can support transmission rates of up to 10 megabaud at distances of 200 to 4000 feet. Additional information on these standards can be found in Section 6 of Chapter 8.

MAP

The Manufacturing Automation Protocol, generally referred to as MAP, is a communication specification spearheaded by General Motors Corporation to allow different types of computerized equipment to communicate with each other. The intention of this protocol is to have an international definition for each of the functions required for data communication. These functions are organized according to the OSI reference model developed by the ISO.

The MAP definition of each of the OSI layers will allow the support of a wide spectrum of applications. Once the majority of PLC and other computer based machine manufacturers decide on one protocol to follow, such as MAP, will real computer integrated manufacturing (CIM) take place. Devices in network nodes, although strange to each other, would be able to communicate and exchange information through a common protocol.

14-7 NETWORK TESTING AND TROUBLE-SHOOTING

As mentioned in the previous section, before a local area network is installed it should be tested to insure that it not only performs the desired function, but also provides the required response time. Application programming should be provided to monitor continuously the response time and take appropriate action if the response time goes beyond the maximum that the process will tolerate. This test may be performed by programming a *buzzer* circuit in which the contact closure is passed through every critical node on the network before it is returned. In this way, the pulse width of the created pulse is equal to the response time. This pulse width can be applied to a timer which is set to the maximum allowable response time. When the timer times out, the response time has been exceeded.

Trouble-shooting can be quite difficult unless both the manufacturer and the user take steps to simplify this task. The manufacturer can provide error counts and a self-test for each node. The user can provide application programming to detect the failure of each node. The extreme case of this would be to provide a "buzzer" and timer between each node and each other node. If the entire network goes down, it is probably due to a node with a transmitter which is shorted or constantly transmitting. The faulty node can be found by disconnecting each node, one at a time, and observing if communication is restored. Some manufacturers provide a network monitor which can be used to detect a failed node, open cable, or excessive electrical interference.

Chapter

—15—

GUIDELINES FOR INSTALLATION, START-UP, AND MAINTENANCE

"If I had been present at the creation, I would have given some useful hints for the better arrangement of the Universe."

—Alfonso the Wise, King of Castille

15-1 PLC SYSTEM LAYOUT

The design nature of programmable controllers includes a number of rugged design features that allow the PLC to be installed in almost any industrial environment. However, a little foresight during the installation will go a long way towards assuring proper system operation.

The system layout is a conscientious approach in placing and interconnecting the components not only to satisfy the application, but also to insure that the controller will operate troublefree in the environment in which it will be placed. With a carefully constructed layout, components are accessible and easily maintained.

In addition to the programmable controller equipment, the system layout also takes into account other components that form part of the total system. These other types of equipment include isolation transformers, auxiliary power supplies, safety control relays, and incoming line noise suppressors.

Although it is not necessary to install the controller in a controlled environment because it is designed to work on the factory floor, certain considerations should be followed for the proper installation of the system components. The best place for the PLC to be located is near the machine or process that it will control, if temperature, humidity, and electrical noise are not problems. Placing the controller near the equipment and using remote I/O where possible will minimize wire runs and simplify start-up and maintenance. Figure 15-1 shows a typical PLC installation.

Panel Enclosures and System Components

PLCs are generally placed in a NEMA-12 or JIC enclosure. The enclosure size depends on the total space required. Mounting the controller components in an enclosure is not always required, but is recommended for most applications to protect against atmospheric contaminants, such as conductive dust, moisture, and any other corrosive or harmful airborne substances. Metal enclosures also help to minimize the effects of electromagnetic radiation that may be generated by other equipment.

The enclosure layout should conform to NEMA standards, and placement and wiring of the components should take into consideration the effects of heat, electrical noise, vibration, maintenance, and safety. A typical enclosure layout is illustrated in Figure15-2 and can be used for reference with the following layout discussions.

General. The following recommendations are concerned with preliminary considerations for the location and physical aspects of the enclosure.

- The enclosure should be placed in a position that allows the doors to be opened fully for easy access to wiring and components for testing or troubleshooting.
- The enclosure depth should be enough to allow clearance between a closed enclosure door and any print pockets mounted on the door or the enclosed components and related cables.
- The enclosure's back panel should be removable to facilitate mounting of the components and other assemblies.

Figure 15-1. Installation of a Texas Instrument PM550 PLC based control system (Courtesy Columbia Gas Transmission Corp.)

- An emergency disconnect device should be mounted in the cabinet in an easily accessible location.
- Accessories, such as AC power outlets, interior lighting, or a gasketed plexiglass window to allow viewing of the processor or I/O indicators should be considered for installation and maintenance convenience.

Environmental. The effects of temperature, humidity, electrical noise, and vibration are important factors that will influence the actual placement of the controller, the inside layout of the enclosure, and the necessity for other special equipment. The following points are concerned with insuring favorable environmental conditions for the controller.

- The temperature inside the enclosure must not exceed the maximum operating temperature of the controller (typically 60° C max.).
- If "hot spots" such as those generated by power supplies or other electrical equipment are present, a fan or blower should be installed to help dissipate the heat.
- If condensation is anticipated, a thermostatically controlled heater may be installed inside the enclosure.
- If the area in which the system will be located contains equipment that generates excessive electromagnetic interference (EMI) or radio frequency interference (RFI), the enclosure should be placed well away from these sources. Examples of such equipment are welding machines, induction heating equipment, and large motor starters.

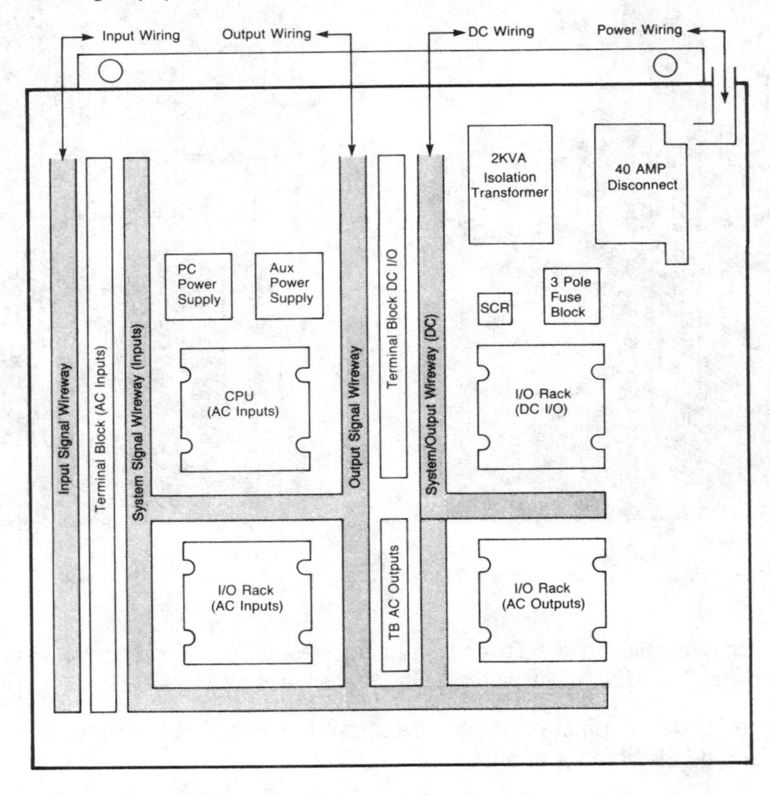

Figure 15-2. Typical enclosure layout diagram with terminal blocks and internal conduits.

- In cases in which the PLC enclosure must be mounted on the equipment to be controlled, careful attention should be given to insure that vibrations caused by that equipment do not exceed the PLC's vibration specifications.

Placement of PLC Components . The placement of the major components of a specific controller is dependent on the number of system components and the physical design or modularity of each component (see Figure 15-3). Although various controllers may have different mounting and spacing requirements, the following considerations and precautions are applicable when placing any PLC inside an enclosure.

- To allow maximum convection cooling, all controller components should be mounted in a vertical (upright) position. Some manufacturers may specify that the controller components can be mounted horizontally. However, in most cases, components mounted horizontally will obstruct air flow.
- The power supply (main or auxiliary) has a higher level of heat dissipation than any other system component and, therefore, should not be mounted directly underneath any other equipment. Typically, the power supply is installed above all other equipment and at the top of the enclosure with adequate spacing (at least ten inches) between the power supply and the top of the enclosure. The power supply may also be placed adjacent to other components, but with sufficient spacing.
- The CPU should be placed at a comfortable working level (e.g. at sitting or standing eye level) that is either adjacent to or below the power supply. If the CPU and power supply are contained in a single unit, then it should be placed towards the top of the enclosure with no other components directly above it, unless there is sufficient spacing.

Figure 15-3. Actual Allen-Bradley PLC enclosure layout (Courtesy Grinnell Corp.)

- Local I/O racks (same panel enclosure as CPU) are placed in any desired arrangement within the distance allowed by the I/O rack interconnection cable. Typically, the racks are placed below or adjacent to the CPU, but not directly above the CPU or power supply.
- Remote I/O racks (away from the CPU) and their auxiliary power supplies are generally placed inside an enclosure at the remote location. The same placement practices described for local racks can be followed.
- Spacing of the controller components to allow proper heat dissipation should adhere to the manufacturer's specifications for vertical and horizontal spacing between major components.

Placement of Other Components. In general, the placement of other components inside the enclosure should be away from the controller components, so as to minimize the effects of noise or heat generated by these devices. The following list outlines some common practices for locating other equipment inside the enclosure.

- Incoming line devices, such as isolation or constant voltage transformers, local power disconnects, and surge suppressors, are normally located near the top of the enclosure and alongside the power supply. This placement assumes that the incoming power enters at the top of the panel. The proper location of incoming line devices keeps power wire runs as short as possible, thus minimizing the chances of transmitting electrical noise to the controller components.
- Magnetic starters, contactors, relays, and other electro-mechanical components should be mounted near the top of the enclosure in an area segregated from the controller components. A good practice is to place a barrier with at least six inches separation between the magnetic area and the controller area. Typically, magnetic components are placed adjacent and opposite to the power supply or incoming line devices.
- If fans or blowers are used for cooling the components inside the enclosure, they should be placed in a location close to the heat generating devices (generally power supply heat sinks). When using fans, be sure that outside air is not brought inside the enclosure unless a fabric or other reliable filter is also used. This filtration will prevent conductive particles or other harmful contaminants from entering the enclosure.

Grouping Common I/O Modules. The grouping of the I/O will allow signal and power lines to be routed properly through the ducts, such that crosstalk interference will be minimized. The following are recommendations for grouping I/O modules.

- I/O modules should be segregated into groups such as AC input modules, AC output modules, DC input modules, DC output modules, analog input modules, and analog output modules whenever possible.
- If possible, a separate I/O rack should be reserved for common input or output modules. If it is not possible to reserve a complete rack for common modules, then the modules should be separated as much as possible within a rack. As an example, a suitable partitioning would involve placing all AC

together or all DC together, and if space permits, allowing an unused slot between the different groups.

Duct and Wiring Layout. The duct and wiring layout defines the physical location of wireways and the routing of field I/O signals, power, and controller interconnections within the enclosure. The enclosure's duct and wiring layout is dependent on the placement of I/O modules within each I/O rack (see Grouping Common I/O). The placement of these modules is determined during the design stages, when the I/O assignment takes place (see Chapter 8). Prior to defining the duct and wiring layout and assigning the I/O, the following practices should be considered to minimize electrical noise caused by crosstalk between I/O lines.

- All incoming AC power lines should be kept separate from low-level DC lines, I/O power supply cables, and all I/O rack interconnect cables.
- Low-level DC I/O lines, such as TTL and analog, should not be routed in the same duct in parallel with AC I/O lines. Whenever possible, always keep AC signals separate from DC signals.
- I/O rack interconnect cables and I/O power cables can be routed together in a common duct not shared by other wiring. When this arrangement is impractical or it is impossible to separate these cables from all other wiring, then they can be routed with low-level DC lines or routed externally to all ducts and held in place using tie wraps or some other fastening method.
- If I/O wiring must cross the AC power lines, it should do so only at right angles. This routing practice will minimize the possibility of electrical noise pick-up.
- When designing the duct layout, at least two inches should be allowed between the I/O modules and any wire duct. If terminal strips are used, then two inches should be allowed between the terminal strip and wire duct as well as between the terminal strip and I/O modules.

Grounding. Proper grounding is an important safety measure in all electrical installations. When installing electrical equipment, users should refer to the National Electric Code (NEC), article 250, which provides data such as the size and types of conductors, color codes, and connections necessary for safe grounding of electrical components. The code specifies that a grounding path must be permanent (no solder), continuous, and able to safely conduct the ground-fault current in the system with minimal impedance. The following grounding practices will have significant impact on the reduction of noise caused by electromagnetic induction.

- Ground wires should be separated from the power wiring at the point of entry to the enclosure. To minimize the ground wire length within the enclosure, the ground reference point should be located as close as possible to the point of entry of the plant power supply.
- All electrical racks or chassis and machine elements should be grounded to a central ground bus, normally located in the magnetic area of the enclosure. Paint or other non-conductive materials should be scraped away from the area where a chassis makes contact with the enclosure. In addition to the

ground connection made through the mounting bolt or stud, a one inch metal braid or size #8 AWG wire (or manufacturer's recommendation) can be used to connect each chassis and the enclosure at the mounting bolt or stud.

- The enclosure should be properly grounded to the ground bus; making sure that good electrical connections are made at the point of contact with the enclosure.
- The machine ground should be connected to the enclosure and to earth ground.

15-2 SYSTEM POWER REQUIREMENTS AND SAFETY CIRCUITRY

PLC power supplies operate, in general, with sources of 120 or 240 VAC single phase. If the controller is installed in an enclosure, the two power leads (L1 hot, L2 common) normally enter the enclosure through the top part of the cabinet to minimize interference with other control lines. It is important to supply the cleanest possible power line so that problems related with line interference to the controller and I/O system are avoided.

Common AC Source. It is always good practice to use a common AC source to the system power supply and the I/O devices. This practice will minimize line interference and prevent the possibility of reading faulty input signals if the AC source to the power supply and CPU is stable, but the AC source to the I/O devices is unstable. By keeping both the power supply and I/O devices on the same power source, the user can take full advantage of the power supply's own line monitoring feature. If line conditions fall below the minimum operating level, the power supply will detect the abnormal condition and signal the processor, which will stop reading input data and turn off all outputs.

Isolation Transformers. Using isolation transformers on the incoming AC power line to the controller is another good practice. An isolation transformer is especially desirable in cases where heavy equipment will be likely to introduce noise onto the AC line. The isolation transformer can also serve as a step-down transformer to reduce the incoming line voltage to a desired level. The transformer should have a sufficient power rating (units of volt-amperes) to supply the load adequately. Users should consult the manufacturer for the recommended transformer rating for a particular application.

Safety Circuitry

Sufficient emergency circuits should be provided to stop, either partially or totally, the operation of the controller or the controlled machine or process. These circuits should be routed outside the controller, so in the event of total controller failure, independent and rapid shutdown means are available. Devices, such as emergency pull rope switches or end of travel limit switches, should operate motor starters, solenoids, or other devices without being processed by the controller. These emergency circuits should be implemented using simple logic with a minimum number of highly reliable, preferably electromechanical, components.

Emergency Stops. It is recommended that emergency stop circuits be incorporated in the system for every machine directly controlled by the PLC. To provide maximum safety in programmable controller systems, these circuits must not be wired into the controller, but should be hardwired. Emergency stop switches should be placed in locations that are easily accessible to the operator and there should be as many as are required. These emergency stop switches are generally wired into master control relay (MCR) or safety control relay (SCR) circuits that will remove power from the Input/Output system in an emergency.

Master or Safety Control Relays. MCRs or SCRs provide a convenient means for removing power from the I/O system during an emergency situation. By de-energizing the MCR (or SCR) coil, power to the input and output devices is removed. This event occurs when any emergency stop switch opens. However, the CPU continues to receive power and to operate even though all its inputs and outputs are disabled.

The MCR circuit may be extended by placing the PLC fault relay (closed during normal PLC operation) in series with any other emergency stop condition. This enhancement will cause the MCR circuit to drop the I/O power in case of a PLC failure (memory error, I/O communications error, etc.). Figure 15-4 illustrates typical wiring of safety circuits.

Emergency Power Disconnect. A properly rated emergency power disconnect should be used in the power circuit feeding the power supply as a means of removing power from the entire programmable controller system (see Figure 15-4). A capacitor (0.47 μF for 120 VAC, 0.22 μF for 220 VAC) is sometimes placed across the disconnect to protect against a condition known as “outrush”. Outrush occurs when the output Triacs are turned off by throwing the power disconnect, thus causing the energy stored in the inductive loads to seek the nearest path to ground, which is often through the Triacs.

15-3 NOISE, HEAT AND VOLTAGE CONSIDERATIONS

If the recommendations that have been previously outlined are closely followed, they should provide favorable operating conditions for most programmable controller applications. However, in certain applications the operating environment may not be considered normal and may have extreme conditions that require special attention. These adverse conditions include excessive noise and heat and nuisance line fluctuations. The following discussions describe these conditions and possible measures that will minimize their effects.

Excessive Noise. Electrical noise is seldom responsible for damaging components, unless extremely high energy or high voltage levels are present. However, the malfunctions due to noise are temporary occurrences of operating errors that can result in hazardous machine operation in certain applications. Noise may be present only at certain times, may appear at widely-spread intervals, or in some cases may exist continuously. The former case is the most difficult to isolate and to correct.

Noise usually enters through input, output, and power supply lines and may be coupled onto the lines electrostatically through the capacitance between these lines and noise signal carrier lines. This effect generally results from the presence of high voltage or long, closely-spaced conductors. Coupling of magnetic fields can also occur when control lines are closely spaced with lines carrying large currents. Devices that are potential noise generators include relays, solenoids, motors, and motor starters, especially when operated by hard contacts such as pushbuttons or selector switches.

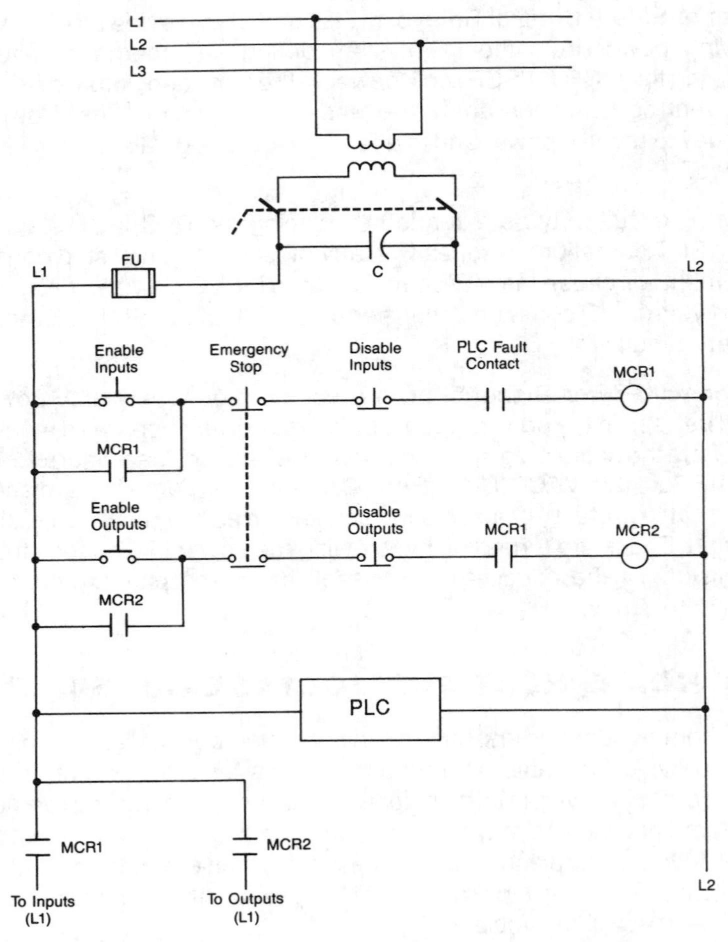

Figure 15-4. Safety wiring diagram example with separate MCR control for inputs and outputs.

Although a great deal of effort has been expended in the design of solid-state controls to achieve reasonable noise immunity, special consideration must be given to minimize noise, especially when the anticipated noise signal has characteristics similar to the desired control input signals. To increase the operating noise margin, the controller must be located away from noise generating devices such as large AC motors and high-frequency welding machines. All inductive loads must be suppressed. Three-phase motor leads should be grouped together and routed separately from low-level signal leads. If the situation is critical, it may be necessary to suppress all three-phase motor leads, as shown in Figure 15-5. Suppression techniques for small and large inductive devices are described in the next section.

Excessive Heat. Programmable controllers are designed to withstand temperatures in the range of 0° to 60° C and are normally cooled by convection, in which a vertical column of air is drawn in an upward direction over the surface of the components. To keep the temperature within limits, the cooling air at the base of the system must not exceed 60° C. Based on these specifications, proper spacing must be allocated between components when they are installed. The spacing recommendations from manufacturers are, in general, determined for typical conditions which exist for most applications. Conditions are considered typical if, usually, 60% of the inputs are ON at any one time, 30% of the outputs are ON at any one time, current supplied by all the modules combined an average specified by the manufacturer, and the air temperature is around 40° C.

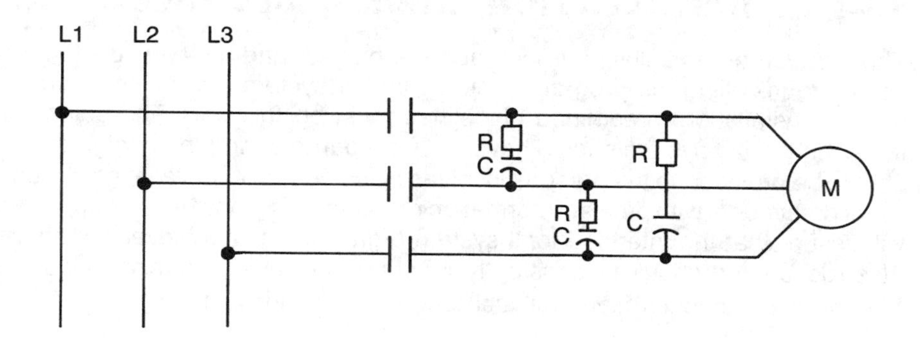

Figure 15-5. Three-phase motor lead suppression example.

Situations in which most of the I/O are ON at the same time and the air temperature is higher than 40° C are not typical. Spacing between components must be larger to provide better convection cooling. If equipment inside or outside the enclosure generates substantial amounts of heat and the I/O system is switched ON continuously, a fan should be utilized inside the enclosure to provide good air circulation to reduce hot spots created near the PLC system. When using a fan, the air being brought in should first pass through a filter to prevent dirt or other contaminants from entering. Dust will obstruct the heat dissipation from components and can be especially harmful on heat sinks when thermal conductivity to the surrounding air is lowered. In cases of extreme heat, an air conditioning unit should be used to prevent heat build-up inside the enclosure.

Excessive Line Voltage Variation. The power supply section of the PLC system is built to sustain line fluctuations and still allow the system to function within its operating margins. As long as the incoming voltage is adequate, the power supply will continue to provide all the logic voltages necessary to support the processor, memory, and I/O. If the voltage drops below the minimum acceptable level, the power supply will signal the processor and a system shut-down will be executed.

In cases where the installation is subject to "soft" AC lines and unusual line variations, a constant voltage transformer can be used to prevent the system from shutting-down too often. However, a first step towards the solution of the line variations is to correct any possible feeder problem in the distribution system. If this correction does not solve the problem, a constant voltage transformer must be used.

The constant voltage transformer stabilizes the input voltage to the power supply and input field devices by compensating for voltage changes at the primary in order to maintain a steady voltage at the secondary. When using a constant voltage transformer, the user should be sure to check that its power rating is sufficient to supply the input devices and the power supply. The output devices are generally connected to the line in front of the constant voltage transformer, instead of providing power to the outputs from the transformer. This arrangement will lessen the load supported by the transformer and allow a smaller rating. The information regarding power rating requirements can be obtained from the manufacturer.

15-4 I/O INSTALLATION, WIRING, AND PRECAUTIONS

The input/output installation is perhaps the biggest and most critical job when it comes to installing the programmable controller system. To minimize errors and simplify installation, predefined guidelines must be followed. The guidelines for installing the I/O system should have been prepared during the design phase and should be provided to the persons involved in the controller installation. A complete set of documents with precise information regarding I/O placement and connections will assure that the intended total system organization is achieved. Furthermore, these documents should be constantly updated during every stage of the installation. The following considerations will facilitate an orderly installation.

I/O Module Installation

Placement and installation of the I/O modules is simply a matter of inserting the correct modules in their proper locations. This procedure involves verifying the type of module (115 VAC output, 115 VDC input, etc.) and the slot address as defined by the I/O address assignment document. Each terminal in the module will be wired to field devices that have been assigned to that termination address. The user should be sure that power to the module (or rack) is removed before installing and wiring any module.

Wiring Considerations

Wire Size. Each I/O terminal has been designed to accept one or more conductors of a particular wire gauge. This wire size should be checked to insure that the correct gauge is being used and that it is properly sized to handle the maximum possible current.

Wire and Terminal Labeling. A label should be used to identify each field wire and its termination point. Reliable labeling methods such as shrink-tubing or tape should be used on each wire, and tape or stick-on labels should be used on each terminal block. Color coding of similar signal characteristics (e.g. AC-red, DC-blue, common-white, etc.) can be used in addition to wire labeling. Typical labeling nomenclature includes wire numbers, device names or numbers, or the input or output address assignment. Good wire and terminal identification will simplify maintenance and trouble-shooting.

Wire Bundling. Wire bundling is a technique commonly used to simplify the connections to each I/O module. When using this method, the wires that will be connected to a single module are bundled, generally using a tie-wrap, and then routed through the duct with other bundles of the same signal characteristics. The input, power, and output bundles carrying the same type of signals should be kept in separate ducts when possible to avoid interference.

Wiring Procedures

Once I/O modules have been placed in the correct slots and wire bundles have been made for each module, the wiring to the modules can take place. The following procedures are recommended for I/O wiring:

- Remove and lock-out input power from the controller and I/O before any installation and wiring begins.
- Verify that all modules are in the correct slots. Check module type and model number by inspection, and I/O wiring diagram. Check the slot location according to the I/O address assignment document.
- Loosen all terminal screws on each I/O module.
- Locate the wire bundle corresponding to each module and route it through the duct to the module location. Identify each of the wires in the bundle and be sure that they correspond to that particular module.
- Starting with the first module, locate the wire from the bundle that connects to the lowest terminal. At the point where the wire is at a vertical height equal to the termination point, bend the wire at a right angle towards the terminal.
- Cut the wire at a length that extends 1/4 inch past the edge of the terminal screw. Strip the insulation from the wire approximately 3/8 of an inch. Insert the uninsulated end of the wire under the pressure plate of the terminal and tighten the screw.
- If two or more modules share the same power source, the power wiring can be jumpered from one module to the next.
- If shielded cable is being used, connect only one end to ground, preferably at the rack chassis. This connection will avoid possible ground loops. The other end should be left cut-back and unconnected, unless specified otherwise.
- Repeat the wiring procedure for each wire in the bundle until the module wiring is complete.
- After all wires are terminated, check for good terminations by gently pulling on each wire.
- Repeat the wiring procedure until all modules are completed.

Special I/O Connection Precautions

Typical connection diagrams for the various types of I/O modules have been illustrated in Chapter 5. However, certain field device wiring connections may need special attention. These connections include leaky inputs, inductive loads, output fusing, and use of shielded cable.

Connecting Leaky Inputs. Some field devices will have a small leakage current even when they are in the OFF state. Both triac and transistor outputs will exhibit this leakage characteristic, although the transistor leakage current is much lower. Often, the leaky input will only cause the module's input indicator to flicker. The leakage could, however, result in misoperation by falsely triggering an input circuit. This situation can be corrected by placing a "bleeding" resistor across the input. A typical device that exhibits this situation is a proximity switch. This leakage may also be observed when driving an input module with an output module when there is no other load. Figure 15-6 illustrates both of these cases along with the protective action taken.

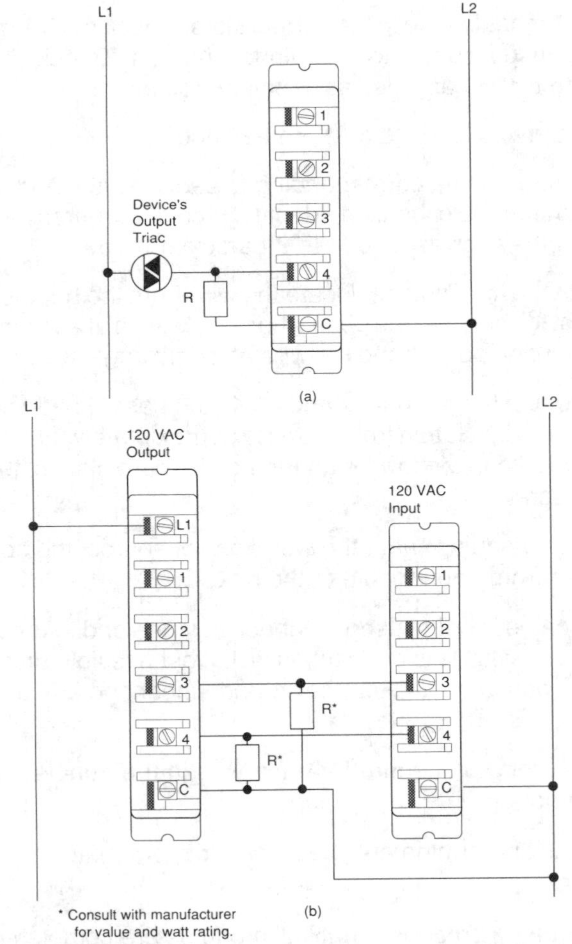

Figure 15-6. (a) Typical connection for leaky input devices.
(b) Connection of output module to input module.

Suppression of Inductive Loads. When the current in an inductive load is interrupted (by turning the output OFF), a very high voltage spike is generated. These spikes, if not suppressed, can reach several thousand volts across the leads which feed power to the device or between both power leads and chassis ground, depending on the physical construction of the device. This high voltage causes erratic operation and, in some cases, may damage the output module. To avoid this situation, a snubber circuit, typically an RC or MOV, should be installed to limit the voltage spike as well as the rate of change of current through the inductor.

Generally, output modules are designed to drive inductive loads and often include suppression networks; however, under certain loading conditions, it may be found that the triac is unable to turn OFF as current passes through zero (commutation), thus requiring external suppression. If this is the case, additional suppression must be installed.

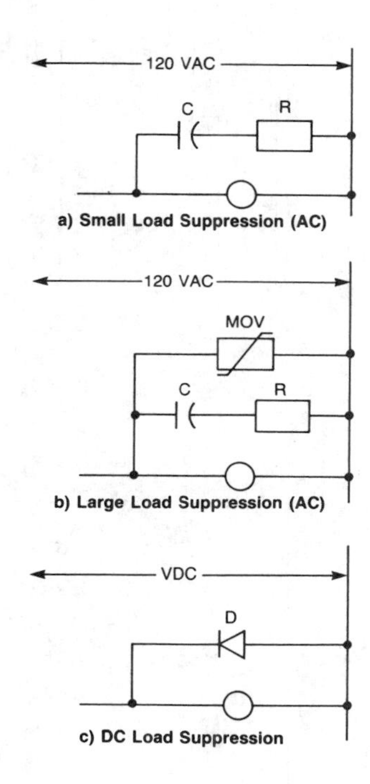

Figure 15-7. Suppression circuit networks for AC and DC loads.

Suppression for small AC devices, such as solenoids, relays, and motor starters up to Size 1, can be accomplished by placing an RC snubber circuit across the device. Larger contactors of Size 2 and above will require an MOV in addition to the RC network (see Figure 15-7). DC suppression is accomplished by using a free-wheeling diode across the load. Figure 15-8 illustrates several examples of suppressing inductive loads.

Fusing Outputs. Solid state outputs are normally provided with fusing on the module, to protect the triac or transistor from moderate overloads. If fuses are not provided internally, they should be installed (normally at the terminal block) externally, during the initial installation. When adding fuses to the output circuit, be sure to adhere to the manufacturer's specification for a particular module. Only a properly rated fuse will insure that, in an overload condition, the fuse will open quickly to avoid overheating of the output switching device.

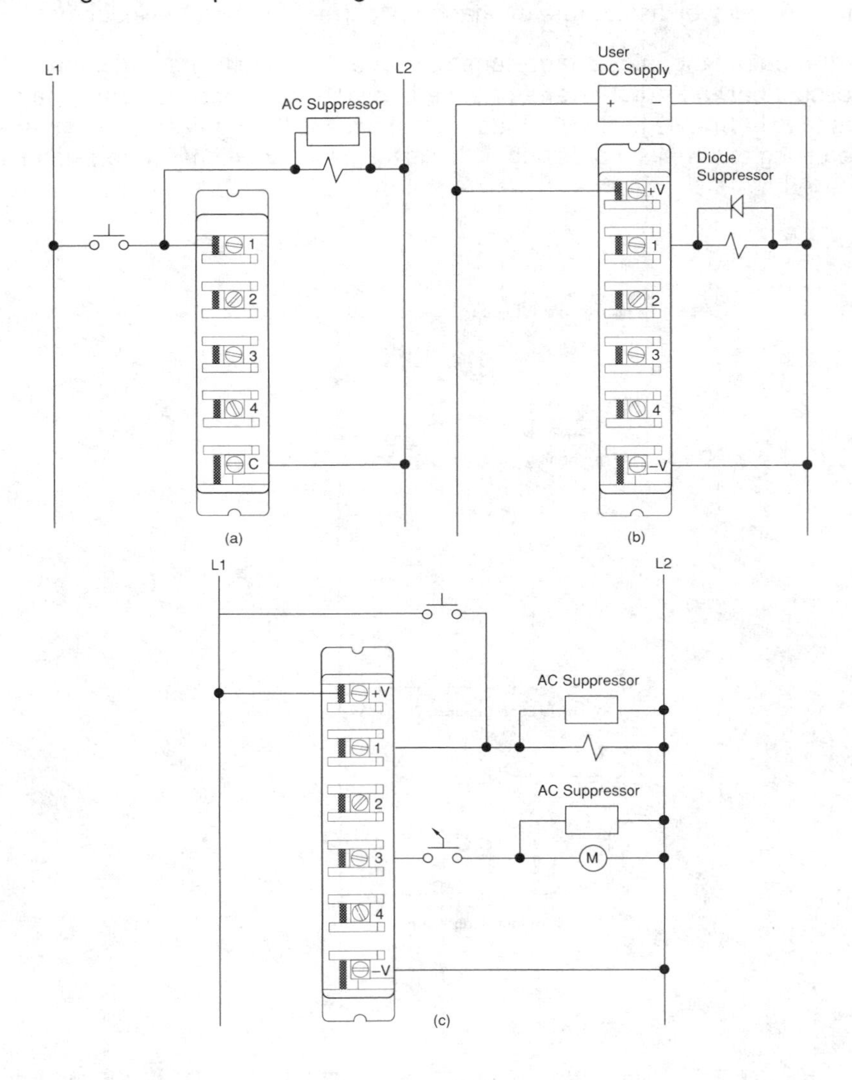

Figure 15-8. a)Suppression of inductive loads in parallel with PLC input; b) suppression of DC load; and c) suppression of loads with switch in parallel and series with PLC output .

Shielding. Control lines such as TTL, analog, thermocouple, and other low-level signals are normally routed in a separate wireway to reduce the effects of signal coupling. For further protection, shielded cable is used to protect the low-level signals from electrostatic and magnetic coupling with lines carrying 60 Hz power and other lines that carry rapidly changing currents. The twisted, shielded cable should have at least a one inch lay, or approximately twelve twists per foot, and be protected on both ends by means such as shrink tubing. The shield should be connected to control ground at only one point, as shown in Figure 15-9, and shield continuity must be maintained for the entire length of the cable. Care should be taken to insure that the shielded cable is routed away from high noise areas and is insulated over its entire length, and makes a connection at ONLY one point.

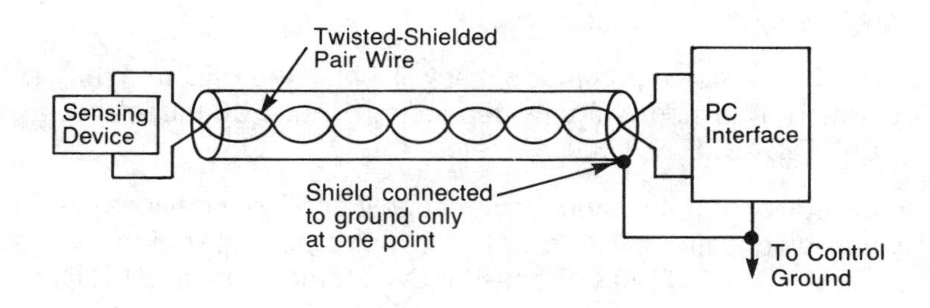

Figure 15-9. Example of shielded cable ground connection.

15-5 PLC START-UP AND CHECKING PROCEDURES

Prior to applying power to the system, several final inspections of the hardware components and interconnections are recommended. These recommended procedures will undoubtedly require time up front; however, this invested time will almost always insure a reduced total start-up time, especially for large systems with many I/O devices. The following checklist can be used for pre-start-up procedures:

- Visually inspect to insure that all PLC hardware components are present. Verify correct model numbers for each component.
- Inspect all CPU and I/O modules to insure that they are installed in the correct slot locations and are placed securely in position.
- Check that the incoming power is wired correctly to the power supply (and transformer) and that the system power is properly routed and connected to each I/O rack.
- Verify that each I/O communication cable that links the processor and each I/O rack is correct, according to the I/O rack address assignment.
- Verify that all I/O wiring connections at the controller end are in place and are securely terminated. This check involves using the I/O address assignment document to verify that a wire is terminated at each point as specified by the assignment document.
- Verify that output wiring connections are in place and properly terminated at

- For maximum safety, the system memory should be cleared of any control program that has been previously stored. If the control program is stored in EPROM, then the chips should be removed temporarily.

Check of Static Input Wiring

The static input wiring check is performed with power applied to the controller and input devices. When performed, it will verify that each input device is connected to the correct input terminal and that the input module or point is functioning properly. Since these tests are performed first, they will also verify that the processor and the programming device are in good working condition. Proper input wiring can be verified using the following procedures:

- Place the controller in a mode that will inhibit the PLC from any automatic operation. This mode will vary depending on the PLC model, but is typically: STOP, DISABLE, PROGRAM, etc.
- Apply power to the system power supply and input devices. Verify that all system diagnostic indicators are indicating proper operation. Typical indicators are: AC OK, DC OK, Processor OK, Memory OK, and I/O Communication OK.
- Verify that the emergency stop circuit will de-energize power to the I/O devices.
- Manually activate each input device and observe the corresponding LED status indicator on the input module and/or monitor the same address on the programming device, if used. If properly wired, the indicator will turn ON. If an indicator other than the expected one turns ON when the input device is activated, then it is possible that the wiring is to the wrong input terminal. If no indicator turns ON, then a fault possibly exists in either the input device, field wiring, or the input module (see Section 15-4).
- Precautions should be taken when activating input devices that are connected in series with loads that are external to the PLC since these could cause injury or damage.

Check of Static Output Wiring

The static output wiring check is performed with power applied to the controller and output devices. A good safe practice, however, is to first locally disconnect all output devices that involve mechanical motion (e.g. motors, solenoids, etc.). When performed, the static output wiring check will verify that each output device is connected to the correct terminal address and that the device and output module are functioning properly. The output wiring can be verified using the following procedures:

- Locally disconnect all output devices that will cause mechanical motion.
- Apply power to the controller and to the input/output devices. If power to the outputs can be removed by an emergency stop, verify that the circuit does remove power when activated.

- Static checkout of the outputs should be performed one at a time. If the device is a motor or other output that has been locally disconnected, re-apply power to that device only, prior to checking. The output operation inspection can be performed using one of the following methods:

  1) Assuming the controller has a forcing function, each output can be tested with the use of the programming device by forcing the output ON and by setting the corresponding terminal address (point) to 1. If properly wired, the corresponding LED indicator will turn ON and the device will energize. If an indicator other than the expected one turns ON when the terminal address is forced, then it is possible that the wiring is to the wrong output terminal. Inadvertent machine operation is avoided since rotating and other motion producing outputs are disconnected. If no indicator turns ON, then a fault possibly exists in either the output device, field wiring, or the output module (see Section 15-4).

  2) An alternative to forcing involves programming a "dummy" program rung that can be used repeatedly for testing each output. Program a single rung with a single normally-open contact (a conveniently located push-button) controlling the output. To test the output, the CPU must be placed in either the RUN or Single-Scan mode, or a similar mode, depending on the controller. With the controller in the RUN mode, the test is performed by depressing the pushbutton. If the Single-Scan is used, it is performed by continuously depressing the pushbutton while the Single-Scan is executed. Observe the output device and LED indicator, as described in the previous procedure.

Control Program Review

The program checkout is simply a final review of the control program. This check can be performed at any time, but should be done prior to loading the program into memory for the dynamic system checkout.

To perform the check will require a complete documentation package that relates the control program to the actual field devices. Documents such as address assignments and wiring diagrams should reflect any modifications that may have occurred during the static wiring checks. When performed, this final program review will verify that the final hardcopy, which will be loaded into memory, is error-free or at least agrees with the original design documents. The following is a checklist for final program checkout:

- Using the I/O wiring document, verify, against the hardcopy program printout, that every controlled output device has a programmed output rung of the same address.

- Inspect the hardcopy printout for any entry errors that may have occurred while entering the program. Verify that all program contacts and internal outputs have valid address assignments.

- Verify that all timer, counter, and other preset values are as intended.

Dynamic Operation

The dynamic operation checkout is a procedure by which the logic of the control program is verified for correct operation of the outputs. This checkout assumes that all static checks have been performed, the wiring is correct, the hardware components are operational and functioning correctly, and the software has been thoroughly reviewed. At this point, it can be assumed that it is safe to gradually bring the system under full automatic control.

Although it may not be necessary to start up a small system partially it is always a good practice to start large systems in sections. Large systems generally use remote subsystems that control different sections of the machine or process. Bringing one subsystem on-line at a time will allow the total system start-up to be performed with maximum safety and efficiency. Remote subsystems can be temporarily disabled either by locally removing power or by disconnecting the communications link with the CPU. The following practices outline possible procedures for the dynamic system checkout:

- Load the control program into the PLC memory.
- The control logic can be tested, in most cases, using one of the following methods:
  1) A mode such as TEST, if available, will allow the control program to be executed and debugged while the outputs are disabled. A check of each rung can be done by observing the status of the output LED indicators or by monitoring the corresponding output rung on the programming device.
  2) If the controller must be in the RUN mode to update outputs during the tests, outputs that are not being tested and could cause damage or harm should be locally disconnected until they are tested. If an MCR or similar instruction is available, it can be used to bypass execution of the outputs that are not being tested, so that disconnection of the output devices is not necessary.
- Check each rung for correct logic operation and modify the logic if necessary. A useful tool for debugging the control logic is the single scan. This procedure will allow users to observe each rung as every scan is executed under their command.
- When all the logic has proven to control the outputs satisfactorily, remove all temporary rungs that may have been used (MCRs, etc.). Place the controller in the RUN mode and test the total system operation. If all procedures have checked correctly, then full automatic control should operate smoothly.
- All modifications to the control logic should be documented immediately and revised on the original documentation. A reproducible copy (e.g. a cassette recording, etc.) of the program should be obtained as soon as possible.

The start-up recommendations and practices that have been presented in this section are considered good procedures that will aid in the safe and orderly start-up of any programmable control system. However, depending on the controller that is being installed, there may be specific start-up requirements that are outlined in the manufacturer's product manual. You should be aware of these specific start-up procedures, prior to attempting to start-up the controller.

15-6 PLC SYSTEM MAINTENANCE

Although programmable controllers have been designed in such a way as to minimize maintenance for troublefree operation, there are several maintenance aspects that should be taken into consideration once the system has been installed and is operational. Certain maintenance measures, if performed periodically, will minimize the chances of system malfunction. This section outlines some of the practices that should be observed to keep the system in good operating condition.

Preventive Maintenance

Preventive maintenance of programmable controller systems includes only a few basic procedures or checks that will greatly reduce the failure rate of system components. Preventive maintenance for the PLC system could be scheduled with the regular machine or equipment maintenance so that the equipment and controller are down for a minimum amount of time. However, depending upon the environment in which the PLC is located, the required preventive maintenance may be more frequent than in other surroundings. The following preventive measures should be taken:

- Any filters that have been installed in enclosures should be cleaned or replaced periodically. This practice will insure that clean air circulation is present inside the enclosure. Filter maintenance should not be put-off until the scheduled machine maintenance, but should be checked periodically at a frequency dependent on the amount of dust in the area.

- Dirt and dust should not be allowed to accumulate on the PLC components. To allow heat dissipation, the CPU and I/O system are generally not designed to be dust-proof. If dust is allowed to build-up on heat sinks and electronic circuitry, an obstruction of heat dissipation could occur and cause circuit malfunction. Furthermore, if conductive dust reaches the electronic boards, a short circuit could result and cause permanent damage to the circuit board.

- The connections to the I/O modules should be periodically checked to insure that all plugs, sockets, terminal strips, and modules are making good connections, and that the module is securely installed. This type of check should be done more often in situations in which the PLC system is located in areas that experience constant vibration that could loosen terminal connections.

- Care should be taken to insure that heavy noise generating equipment is not moved too closely to the PLC.

- The personnel performing the maintenance should make sure that unnecessary articles are kept away from the equipment inside the enclosure. Leaving articles such as drawings, installation manuals, or other booklets on top of the CPU rack or other rack enclosures could obstruct the air flow and create hot spots, which can result in system malfunction.

Spare Parts

It is a good practice to keep on hand a stock of replacement parts. This practice will minimize any down time resulting from component failure. In a failure situation, having the right spare in stock could mean a shutdown of only minutes, instead of hours or days. As a rule of thumb, the spares stocked should be 10% of the number of each module used. If a module is used infrequently, then less than 10% of that particular module can be stocked.

Main CPU board components should have one spare each, regardless of how many CPUs are being used. Each power supply, whether main or auxiliary, should have a backup. There are certain applications that may require a complete CPU rack as a stand-by spare. This extreme case exists when there is no time during a failure for determining which CPU board has failed, and the system must be brought into operation immediately.

Replacement of I/O Modules

If a module has to be replaced, the user should make sure that the module being installed is of the correct type. Some I/O systems allow modules to be replaced while power is still applied, but others may require that power be removed. If replacing a module solves the problem, but the failure reoccurs in a relatively short period, the user should check inductive loads that may be generating voltage/current spikes that may require external suppression. If the module fuse blows again after it is replaced, it may be that the module's output current limit is being exceeded or that the output device is shorted.

15-7 TROUBLE-SHOOTING THE PLC SYSTEM

Diagnostic Indicators

The LED status indicators can provide much information regarding the field device, wiring, and the I/O module. Most input/output modules will have at least a single indicator. Input modules will normally have a "power" indicator, while output modules will normally have a "logic" indicator.

For an input module, the power LED ON indicates that the input device is activated and a signal is present at the module. This indicator alone cannot isolate malfunctions to the module, so some manufacturers provide an additional diagnostic indicator, the logic indicator. The logic LED ON indicates that the input signal has been recognized by the logic section of the input circuit. If the logic and power indicators do not match, the module is unable to transfer the incoming signal to the processor correctly.

The output module's logic indicator functions similarly to the input logic indicator. When it is ON, it indicates that the module's logic circuitry has recognized a command from the processor to turn ON. In addition to the logic indicator, some output modules will incorporate a blown fuse indicator, a power indicator, or both. The blown fuse indicator simply indicates the status of the protective fuse in the output circuit. When ON, the output power indicator shows that power is being applied to the load. Like the power and logic indicators of the input module, if both are not ON simultaneously, the output module is malfunctioning.

LED indicators provide much assistance in trouble-shooting. With both power and logic indicators, a malfunctioning module or circuit can be immediately pinpointed. LED indicators, however, cannot diagnose all possible problems, but instead serve as preliminary signs of system malfunctions.

Diagnosing Input Malfunctions

In the case of an input malfunction, the first check is to see if the LED power indicator is responding to the field device (i.e. pushbutton, limit switch, etc.). If the input device is activated but the indicator does not energize, then the next test is to take a voltage measurement across the input terminal to check for the proper voltage level. If the voltage level is correct, then the input module should be replaced. If an LED logic indicator is illuminated and according to the programming device monitor the processor is not recognizing the input, then the input module may have a fault. If a replacement module does not eliminate the problem and wiring is assumed correct, then the I/O rack or communication cable should be suspected.

Diagnosing Output Malfunctions

In the case of output malfunctions, the first check is to see if the output device is responding to the LED status indicators. If an output rung is energized, the module indicator is ON, and the output device is not responding, then the field wiring should be suspected; but first, a check should be made for a blown fuse or the module should be replaced. If the fuse is OK and the replacement module does not solve the problem, then the field wiring should be checked. If according to the programming device monitor, an output device is being commanded to turn ON, but the indicator is OFF, then the module should be replaced.

When diagnosing input/output malfunctions, the best method is to isolate the problem either to the module itself or to the field wiring. If both power and logic indicators are available, then module failures become readily apparent. Normally, the first test is to replace the module or take a voltage measurement for the proper voltage level at the input or output terminal. If the proper voltage is measured at the input terminal and the module is not responding, then the module should be replaced. If the replacement module has no effect, then field wiring should be suspected. A predetermined voltage level at the output terminal, while the output device is OFF, indicates an error in the field wiring. If an output rung is activated and the LED indicator is OFF, then the module should be replaced.

If a malfunction cannot be traced to the I/O module, then the module connectors should be inspected for poor contact or misalignment. Finally, check for broken wires under connector terminals and cold solder joints on module terminals.

CHAPTER

—16—

PLC SYSTEM SELECTION GUIDELINES

"This is not the end. It is not even the beginning of the end. But it is, perhaps, the end of the beginning."

—Winston Churchill

16-1 INTRODUCTION

Programmable controllers are available in all shapes and sizes, covering a wide spectrum of capability. On the low end are "relay replacers", with minimum I/O and memory capability. At the high end are large supervisory controllers, which play an important role in hierarchical systems by performing a variety of control and data acquisition functions. In between these two extremes are multifunctional controllers with communication capability, which allows integration with various peripherals, and expansion capability, which allows the product to grow as the application requirements change.

Deciding on the right controller for a given application has become increasingly difficult. Having been complicated by an explosion of new products, including general and special purpose programmable controllers, this process places an even greater demand on the system designer to take a system approach to selecting the best product for each task. Many factors are affected by programmable controller selection; it is up to the designer to determine what characteristics are desirable in the control system and which controller best fits the present and future needs of the application.

This chapter covers the range of PLC capabilities, several guidelines for defining and configuring the control system, and also other factors that will affect the final selection.

16-2 PLC SIZES AND SCOPE OF APPLICATIONS

Prior to evaluating the system requirements, it might be helpful to understand the various ranges of programmable controller products and typical features found within those ranges. This understanding will enable the user to identify quickly the general area in which the specifications for an application may be found and to select the product that comes closest to matching these requirements.

Figure 16-1 illustrates the product ranges, divided into four major areas showing overlapping boundaries. The basis for segmentation of the product areas is the number of possible inputs and outputs the system can accommodate (I/O count), the amount of memory available for the application program, and the system's general hardware and software structure. As I/O count increases, the complexity and cost of the system also increases. Similarly, as the system complexity increases, so does the memory capacity, the variety of I/O modules, and the capabilities of the instruction set.

Shaded areas A,B, and C reflect the possibility of a controller with enhanced (not standard) features for a particular range. These enhancements place the product in a grey area that overlaps the next higher range. For example, because of its I/O count, a small PLC falls into area 2, but it has analog control functions that are usually standard in medium size controllers. This fact will place that product in area A. Products that fall into these overlapping areas will always allow the user to select the product that matches the requirements closest, without having to select the larger product unless it is necessary.

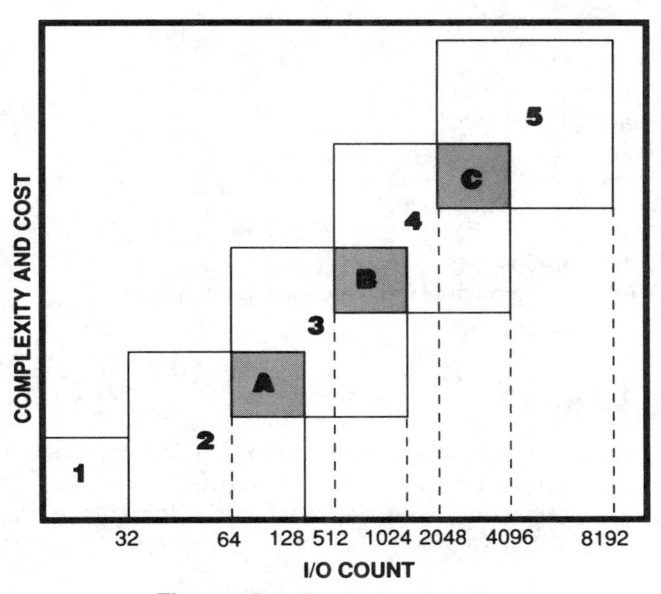

Figure 16-1. Illustration of product ranges.

The typical characteristics of the various product segments are explained in the following discussions.

Segment 1—Micro PLCs

Micro PLCs ar eused in applicaitons which require in general the control of a few discrete I/O devices such as small conveyor controls. Some Micro PLCs may be found to handle some analog I/O for monitoring (e.g. a temperature setpoint or activating an output). Figure 16-2 illustrates one of these micro controllers. The standard features generally found in this segment are shown Table 16-1.

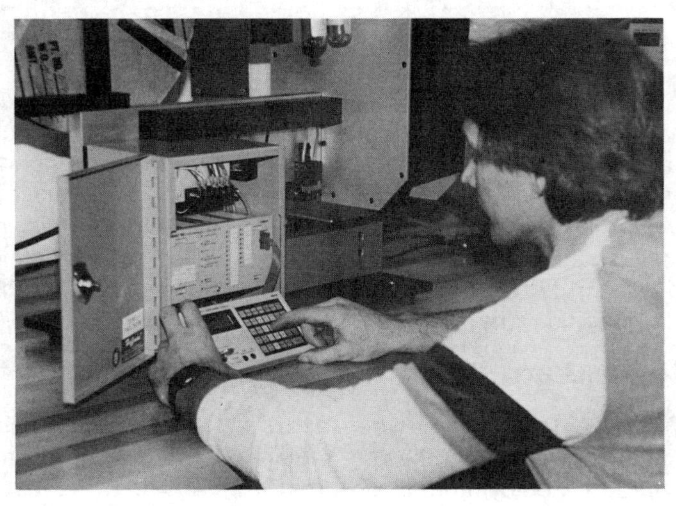

Figure 16-2. Micro PLC application (Courtesy Omron Electronics, Schaumburg, IL).

- Up to 32 I/O
- 8- bit processor
- Relay replacing mostly
- Memory up to 1K
- Digital I/O
- Built-in I/Os in compact unit
- MCRs
- Timers and Counters
- Generally programmed with hand-held programmer

Table 16-1

Segment 2—Small PLCs

These small controllers are usually found in applications which logic sequencing and timing functions are required for ON/OFF control. Micro-controllers and small PLCs are widely used for individual control of small machines. Often, these products are single board controllers. Table 16-2 lists standard features found in segment 1.

- Up to 128 I/O
- 8 bit processor
- Relay replacing mostly
- Memory up to 2K words
- Digital I/O
- Local I/O only
- Ladder or Boolean language only
- Timers/Counters/Shift registers (TCS)
- Master Control Relays (MCR)
- Drum Timers or Sequencers
- Generally programmed with hand-held programmer

Table 16-2. Typical standard features for small PLCs.

Area A. This shaded area includes controllers that are capable of having up to 64 or 128 I/O and also includes products having features normally found in medium sized controllers. The enhanced capabilities of these small controllers allow them to be used effectively in applications in which a small number of I/O is needed, yet analog control, basic math, LANs, remote I/O, and/or limited data handling may be required (Figure 16-3). A typical case is a transfer line application in which several small machines, under individual control, must be interlocked (through LAN).

Segment 3—Medium PLCs

Medium PLCs (Figure16-4) are applied when more than 128 I/O, analog control, data manipulation, and arithmetic capabilities are required. In general, the controllers in segment 2 are characterized by more flexible hardware and software features than those previously mentioned. These features are listed in Table 16-3.

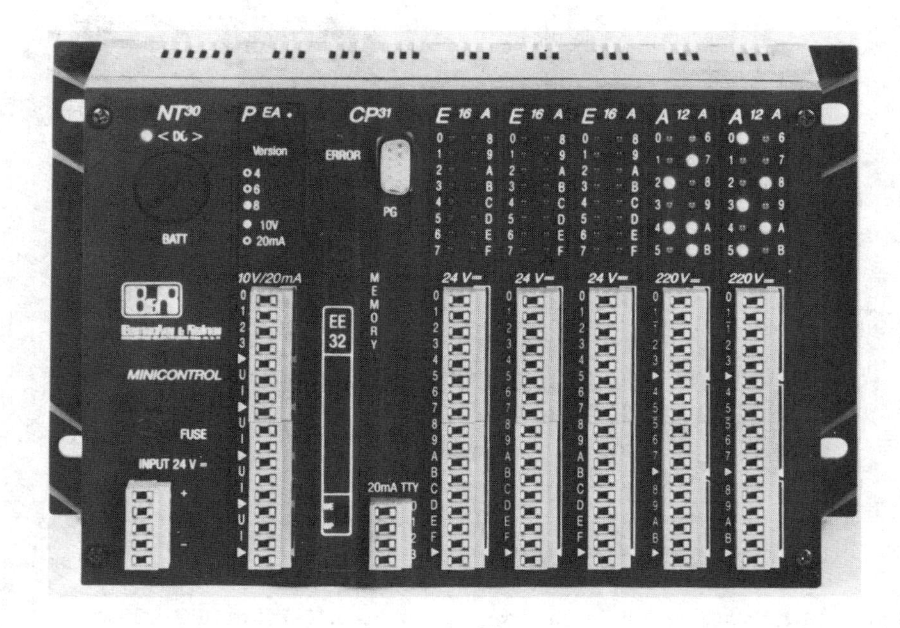

Figure 16-3. Small PLC that falls in the A category. This small controller (10W x 6H x 3D) can control up to 96 I/O) and is capable of handling analog I/O (Courtesy B & R Industrial Automation, Stone Moutain, GA).

- Up to 1024 I/O
- 8 bit processor
- Relay replacing and analog control
- Typical memory up to 4K words. Expandable to 8K.
- Digital I/O
- Analog I/O
- Local and remote I/O
- Ladder or Boolean language
- Functional block/high level language
- TCSs
- MCRs
- Jump
- Drum Timers or Sequencers
- Math Capabilities
  —Addition
  —Subtraction
  —Multiplication
  — Division
- Limited data handling
  —Compare
  —Data conversion
  —Move register/file
  —Matrix functions
- Special function I/O modules
- RS 232 communication port
- Local Area Networks (LANs)
- CRT programmer

Table 16-3. Typical standard features for medium size PLCs.

Area B. In general, the shaded area B contains products in segment 2 that have more memory, table handling, PID, subroutine capability, and more arithmetic or data handling instructions. These feature are typically standard in segment 3 and basically are enhanced instruction sets. The Allen-Bradley PLC 2/30, shown in Figure 16-5, falls into this category.

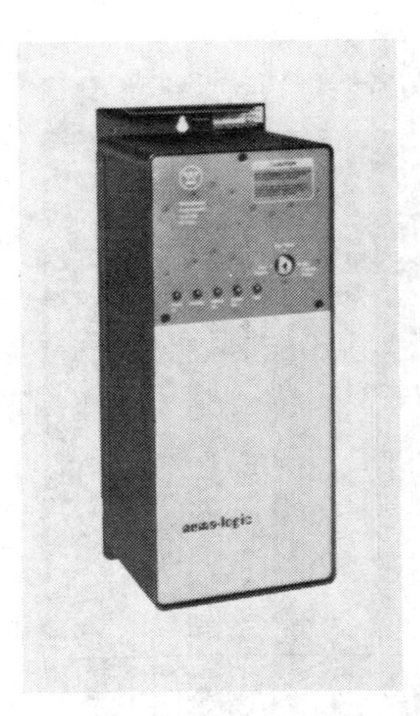

Figure 16-4. Westinghouse Numa-Logic PC-700 with capacity of 512 I/O (Courtesy of Westinghouse Electric, Pittsburgh, PA).

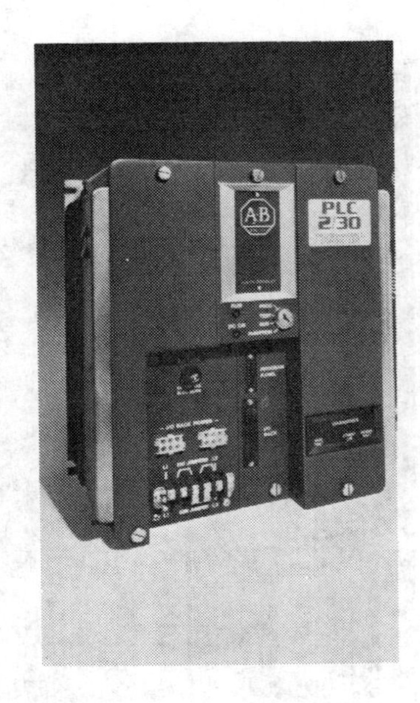

Figure 16-5. The PLC-2/30 from Allen-Bradley with 896 I/O capacity (Courtesy of Allen-Bradley, Highland Heights, OH).

Segment 4—Large PLCs

Large controllers (Figure 16-6) are used in more complicated control tasks that require extensive data manipulation, data acquisition, and reporting. Further software enhancements allow these products to perform more complex numerical computations. These standard features are summarized in Table 16-4.

- Up to 2048 I/O
- 8 or 16 bit processor
- Relay replacing and analog control
- Typical memory up to 12K words. Expandable to 32K
- Digital I/O
- Analog I/O
- Local and remote I/O
- Ladder or Boolean language
- Functional block/high level language
- TCSs
- MCRs
- Jump
- Subroutines, interrupts
- Drum Timers or Sequencers
- Special function I/O modules
- PID modules or system software PID
- One or more RS 232 communication ports
- Local Area Networks (LANs)
- Math Capabilities
  —Addition
  —Subtraction
  —Multiplication
  —Division
  —Square root
  —Double precision
- Extended Data Handling
  —Compare
  —Data conversion
  —Move register/file
  —Matrix functions
  —Block transfer
  —Binary tables
  —ASCII tables
- Host computer communication modules
- CRT programmer

Table 16-4. Typical standard features for large PLCs.

Figure 16-6. General Electric's Series Six model 600 with capacity of 2000 I/O (Courtesy of General Electric, Charlottesville, VA).

Area C. The shaded area C may include some of segment 3 PLCs that have a large amount of application memory and more I/O capacity. Greater math and data handling capabilities can also be found in this area. The Gould 984 mainframe PLC shown in Figure 16-7 is an example of a controller in this area.

Figure 16-7. Gould 984 PLC mainframe. (Courtesy Gould, Andover, MA).

Segment 5—Very Large PLCs

Very large PLCs (Figure 16-8) are utilized in sophisticated control and data acquisition applications when basic requirements are large memory and I/O capacity. Remote and special I/O interfaces are also a standard requirement. Typical application areas include steel mills and refineries. These PLCs usually serve as supervisory controllers in large distributed control applications. Table 16-5 lists standard features found in segment 4.

Figure 16-8. The PLC-3 from Allen-Bradley is capable of handling up to 8190 I/O (Courtesy of Allen-Bradley, Highland Heights, OH).

- Up to 8192 I/O
- 16 bit or 32 bit processor or multi-processors
- Relay replacing and analog control
- Typical memory up to 64K words. Expandable to 1 Meg.
- Digital I/O
- Analog I/O
- Remote analog I/O
- Remote special modules
- Local and remote I/O
- Ladder or Boolean language
- Functional block/high level language
- TCSs
- MCRs
- Jump
- Subroutines, interrupts
- Drum Timers or Sequencers
- Special function I/O modules
- PID modules or system software PID
- Two or more RS 232 communication ports
- Local Area Networks (LANs)
- Host computer communication modules
- Math Capabilities
  —Addition
  —Subtraction
  —Multiplication
  —Division
  —Square root
  —Double precision
  —Floating point
  —Cosine functions
- Powerful Data Handling
  —Compare
  —Data conversion
  —Move register/file
  —Matrix functions
  —Block transfer
  —Binary tables
  —ASCII tables
  —LIFO
  —FIFO
- Machine diagnostics
- CRT programmer

Table 16-5. Typical standard features for very large PLCs.

16 -3 PROCESS CONTROL SYSTEM DEFINITION

Selecting the right programmable controller to control a machine or process requires several preliminary considerations regarding not only current needs of the application, but perhaps needs that will involve future plant goals. Keeping the future in mind will allow changes and additions to the system at minimal cost. With proper considerations, the need for memory expansion may only require that a memory module be installed; the need for an additional peripheral can be accommodated if the communication port is available. A local area network consideration will allow future integration of each controller into a plantwide communication scheme. If present and future goals are not properly evaluated, the control system may quickly become inadequate and obsolete.

Once basic considerations have been investigated, you should be prepared to begin defining specific controller requirements. The following items need to be evaluated and defined:

- Input/Output
- Type of Control
- Memory
- Software
- Peripherals
- Physical and Environmental

Input/Output Considerations

Determining the amount of I/O is typically the first problem that must be addressed. Once the decision has been made to automate some machine or process either totally or partially, determining the amount of I/O is simply a matter of counting the discrete and/or analog devices that will be monitored or controlled. This determination will help to identify the minimum size constraints for the controller. Remember to allow for future expansion and spares (typically 10% to 20% spares). Spares do not effect the choice of PLC size.

Discrete Input/Output. Input/Output interfaces of standard ratings are available for accepting signals from sensors and switches (e.g.pushbuttons, limit switches, etc.) as well as control (ON/OFF) devices (e.g. pilot lights, alarms, motor starters, etc.). Typical AC inputs/outputs range from 24 volts to 240 volts, and DC inputs/outputs from 5 volts to 240 volts.

Although input circuits vary from one manufacturer to another, certain characteristics are desirable. Look for debouncing circuitry to protect against false signals and surge protection to guard against large transients. Most input circuits will have optical or transformer isolation between the high power input and the control logic circuitry of the interface circuit.

When evaluating discrete outputs, look for the following: fuses, transient surge protection, and isolation between the power and logic circuits. Fused circuits usually cost more initially, but will probably cost less than having the fuse installed externally. Check for fuse accessibility; replacing fuses may mean shutting down several other devices. Most output circuits with fuses will also have blown fuse indicators, but make sure to check. Finally, check the output current ratings and the specified operating temperature. Typically, the temperature rating is at 60° C.

Interface circuits with isolated commons (return lines) will be required if the input/output devices are powered from separate sources.

Analog Input/Output. Analog input/output interfaces are used to sense signals generated from transducers. These interfaces measure quantity values such as flow, temperature, and pressure and are used for controlling voltage or current output devices. Typical interface ratings include -10V to +10V, 0V to +10V, 4 to 20mA, or 10 to 50mA.

Some manufacturers provide special analog interfaces to accept low level signals (e.g. RTD, thermocouple). Typically, these interface modules will accept a mix of various thermocouple or RTD types on a single module. Users should consult the vendor concerning specific requirements.

Special Function Input/Output. When selecting a programmable controller, the user may be faced with a situation requiring some special type of I/O conditioning (e.g. positioning, fast input, frequency, etc.) that may be impossible to implement using standard I/O. The user should find out whether or not the vendors under consideration have special modules that would help minimize control efforts. Smart modules, a subset of special interfaces, should also be considered. Typically, these interfaces perform all the field data processing in the module itself, thus relieving the CPU from performing time consuming tasks. For example, PID, 3 axis positioning, and stepper motor control modules would make the control implementation much easier and feasible, thus reducing programming and implementation time.

Remote Input/Output. Remote I/O should always be considered, especially in large systems. I/O subsystems, located a distance (over twisted pairs) away from the CPU, can mean a dramatic reduction in wiring cost, both from a labor and material standpoint. Another advantage of remote I/O subsystems is that the inputs/outputs can be strategically grouped to control separate machines or sections of a machine or process. This grouping allows easy maintenance and start-up without involving the entire system. Most controllers that have remote I/O will also have remote digital I/O. However, users who require remote analog I/O should check to see if it is available in the products being considered.

Control System Organization

With the advent of new, smarter programmable controllers, the decision on the type of control has become a very important initial step. Questions such as, "what type of control should I use?" are more often asked when automating a process. Knowing process application and future automation requirements will help the user to decide what type of control will be required. This knowledge will ease the selection of a particular PLC. Possible control configurations include individual or segregated control, centralized control, or distributed control. Figure 16-9 illustrates these configurations.

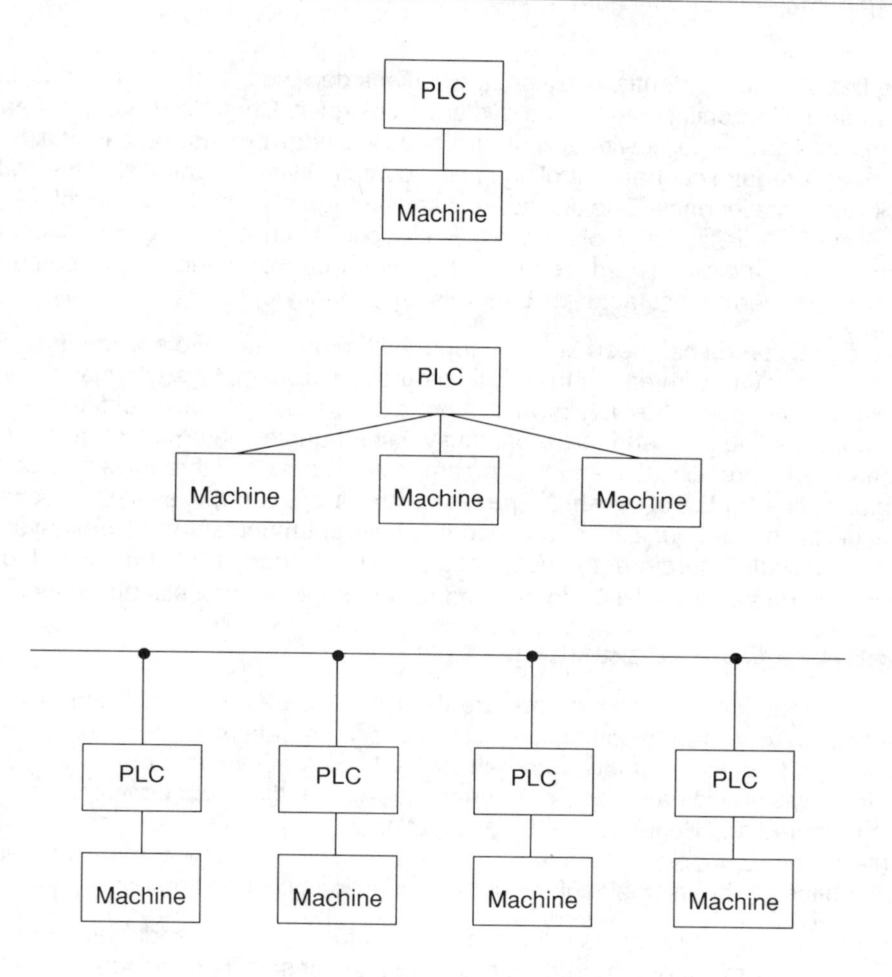

Figure 16-9. Control configuration.

Individual or segregated control is used when a PLC controls a single machine with only local I/O or with local and a few remote I/O. This type of control does not normally require communication with any other controllers or computers. Individual control is primarily applied to OEM and end-user equipment, such as injection molding machines, small machine tools, and small dedicated batching processes. When deciding on this approach, the user should take into consideration whether future intercontroller communication will be desired. If so, the appropriate controller can be selected then to avoid extra design expenses at a later date.

Centralized control is used when several machines or processes are controlled with one central programmable controller. This type of control can have many subsystems spread out through the factory, each one interfaced with specific I/O devices that may or may not be related to the same control. The communication that exists is only to subsystems and/or peripherals; no exchange of PLC status or data is sent to other PLCs.

The flexibility and potential of these applications depend greatly on the PLC used and the design philosophy of the system's designer. Centralized control can be interpreted as a large individual control applied to a large process or machine. Some processes require central control due to the complexity of decentralizing the control tasks into smaller ones. One distinctive disadvantage of centralized control is that if the main PLC fails, the whole process is stopped. In critical, large central control, when a back-up is needed, redundant systems can be used to overcome this problem. Several manufacturers offer this redundancy option.

Distributed control has been brought about by the need to have several main PLC s communicating with each other. This type of control employs what is known as Local Area Networks (LANs), in which several PLC s are controlling different stages or processes locally and are constantly interchanging information and status regarding the process. Communication among PLC s is done through single coaxial cables or fiber-optics at very high speeds (up to 1 Megabaud). Despite this powerful configuration, communication between two different manufacturer's LAN systems can be difficult. Therefore, the user should define properly from the beginning the functional needs of the PLCs to be used according to the process application.

Memory Considerations

The two main considerations here are the type and the amount of memory. The application may require both nonvolatile memory (i.e. retains contents upon loss of power) and volatile memory with battery backup. A nonvolatile memory such as EPROM will provide a reliable permanent storage medium, once the program has been created and debugged. If the application will require on-line changes, Read/Write memory supported by battery should be considered. Some controllers will offer both options to be used singularly or in conjunction with each other.

The memory capacity of small PLC s is normally fixed (non-expandable) between 1/2 K and 2K words and is, therefore, not really a major selection concern. In medium and large controllers, however, memory can be expanded incrementally in units of 1K, 2K, 4K, etc. Although there are no fixed rules for determining the amount of memory, certain guidelines can be used to evaluate memory requirement.

The amount of memory required for a given application is a function of the total number of inputs and outputs to be controlled and the complexity of the control program. The complexity refers to the amount and type of arithmetic and data manipulation functions that will be performed. Manufacturers will normally have a rule-of-thumb formula for each of their products that will aid in making an approximation of the memory requirement. This formula involves multiplying the total number of I/O by some constant (usually a number between 3 and 8). An additional 25% to 50% should be added to the first approximation, if the program involves arithmetic or data manipulation.

Finally, the best way to obtain the memory requirement is to create the program and count the number of words used. Knowledge of the number of words required to store each instruction will allow the user to determine exact memory requirements.

Software Considerations

During the implementation of the system, the user will be faced with the programming of the PLC . Because the programming is so important, users should be aware of the software capabilities of the product they choose. Generally, the software capability of a system is tailored to the handling of the control hardware that is available with the controller. However, an application may require special software functions beyond the control of the hardware components. For instance, an application may involve special control or data acquisition functions that require complex numerical calculations and data handling manipulations. The instruction set selected will determine the ease with which the software task can be implemented. The available instruction set will directly affect the time needed to implement the control program and the program execution time.

Peripherals

The first peripheral to consider is a programming device. In general, programmers are available in two types the small (low cost) handheld units or CRT display type. The handheld programmer, typically used with small PLCs, can display either a single program element or in some cases a single program rung.

Both handheld and CRT programmers provide programming and monitoring capability, but the CRT allows greater flexibility. Some CRT programmers also provide program storage capability, as well as limited graphics. Intelligent CRTs, as they are labelled, are microprocessor based devices that can be programmed for various functions. The capability of these units differs depending on the manufacturer. Typical features are listed later in this section.

In addition to the programming device, users may require peripherals at certain control stations to provide interface between the controller and an operator. Two of the most common peripherals are the line printer, used for obtaining a hardcopy printout of the program, and a cassette recorder, used to store the program. Messages or alarms can be sent to graphic or alphanumeric display devices, while hourly or monthly production reports can de stored on a floppy diskette and later sent to a line printer.

Peripheral requirements should be evaluated along with the CPU, since the CPU will determine the type and number of peripherals that can be interfaced, as well as the method of interfacing. The distance that peripherals can be placed from the CPU is also affected. Typical peripherals are printers, CRTs, diskette drives, color displays, alphanumeric displays, cassette drives.

Physical and Environmental

The physical and environmental characteristics of the various controller components will have significant impact on the total system reliability and maintainability. Consider the ambient conditions in which the controller will be operating. Conditions such as temperature, humidity, dust level, and corrosion can all affect the controller's ability to operate properly. It is important to determine operating parameters (i.e.

temperature, vibration, EMI/RFI, etc.), and packaging (i.e. dust proof, drip proof, ruggedness, type of connections, etc.) when specifying the controller and its I/O system. Most programmable controller manufacturers provide products that have undergone certain environmental and physical tests (e.g. temperature, EMI/RFI, shock, vibration, etc.). Users should be aware of the test performed and whether or not the results meet the demands of the operating environment.

A checklist with typical sample answers is shown in Table 16 -6. This list will help to examine most of the features a user should look for when evaluating PLC requirements. Note that all product ranges are covered in the list, from small to very large. Some PLC s may not answer all checks, primarily due to their range characteristics.

I/O SYSTEM CHECKLIST	TYPICAL ANSWERS
I/O Count	
• Digital count	Maximum of 128 I/O mixable
• Analog count	Maximum of 16 I/O mixable
Digital I/O	
• Inputs	
Number of points/module	4 points/module, etc
Input Type	AC,DC,non-voltage,etc.
Input ratings	110 VAC, 220 VAC, 5-24 VDC, contact,etc.
Maximum inputs/channel	64 points/channel
Input status indicators	Power, logic
Isolation	1500 Volts optical
• Outputs	
Number of points/module	16 points/module
Output type	AC,DC, contact, etc.
Output ratings	110 VAC, 220 VAC, contact
Output current Amps/point	1 Amp/point with all outputs on @ 115 VAC
Maximum outputs/channel	64 points/channel
Output status indicators	Power, individual blown fuse, logic
Output protection	Fuse, suppression on contact output
Analog I/O	
• Inputs	
Number of points/module	4 analog inputs/module
Resolution	11 bits
Input type	current, voltage
Input ratings	4-20 ma., 0-5 volts, 0-10 volts
Built-in transducer	Yes, thermocoupler input
Maximum analog inputs/channel	32
Supply of power	Internal to PLC
• Outputs	
Number of points/modules	2 analog outputs/module
Output type	current, voltage
Output ratings	4-20 ma., 10-50 ma., 0-10 volts
Maximum analog outputs/channel	16
Supply of power	+15 VDC and -15 VDC

Table 16 -6. System Checklists.

Table 16-6 continued.

Remote I/O	
• Digital	
Distance	1500 ft.
Number of I/O per remote	32 I/O per remote
Communication Link	Twisted pair, 100 ohms impedance
• Analog	
Distance	5000 ft. with receiver/transmitter
Number of I/O per remote	16 analog inputs/outputs per remote
Communication Link	Coaxial
Special I/O	
• High-speed pulse counter	Local and remote, 50 KHz
• Fast electronic input	Yes, 5 microsecond pulse width minimum
• Interrupt module	Yes
• Absolute encoder	Direct connection to encoder
• Incremental encoder	Not available
• BCD input/output module	Remote, 4 and 8 BCD digits
• Stepper motor	Yes
• ASCII communications module	Yes, full ASCII, 300-4800 baud
• Host computer	Yes, protocol decode on module
• LAN I/O module	Extra board in CPU
• PID module	Local and remote, 2 loops per module
• Language module	Yes, basic interpreter module
Physical	
• Wire size to I/O	20 AWG, can handle two wires per I/O
• Separate commons	Yes for 4 points/module, no for 16
• Removable under power	Yes
• Disturb wiring to remove I/O	Disconnect screw from I/O module
CENTRAL PROCESSING UNIT CHECKLIST	**TYPICAL ANSWERS**
Processor	
• Microprocessor	8 bit micro, and multiprocessor board
• Scan time	10 msec/K of memory
• Communication ports	Two RS-232C ports
Memory	
• Memory type	RAM, EEPROM
• Total system memory	64K
• Application memory size	8K for user
• Word size	8 bits
• Memory utilization	1 word per element (coil, contact)
• Memory protect	Yes, key switch
Power Supply	
• Incoming Power	120/240 VAC, 24 VDC
• Frequency	50/60 Hz
• Voltage variation	+15%, -10%
• Overvoltage protection	Yes
• Current limiting	Yes
• Maximum current supply	100 ma. @ 14 VDC, 2.5 amps @ 5 VDC

Table 16-6 continued.

• Isolation	1500 volts
• Location	Built-in CPU
Environmental	
• Operating temperature	0-60 degrees C
• Humidity	5%-90% relative humidity non-condensing
• EMI/RFI	Satisfies NEMA and IEEE tests
• Noise	1,000 volts peak-peak, 1 microsecond
• Vibration	Withstands 16.7 Hz, double amplitude
• Shock	10g X,Y, and Z direction
SOFTWARE CHECKLIST	**TYPICAL ANSWERS**
Language	
• Ladder or Boolean	Ladder language
• High level	Functional blocks
Software Coils	
• Number of internals	128
• Number of timers	32 timers, maximum count 9999 sec. BCD
• Number of counters	166
• Number of shift registers	32, 16 bit each
• Number of drum timers	16
• Timer's time base	0.1, 1.0 seconds
• Timer type	on delay and off delay
• Counter type	up/down count
• Latch coil	32
• Transitional coil	16, OFF-ON, and ON-OFF
• MCR	8
• Global coil	256 in LAN
• Global register	128 in LAN
• Fault coil	Yes, detection of CPU failure
• Interrupt coil	Yes
Math	
• Addition	Yes, double precision
• Subtraction	Yes, double precision
• Multiplication	Yes
• Division	Yes
• Square root	Yes
• Floating point	Yes, 1E+38, 1E-38
• Trigonometric functions	Yes, sine and cosine
Data Handling	
• Number of registers	128, 16 bits each
• Data size in registers	+32767, -32767, and 9999 BCD
• Compare	Yes, greater/less than or equal to
• Conversions	Binary-BCD, BCD-Binary
• Move	Registers and single files
• Matrix	AND, OR, EXOR, NAND
• Tables	Move to ASCII or binary tables
• PID	Software functional block, 20 loops
• LIFO	Yes
• FIFO	Yes

Table 16-6 continued.

• Jump	Conditional and direct
• Subroutines	Yes
• ASCII instructions	Yes, print and read
• Sort	No
• Machine diagnostics	Yes
PROGRAMMING AND STORAGE DEVICES CHECKLIST	**TYPICAL ANSWERS**
CRT or Computer	
• Physical	
CRT or computer	Both
Display size	9" screen
Graphics	No
Ladder matrix size	10x7 elements
Built-in storage	Yes, tape
Local Area Network	Yes
Communication	RS-232C and 20 ma current loop
Incoming power	115/230 VAC
Operating temperature	0-40 degrees C
Keyboard type	Mylar or standard keys
• Functional	
Intelligent	Yes
Single scan	No
Power flow	Yes, element intensified on screen
OFF-LINE programming	Yes
Monitor function	Yes
Modify function	Yes
Force I/O	Yes, indicates forcing on mainframe
Search	No
Mnemonics	Yes
Manual Programmer	
• Physical	
Display type	LCD or LED
Ladder matrix size	7x4 elements
Communication	RS-232C
Incoming power	From unit
Operating temperature	0-40 degrees C
Keyboard type	Mylar
• Functional	
Intelligent	No
Single scan	No
Power flow	Yes
Monitor function	Yes
Modify function	Yes
Force I/O	Yes
Search	No
Mnemonics	Yes, also messages

Table 16-6 continued.

Storage Devices	
• Digital cassette	Yes, Built-in
• Floppy disk	No
• Computer	Yes, thru computer module
• Electronic memory module	Yes, for small PLC
SYSTEM DIAGNOSTICS CHECKLIST	**TYPICAL ANSWERS**
Power Supply	
• Power loss detection	Yes, after 3 cycles
• Voltage level detection	Yes, DC levels for CPU
• Diagnostic monitoring	Continuously
Memory	
• Memory OK	Yes, checksum, LED indicator
• Battery OK	Yes, LED indicator
• Diagnostic monitoring	At power up only
Processor	
• Local	Yes, watch dog timer, and LED indicator
• Remote	Yes, indicator in CPU
• Diagnostic monitoring	Continuously
Communication	
• Local I/O	Yes
• Remote I/O	Yes, check-sum
• Programming device	CRT port OK, and RS-232C OK
• Diagnostic monitoring	During transmission
Fault Indications	
• CPU	Yes, external relay contacts
• Remote	Yes, external contacts at remote driver
• LAN	Yes, internal coil
• I/O	Yes, detects presence of I/O module

16-4 OTHER CONSIDERATIONS

An evaluation of the hardware and software requirements will narrow the selection down to one of a few possible candidates. Eventually, two or more products will meet all the requirements of the preliminary system design, and then a final decision must be made. At this point, there are still several other factors that can lead to a final selection of the product that best fits the system specifications and the in-house requirements. The following considerations should be made and discussed with the potential vendors.

Product Proven Reliability

The reliability of the controller plays an important role in overall system performance. Lack of reliability usually translates into downtime, poor quality product, and higher scrap levels.

Several factors can be investigated to determine the proven reliability of a particular product. Mean Time Between Failure studies (MTBF) can be helpful if the user can evaluate the data. Knowledge of a similar application in which the product has been sucessfully applied is also useful. A sales representative can provide such information or even on occasion arrange an on-site visit. If there are any unique or peculiar specifications that must be met (e.g. EMI and vibration specs), be sure that the vendor can truly satisfy these requirements. Finally, burn-in procedures of the product should be investigated (e.g. total system burn-in process or parts burn-in process). In general, any of this information can be obtained from the vendor upon request.

Standardization of PLC Equipment

A final consideration that may help to finalize the decision on a product is the possibility of future plant goals to standardize on a given manufacturer and product line. This route is being taken by several companies today for many good reasons. The fact that several vendors are creating complete product families of PLCs that cover the entire range of capabilities is making standardization more feasible.

A current trend by manufacturers is to build completely intercompatible product families, with products ranging from the very small to the very large. These famalies share the same I/O structure, programming device, and elementary instruction set. They also have similar memory organization and structure. Most families of products can also be linked in a network configuration. Family traits, such as the following, provide several important benefits that should be considered:

- Training on a new family member is usually a progression instead of totally new training on a new product.
- Standard products can mean better plant maintenance in emergency situations.
- I/O spares can be used for all family products (mimimum spare inventory).
- An outgrown product can be replaced by the next larger product by simply removing the smaller CPU, installing the larger CPU, and reloading the old program.

16-5 SUMMARY

This chapter has presented a general approach for making a programmable controller selection. We have seen that this selection relates not only to such obvious factors as I/O capacity, memory capacity, or sophistication of control, but also to intangible factors that have a significant impact on final system results. The following steps are a summary of the major considerations involving PLC selection.

Step 1: Know the process to be controlled.

Step 2: Determine the type of control.

a. Distributed control
b. Centralized control
c. Individual control

Step 3: Determine the I/O interface requirement.

a. Estimate digital and analog inputs and outputs
b. Check for input/output specifications
c. Determine if remote I/O is required
d. Special I/O requirements
e. Allow for future expansion

Step 4: Determine software language and functions.

a. Ladder, Boolean, and/or high level
b. Basic instructions (timers, counters, etc.)
c. Enhanced instructions/functions (math, PID, etc.)

Step 5: Consider the type of memory.

a. Volatile (R/W)
b. Non-volatile (EEPROM, EPROM, NOVRAM, Core, etc.)
c. Combination of volatile and nonvolatile

Step 6: Consider memory capacity.

a. Approximate memory usage based on memory utilization per instruction
b. Allow extra memory for complex programming and future expansion

Step 7: Evaluate processor scan time requirements.

Step 8: Define programming and storage device requirements.

a. CRT
b. Computer
c. Cassette or disk storage
d. Manual programmer
e. Consider the functional capabilities of programming devices

Step 9: Define peripheral requirements.

a. Graphic display
b. Operator interface
c. Line printers
d. Documentation systems
e. Report generation systems

Step 10: Determine any physical and environmental constraints.

a. Available space for system
b. Ambient conditions

Step 11: Evaluate other factors that can affect selection.

a. Vendor support
b. Product proven reliability
c. Plant goals for standardization

APPENDIX

Glossary

-A-

AC input interface: An input circuit that conditions various AC signals from connected devices, to logic levels that are required by the Processor. This interface may or may not be a module type. Typical voltages for AC inputs include 24 VAC, 115 VAC, and 230 VAC.

AC output interface: An output circuit that switches the User supplied control voltage required, to control connected AC device. This interface may or may not be a module type. Typical voltages for AC outputs are 24 VAC,115 VAC, and 230 VAC.

acknowledgement (ACK):A control signal that indicates acceptance of data, in an I/O transmission process. An acknowledgement can be implemented with hardware or software.

ADCCP: Advanced Data Communication Control Procedures. An ANSI standard communication protocol for synchronous links, in which various combinations of primary and secondary link control functions are defined.

address field: The sequence of eight (or any multiple of eight, if extended) bits immediately following the opening flag sequence of a frame. The address field identifies which secondary station is sending (or is designated to receive) the frame.

addressability: The total number of devices that can be connected to the network. The ability of a network to accommodate device expansion.

address: 1) An alphanumeric value that uniquely identifies where data is stored. 2) An alphanumeric value used to identify a specific I/O rack, module group, and terminal.

algorithm: A set of procedures for solving a problem. More generally used in reference to a software program.

alphanumeric: Character strings consisting of any combination of alphabets, numerals, and/or special characters (e.g. A15$), for representing text, commands, numbers, and/or code groups.

ambient temperature: The temperature of the air surrounding a module or system.

American Wire Gauge (AWG): A standard system used for designating the size of electrical conductors. Gauge numbers have an inverse relationship to size; larger numbers have a smaller diameter.

analog: An expression of values which can vary continuously between specified limits.

analog device: Apparatus that measures continuous information (e.g. voltage–current). The measured analog signal has an infinite number of possible values. The only limitation on resolution is the accuracy of the measuring device.

analog input interface: An input circuit that employs an analog–to–digital converter to convert an analog value, measured by an analog measuring device, to a digital value that can be used by the Processor.

analog output interface: An output circuit that employs a digital–to–analog converter to convert a digital value, sent from the Processor, to an analog value that will control a connected analog device.

analog signal: One having the characteristic of being continuous and changing smoothly over a given range, rather than switching suddenly between certain levels as with discrete signals.

AND: A Boolean operation that yields a logic "1" output if all inputs are "1," and a logic "0" if any input is "0."

ANSI: American National Standards Institute. A clearing-house and coordinating agency for voluntary standards in the United States.

application: 1) A machine or process monitored and controlled by a PLC. 2) The use of computer- or processor-based routines for specific purposes.

application memory: That part of the total system memory that is available for the storage of the application program and associated data.

application program: The set of instructions written by the User for control, data acquisition, or report generation. This software is stored in the application memory.

arithmetic capability: The ability to perform such math functions as addition, subtraction, multiplication, division, square roots, and other advanced math functions with a PLC processor. A given controller may have some or all of these functions.

arithmetic logic unit: A processor subsystem that performs arithmetic operations like addition and subtraction, and logic operations such as Exclusive–OR, AND, and OR functions. Also called ALU.

artificial intelligence (AI): A subfield of computer science which encompasses the creation or development of (computer) programs to solve tasks that require extensive knowledge.

ASCII: American Standards Code for Information Interchange. It is a seven-bit code with an optional parity bit used to represent alphanumerics, punctuation marks, and control code characters.

assembly language: A symbolic programming language that can be directly translated into machine language instructions.

asynchronous response mode (arm): A mode in which a secondary station may initiate transmission without explicit permission from the primary.

asynchronous: Recurrences or repeated operations that take place in patterns unrelated over time.

asynchronous shift register: A shift register that is loaded and/or unloaded based on external conditions and/or timing functions.

asynchronous transmission: Transmission controlled by start and stop elements at the beginning and end of each character. Time intervals between characters may be of unequal length.

attenuation: The decrease in magnitude of a signal.

AWG: See American Wire Gauge.

-B-

backplane: A printed circuit board, located in the back of a chassis, that contains a data bus, power bus, and mating connectors for modules to be inserted in the chassis.

backup: A device or system that is kept available to replace something that may fail in operation. Also see Hot Backup.

backward chaining: A way of finding the causes of an outcome by analyzing the consequents to obtain the antecedents.

bandwidth: The range of frequencies over which a system is designed to operate. The bandwidth is expressed in Hertz between the highest and lowest frequencies.

baseband: Transmission of one signal at its original frequency.

battery backup: A battery or set of batteries that will provide power to processor memory only in case of a system power outage.

battery low: A condition that exists when the backup battery voltage drops low enough to require battery replacement. Generally, an indicator or status bit signals this condition.

baud: Officially defined as the reciprocal of the shortest pulse width in a data communication stream, but usually used to refer to the number of binary bits transmitted per second during a serial data transmission.

BCD: See Binary Coded Decimal.

binary: A numbering system using only the digits 0 and 1. Also called 2 base.

Binary Coded Decimal: A binary number system in which each decimal digit from 0 to 9 is represented by four binary digits (bits). The four positions have a weighted value of 1, 2, 4, and 8 respectively starting with the least significant bit. A thumbwheel switch is a BCD device, and when connected to a programmable controller, each decade requires four wires. Decimal 9 = 1001 BCD

Binary Number System: A number system that uses two numerals (binary digits) "0" and "1." Each digit position for a binary number has a place value of 1,2,4,8,16,32,64,128, and so on beginning with the least significant (right–most) digit. Base 2. Example: 1101 = 1(1) + 0 (2) + 1 (4) + 1(8) = 13

binary word: A related grouping of ones and zeros having coded meaning assigned by position, or as a group, has some numerical value. A 10010010 is an eight bit binary word, in which each bit could have coded significance or as a group represent the number 146 in decimal.

bit: One binary digit. The smallest unit of binary information (Abbreviation of **B**inary dig**IT**). A bit can have a value of "1" or "0."

bit manipulation: The process of controlling data table bits (on or off) through user instructions or keyboard entry.

bit-oriented protocol: A type of data link control that uses a minimum of control characters, relying instead on bit position to determine the message. A resulting advantage is code–transparency.

bit rate: See Baud.

bit storage: A user defined data table area in which bits can be set or reset without directly affecting or controlling output devices. However, any storage bit can be monitored as necessary in the user program.

blackboard (architecture): The distribution of knowledge inferencing as well as global and knowledge data bases in a control system. It utilizes several subsystems containing local global and knowledge data bases that may work independently from each other.

block: A group of words transmitted as a unit.

Block Check Character (BCC): A transmission verification algorithm where the result accumulates over a transmission block, which is normally appended at the end of the block or frame.

block diagram: A schematic drawing.

block length: The total number of words transferred at one time.

block transfer: A programming technique used to transfer up to 64 words of data to, or from, an intelligent I/O module.

Boolean operators: Logical operators such as AND, OR, NAND, NOR, NOT, and Exclusive–OR, that can be used singly or in combination to form logical statements or circuits that can have an output response be true or false.

Bourdon Tube: A pressure transducer which converts pressure measurements into displacement. The different types of Bourbon tubes are spiral, helical, twisted, and C-tube.

branch: A parallel logic path within a rung.

breadth first search: A method which evaluates each rule in the same level of a tree before expanding downward.

bridge circuit: a mechanism used in the implementation of transducer circuits primarily when the changing parameter in the measurement is a resistive element. Examples: voltage sensitive bridge, current sensitive bridge.

broadband: Transmission of two or more channels simultaneously via frequency division multiplexing.

buffer: 1)In software terms a register or group of registers used for temporary storage of data, to compensate for transmission rate differences between the transmitting and receiving device. 2) In hardware terms, a circuit that restores a signal to a proper drive level.

bugs: 1) Software errors that cause unwanted behavior. 2) Functional problems due to hardware.

build: A programming concept that takes a user keyboard command (source code) and converts it into hexadecimal format to generate an object code for program execution.

bulk memory: A supplementary large volume memory.

burn: The process by which information is entered into PROM memory.

burn-in: The process of operating a device at an elevated temperature to improve the probability that any component weakness will cause a failure. The intent is to isolate any early–failing parts.

bus: 1) A group of lines used for data transmission or control. 2) Power distribution conductors.

bus topology: A network topology in which all stations are connected in parallel with the medium. These stations are capable of concurrently receiving a signal transmitted by any other station connected to the medium.

byte: A group of adjacent bits usually operated upon as a unit, such as when moving data to and from memory. There are eight bits/byte.

-C-

carrier: A continuous frequency capable of being modulated or impressed with a signal.

carrier system: A means of operating a number of channels over a single path by modulating each channel with a different carrier frequency at the transmitting end, and by demodulating each channel at the receiving end.

Cartesian Coordinate System: A coordinate system in 2 or 3 axes that intersect each other at right angles at an origin, enabling any point to be identified by the distance from the axes.

cathode ray tube: See CRT

CCITT: International Telegraph and Telephone Consultative Committee. International standards committee responsible for Recommendation X25.

Central Processing Unit: That part of the programmable controller that governs system activities, including the interpretation and execution of programmed instructions. In general the CPU consists of the arithmetic–logic unit, timing/control circuitry, accumulator, scratch pad memory, program counter and address stack, and an instruction register. The Central Processing Unit is sometimes referred to as the Processor or the CPU.

channel: A designated path for a signal.

channel capacity: The amount of information that can be transmitted per second on a given communications channel. A function of medium, line length and modulation rate.

character: One symbol of a set of elementary symbols, such as a letter of the alphabet or a decimal number. Characters may be expressed in many binary codes. An ASCII character can be represented by a group of seven or eight (with parity) bits

character stuffing: In character–oriented protocols, a technique for distinguishing control characters from their corresponding bit sequences by the insertion and deletion of special control characters into the data stream.

character-oriented protocol: A type of data link control in which transmitted data is encoded using an alphabet of (multi–bit) characters, such as ASCII. Compare to "bit–oriented."

chassis: A hardware assembly used to house PLC devices such as I/O modules, adapter modules, processor modules, power supplies, and processors.

checksum: A character placed at the end of a data block, that corresponds to the binary sum of all the characters in the block. This is one technique used for error detection.

chip: A very small piece of semiconductor material, on which electronic components are formed. Chips are normally made of silicon and are typically less than 1/4 inch square and 1/100 inch thick.

clear: To remove data from a single memory location or all memory locations, and return to a nonprogrammed state or some initial condition (normally "0").

clear-to-send (CTS): A signal that tells the transmitting device to start transmitting data.

clock: 1) A pulse generator which synchronizes the timing of various logic circuits. 2) Circuitry used to measure time.

clock rate: The speed (frequency) at which the processor operates, as determined by the time elapsed as words or bits are transferred through internal logic sequences.

clock signal: A clock pulse that is periodically generated and used throughout the system to synchronize equipment operation.

CMOS: Abbreviation for Complementary Metal Oxide Semiconductor. An integrated circuit family characterized by low power consumption and high noise immunity.

coaxial cable: A transmission line consisting of a central conductor surrounded by dielectric materials and an external conductor, and possessing a predictable characteristic impedance.

code: 1)A binary representation of numbers, letters, or symbols which has some meaning. 2)A set of programmed instructions.

combined error: See Propagation Error.

command: In data communications, an instruction represented in the control field or a frame and transmitted by the primary device. It causes the addressed secondary to execute a specific data link control function.

common carrier: A public utility company (e.g. Bell, Western Union), that supplies communication services to the general public.

communication control character: A function character intended to control or facilitate transmission over data networks.

compatibility: 1) The ability of various specified units to replace one another, with little or no reduction in capability. 2) The ability of units to be interconnected and used without modification.

complement: A logical operation that inverts a signal or bit. The complement of 1 is 0, and the complement of 0 is 1.

computer interface: Circuitry designed to allow communication between a computer and some other processor, such as the programmable controller's Central Processing Unit.

conditional probability inferencing: The conditional probability in an AI system.

contact: 1) One of the conducting parts of a relay, switch, or connector that are engaged or disengaged to open or close the associated electrical circuits. 2) In reference to software: the juncture point that provides a complete path when closed. See ladder contact.

contact histogram: A feature which allows a display (or printout) of the on and off times for any selected data table bit.

contact symbology: A set of symbols used to express the control program using conventional relay symbols (e.g. -][- normally open contact, -]/[- normally closed contact, and -()- coil).

contention: A condition on a communication channel where two or more stations try to transmit at the same time.

contention system: A communications network in which two or more stations have equivalent status and contend for access to the bus.

control: 1) A unit, such as a PLC or relay panel, which operates an industrial application. 2) To cause a machine or process to function in a predetermined manner. 3) To energize or de-energize a PLC output, or to set a data table bit to on or reset it to off, by means of the user program.

control field: The sequence of eight (or sixteen, if extended) bits immediately following the address field of a frame. The content of the control field is interpreted by: a) the receiving secondary, designated by the address field, as a command instructing the performance of some specific function. b) the receiving primary, as a response from the secondary, designated by the address field, to one or more commands.

control logic: The control plan for a given system. The program.

control panel: 1) A panel which may contain instruments or pushbutton switches. 2) In the Advisor system, a device which allows an operator to access and control plant operations through manipulation of the PLC data table.

core memory: A type of memory that uses tiny magnetic rings to store each bit as a permanent magnetic field. This nonvolatile memory was utilized in many of the earlier programmable controllers.

counter: 1) An electro–mechanical relay–type device that counts the occurrence of some event. The event to be counted may be pulses developed from operations such as switch closures, interruptions of light beams, or other discrete events. 2) A programmable controller eliminates the need for hardware counters by using software counters. The software counter can be given a preset count value, and will count up or down, depending on program, whenever the counted event occurs. The software counter has greater flexibility than the hardware counter.

CPU: See Central Processing Unit.

crosstalk: The intermixing of signals between adjacent channels or circuits.

CRC: See Cyclic Redundancy Check.

CRT: Abbreviation for cathode–ray tube, a vacuum tube with a viewing screen as an integral part of its envelope.

CRT programmer: A programming device containing a cathode ray tube. The CRT programmer is primarily used to create and monitor the control program, but can also be used display data and on–line reports. Some CRT programmers incorporate storage devices to record the control program.

CTS: See Clear-to-Send.

current loop: A two wire communication link in which the presence of a 20 milliamp current level indicates a binary "1" (mark), and its absence indicates no data, a binary "0" (space).

current sink: A device that receives current.

current source: A device that supplies current.

cursor: An illuminated position indicator on the display of the programming device, that indicates where the next typed character will appear.

cycle: 1) A sequence of operations that is repeated regularly. 2) The time it takes for one such sequence to occur.

Cyclic Redundancy Check (CRC): An error detection scheme in which all the bits in a block of data are divided by a predetermined binary number. A check character is determined by the remainder.

-D-

data: A general term for any type of information.

data address: A location in memory where data can be stored.

Data Carrier Detect (DCD): A signal that indicates the carrier is being received.

Data Communication Equipment (DCE): 1) Equipment that provides the functions required to establish, maintain, or terminate a connection. 2) The signal conversion and coding required for communication between data terminal equipment and data circuits. DCE may or may not be an integral part of a computer.

data files: Groups of data values stored in the data table. These files are manipulated by the user program as required by the application.

data highway: The means of transmitting frames between stations interconnected by a data transmission line. A data highway consists of a data circuit and the Physical and Data Link Layers of the stations connected to the data circuit.

data initialization: A function performed by the processor that sets starting data values.

data integrity: The ability of a communication system to deliver data from its origin to its destination with an acceptable residual error rate.

data link layer: Layer 2 of the OSI architecture. It provides functional and procedural means to establish, maintain and release data–link connections among network–entities. A data–link–connection is built upon one or several physical–connections. The objective of this layer is to detect and possibly correct errors which may occur in the physical layer. In addition, the Data Link Layer conveys to the Network Layer the capability to request assembly of data circuits within the physical layer (i.e., the capability of performing control of circuit switching).

data link: Equipment that permits the transmission of information.

Data Link Escape (DLE): An ASCII control character used to provide supplementary line communication signals. There are two characters in the sequence: the first character is DLE, the second character varies according to the function desired and the code used.

data network: The means of transmitting messages between a data source and one or more data sinks. A data network may contain one or more data highways interconnecting the same or different sets of devices. A data network consists of these highways and the Network Layers of the stations interconnected by these highways.

Data Set Ready (DSR): A signal that indicates the modem is connected, powered up, and ready for data transmission.

data signalling rate: The rate, expressed in bits per second, at which data are transmitted or received by a data terminal equipment.

data table: The part of processor memory that contains I/O values and files where data is monitored, manipulated, and changed for control purposes.

data terminal: 1) A device used only to send or receive data. 2) A peripheral device which can load, monitor, or dump PLC memory data, including the data table or data files. This also includes CRT devices and line printers.

Data Terminal Equipment (DTE): Equipment which is attached to a network to send or receive data, or both.

Data Terminal Ready (DTR): A signal which indicates that the transmission device (terminal) is connected, powered up, and ready to transmit.

data transfer: The process of moving information from one location to another, register–to–register, device–to–device, etc.

data transmission line: A medium for transferring signals over a distance.

Data Transmission Rate: See Baud.

DCD: See Data Carrier Detect.

DCE: See Data Communication Equipment.

debouncing: The act of removing intermediate noise states from a mechanical switch.

debug: The process of locating and removing mistakes from a software program or from hardware interconnections.

DCE: Data Communications Equipment. Equipment designed to establish, maintain, and terminate a connection.

DDCMP: Digital Data Communication Message Protocol. Logic that controls the transmission of data between stations in a point-to-point or multipoint data communications system. The method of physical data transfer used may be parallel, series synchronous, or series asynchronous.

debugging: The process of detecting, locating, and correcting errors in hardware or software.

decentralized connection: A multi–endpoint connection in which data sent by any entity associated with a connection–endpoint is received by all other entities.

Decimal Number System: A number system that uses ten numeral digits (decimal digits, 0,1,2,3,4,5,6,7,8,9. Each digit position has a place value of 1,10,100,1000, and so on, beginning with the least significant (right–most) digit. Base 10.

delimiter: A character that, when placed before and/or after a string of data, causes the data to be interrupted in a predetermined manner.

democratic system: A distributed system which attempts to maintain equal access times for all stations.

depth first search: A method that evaluates rules in a prioritized basis that is expanding downward.

derivative control: See Proportional, Integral, Derivative Module.

device driver: A program that services a peripheral device by controlling its hardware activities. The interface between a device and the I/O code of a network.

diagnostics: Pertains to the detection and isolation of an error or malfunction.

diagnostic program: A system program that checks a specific system operation, or a user application program that detects certain malfunctions of the controlled machine or process. Typical system diagnostics are CPU, Memory, Input/Output communication to processor program scan loss etc.. An application diagnostic program may detect and in some cases prevent certain machine or process failures.

diagnostic system: The lowest level of artificial intelligence; based primarily on the detection of faults within an application without regard to how a solution to the problem can be accomplished.

differential transmission: A method of signal transmission through two wires which always has opposite states. The signal data is the polarity difference between the wires; if one is high, the other is low. Neither wire is grounded. Usually used in reference to encoders, I/O modules, and communication systems.

digital: Information presented in a discrete number of codes.

digital signal: One having the characteristic of being discrete or discontinuous in nature. One that is present or not present, can be counted and represented directly as a numerical value.

digital device: One that processes discrete electrical signals.

direct memory access (DMA): A process in which a direct transfer of data to or from the memory of a processor–based system can take place without involving the central processing unit.

disk cartridge: Some diskettes are encased in a cartridge–like container for protection. As a result, it is often called a disk cartridge.

disk drive: The input/output device that writes data on and reads data from a magnetic diskette.

disk file: An organized collection of records stored on a magnetic disk.

disk storage: A supplementary data storage area.

diskette: A thin flexible sheet of Mylar, coated with a magnetic–oxide surface on which data is stored in tracks.

display: The image which appears on a CRT screen or on other image projection systems.

display menu: The list of displays from which the user selects specific information for viewing.

distributed application: A function which is implemented using cooperating application–processes, characterized by multiple procedures which may be executed in different systems.

distributed control: A design approach in which factory or machine control is divided into several subsystems, each managed by a unique programmable controller, yet all interconnected to form a single entity. Various types of communications busses are used to connect the controllers.

distributed media access control: A Data Link Function in which the responsibility for media access control is distributed among more than one station.

distributed processing system: A design approach in which the processing task of a single programmable controller is disbursed among several processors all within the same system.

DLE: See Data Link Escape

DMA: See Direct Memory Access.

documentation: An orderly collection of recorded hardware and software information concerning the control system. These records provide valuable reference data for installation, debugging, and maintenance of the programmable controller. Examples: Input/Output address assignments, Hardware arrangement (location) diagram, Program printout, etc.

double precision: A method of increasing the range of expressible numbers by using multiple bytes or words to represent single numbers.

download: The process of transferring data tables from a network controller to the user memory of an end–device, in order to make that device network–compatible.

downtime: The time when a system is not available for production due to required maintenance (scheduled or unscheduled).

driver: A transistor output circuit which has an emitter follower configuration.

drop line: A flexible coaxial cable, normally type RG–6U, which usually drops from a overhead tap in the coaxial network. The end of the drop line has the network outlet connector which is used to couple an external device.

drop line device: Any external device attached to the coaxial network through a drop line (e.g. RF modem, TV set, audio modular, etc.).

DSR: See Data Set Ready.

DTE: Data Terminal Equipment. Equipment that includes the data source, data sink, or both.

DTR: See Data Terminal Ready.

dump: The process of printing out or externally storing the entire contents of the controller's memory.

duplex: A communication link in which data transmission can take place in both directions simultaneously.

dynamic start–up: That part of the controller start–up procedure in which the system is checked under program control.

-E-

EAROM: Electrically Alterable Read Only Memory.

echo: A portion of the transmitted signal returned to the source with sufficient magnitude and delay to cause interference.

ECMA: European Computer Manufacturers' Associated.

EDC: Error Detection and Correction. A memory system that uses error detection and correction diagnostic routines.

edit: To deliberately modify a stored program.

EEPROM: Electrically Erasable Programmable Read Only Memory. A type of PROM that is programmed and erased by electrical pulses.

EIA: Abbreviation for electronic Industries Association. An agency that sets electrical/electronic standards.

Electrical-optical isolator: A device which provides electrical isolation using a light source and detector in the same package.

enable: The ability to respond to program instructions.

encoder: 1) A rotary device which transmits position information. 2) A device which transmits a fixed number of pulses for each revolution.

end-device: A device located at the end of a particular network drop. An end-device may have direct control over a real application (PC, robot, NC machine), or it may monitor such an application (terminal).

End of Transmission (EOT): An ASCII control character that indicates the end of a transmission.

environment: In a systems context, the environment is anything that is not a part of the system itself. Knowledge about the environment is important because of the affect it can have on the system or because of possible interactions between the system and the environment.

EOT: See End of Transmission.

EPROM: Erasable Programmable Read Only Memory. A PROM that can be erased with ultraviolet light, and then reprogrammed.

error control procedure: That part of a protocol controlling the detection, and possibly the correction, of transmission errors.

error correcting code (ECC): A code in which each acceptable expression conforms to specific rules of construction that also define one or more equivalent non–acceptable expressions, so that if certain errors occur in an acceptable expression the results will be one of its equivalents and thus the error can be corrected.

error detecting code (EDC): A code in which each expression conforms to specific rules of construction so that if certain errors occur in an expression the resulting expression will not conform to the rules of construction and thus the presence of errors is detected. Synonymous with self–checking code.

error signal: A signal proportional to the difference between the actual output and the desired output.

Exclusive–OR (XOR): A logical operation that has only two inputs, and yields a logic "1" output if either of the two inputs is "1," and a logic "0" output if both inputs are "1" or "0."

execution: The performance of a specific operation such as would be accomplished through processing one instruction, a series of instructions, or a complete program.

execution time: The time required to perform one specific instruction, a series of instructions, or a complete program. The execution time for a given instruction may vary depending on the outcome of the instruction (i.e. true or false), and the parameters involved.

Executive Memory: That portion of the System memory that is responsible for supervisory functions that govern system operation. It allocates and controls system resources such as routines that allow the user to communicate with the processor, and the execution of the application program.

expert system: The top of the line artificial type application. It performs all the capabilities a knowledge system does and provides additional capability of analyzing process data using statistical analysis.

-F-

false: As related to PLC instructions, a reset logic state.

fault: Any malfunction which interferes with normal application operation.

feedback: The signal or data transmitted to the PLC from a controlled machine or process to denote its response to the command signal.

FIFO: See First-In-First-Out.

file: 1) One or more data table words used to store related data. See File Organization. 2) A collection of related records treated as a unit. The records are organized or ordered on the basis of some common factor called a key. Records may be fixed or vary in length and can be stored in different devices and storage media.

file organization: A method of ordering data records stored as a file, while also providing a way to access stored records.

firmware: A series of instructions in a PROM. These instructions are for internal processor functions only, and are transparent to the user.

First-In-First-Out (FIFO): The order that data is entered into, and retrieved from, a file.

flag: A programming technique of using a single bit in memory to detect and remember some event. A storage bit.

flag bit: A processor memory bit, controlled through firmware or a user program, used to signify a certain condition. Example: battery low.

flag sequence: The unique sequence of eight bits (01111110) employed to mark the beginning and ending of a frame.

floating master: A communications system where mastership changes from station to station on an event basis. Mastership contention is resolved using a polling technique.

floating point: A data manipulation format used to locate the point by expressing the power of the base, and which involves the use of two sets of digits. For example, for a floating decimal notation, the base is 10, so the number 8,700,000 would be expressed as 8.7(10)6 or 8.7E6.

floppy disk: A low cost mass storage medium usually consisting of a thin flexible sheet of Mylar with a magnetic oxide surface. Sometimes called diskette.

flow chart: A graphical representation of a definition or method of solution to some task or problem. Example: One might flow chart the step–by–step procedure for load shedding during peak demand hours.

foreground: The area in Network Controller memory that contains high–priority application programs.

forward chaining: The method used to arrive at possible outcomes from given data inputs.

forward channels: In a CATV system, the frequency bands assigned for transmission from the head end into the cable. In a mid–split system, these frequencies are in the range of 150-300 MHz.

frame: A sequence of bits which is delimited or otherwise defined by some distinguished bit sequence; most often applied to sequences processed at the data link layer.

frame check sequence (FCS): The field immediately preceding the ending flag sequence of a frame, containing the bit sequence that provides for the detection of transmission errors by the receiver.

frequency shift keying (FSK): A signal modulation technique, in which a carrier frequency is shifted high or low to represent a binary one or zero, respectively. Offers a high degree of noise immunity.

FSK: See Frequency Shift Keying.

front-end processor (FEP): A single-user computer operating system (e.g. DEC RT–11 or 624–30 Translator) that may act as Network Controller. Connected to a host computer, the FEP preprocesses data between the host and the network.

Full-Duplex (FDX): A bidirectional mode of communication where data may be transmitted and received simultaneously.

full-duplex line: A communication line used to simultaneously transmit data to and from the central processing unit. Contrast with Half-Duplex Line.

functional block instruction set: A set of instructions that moves, transfers, compares, or sequences blocks of data.

-G-

gate: A circuit having two or more input terminals and one output terminal, where an output is present when and only when the prescribed inputs are present.

gateway: A device (or pair of devices) that connects two or more communications networks. This device might appear to each network as a host on that network. A gateway may transfer messages between networks by translating protocols.

global address: An 8-bit address field consisting of all ones. A message with this address will be received by all stations on the network.

global output: An internal output used for inter-processor communication.

global data base: The part of the system where data measurements from the process being performed are stored.

gross errors: Errors resulting from human miscalculation.

guarantee error: Value that comes from a known specification which defines that a product or material will be between a specified arithmetic deviation from the mean.

-H-

Half Duplex (HDX): A communication link in which data transmission is limited to one direction at a time.

half-duplex line: A communication line used to transmit data to and from the central processing unit, but only in one direction at a time.

handshaking: Two-way communication between two devices in order to ensure successful data transfer. Based on a data ready/data received scheme, one device alerts the other that it is ready to send. The other device signals when ready to receive, and acknowledges when the data is received. Handshaking can be accomplished through hardware using special lines, or through software using special codes.

handshaking signals: Special interface lines used for synchronous (not timed) data transfers.

hard contacts: Any type of physical switch contacts.

hard copy: A printed document of what is stored in memory. Examples: ladder program listing, Input/Output cross reference, data table contents, etc.

hardware: Includes all the physical components of the programmable controller, including peripherals, as contrasted to the programmed software components that control its operation.

hardwired logic: Logic control functions that are determined by the way devices are interconnected, as contrasted to programmable control in which logic control functions are programmable and easily changed.

HDLC (High Level Data Link Control): A communications protocol developed by the International Standards Organization (ISO), which defines procedures for the Data Link and Physical protocol layers (.2 and .1).

HDX: See Half-Duplex.

head end: An electronic control center for a CATV network. Also called a "Hub" for bi–directional (e.g. mid–split) systems.

header: The control prefix in an asynchronous message frame. The prefix includes source/destination code, messages type, and priority.

hexadecimal number system: A number system that uses the numerals 0,1,2,3,4,5,6,7,8,9, and the letters A,B,C,D,E,F to represent numbers and codes. Base 16.

high-level language: A powerful set of user-oriented instructions, in which each statement may translate into a series of instructions or subroutines in machine language.

High = True: A signal type where the higher of two voltages indicates a logic state of on (1). See Low = True.

host: A central-controlling computer in a network system.

host computer: A computer attached to a network supplying such services as computation, data base access, or special programs.

host interface: The communications interface to the host computer.

hot backup: A standby processor in a programmable controller system. It consists of a primary and backup processor. If the primary processor fails, the backup processor takes over PLC operations.

-I-

IEEE 802: A family of standards specified by the Institute of Electrical and Electronic Engineers for data communication over local and metropolitan area networks.

image table: An area in PLC memory dedicated to I/O data. Ones and zeros (1 and 0) represent on and off conditions, respectively. During every I/O scan, each input controls a bit in the input image table; each output is controlled by a bit in the output image table.

inference engine: The section of an AI system where all the decisions are made. It uses the knowledge stored in the knowledge data base to arrive at the decision.

information field: The sequence of bits, occurring between the last bit of the control field and the first bit of the frame check sequence. The information field contents are not interpreted at the link level.

information rate: A network measure defined as the amount of data sent per unit time, with units in bits per second.

input: Information sent to the processor from connected devices, via some input interface.

input device: Any connected equipment that will supply information to the central processing unit such as control devices (e.g. switches, buttons, sensors), or peripheral devices (e.g. CRT, manual programmer). Each type of input device has a unique interface to the processor.

instruction set: The set of general–purpose instructions available with a given controller. In general, different machines have different instruction sets.

integer: Any positive or negative whole number or zero.

integral control: See Proportional, Integral, Derivative Module.

intelligent I/O module: Microprocessor based modules that perform processing or sophisticated closed-loop, application functions.

intelligent terminal: An input/output device with built–in intelligence in the form of a microprocessor, and able to perform functions that would otherwise require use of the main memory, and the central processor (e.g. an intelligent CRT programmer).

interface: A circuit that permits communication between the central processing unit and a field input or output device. Different devices require different interfaces. Typical interfaces are: AC inputs, DC inputs, DC outputs, analog inputs, analog outputs, etc.

interference: Any undesired electrical signal induced into a conductor by electrostatic or electromagnetic means.

interframe time fill: The series of flag sequences transmitted between frames.

interlock: A device actuated by the operation of some other device to which it is associated, to govern the succeeding operation of the same or allied devices.

internal output: A program output that is used strictly for internal purposes (does not drive a field device). It provides interlocking functions like a hardwired control relay, however normally–closed and normally open contacts from an internal output may be used as often as required. Internal outputs also provide a solution to ladder diagram format restrictions imposed by the programming unit. Also called internal storage bit, or internal coil.

International Standards Organization (ISO): An organization established to promote the development of international standards.

interrupt: The act of directing a program's execution to a more urgent task.

I/O: Abbr. for Input/Output.

I/O address: A unique number assigned to each input and output. The address number is used when programming, monitoring, or modifying a specific input of output.

I/O channel: A single input or output circuit.

I/O chassis: See chassis.

I/O module: A plug–in type assembly that contains more than one input or output circuit. A module usually contains two or more identical circuits. Normally 2,4,8, or 16 circuits.

I/O rack: See rack.

I/O scan time: The time required for the processor to monitor inputs and control outputs.

I/O update: The continuous process of revising each and every bit in the I/O tables, based on the latest results from reading the inputs and processing the outputs according to the control program.

I/O update time: The time required update all local and remote I/O. Also called I/O scan.

ISO: International Standards Organization.

isolated I/O: Input and output circuits that are electrically isolated from any and all other circuits of a module. Isolated I/O are designed to allow for connecting field devices that are powered from different sources to one module.

-J-

jump: Change in normal sequence of program execution, by executing an instruction that alters the program counter. Sometimes called a branch. In ladder programs a JUMP (JMP) instruction causes execution to jump forward to a labelled rung. In a high–level language, execution is typically passed to a statement line number or an address label like START.

-K-

K: 2^{10} = 1K = 1024. Used to denote size of memory and can be expressed in bits, bytes, or words. Example: 2K = 2048.

k: Kilo. A prefix used with units of measurement to designate quantities a 1000 times as great.

keying: Keying bands installed on backplane connectors to ensure that only one type of module can be inserted into a keyed connector.

knowledge AI system: A progression that detects faults and process behavior based on knowledge introduced, and makes decisions on the process and/or a probable cause of a fault.

knowledge data base: A section of the AI system where the information extracted from the expert is stored.

knowledge inference: Decision -making dealing with the methodology used for the gathering and analyzing of data to draw up a conclusion.

knowledge representation: The method in which a complete artificial intelligence system strategy is organized.

-L-

ladder diagram: An industry standard for representing relay–logic control systems.

ladder diagram programming: A method of writing a user program in a format similar to a relay ladder diagram.

ladder element: Any one of the elements that can be used in a ladder program. The elements include contacts, coils, shunts, timers, counters, etc.

ladder program: A type of control program that uses relay–equivalent contact symbols as instructions (e.g. normally–open, normally–closed contacts, coils).

ladder matrix: A rectangular array of programmed contacts, that defines the number of contacts that can be programmed across a row and the number of parallel branches allowed in a single ladder rung.

LAN: Local Area Network.

language: A set of symbols and rules for representing and communicating information among people, or between people and machines. The method used to instruct a programmable device to perform various operations. Examples are: Boolean, ladder contact symbology, Basic.

latch: A ladder program output instruction that retains its state even though the conditions which caused it to latch ON may go OFF. A latched output must be unlatched. A latched output will retain its last state (ON or OFF) if power is removed.

latching relay: A relay that maintains a given position by mechanical or electrical means until released mechanically or electrically.

layered architecture: A system that breaks network protocol into seven distinct layers. Each layer provides services to the next higher layer, until the highest layer is capable of performing distributed applications.

LCD: See Liquid Crystal Display.

leading edge one-shot: A programming technique that sets a bit for one scan when its input condition has made a false-to-true transition. The false-to-true transition represents the leading edge of the input pulse.

leased line: A private communications line, reserved solely for the customer without interexchange switching.

Least Significant Bit (LSB): The bit that represents the smallest value in a nibble, byte, or word.

Least Significant Digit (LSD): The digit that represents the smallest value in a byte or word.

LED: Abbreviation for light–emitting diode. A semiconductor diode, the junction of which emits light when passing a current in the forward direction. LEDs are used as diagnostic indicators on various controller hardware components.

limit switch: An electrical switch actuated by some part and/or motion of a machine or equipment.

line: 1) A component part of a system used to link various subsystems located remotely from the processor. 2) The source of power for operation. Example: 120V AC line.

Linear Variable Differential Transformer (LVDT): An electromechanical mechanism which provides a voltage reference which is proportional to a movement or displacement of a core inside of a coil.

line printer: A hard–copy device that prints one line of information at a time.

link: The data path established between two or more stations.

liquid crystal display (LCD): A display device consisting basically of a liquid crystal hermetically sealed between two glass plates. One type of LCD depends upon a backlighting source. The readout is either dark characters on a dull white background or white characters on a dull black background. LCDs are used on many hand–held programmers.

load: 1) The power used by a machine or apparatus. 2) To place data into an internal register under program control. 3) To place a program from an external storage device into central memory under operator control.

load cell: A force or weight transducer which is based on a direct application of bonded strain gauges.

load resistor: A resistor connected in parallel with a high impedance load so the output circuit driving the load can provide at least the minimum current required for proper operation.

local area network (LAN): An ensemble of interconnected processing elements (nodes) which are typically confined to fall within a radius not exceeding a few miles.

local I/O PLC: A PLC where I/O distance is physically limited and must be located near the processor. However, it may still be mounted in a separate enclosure. See Remote I/O PLC.

location: In reference to memory, a storage position or register identified by a unique address.

logic: A process of solving complex problems through the repeated use of simple functions that can be either true or false. the three basic logic functions are AND, OR, and NOT.

logic diagram: A drawing which shows AND, OR, and NOT logic symbols, interconnected to graphically describe system operation or control.

logic level: The voltage magnitude associated with signal pulses that represent ones and zeros in digital systems.

Longitudinal Redundancy Check (LRD): An error checking technique based on an accumulated "exclusive or" of transmitted characters. An LRC character is accumulated at both the sending and receiving stations. Similar to CRC.

loop: A sequence of instructions which is executed repeatedly until a terminating condition is satisfied.

loop resistance: The total resistance of two conductors measured at one end (conductor and shield, twisted pair, conductor and armor).

Low = True: A signal type where the lower of two voltages indicates a logic state of on (1). See High = True.

LRC: See Longitudinal Redundancy Check.

LSB: See Least Significant Bit.

LSD: See Least Significant Digit.

-M-

machine language: A program written in binary form.

macro: An instruction set made up of several micro instructions.

magnetic core memory: See Core Memory.

magnetic disk: A flat, circular plate with a magnetic surface on which data can be stored by selective polarization.

main memory: The block of data storage location connected directly to the CPU.

malfunction: Any incorrect function within electronic, electrical, or mechanical hardware. See Fault.

management information: A type of data derived from operational conditions. It may be displayed or printed as reports, which may be used in making decisions relative to the application. This data can include uptime records, production summaries, operating conditions, or a variety of other categories to aid in the MIS (Management Information System) effort.

manipulation: The process of controlling and monitoring data table bits, bytes, or words by means of the user program to vary application functions.

mask: A logical function used to always set certain bits in a word to an established state.

master: A device used to control secondary devices.

Master Control Relay (MCR): A mandatory hardwired relay that can be de-energized by any series-connected emergency stop switch. Whenever the master control relay is de-energized, its contacts open to de-energize all application I/O devices.

master file: A file of data containing the history or current status of a factor or entity of interest to an organization. A master file must be updated periodically to maintain its usefulness.

master station: A data station that has accepted the nomination to ensure a data transfer to and/or from one or more slave stations. A responder or listener cannot be a master under this definition.

MCR Zones: User program areas where all non-retentive outputs can be turned off simultaneously. Each MCR zone must be delimited and controlled by MCR fence codes (MCR instructions).

mean: The average value of readings that have been taken from a complete data sample set.

media access control: A data link function which determines which station on a data highway shall be enabled to transmit at a given time.

median: The middle value of all the readings, in ascending order, that have been taken from a complete data sample set.

medium: 1) The material used to transmit data in a computer network. Example: twisted-pairs, coaxial cables, and optical fibers. 2) The material used to store data. Example: a disk.

memory: That part of the programmable controller where data and instructions are stored either temporarily or semipermanently. The control program is stored in memory.

memory map: A diagram showing a system's memory addresses and what programs and data are assigned to each section of memory.

memory mapped I/O: A method of interfacing by assigning memory addresses to I/O ports as well as memory. data is then sent to and read from external devices by simply reading or writing to certain memory locations.

message: A group of data and control bits transferred as an entity from a data source to a data sink, whose arrangement of fields is determined by the data source.

message mode: A manner of operating a data network by means of message switching.

memory protect: The capability of preventing unauthorized memory entries or program changes. Usually provided by a keylock switch, or software access code.

menu: In an interactive system (e.g. programming device), a list of system programs, (typically displayed on a viewing screen or other display) of which the user can select which operation he wishes to perform. The menu system provides prompts to guide the user's input and response.

message switching: The process of routing messages by receiving, storing, and forwarding complete messages within a data network. Compare with "Packet."

MOS: Metal Oxide Semiconductor. A semiconductor device in which an electric semiconductor device in which an electric field controls the conductance of a channel under a metal electrode called a gate.

motion detection flow meter: A mechanism which measures fluid flow by detection of fluid motion (e.g. turbine flow meter).

micro-floppy: A 3-1/2 inch, magnetic disk used for data storage. Also called a diskette.

microprocessor: A digital electronics–logic package (usually on a single chip), capable of performing the program execution, control, and data processing functions of a central processing unit. The microprocessor usually contains an arithmetic logic unit, temporary storage registers, instruction decode circuitry, a program counter and bus interface circuitry.

microsecond: One millionth of a second. 1×10^{-6} second or 0.000001 seconds.

millisecond: One thousandth of a second. 1×10^{-3} or 0.001 seconds.

mnemonic codes: Symbolic names, usually alphanumeric, for referencing instructions, registers, addresses, etc., to eliminate the need for remembering numbers by substituting meaningful codes.

mode: The sample value that is most consistent in a complete data sample set.

modem: Abbreviation for modulator/demodulator. A device that modulates digital signals and sends them across a telephone line. It also demodulates incoming signals and converts them into digital signals.

modem handshaking: A signalling protocol used for transferring information between devices in a synchronized manner at a rate acceptable to both devices. It may be hardware or software.

modulation rate: A network measure defined as the rate at which a signal changes, with units in baud. Differs from information rate when more than one symbol defines a cell.

module: An interchangeable, plug-in item containing electronic components.

module addressing: A method of identifying the I/O modules installed in a chassis.

monitor: 1) A CRT display. 2) To observe an operation.

Most Significant Bit (MSB): The bit representing the greatest value of a nibble, byte, or word.

Most Significant Digit (MSD): The digit representing the greatest value of a byte or word.

motherboard: A part of the system hardware that contains the connectors and bus wiring for the system's circuit boards (e.g. processor boards, memory boards, I/O modules). Also called a backplane.

motor controller: A device or group of devices that serve to govern, in a predetermined manner, the electrical power delivered to a motor.

motor starter: See Motor Controller.

MOV: Metal Oxide Varistor

MSB: See Most Significant Bit.

MSD: See Most Significant Digit.

multidrop link: A modem cable that terminates at more than one point or location.

multiplexing: 1) The time-shared scanning of a number of data lines into a single channel. Only one data line is enabled at any time. 2) The incorporation of 2 or more signals into a single wave from which the individual signals can be recovered.

multipoint configuration: A network in which connections are made between more than two statins (end–devices).

multiprocessing: Concurrent execution of two or more tasks residing in memory. Available in both RT–11 and RSX–11M operating systems.

multiplex: The act of channelling two or more signals to one source, or sharing a system resource.

-N-

NAK: See Negative Acknowledgement.

NAND: A logical operation that yields a logic "1" output if any input is "0," and a logic "0" if all inputs are "1." The negated AND function. Result of negating the output of an AND gate by following it with a NOT symbol.

narrowband channels: A channel characterized by a communication rate of 100 to 200 bits per second.

National Bureau of Standards (NBS): An organization under the United States Department of Commerce responsible for developing and disseminating federal standards in many areas.

National Electrical Code (NEC): A set of regulations governing the construction and installation of electrical wiring and apparatus, established by the National Fire Protection Association and suitable for mandatory application by governing bodies exercising legal jurisdiction. It is widely used by state and local authorities within the United States.

NBS: See National Bureau of Standards.

NEC: See National Electrical Code.

Negative Acknowledgement (NAK): An ASCII control character transmitted by a receiver as a negative response to the sender.

negative logic: The use of binary logic in such a way that "0" represents the voltage level normally associated with logic 1 (e.g. 0 = +5V, 1 = 0V). Positive logic is more conventional (i.e. 1 = +5V, 0 = 0V).

NEMA Standards: Consensus standards for electrical equipment approved by the members of the National Electrical Manufacturers Association (NEMA).

NEMA Type 12: A category of industrial enclosures intended for indoor use to provide a degree of protection against dust, falling dirt, and dripping non-corrosive liquids. They do not provide protection against conditions such as internal condensation.

network: A series of points (or devices) connected by some type of communications medium.

network access time: The time it takes for a station to gain access to a medium.

network communication time: The length of time transmissions are active.

network controller: A host computer or front–end processor that establishes and maintains the flow of data in the network. It gathers data from the end–devices for presentation to an application program.

Network Layer: Layer 3 of the OSI architecture. It provides to the transport–entities independence from routing and switching considerations associated with the establishment and operation of a given network connection. This includes the case where several transmission resources are used in tandem or in parallel. It makes invisible to transport–entities the method in which the Network Layer uses underlying resources such as data–link–connections to provide network–connections.

network response time: The elapsed time between the point a communication command from the application is interpreted to the point a communication complete reply is available to the application.

node: A point on the network bus, where it connects to a secondary station, at which network messages are received and responses placed.

node processing time: 1) The time it takes for a station to prepare a message for transmission across the network once it interprets the command. 2) The time it takes to transmit a message to the application after it is received by the station.

node response time: The time required to receive and reply to a message on the network.

node throughput: The number of messages of a given size and/or type passing through the station per unit of time.

noise: Random, unwanted electrical signals, normally caused by radio waves or electrical or magnetic fields generated by one conductor and picked up by another.

noise immunity: The ability to function in the presence of noise.

noise spike: A noise disturbance of relatively short duration.

non-retentive output: An output which is continuously controlled by a program rung. Whenever the rung changes state (true or false), the output turns on or off. Contrasted with a retentive output which remains in its last state (on or off) depending on which of its two rungs, latch or unlatch, was last true.

nonvolatile memory: A type of memory whose contents are not lost or disturbed if operating power is lost. Examples: EPROM, Core NOVRAM.

Nonvolatile Random Access Memory: A special kind of RAM that will not lose its contents due to a loss of power.

NOR: A logical operation that yields a logic "1" output if all inputs are "0" and a logic "0" output if any input is "1." The negated OR function. Result of negating the output of an OR gate, by following it with a NOT symbol.

normally-closed contact: 1) A relay contact-pair which is closed when the coil of the relay is not activated, and opens when the coil is activated. 2) A ladder program symbol that will allow logic continuity (flow) if the referenced input is logic "0" when evaluated.

normally-open contact: 1) A relay contact-pair which is open when the coil of the relay is not activated, and closes when the coil is activated. 2) A ladder program symbol that will allow logic continuity (flow) if the referenced input is logic "1" when evaluated.

normal response mode (NRM): A mode in which a secondary station may only initiate transmission after receiving permission to do so from the primary.

NOT: A logical operation that yields a logic "1" at the output if a logic "0" is entered at the input, and a logic "0" at the output if a logic "1" is entered at the input. The NOT, also called the inverter, is normally used in conjunction with the AND and OR functions.

NOVRAM: See Non-Volatile Random Access Memory.

NRZI (Non-Return to Zero Inverted): A self-clocking pulse code used to establish reliable synchronous transmission. The industry standard for serial communications.

null modem: A cable that interconnects two RS-232C devices by acting as a dummy piece of data communication equipment.

-O-

Octal Number System: A number system that uses eight numeral digits (octal digits, 0,1,2,3,4,5,6,7. Base 8.

off-line: Refers to not being in continuous direct communication with the processor; done independently of the processor (e.g. off-line program generation and storage).

one's complement: A binary representation in which the MSB of the word is assigned the negative value of its normal weight minus one. Negative numbers can thereby be expressed and numbers can be negated by simply complementing them.

one-shot: A programming technique which sets a storage bit or output for only one program scan. See Leading Edge One-Shot and Trailing Edge One-Shot.

one-way interaction: Use of a session wherein one presentation-entity always sends and the other presentation-entity always receives.

on-line: Refers to being in continuous communication with the processor while it is running or stopped (e.g. on-line programming).

on-line data change: Allows the user to change various data table values using a peripheral device while the application is operating normally.

on-line editing: Allows the user to modify a program using a peripheral device while the application is operating normally.

on-line operation: Operations where the programmable controller is directly controlling a machine or process.

open system: A system which can be interconnected to others according to established standards.

Open System Interconnect Reference Model (OSIRM): A standard means of communication between open systems proposed by the ISO. It is a seven layer model which represents network architecture. The seven layers are described as follows:

- **Layer 7 - Application:** Provides services for user application processes and management functions.
- **Layer 6 - Presentation:** Provides services for data interpretation, format, and code transformation.
- **Layer 5 - Session:** Provides services for administration and control of sessions between two entities.
- **Layer 4 - Transport:** Provides services for transparent data transfer, end-to-end control, multiplexing, and mapping.

- **Layer 3 - Network:** Provides services for routing, switching, segmenting, blocking error recovery, and flow control.
- **Layer 2 - Data Link:** Provides services to establish, maintain, and release data links; error detection and flow control.
- **Layer 1 - Physical:** Provides services for electrical, mechanical, and functional control of data circuits.

operating system: A collection of programs that organizes a set of hardware devices into a working unit that people can use.

optical coupler: A device that couples signals from one circuit to another by means of electromagnetic radiation, usually infrared or visible. A typical optical coupler uses a light-emitting diode (LED) to convert the electrical signal of the primary circuit into light, and a phototransistor in the secondary to reconvert the light back into an electrical signal. Sometimes referred to as optical isolation.

optical isolation: Electrical separation of two circuits with the use of an optical coupler.

OR: A logical operation that yields a logic "1" output if one of any number of inputs is "1," and a logic "0" if all inputs are "0."

OSI Model: A description of communications functions and services organized as seven layers in such a way as to help promote "open systems interconnection."

OSIRM: See Open System Interconnect Reference Model.

output: Information sent from the processor to a connected device via some interface. The information could be in the form of control data that will signal some device such as a motor to switch ON or OFF, or vary the speed of a drive. It could also be pure data such as a string of ASCII characters that will generate a report.

output device: Any connected equipment that will receive information or instructions from the central processing unit, such as control devices (e.g. motors, solenoids, alarms, etc.) or peripheral devices (e.g. line printers, disk drives, color displays, etc.) Each type of output device has a unique interface to the processor.

overflow: The act of exceeding an ALU's or ladder's numerical capacity in the positive or negative direction.

overtemperature (OVTEMP): An indicator that lights whenever DC circuits are disabled due to a detected overtemperature condition.

-P-

packet: Data and a sequence of control bits arranged in a specified format and transferred as an entity that is determined by the process of transmission. Compare with "Message."

packet mode: A manner of operating a data network by means of packet switching.

packet size: The number of bytes which can be transmitted as one independent group on a physical medium. The size may vary, with the maximum packet size being determined by the implementation.

packet switching: The process of routing and transferring data by means of addressed packets so that a connection is occupied during the transmission of the packet only, and upon completion of the transmission, the connection is made available for transfer of other packets. Packet switching is a combination of multiplexing, segmenting, and routing, whereas message switching involves routing and multiplexing only.

parallel circuit: A circuit in which two or more of the connected components or contact symbols in a ladder program are connected to the same pair of terminals so that current may flow through all the branches, as contrasted with a series connection, where the parts are connected end–to–end so that current flow has only one path.

parallel communication: A form of communication in which groups of bits or entire words are transmitted simultaneously.

parallel output: Simultaneous availability of two or more bits, channels, or digits.

parallel transmission: The simultaneous transmission of a group of bits that constitute a character or a frame of data. Compare "serial transmission."

parity: The even or odd characteristic of the number of 1's in a byte or word of memory.

parity bit: An additional bit added to each memory word as a means of error detection. See parity check.

parity check: A check for the number of 1's in a memory word, by performing an exclusive–OR. If the result is odd, then parity is odd, and if the result is even, then parity is even. The test will be for either odd or even parity, and the wrong result is a parity error. The parity check is to improve reliability on memory and interface transfers.

partition: A logical division of main memory. In a multiprogramming system, tasks are executed in a specific partition and in parallel with other tasks.

passing: See Token Passing.

PC: See Programmable Controller.

peer-to-peer: A form of communications in which messages are exchanges between entities with comparable functionality in different systems.

peripherals: External devices that are connected to the programmable controller (e.g. line printers, cassette recorders, CRT programmers, disk drives etc.).

permanent virtual circuit (PVC): A network-facility providing a permanent association between two network-connection-endpoints as specified in CCITT Recommendation X.25. A PVC is analogous to a point-to-point private line; hence, no call setup or call clearing action is required or allowed.

PID: Proportional-Integral Derivative. A mathematical formula that provides a closed loop control of a process. Inputs and outputs are continuously variable and typically will be analog signals. See Proportional, Integral, Derivative Module.

physical interface: A shared boundary defined by common physical interconnection characteristics, signal characteristics, and functional characteristics of the interchange circuits.

pilot-type device: A device used in a circuit as a control apparatus to carry electrical signals for directing performance. This device does not carry primary current.

PLC: Programmable Logic Controller. See Programmable Controller.

point-to-point configuration: A network in which a connection is made between two and only two terminal installations.

polling: After an instruction message has been sent, the Network Controller sends another message (polls) asking for a result. May be addressed to a single station or to all stations.

port: A signal input or output point (e.g. communication port).

positive logic: The use of binary logic in such a way that "1" represents a positive logic level (e.g. 1 = +5V, 0 = 0V). This is the conventional use of binary logic.

potentiometer: A very simple displacement transducer which works based by a principle based on resistance changes due to movement of a potentiometer wiper.

power supply: The unit that supplies the necessary voltage and current to the system circuitry.

pressure based flow meter: A mechanism used to measure fluid flow by means of measuring pressure differentials (e.g. Venturi tube, orifice plate).

presentation layer: Layer 6 of the OSI architecture. The purpose of this layer is to represent information to communicating application-entities in a way that preserves meaning while resolving syntax differences. The presentation layer provides syntax differences. The presentation layer provides the services of data transformation and formatting, syntax selection, and presentation-connections to the application layer.

Preset Value (PR): The number of time intervals or events to be counted.

pressure switch: A switch that is activated at a specified pressure.

primary processor: A PLC that controls all the I/O, but has a backup PLC to take over system operation in case it fails.

primary station: That part of the data station that supports the primary control functions of the data link. The primary generates commands for transmission and interprets received responses. Specific responsibilities assigned to the primary include: a) initialization of control signal interchange b) organization of data flow c) actions regarding error control and error recovery functions at the link level.

priority: The importance of a device or peripheral in an interrupt system. Each interrupt device has an interrupt service routine. The interrupt will be serviced according to priority in the case of a simultaneous occurrence.

process: 1) Continuous and regular production executed in a definite uninterrupted manner. 2) One or more entities threaded together to perform a requested service.

processor: See CPU.

program: A planned set of instructions stored in memory, and executed in an orderly fashion by the central processing unit.

programming device: A device for inserting the control program into memory. The programming device is also used to make changes to the stored program. Most programmers have displays that allow the program to be monitored while the machine or process is in operation.

program scan: The time required by the processor to evaluate and execute the control logic. This time does not include the I/O update time. The program scan repeats continuously while the processor is in the run mode.

program storage: The portion of memory reserved for saving programs, routines, and subroutines.

programmable controller: A solid-state control device that can be programmed to control process or machine operation. The programmable controller consists of five basic components (i.e. processor, memory, input/output, power supply, programming device). A PLC is designed as an industrial control system.

PROM: Programmable Read Only Memory. A Read Only Memory that can be programmed once, and cannot be altered after that.

propagation error: A final error which is caused by the interaction of two or more independent variables, each one causing a different error. Also called a combined error.

Proportional, Integral, Derivative Module (PID): An intelligent I/O module or ladder diagram instruction which provides automatic closed-loop operation of multiple continuous process control loops. For each loop, this module can perform any or all of the following control functions:

- **Proportional** control causes an output signal to change as a direct ratio of the error signal variation.
- **Integral** control causes an output signal to change as a function of the error signal and time duration.
- **Derivative** control causes an output signal to change as a function of the rate of change of the error signal.

protocol: A formal definition that describes how communications will take place (e.g. timing considerations, data format, what control signals mean and what they do, the pin numbers for specific functions, the meaning and priority of various messages etc.) Handshaking is a communications protocol.

PROWAY: A data highway for process control proposed by IEC TC 65/SC65A.

-Q-

Queue: Tasks or data waiting for processing time.

-R-

RAM: Random Access Memory. Commonly referred to as read/write, because it can be written to as well as read from However, a stricter definition of RAM is a memory that stores memory in such a way that each bit of information can be stored or retrieved within the same amount of time as any other bit (as contrasted with serial memory, in which data is stored and retrieved in a sequential order).

random errors: Errors resulting from unexpected actions in the process line.

read: 1) To acquire data from a storage device. 2) The transfer of data between devices such as a peripheral device and a computer.

Read/Write memory: A type of memory which can be read from or written to. Read/ Write memory can be altered quickly and easily by merely writing over the part to be changed or inserting a new part to be added. See RAM.

Real-Time Clock (RTC): A device that continually measures time in a system without respect to what tasks the system is performing.

Received Data (RXD): A serialized data input to a module. RXD and RXDRET are isolated from the rest of the circuitry on the module.

Received Data Return (RXDRET): The return signal for RXD. It is connected to the isolated receiver, and is isolated from all other circuitry on the module.

redundancy: The number of total characters or bits in a protocol that can be eliminated without a loss in information.

redundant system: Two or more devices that actively control the outputs of a system. Each device in the system votes on every control decision.

redundant transmission: A method of sending each message several times around the network in order to ensure total reception.

register: A temporary storage device for various types of information and data (e.g. timer/counter preset values). In PLCs a register is normally 16 bits wide.

relay: An electrically operated device that mechanically switches electrical circuits.

relay logic: A representation of the program or other logic in a form normally used for relays.

remote I/O PLC: A PLC system where some or all of the I/O racks are remotely mounted from the PLC. The location of remote I/O racks from the PLC may vary depending on the the application and the PLC used. Also see Local I/O PLC.

remote programming: A method of performing PLC programming by connecting the programming device to the network rather than to the PLC.

repeater: A device whereby signals received over one circuit are automatically repeated in another circuit or circuits, generally in an amplified and/or reshaped form.

reply: Data transmitted in response to a request.

report: An application data display or printout containing information in a user-designed format. Reports could include operator messages, part records, and production lists. Initially entered as messages, reports are stored in a memory area separate from the user program.

report generation: The printing or displaying of user-formatted application data by means of a data terminal. Report generation can be initiated by means of either a user program or a data terminal keyboard.

Request-To-Send (RTS): A request from the module to the modem to prepare to transmit. It typically turns on the data carrier.

Resistance Temperature Detector (RTD): Temperature transducer made of wire conductive elements. The most common elements used are platinum, nickel, copper, and nickel-iron.

resolution: The smallest detectable increment of measure. Analog-to-Digital resolution is usually principally limited by the number of bits used to quantify the analog input signal. (e.g. a 12–bit analog-to-digital converter has a resolution of one part in 4096 ($2^{12} = 4096$).

response: In data communications, a reply represented in the control field of a response frame. It advises the primary with respect to the action taken by the secondary to one or more commands.

response time: The elapsed time between the generation of the last character of a message at a terminal and the receipt of the first character of the reply. It includes terminal delay, network delay, and service node delay.

RFI: Radio Frequency Interference.

ring network: A network topology where each device is physically connected to adjacent devices.

ring topology: A network where signals are transmitted from one station and relayed through each subsequent station in the network.

ROM: Read Only Memory. A type of memory which permanently stores information (e.g. a math function or a microprogram). A ROM is programmed during fabrication according to the User's requirements, and cannot be reprogrammed.

routine: A sequence of PLC instructions which monitors and controls a specific application function.

routing: A function within a layer to translate the title or address of an entity into a path by which the entity is to be reached.

RS-232-C: An EIA standard, originally introduced by the Bell System, for the transmission of data over a twisted-wire pair, less than 50 feet in length. It defines pin assignments, signal levels, etc. for receiving and transmitting devices. Other RS-standards cover the transmission of data over distances in excess of 50 feet.

RS-422: An EIA standard for the electrical characteristics of balanced voltage digital interface. Maximum distance-4000 ft.

RTC: See Real-Time Clock.

RTS: See Request-To-Send.

rule based knowledge representation: The expert's knowledge (of an AI system) transformed into IF and THEN/ELSE situations where action and decisions are made.

rung: A ladder program term that refers to the programmed instructions that drive one output. A single rung may be only a portion of a complete control program which in turn may be several rungs.

R/W: See Read/Write memory.

RXD: See Received Data.

RXDRET: See Received Data Return.

-S-

scan time: The time required to read all inputs, execute the control program, and update local and remote I/O. This is effectively the time required to activate an output that is controlled by programmed logic. See Program Scan Time and I/O Scan Time.

SCR: Abbreviation for Silicon Controlled Rectifier. A semiconductor device that functions as an electrically controlled switch for DC loads. A component of DC output circuits.

scratch pad memory: A temporary storage area used by the CPU to store a relatively small amount of data used for interim calculations or control. Data that is needed quickly is stored in this area to avoid the access time that would be involved if stored in the main memory.

screen: The viewing surface of a CRT where data is displayed.

scrolling: The vertical movement of data on a CRT display caused by the dropping of one line of displayed data for each new line added.

SDLC: Synchronous Data Link Control. A method of controlling the transfer of data between stations in a point-to-point, multipoint, or loop arrangement using synchronous data transmission techniques.

secondary station: That part of the network that executes data link functions as instructed by the primary. A secondary interprets received commands and generates responses for transmission. A slave station in HDLC and ADCCP unbalanced configurations.

self-diagnostic: The hardware and firmware within a controller that monitors its own operation and indicates any fault which it can detect.

sequencer: A function of an (N)-entity to provide the (N)-service of delivering data in the same order as it was submitted.

serial communication: Type of communication in which bits are transmitted sequentially rather than simultaneously as with parallel communication.

serial transmission: The sequential transmission of a group of bits that constitute a character or a frame of data. Compare with "parallel" transmission.

series circuit: A circuit in which the components or contact symbols are connected end-to-end, and all must be closed to permit current flow.

session: Synonym for session-connection. A cooperative relationship between two application-entities characterizing the communication of data between them.

session-dialogue-service: A session-service controlling data exchange, delimiting and synchronizing data operation between two presentation-entities.

session-interaction-unit: A session-service-data-unit used by presentation-entities to control the transfer of turn in an agreed upon interaction mechanism.

session layer: Layer 5 of the OSI architecture. Its purpose is to provide the means for cooperating presentation-entities or organize and synchronize their dialogue and manage their data exchange.

shield: A barrier, usually conductive, that substantially reduces the effect of electric and/or magnetic fields.

signal level: The RMS voltage measured during an RF signal peak, usually expressed in microvolts, referred to an impedance of 75 ohms, or in dBmV, the value in decibels with respect to a reference level of 0 dBmV, which is 1 millivolt across 75 ohms.

significant digit: A digit that contributes to the precision of a number. The number of significant digits is counted beginning with the digit contributing the most value, called the Most Significant Digit (left-most), and ending with the digit contributing the least value, called the Least Significant Digit.

single-scan: A supervisory type command initiated by the user while the controller is in the Stop mode. This command causes the control program to be executed for one scan, including I/O update. This troubleshooting function allows step-by-step inspection of occurrences while the machine is stopped.

slave: A remote system or terminal whose functions are controlled by a central "master" system.

slave station: A station that is selected by a master station to receive data and/or respond with data.

snubber: A circuit generally used to suppress inductive loads; it consists of a resistor in series with a capacitor (RC snubber) and/or a MOV placed across the AC load.

software: The term applied to programs that control the processing of data in a system.

solenoid: A transducer that converts a current into linear motion; it consists of one or more electromagnets that move a metal plunger. The plunger is sometimes returned to its original position after excursion with a spring or permanent magnet.

solid-state: Circuitry designed using only integrated circuits, transistor, diodes, etc.; no electro-mechanical devices such as relays are utilized.

software: 1) Any written documents associated with the system hardware. 2) Stored instructions (the program).

splitter: A passive, 5-300 MHz or 800 MHz band pass device. The device is coupled in-line to a main trunk or branch of a broadband network for splitting the power and the information signal two or more ways on a coaxial network. Splitters always pass through 60 Hz power to all outlets.

standard deviation: The relationship between any reading and the mean of a complete data sample set. See Chapter 12 for equation.

star topology: A network where all devices are connected to a central or master communications device which routes messages.

start bit: The first bit sent in an asynchronous word transmission. The start bit is for control purposes and does not convey data.

state: The logic 0 or 1 condition in PLC memory or at a circuit input or output.

station: Any PLC, computer, or data terminal connected to, and communicating by means of, a data highway. See Node.

status: The condition or state of a device (e.g. ON/OFF).

stop bit: The last bit in an asynchronous word transmission. The stop bit is for control purposes and does not convey data.

storage: See Memory.

storage bit: A bit in a data table word which can be set or reset, but is not associated with a physical I/O terminal point.

storage media: Materials on which data may be recorded. See Disk Storage.

strain gauge: A mechanical transducer which is used to measure body deformation, or strain, due to the force applied to the area of a rigid body. Strain gauges can be either bonded or unbonded.

subroutine: A program segment in the ladder diagram that performs a specific task and is available for use.

subsystem: 1) A part of a larger system having the properties of a system in its own right. 2) A system within another system.

synchronous: A type of serial transmission that maintains a constant time interval between successive events.

synchronous shift register: A shift register where only one change of state occurs per control pulse.

synchronous transmission: Transmission in which data is sent at a fixed rate in synchronization with a timing signal or clock pulse. This eliminates the need for start/stop bits.

syntax: Rules governing the structure of a language.

system: A set of one or more PLCs, I/O devices and modules, computers, the associated software, peripherals, terminals, and communication networks, that together, provide a means of performing information processing for controlling machines or processes.

system errors: Errors resulting from the instrument itself or the environment.

-T-

tap: A device which provides mechanical and electrical connection to a trunk cable. The tap allows part of the signal on the trunk to be passed to a station, and the signal transmitted by the station to be passed to the trunk.

tasks: A set of instructions, data, and control information capable of being executed by a CPU to accomplish a specific purpose.

termination: 1) The load connected to the output end of a transmission line. 2) A provision for ending a transmission line and connecting to a bus bar or other terminating device (e.g. screw termination).

thermal transducer: A mechanism used to sense and monitor changes in temperatures (e.g. RTDs, thermistors, thermocouples.

thermistor: A temperature transducer which exhibits changes in internal resistance that are proportional to changes in temperature. Thermistors are made of semiconductor material such as oxides of cobalt, nickel, manganese, iron, and titanium.

thermocouple: A temperature measuring device which is made of two metals (bimetallic). It produces its own voltage differential (the Seebeck effect) by joining together two metals of different temperatures.

thermopiles: The connection of several thermocouples in series to produce increases in thermocouple resolution.

throughput: The speed at which an application or part of an application is performed. Throughput is dependent on speed, medium, protocol, packet size and amount of data handled by a network.

thumbwheel switch: A rotating switch used to input numeric information into a controller.

time base: A unit of time generated by the system clock and used by software timer instructions. Normal time bases are 0.01, 0.1, and 1.0 second.

time-division multiplex: A means of supporting more than one data link-connection on single data circuit by enabling data transmission by each station intermittently, generally at regular intervals and by means of an automatic distribution.

token: A frame that grants bus mastership to a station from the present bus master. Mastership is required for a station to originate communication.

token bus: A token access procedure used with broadcast topology.

token passing: In this message transmission technique a token is passed along the bus and each node has a set amount of time to receive and/or respond to it. Each station must bear a part of the transmission load.

token ring: A token access procedure used with physical ring topology.

topology: The way in which a network is physically structured, such as in a ring, bus, or star configuration.

trailing edge one-shot: A programming technique that sets a bit for a scan when its input condition has made a true-to-false transition. The true-to-false transition represents the trailing edge of the input pulse.

transducer: A device used to convert physical parameters such as temperature, pressure, and weight into electrical signals.

transitional contact: A contact, depending on how it is programmed, will be energized for one program scan every 0 to 1 transition, or every 1 to 0 transition of the referenced coil.

transmission line: A physical means of connecting two or more locations to each other for the purpose of transmitting and receiving data.

transmission medium: The technology used (coaxial cable, fiber optics, etc.) in a given transmission line.

Transmitted Data (TXD): An output from the module that carries serialized data.

Transmitted Data Return (TXDRET): The return signal for TXD. It is connected to module logic ground through a resistor.

transparency: A feature, usually provided by an interface unit, which allows the end-device to operate without "knowledge" of network protocol.

transport layer: Layer 4 of the OSI architecture. The transport-service relieves its users from any concern with the detailed way in which reliable and cost effective transfer of data is achieved. Transport-functions allow the Network Layer to be composed of more than one communication resource in tandem (e.g. a public packet switched network used in tandem with a circuit-switched network).

triac: A semiconductor device that functions as an electrically controlled switch for AC loads. A component of AC output circuits.

true: As related to PLC instructions, an enabled logic state.

truth table: A table listing that shows the state of a given output as a function of all possible input combinations.

TTL: Abbreviation for Transistor-Transistor Logic. A semiconductor logic family in which the basic logic element is a multiple-emitter transistor. This family of devices is characterized by high speed and medium power dissipation.

two's complement: A numbering system used to express positive and negative binary numbers in which the MSB takes on the negative value of its normal weight.

TXD: See Transmitted Data.

TXDRET: See Transmitted Data Return.

-U-

UART: Abbreviation for Universal Asynchronous Receiver/Transmitter. Interface device for serial/parallel conversion, buffering, and adding check bits.

unbalanced configuration: An HDLC or ADCCP configuration involving a single primary station and one or more secondary stations.

unit interval: A network measure defined as the shortest code symbol. The reciprocal of baud rate.

up-load: The process of transferring end-device data from a network interface to the network controller. Control programs can be up-loaded to be verified against a master copy in memory.

USART: Universal Synchronous/Asynchronous Receiver/Transmitter. A UART with the added capability for synchronous data communications.

user memory: The memory where the application control program is stored.

utility: A general-purpose program that executes common functions for the central processor.

-V-

VA: Voltamperes

variable: A factor which can be altered, measured, or controlled.

variable data: Numerical information which can be changed during application operation. It includes timer and counter accumulated values, thumbwheel settings, and arithmetic results.

varistor: A resistor whose resistance varies proportionately with the voltage applied to it.

Vertical Redundancy Check (VRC): A check or parity bit added to each character in a message so the number of bits in each character, including the parity bit, is odd (odd parity) or even (even parity).

virtual circuit: An association between two DTE's (a source and a sink) over which data packets are transmitted.

volatile memory: A memory whose contents are irretrievable when operating power is lost.

VRC: See Vertical Redundancy Check.

-W-

WACK: Wait Before Transmitting Positive Acknowledgement. In binary synchronous communications, this DLE sequence is sent by a receiving station to indicate that it is temporarily not ready to receive.

watchdog timer: A timer that monitors logic circuits controlling the processor. If the watchdog timer, which is reset every scan, ever times out, the processor is assumed faulty and is disconnected from the process.

weighted value: The numerical value assigned to any single bit as a function of its position in the code word.

wideband: A channel characterized by a communication rate greater than 200 bits per second.

word: The unit number of binary digits (bits) operated on at a time by the central processing unit when it is performing an instruction or operating on data. A word is usually composed of a fixed number of bits.

word length: The number of bits in a word. In a PLC, these are generally only data bits. One PLC word equals 16 data bits.

write: The process of putting information into a storage location.

-X-

X.25: A CCITT recommendation that establishes procedures for the first three layers of network protocol (physical, data link, and network.) X.25 provides a link service, but does not provide network control functions.

AND	OR	A	B	Y
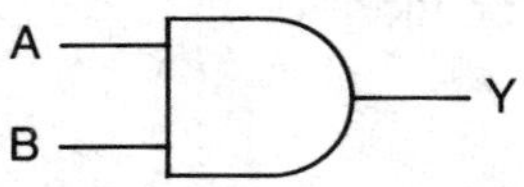 A B Y	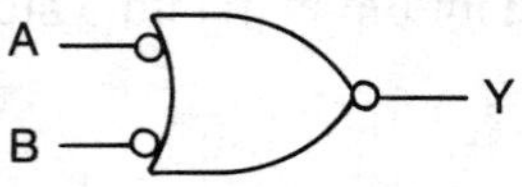 A B Y	1 1 0 0	1 0 1 0	1 0 0 0
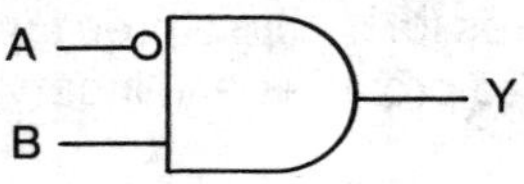 A B Y	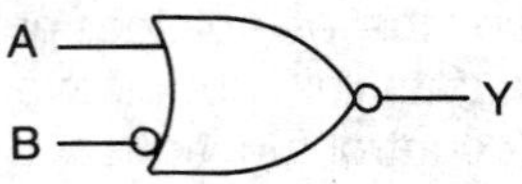 A B Y	1 1 0 0	1 0 1 0	0 0 1 0
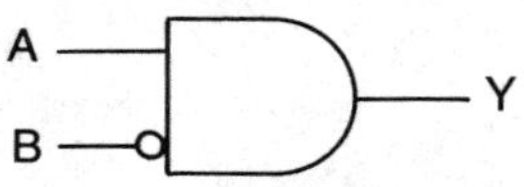 A B Y	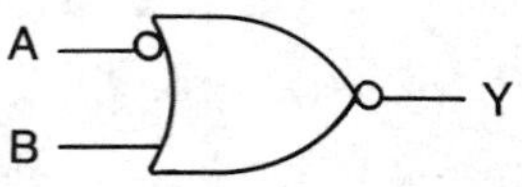 A B Y	1 1 0 0	1 0 1 0	0 1 0 0
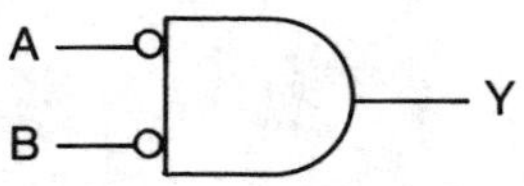 A B Y	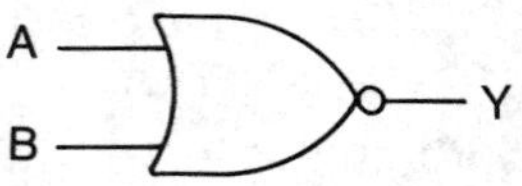 A B Y	1 1 0 0	1 0 1 0	0 0 0 1
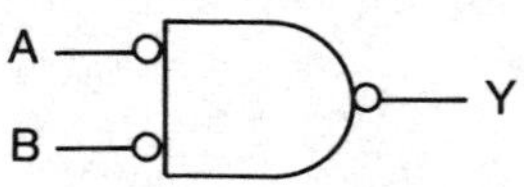 A B Y	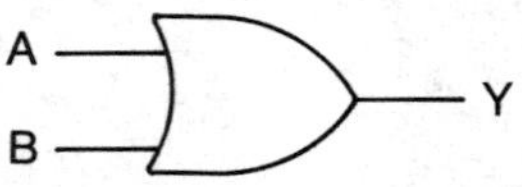 A B Y	1 1 0 0	1 0 1 0	1 1 1 0
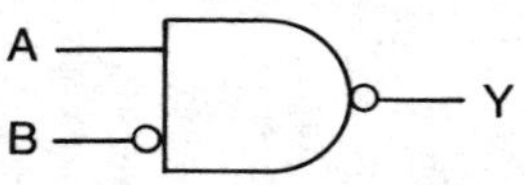 A B Y	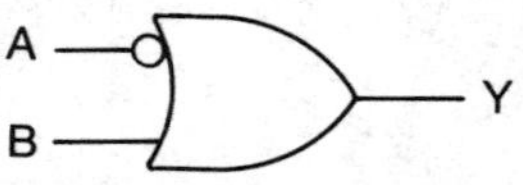 A B Y	1 1 0 0	1 0 1 0	1 0 1 1
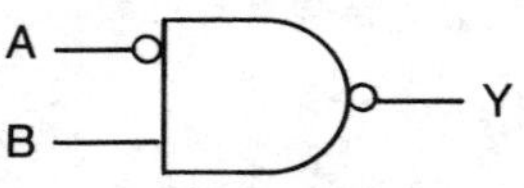 A B Y	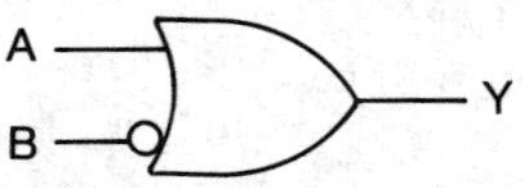 A B Y	1 1 0 0	1 0 1 0	1 1 0 1
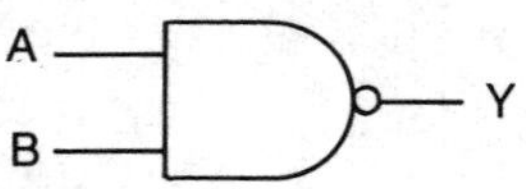 A B Y	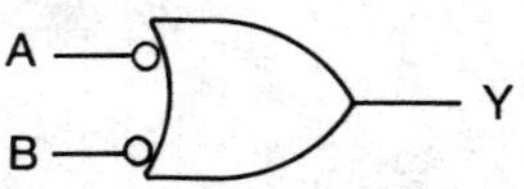 A B Y	1 1 0 0	1 0 1 0	0 1 1 1

Logic Diagrams

Ladder Diagrams

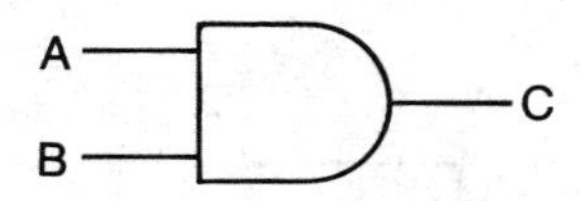

AND Gate

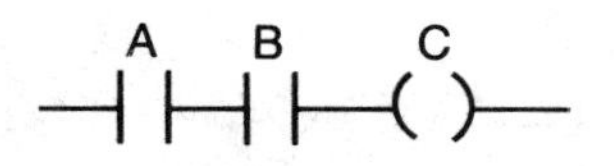

Equivalent Circuit

A	B	C
0	0	0
0	1	0
1	0	0
1	1	1

AND Truth Table

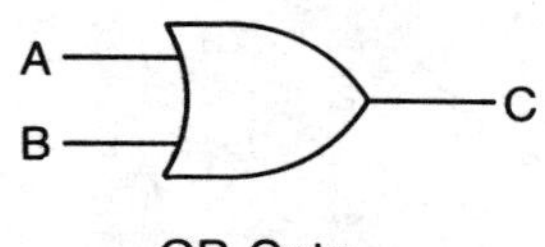

OR Gate

$A + B = C$

Equivalent Circuit

A	B	C
0	0	0
0	1	1
1	0	1
1	1	1

OR Truth Table

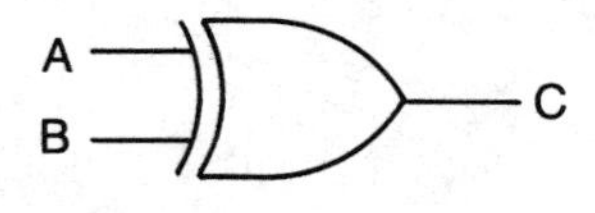

Exclusive-OR Gate

$A \oplus B = C$

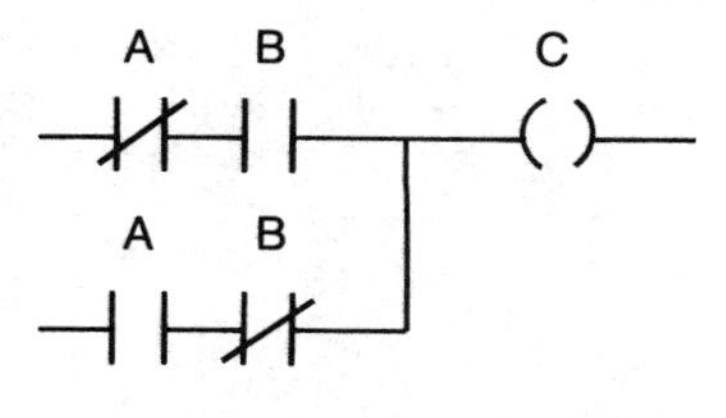

Equivalent Circuit

A	B	C
0	0	0
0	1	1
1	0	1
1	1	0

Exclusive-OR Truth Table

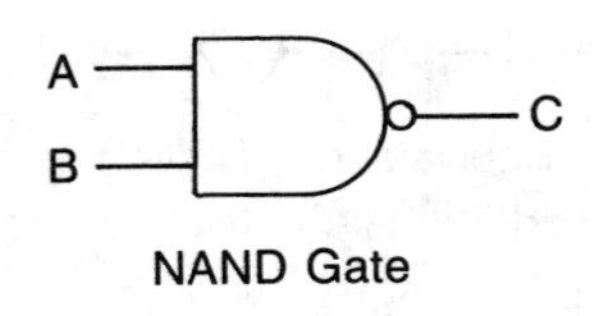

NAND Gate

$A \cdot B = C$

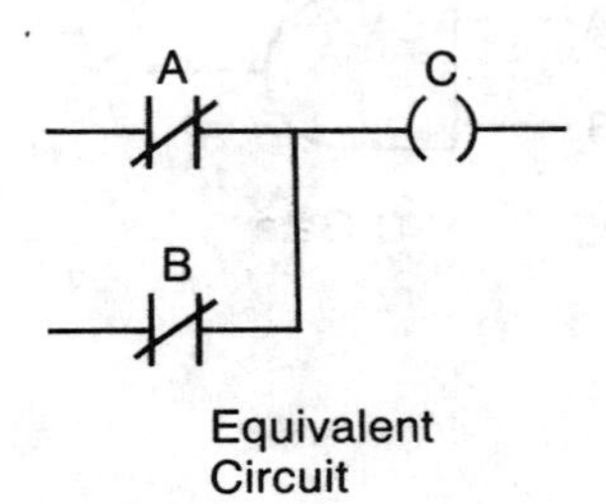

Equivalent Circuit

A	B	C
0	0	1
0	1	1
1	0	1
1	1	0

NAND Truth Table

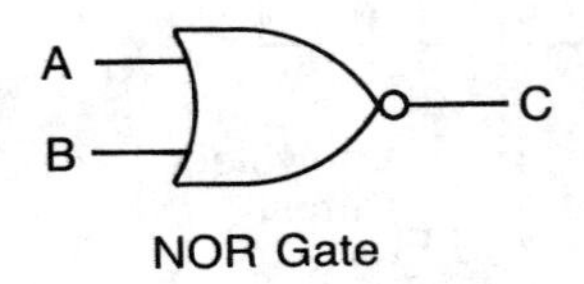

NOR Gate

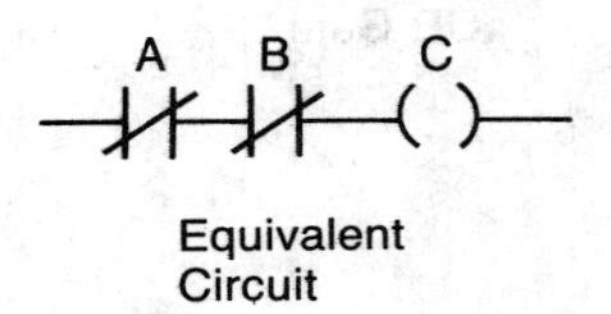

Equivalent Circuit

A	B	C
0	0	1
0	1	0
1	0	0
1	1	0

NOR Truth Table

ALU: Arithmetic Logic Unit
ANSI: American National Standards Institute
ASCII: American Standard Code for Information Interchange
BCC: Block Check Character
BCD: Binary Coded Decimal
BCP: Byte Or Character Controlled Protocol
CMOS: Complimentary Metal Oxide Semiconductor
CPU: Central Processing Unit
CRC: Cyclic Redundancy Check (For Error Detection)
CRT: Cathode Ray Tube
CTS: Clear To Send; Handshake Between DTE & DCE
CSMA/CD: Carrier Sense Multiple Access with Collision Detection
DLE: Data Link Escape (a BCP Control Character)
DMA: Direct Memory Access
DSR: Data Set Ready: Handshake Between DTE & DCE
DTE: Data Terminal Equipment
DTR: Data Terminal Ready; Handshake Between DTE & DTE
EAROM: Electrically Alterable Read Only Memory
EBCDIC: Extended Binary-Coded Decimal Information Code
ECC: Error Correction Code
EEPROM: Electrically Eraseable Programmable Read Only Memory
EIA: Electronic Industries Association
EPROM: Eraseable Programmable Read Only Memory
FIFO: First-In First-Out
HDLC: High (level) Data Link Control
IEEE: Institute of Electrical and Electronics Engineers
I/O: Input/Output
ISO: International Standards Organization
LAN: Local Area Network
LCD: Liquid Crystal Display
LED: Light Emitting Diode
LIFO: Last-In First-Out
LRC: Longitudinal Redundant Check
LSB: Least Significant Bit/Byte
LSI: Large Scale Integration
MCR: Master Control Relay
MODEM: Modulator/Demodulator
MSB: Most Significant Bit/Byte
msec: millisecond
NOVRAM: Non-Volatile Random Access Memory
PC: Programmable Controller
PROM: Programmable Read Only Memory
RAM: Random Access Memory
ROM: Read Only Memory
RTS: Request To Send
R/W: Read/Write
TTL: Transistor-Transistor Logic
UART: Universal Asynchronous Receiver/Transmitter
μsec: microsecond
VDU: Video Display Unit
VPU: Video Programming Unit
VRC: Vertical Redundancy Check
XFER: Transfer
XMIT: Transmit
XOR: Exclusive OR

OCT	PARITY	HEX	ASCII CHAR	CTRL KEYBD EQUIV	ALTERNATE CODE NAMES
000	EVEN	00	NUL	@	NULL, CTRL SHIFT P, TAPE LEADER
001	ODD	01	SOH	A	START OF HEADER,SOM
002	ODD	02	STX	B	START OF TEXT,EOA
003	EVEN	03	ETX	C	END OF TEXT, EOM
004	ODD	04	EOT	D	END OF TRANSMISSION, END
005	EVEN	05	ENQ	E	ENQUIRY,WRU,WHO ARE YOU
006	EVEN	06	ACK	F	ACKNOWLEDGE,RU,ARE YOU
007	ODD	07	BEL	G	BELL
010	ODD	08	BS	H	BACKSPACE,FE0
011	EVEN	09	HT	I	HORIZONTAL TAB,TAB
012	EVEN	0A	LF	J	LINE FEED,NEW LINE,NL
013	ODD	0B	VT	K	VERTICAL TAB,VTAB
014	EVEN	0C	FF	L	FORM FEED,FORM,PAGE
015	ODD	0D	CR	M	CARRIAGE RETURN,EOL
016	ODD	0E	SO	N	SHIFT OUT,RED SHIFT
017	EVEN	0F	SI	O	SHIFT IN,BLACK SHIFT
020	ODD	10	DLE	P	DATA LINK ESCAPE,DC0
021	EVEN	11	DC1	Q	XON,READER ON
022	EVEN	12	DC2	R	TAPE,PUNCH ON
023	ODD	13	DC3	S	XOFF,READER OFF
024	EVEN	14	DC4	T	TAPE,PUNCH OFF
025	ODD	15	NAK	U	NEGATIVE ACKNOWLEDGE,ERR
026	ODD	16	SYN	V	SYNCHRONOUS IDLE,SYNC
027	EVEN	17	ETB	W	END OF TEXT BUFFER,LEM
030	EVEN	18	CAN	X	CANCEL,CNCL
031	ODD	19	EM	Y	END OF MEDIUM
032	ODD	1A	SUB	Z	SUBSTITUTE
033	EVEN	1B	ESC	[	ESCAPE,PREFIX
034	ODD	1C	FS	\	FILE SEPARATOR
035	EVEN	1D	GS	]	GROUP SEPARATOR
036	EVEN	1E	RS		RECORD SEPARATOR
037	ODD	1F	US	_	UNIT SEPARATOR
040	ODD	20	SP		SPACE,BLANK
041	EVEN	21	!		
042	EVEN	22	"		
043	ODD	23	#		
044	EVEN	24	$		
045	ODD	25	%		
046	ODD	26	&		
047	EVEN	27	'		APOSTROPHE
050	EVEN	28	(		
051	ODD	29	)		
052	ODD	2A	*		ASTERISK
053	EVEN	2B	+		
054	ODD	2C	,		COMMA
055	EVEN	2D	-		MINUS
056	EVEN	2E	.		PERIOD
057	ODD	2F	/		
060	EVEN	30	0		NUMBER ZERO
061	ODD	31	1		NUMBER ONE
062	ODD	32	2		
063	EVEN	33	3		
064	ODD	34	4		
065	EVEN	35	5		
066	EVEN	36	6		
067	ODD	37	7		
070	ODD	38	8		
071	EVEN	39	9		
072	EVEN	3A	:		COLON
073	ODD	3B	;		SEMI-COLON
074	EVEN	3C	<		LESS THAN
075	ODD	3D	=		
076	ODD	3E	>		GREATER THAN
077	EVEN	3F	?		

To transmit control codes, depress"CTRL," then the desired character under keyboard equivalent.

OCT	PARITY	HEX	ASCII CHAR	ALTERNATES
100	ODD	40	@	SHIFT P
101	EVEN	41	A	
102	EVEN	42	B	
103	ODD	43	C	
104	EVEN	44	D	
105	ODD	45	E	
106	ODD	46	F	
107	EVEN	47	G	
110	EVEN	48	H	
111	ODD	49	I	LETTER I
112	ODD	4A	J	
113	EVEN	4B	K	
114	ODD	4C	L	
115	EVEN	4D	M	
116	EVEN	4E	N	
117	ODD	4F	O	LETTER O
120	EVEN	50	P	
121	ODD	51	Q	
122	ODD	52	R	
123	EVEN	53	S	
124	ODD	54	T	
125	EVEN	55	U	
126	EVEN	56	V	
127	ODD	57	W	
130	ODD	58	X	
131	EVEN	59	Y	
132	EVEN	5A	Z	
133	ODD	5B	[	SHIFT K
134	EVEN	5C	\	SHIFT L
135	ODD	5D	]	SHIFT M
136	ODD	5E	^	SHIFT N
137	EVEN	5F	—	,SHIFT O, UNDERSCORE
140	EVEN	60		ACCENT GRAVE
141	ODD	61	a	
142	ODD	62	b	
143	EVEN	63	c	
144	ODD	64	d	
145	EVEN	65	e	
146	EVEN	66	f	
147	ODD	67	g	
150	ODD	68	h	
151	EVEN	69	i	
152	EVEN	6A	j	
153	ODD	6B	k	
154	EVEN	6C	l	
155	ODD	6D	m	
156	ODD	6E	n	
157	EVEN	6F	o	
160	ODD	70	p	
161	EVEN	71	q	
162	EVEN	72	r	
163	ODD	73	s	
164	EVEN	74	t	
165	ODD	75	u	
166	ODD	76	v	
167	EVEN	77	w	
170	EVEN	78	x	
171	ODD	79	y	
172	ODD	7A	z	
173	EVEN	7B	[	
174	ODD	7C	!	VERTICAL SLASH
175	EVEN	7D	]	ALT MODE
176	EVEN	7E	~	ALT MODE)
177	ODD	7F	DEL	DELETE,RUBOUT

Table F-1. Powers of Two

2^n	n
1	0
2	1
4	2
8	3
16	4
32	5
64	6
128	7
256	8
512	9
1 024	10
2 048	11
4 096	12
8 192	13
16 384	14
32 768	15

Table F-2. Powers of Eight

8^n	n
1	0
8	1
64	2
512	3
4 096	4
32 768	5
262 144	6
2 097 152	7
16 777 216	8
134 217 728	9
1 073 741 824	10
8 589 934 592	11
68 719 476 736	12
549 755 813 888	13
4 398 046 511 104	14
35 184 372 088 832	15

Table F-3. Powers of Sixteen

16^n	n
1	0
16	1
256	2
4 096	3
65 536	4
1 048 576	5
16 777 216	6
268 435 456	7
4 294 967 296	8
68 719 476 736	9
1 099 511 627 776	10
17 592 186 044 416	11
281 474 976 710 656	12
4 503 599 627 370 496	13
72 057 594 037 927 936	14
1 152 921 504 606 846 976	15

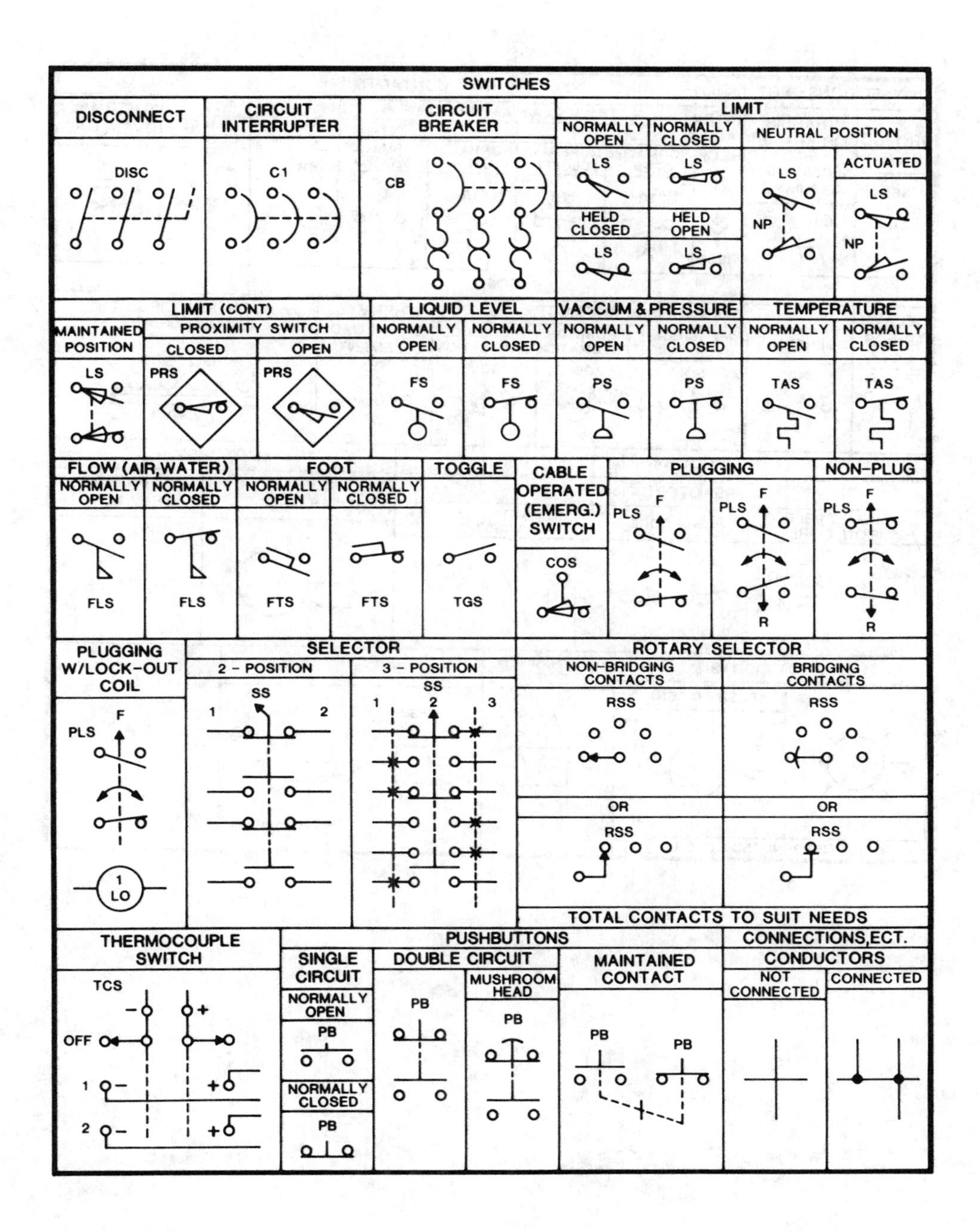

SWITCHES
DISCONNECT
CIRCUIT INTERRUPTER
CIRCUIT BREAKER
LIMIT
NORMALLY OPEN
NORMALLY CLOSED
NEUTRAL POSITION
ACTUATED
HELD CLOSED
HELD OPEN
DISC
C1
CB
LS
NP
LIMIT (CONT)
MAINTAINED POSITION
PROXIMITY SWITCH
CLOSED
OPEN
PRS
LIQUID LEVEL
VACCUM & PRESSURE
TEMPERATURE
FS
PS
TAS
FLOW (AIR,WATER)
FOOT
TOGGLE
CABLE OPERATED (EMERG.) SWITCH
PLUGGING
NON-PLUG
FLS
FTS
TGS
COS
PLS
F
R
PLUGGING W/LOCK-OUT COIL
SELECTOR
2 - POSITION
3 - POSITION
SS
ROTARY SELECTOR
NON-BRIDGING CONTACTS
BRIDGING CONTACTS
RSS
OR
1 LO
TOTAL CONTACTS TO SUIT NEEDS
THERMOCOUPLE SWITCH
TCS
OFF
PUSHBUTTONS
SINGLE CIRCUIT
DOUBLE CIRCUIT
MUSHROOM HEAD
MAINTAINED CONTACT
PB
CONNECTIONS,ECT.
CONDUCTORS
NOT CONNECTED
CONNECTED

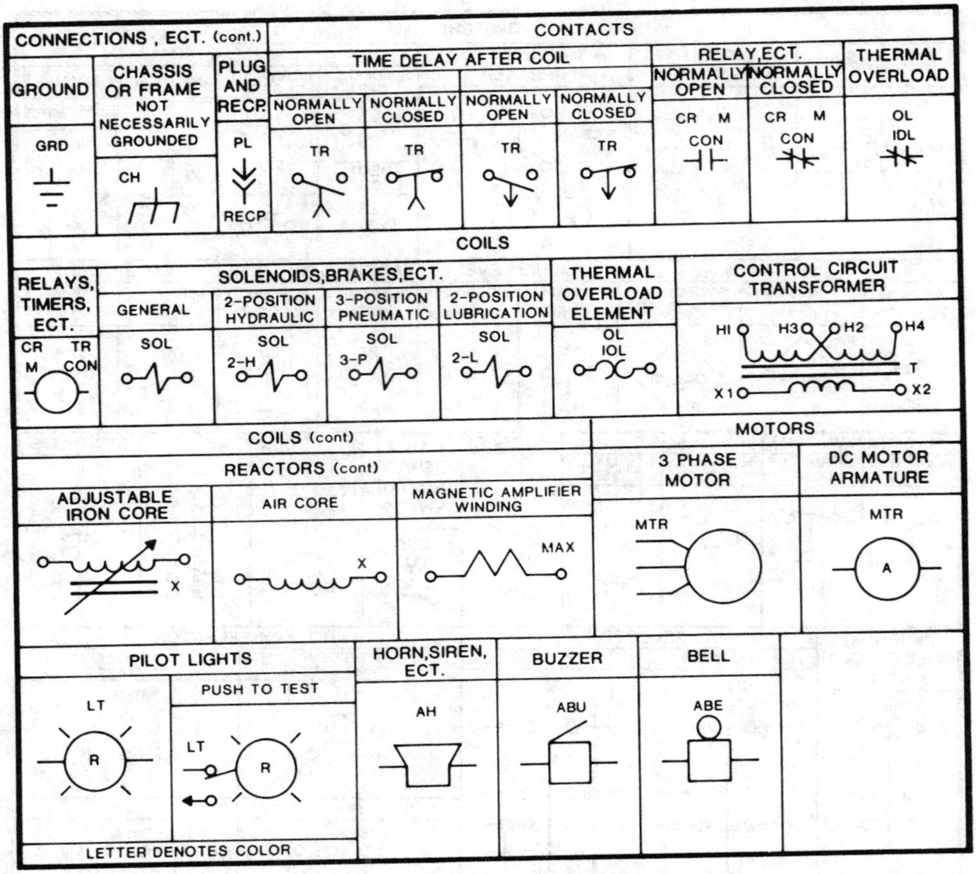

CONNECTIONS, ECT. (cont.)			CONTACTS						
GROUND	CHASSIS OR FRAME NOT NECESSARILY GROUNDED	PLUG AND RECP.	TIME DELAY AFTER COIL: NORMALLY OPEN	TIME DELAY AFTER COIL: NORMALLY CLOSED	TIME DELAY AFTER COIL: NORMALLY OPEN	TIME DELAY AFTER COIL: NORMALLY CLOSED	RELAY, ECT.: NORMALLY OPEN	RELAY, ECT.: NORMALLY CLOSED	THERMAL OVERLOAD
GRD	CH	PL RECP	TR	TR	TR	TR	CR M CON	CR M CON	OL IDL

COILS						
RELAYS, TIMERS, ECT.	SOLENOIDS, BRAKES, ECT.: GENERAL	2-POSITION HYDRAULIC	3-POSITION PNEUMATIC	2-POSITION LUBRICATION	THERMAL OVERLOAD ELEMENT	CONTROL CIRCUIT TRANSFORMER
CR TR M CON	SOL	SOL 2-H	SOL 3-P	SOL 2-L	OL IOL	HI H3 H2 H4 T X1 X2

COILS (cont) — REACTORS (cont)			MOTORS	
ADJUSTABLE IRON CORE	AIR CORE	MAGNETIC AMPLIFIER WINDING	3 PHASE MOTOR	DC MOTOR ARMATURE
X	X	MAX	MTR	MTR A

PILOT LIGHTS	PUSH TO TEST	HORN, SIREN, ECT.	BUZZER	BELL
LT R	LT R	AH	ABU	ABE

LETTER DENOTES COLOR

Appendix H

The equation that describes the line that passes through two given points $P_1(X_1,Y_1)$ and $P_2(X_2,Y_2)$ can be calculated by:

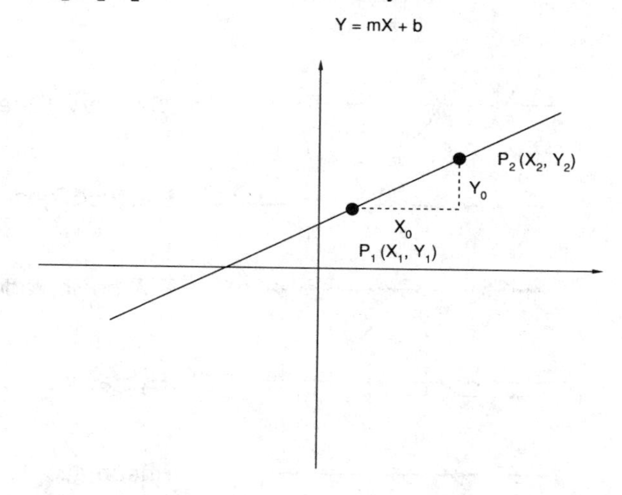

where m represents the slope of the line and b is the value of Y when $X = 0$. The value of m can be calculated as:

$$m = \frac{Y_0}{X_0} = \frac{Y_2 - Y_1}{X_2 - X_1}$$

If the value of b is not given and P_1 and P_2 are known, then Y can be obtained in this method:

$$Y - Y_1 = \frac{Y_2 - Y_1}{X_2 - X_1}(X - X_1)$$

where X_1 and Y_1 are the X and Y values at point P1 and X_2 and Y_2 are the values of X and Y at point P_2.

e.g.:

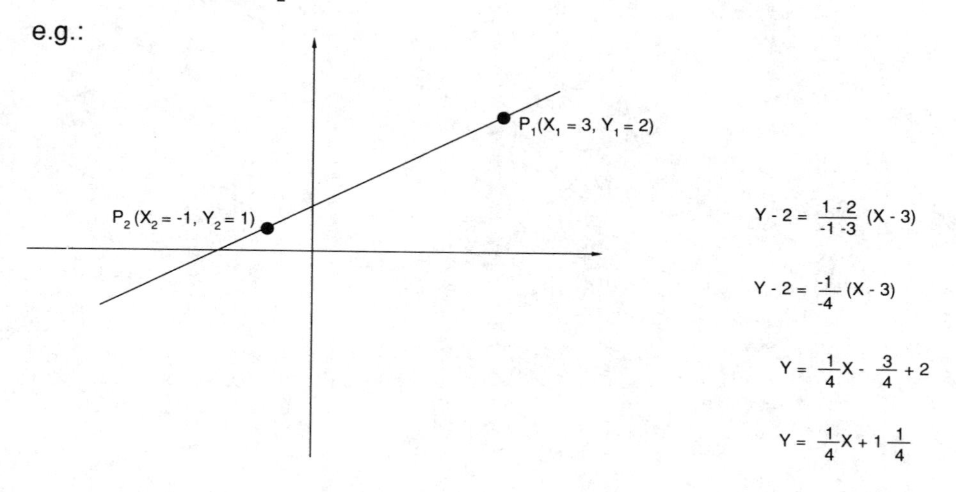

P & ID Symbols

Instrument Line Symbols

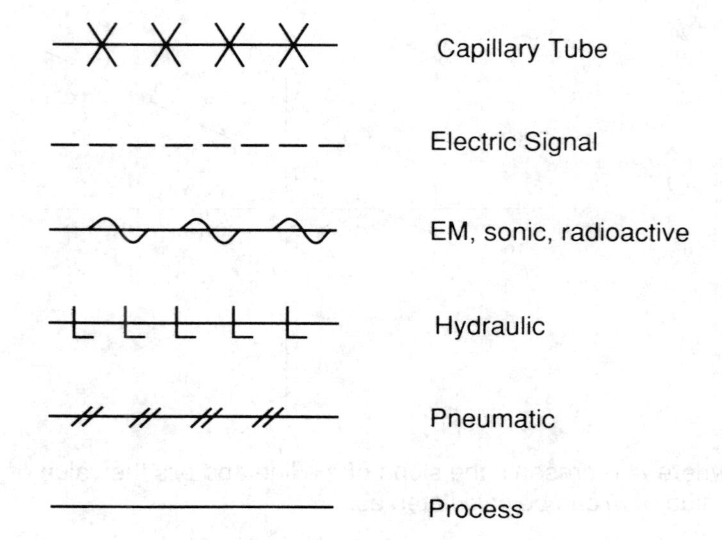

Instrument Identification Letters

	First Letter	Second Letter
A	Analysis	Alarm
B	Burner, Combustion	User's Choice*
C	User's Choice	Control
D	User's Choice	
E	Voltage	Sensor (Primary element)
F	Flow Rate	
G	User's Choice	Glass (sight tube)
H	Hand (manually initiated)	
I	Current (electric)	Indicate
J	Power	
K	Time or Time Schedule	Control station
L	Level	Light (pilot)
M	User's Choice	
N	User's Choice	User's Choice
O	User's Choice	Orifice,Restriction
P	Pressure, Vacuum	Point (test connection)
Q	Quantity	
R	Radiation	Record or Print
S	Speed or Frequency	Switch
T	Temperature	Transmit
U	Multivariable	Multifunction
V	Vibration, Mechanical Analysis	Valve, Damper, Louver
W	Weight, Force	Well
X	Unclassified **	Unclassified
Y	Event, State or Presence	Relay, Compute
Z	Position, Dimension	Driver, Actuator, Unclassified

*User's Choice may be used to denote a particular meaning. Having a meaning as a first letter and another meaning as a second letter. The user must describe the particular meaning in the legend. This letter could be used repetitively in a particular project.

** Unclassified letters may be used only once or used to a limited extent. If used the letter may have a meaning as a first letter and any meaning as a second letter. The user must specify the meaning(s) in the legend.

Reference:

ANSI/ISA –S5.1–1984, Instrumentation Symbols and Identification, ISBN 0-87664-844-8

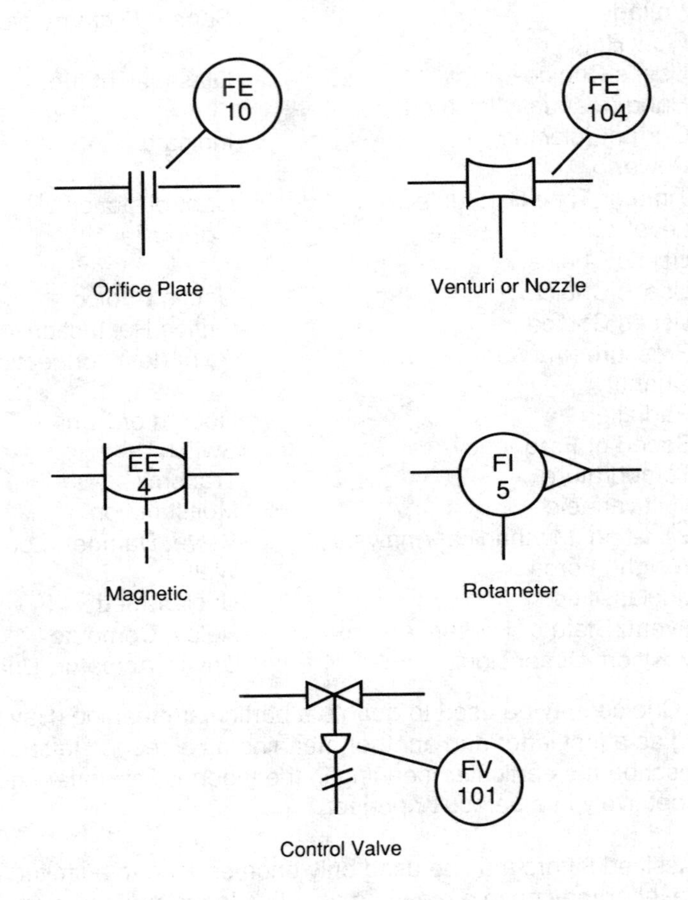

P & ID Symbols for Transducers and Elements

Bibliography

Allocca, J.A. and Stuart, A. *Transducers: Theory and Applications*, Reston Publishing Company Inc., Reston, VA, 1984.

Artwick, B.A. *Microcomputer Interfacing*, Prentice-Hall, Englewood Cliffs, NJ, 1980.

Beckwith, T.G., Buck, N.L., and Marangoni, R.D. *Mechanical Measurements,* 3rd Edition, Addison-Wesley Publishing Company, Reading, MA, 1982.

Beers, Y. *Introduction to the Theory of Error,* 2nd Edition, Addison-Wesley Publishing Company Inc., Reading, MA, 1957.

Boyce, J.C. *Digital Logic and Switching Circuits: Operation Analysis*, Prentice-Hall, Englewood Cliffs, NJ, 1975.

Gilmore, J.F., Pulaski, K., and Howard, C. "A Comprehensive Evaluation of Expert System Tools," Georgia Tech Research Institute, GA Institute of Technology, Atlanta, GA, 1985.

Gilmore, J.F., Roth, S.P., and Tynor, S.D. "GEST—The Anatomy of a Blackboard Expert System Tool," Georgia Tech Research Institute, GA Institute of Technology, Atlanta, GA, 1987.

Handbook of Measurement and Control, Schaevitz Engineering Co., Pennsauken, NJ, 1986.

Harmon, P., and King, D. *Expert Systems*, John Wiley and Sons Inc., New York, NY, 1985.

Hill, F.J. and Peterson, G.R. *Introduction to Switching Theory and Logical Design*, 2nd Edition, John Wiley and Sons, New York, NY, 1974.

Hill, F.J., and Peterson, G.R. *Digital Systems: Hardware Organization and Design*, John Wiley and Sons, New York, NY, 1973.

Johnson, C.D. *Process Control Instrumentation Technology,* 2nd Edition, John Wiley and Sons Inc., New, NY 1982.

Landers, G. "Xicor Replacer Dip Switches and Trimmers with Novram Memories," Xicor Publication AN-103, Milpitas, CA.

Liptak, B.G., and Venczel, K. *Instrument Engineers' Handbook: Process Measurement,* 2nd Edition, Chilton Book Company, Radnor, PA, 1982.

Modicon 584 "Programmable Controller User's Manual," Gould-Modicon Division, Andover, MA, 1982.

Nagle, T.H. Carroll, B.D., and Irwin, J.D. *An Introduction to Computer Logic*, Prentice-Hall, Englewood Cliffs, NJ, 1975.

PLC-2/30 "Programmable Controller: Programming and Operations Manual," Allen Bradley, Highland Heights, OH.

PLC-3 "Programmable Controller: Programming and Operations Manual,"Allen Bradley, Highland Heights, OH.

Rockis, G., and Mazur, G. Electrical Motor Controls, American Technical Publishers, Inc., Alsip, IL, 1982.

Thermistor Sensor Handbook, Thermometrics Inc., Edison, NJ, 1987.

Waterman, D.A. *A Guide to Expert Systems*, Addison-Wesley Publishing Company, Reading, MA, 1986.

Index

-D-

-E-

-F-

-G-

-H-

-N-

-O-

-P-